INSTRUCTOR'S SOLUTIONS MANUAL

Aimee L. Calhoun

Monroe Community College

Richard C. Stewart

Monroe Community College

A SURVEY OF MATHEMATICS WITH APPLICATIONS

SEVENTH EDITION AND EXPANDED SEVENTH EDITION

Allen R. Angel

Monroe Community College

Christine D. Abbott

Monroe Community College

Dennis C. Runde

Manatee Community College

PEARSON

Addison
Wesley

Boston   San Francisco   New York
London   Toronto   Sydney   Tokyo   Singapore   Madrid
Mexico City   Munich   Paris   Cape Town   Hong Kong   Montreal

Reproduced by Pearson Addison-Wesley from electronic files supplied by the authors.

Copyright © 2005 Pearson Education, Inc.
Publishing as Pearson Addison-Wesley, 75 Arlington Street, Boston, MA 02116

ISBN   0-321-20594-4

2 3 4 5 6 VHG 07 06 05 04

PEARSON

Addison
Wesley

Table of Contents

ACKNOWLEDGMENTS

We would like to thank Allen Angel, Christine Abbott, and Dennis Runde, the authors of *A Survey of Mathematics with Applications*, for their support and encouragement; Joe Vetere from Addison Wesley for his computer expertise; Lauren Morse from Addison Wesley for her assistance; and Jane Cummings from Monroe Community College for her technical support.

Aimee L. Calhoun
Richard C. Stewart

To my wonderful husband, Justin, for his love and support throughout the process of co-authoring this manual;
my children, Melanie and Jacob, for being my inspiration;
my incredible parents and the rest of my loving family for always believing in me;
and to the loving memories of my grandparents.

Aimee L. Calhoun

I am grateful to my wife, Christy, who enthusiastically supported my efforts to contribute to this manual, and to my daughters, Sarah and Sheila, for their encouragement to serve as an educator.

Richard C. Stewart

CHAPTER ONE

CRITICAL THINKING SKILLS

Exercise Set 1.1

1. a) 1, 2, 3, 4, 5,
   b) Counting numbers

2. a) If $a \div b$ has a remainder of zero, then a is divisible by b.
   b) 4, 8, 12
   c) 9, 18, 27

3. A **conjecture** is a belief based on specific observations that has not been proven or disproven.

4. **Inductive reasoning** is the process of reasoning to a general conclusion through observation of specific cases.

5. **Deductive reasoning** is the process of reasoning to a specific conclusion from a general statement.

6. A **counterexample** is a specific case that satisfies the conditions of the conjecture but shows the conjecture is false.

7. Inductive reasoning

8. Deductive reasoning

9. Inductive reasoning, because a general conclusion was made from observation of specific cases.

10. Inductive reasoning, because a general conclusion was made from observation of specific cases.

11. $1 \quad 5(1+4) \quad 10(4+6) \quad 10(6+4) \quad 5(4+1) \quad 1$

12. $100,000 = 10^5$

13. $5 \times 9 = 45$

14. $11 \times 14 = 154$

15.

16.

17.

18.

19. 15, 18, 21 (Add 3 to previous number.)

20. 2, - 4, - 10 (Subtract 6 from previous number.)

21. - 1, 1, - 1 (Alternate - 1 and 1.)

22. - 5, - 7, - 9 (Subtract 2 from previous number.)

23. $\dfrac{1}{81}, \dfrac{1}{243}, \dfrac{1}{729}$ (Multiply previous number by $\dfrac{1}{3}$.)

24. 162, - 486, 1458 (Multiply previous number by 3.)

25. 36, 49, 64 (The numbers in the sequence are the squares of the counting numbers.)

26. 21, 28, 36 ($15 + 6 = 21, 21 + 7 = 28, 28 + 8 = 36$)

27. 34, 55, 89 (Each number in the sequence is the sum of the previous two numbers.)

28. $\dfrac{80}{81}, -\dfrac{160}{243}, \dfrac{320}{729}$

(Multiply previous number by $-\dfrac{2}{3}$.)

29. Y: There are three letters in the pattern. $39 \times 3 = 117$, so the 117th entry is the second R in the pattern. Therefore, the 118th entry is Y.

30. a) Answers will vary.
    b) The sum of the digits is 9.
    c) The sum of the digits in the product when a one or two digit number is multiplied by 9 is 9.

31. a) 36, 49, 64
    b) Square the numbers 6, 7, 8, 9 and 10.
    c) $8 \times 8 = 64$ $9 \times 9 = 81$

    72 is not a square number since it falls between the two square numbers 64 and 81.

32. a) 28 and 36
    b) To find the 7th triangular number, add 7 to the 6th triangular number. To find the 8th triangular number, add 8 to the 7th triangular number. To find the 9th triangular number, add 9 to the 8th triangular number. To find the 10th triangular number, add 10 to the 9th triangular number. To find the 11th triangular number, add 11 to the 10th triangular number.
    c) $36 + 9 = 45$ $45 + 10 = 55$ $55 + 11 = 66$ $66 + 12 = 78$

    72 is not a triangular number since it falls between the two triangular numbers 66 and 78.

33. Blue: 1, 5, 7, 10, 12   Purple: 2, 4, 6, 9, 11   Yellow: 3, 8

34. a) 19 (Each new row has two additional triangles.)
    b) $1 + 3 + 5 + 7 + 9 + 11 + 13 + 15 + 17 + 19 = 100$

35. a) ≈ 58 million
    b) ≈ 45 million
    c) We are using observation of specific cases to make a prediction.

36. a) $\approx \$28,000$
    b) $\approx \$61,000$
    c) We are using observation of specific cases to make a prediction.

37.

P	B	P	B
B	P	B	P
P	B	P	B
B	P	B	P

38.

39. a) You should obtain the original number.
    b) You should obtain the original number.
    c) Conjecture: The result is always the original number.
    d) $n, 4n, 4n+8, \dfrac{4n+8}{4} = \dfrac{4n}{4} + \dfrac{8}{4} = n+2, n+2-2 = n$

40. a) You should obtain twice the original number.
    b) You should obtain twice the original number.
    c) Conjecture: The result is always twice the original number.
    d) $n, 10n, 10n+5, \dfrac{10n+5}{5} = \dfrac{10n}{5} + \dfrac{5}{5} = 2n+1, 2n+1-1 = 2n$

41. a) You should obtain the number 5.

 b) You should obtain the number 5.

 c) Conjecture: No matter what number is chosen, the result is always the number 5.

 d) $n, n+1, n+(n+1) = 2n+1, 2n+1+9 = 2n+10, \dfrac{2n+10}{2} = \dfrac{2n}{2} + \dfrac{10}{2} = n+5, n+5-n = 5$

42. a) You should obtain the number 0.

 b) You should obtain the number 0.

 c) Conjecture: No matter what number is chosen, the result is always the number 0.

 d) $n, n+10, \dfrac{n+10}{5}, 5\left(\dfrac{n+10}{5}\right) = n+10, n+10-10 = n, n-n = 0$

43. $999 \times 999 = 998{,}001$ is one example.

44. $11+12+13 = 36$ is one example.

45. Two is a counting number. The sum of 2 and 3 is 5. Five divided by two is $\dfrac{5}{2}$, which is not an even number.

46. One and three are counting numbers. The product of 1 and 3 is 3, which is not divisible by 2.

47. One and two are counting numbers. The difference of 1 and 2 is $1-2 = -1$, which is not a counting number.

48. The sum of the odd numbers 1 and 5 is 6, which is not divisible by 4.

49. a) The sum of the measures of the interior angles should be $180°$.

 b) Yes, the sum of the measures of the interior angles should be $180°$.

 c) Conjecture: The sum of the measures of the interior angles of a triangle is $180°$.

50. a) The sum of the measures of the interior angles should be $360°$.

 b) Yes, the sum of the measures of the interior angles should be $360°$.

 c) Conjecture: The sum of the measures of the interior angles of a quadrilateral is $360°$.

51. 129, the numbers in positions are found as follows: $\begin{matrix} a & b \\ c & a+b+c \end{matrix}$

52. 1881, 8008, 8118 (They look the same when looked at in a mirror.)

53. Counterexample

54. c

Exercise Set 1.2

(Note: Answers in this section will vary depending on how you round your numbers. The answers may differ from the answers in the back of the textbook. However, your answers should be something near the answers given. All answers are approximate.)

1. $431 + 327.2 + 73.5 + 20.4 + 315.9 \approx 430 + 330 + 70 + 20 + 320 = 1170$

2. $3.89 + 402.8 + 156.9 + 189 + 0.23 + 416 \approx 4 + 403 + 157 + 190 + 0 + 416 = 1170$

3. $297{,}700 \times 4087 \approx 300{,}000 \times 4000 = 1{,}200{,}000{,}000$

4. $1854 \times 0.0096 \approx 1900 \times 0.01 = 19$

5. $\dfrac{405}{0.049} \approx \dfrac{400}{0.05} = 8000$

6. $297.521 - 85.964 \approx 300 - 90 = 210$

7. $0.049 \times 1989 \approx 0.05 \times 2000 = 100$

8. 9% of $2164 \approx 10\%$ of $2200 = 0.10 \times 2200 = 220$

9. $51{,}608 \times 6981 \approx 52{,}000 \times 7000 = 364{,}000{,}000$

10. $\dfrac{0.0498}{0.00052} \approx \dfrac{0.05}{0.0005} = 100$

11. $592 \times 2070 \times 992.62$
 $\approx 600 \times 2000 \times 1000 = 1{,}200{,}000{,}000$

12. $296.3 \div 0.0096 \approx 300 \div 0.01 = 30{,}000$

13. $52 \times \$0.37 \approx 50 \times \$0.40 = \$20$

14. $32 \text{ hours} \times \$7.95 \text{ per hour}$
     $\approx 32 \text{ hours} \times \$8 \text{ per hour} = \$256$

15. $1521 + 1897 + 2324 + 2817$
     $\approx 1500 + 1900 + 2300 + 2800 = 8500 \text{ mi}$

16. $6 \times 15.87 \approx 6 \times 16 = 96 \text{ lb}$

17. $\$2.29 + \$12.16 + \$4.97 + \$6.69 + \$49.76 + \$0.47$
     $+ \$3.49 + \$5.65 \approx \$2 + \$12 + \$5 + \$7 + \$50 + \$0.50$
     $+ \$3.50 + \$5.70 = \$85.70$

18. $\dfrac{3.12}{6} \approx \dfrac{3}{6} = 0.5 \text{ lb}$

19. $\dfrac{\$44,569}{5} \approx \dfrac{\$45,000}{5} = \$9000$

20. $32,798 - 14,292 \approx 32,800 - 14,300 = 18,500 \text{ lb}$

21. $9 \times 5.12 \approx 9 \times 5 = 45 \text{ lb}$

22. $8\% \text{ of } \$14,876$
     $\approx 8\% \text{ of } \$15,000 = 0.08 \times \$15,000 = \$1200$

23. $\dfrac{23,663}{12} \approx \dfrac{24,000}{12} = 2000 \text{ mi}$

24. $\dfrac{\$10.87}{3.2} \approx \dfrac{\$11}{3} = \$3.\overline{6} \approx \$3.70 \text{ per pound}$

25. $12(\$29.17 + \$39.95)$
     $\approx 12(\$30 + \$40) = 12(\$70) = \$840$

26. Team A: $189 + 172 + 191 \approx 190 + 170 + 190 = 550$
     Team B: $183 + 229 + 167 \approx 180 + 230 + 170 = 580$
     $580 - 550 = 30 \text{ lb}$

27. $15\% \text{ of } \$38.60 \approx 15\% \text{ of } \$40 = 0.15 \times \$40 = \$6$

28. $3.8 \text{ grubs per square foot} \times (60 \text{ ft} \times 80.2 \text{ ft})$
     $\approx 4 \text{ grubs per square foot} \times (60 \times 80 \text{ square feet})$
     $= 4 \times 4800 \text{ grubs} = 19,200 \text{ grubs}$

29. $100 \text{ Mexican pesos} = 100 \times 0.092 \text{ U.S. dollars}$
     $\approx 100 \times 0.09 \text{ U.S. dollars} = 9 \text{ U.S. dollars}$
     $\$50 - \$9 = \$41$

30. $\$973 + 6(\$41) + 6(\$97) + 6(\$90)$
     $\approx \$970 + 6(\$40) + 6(\$100) + 6(\$90)$
     $= \$970 + \$240 + \$600 + \$540 = \$2350$

31. $\approx 375 \text{ miles}$

32. $\approx 70 \text{ miles}$

33. a) $30.98\% \times 105 \text{ million} \approx 31\% \times 105 \text{ million}$
     $= 0.31 \times 105 \text{ million} = 32.55 \text{ million} \approx 32.6 \text{ million}$
     b) $18.41\% \times 3141$
     $\approx 18\% \times 3100 = 0.18 \times 3100 = 558 \text{ counties}$
     c) The counties that use punch cards could be the
     largest counties with the most voters.

34. a) $39\% \times \$40,075 \approx 40\% \times \$40,000 = \$16,000$
     b) $22.9\% \times \$40,075 \approx 23\% \times \$40,000 = \$9200$

35. a) 4 million
     b) 98 million
     c) $98 \text{ million} - 34 \text{ million} = 64 \text{ million}$
     d) $19,000 + 78,000 + 82,000 + 61,000 + 35,000$
     $= 275 \text{ million}$

36. a) 19%
     b) 25%
     c) $28\% \text{ of } 180 \text{ lb} = 0.28 \times 180 = 50.4 \approx 50 \text{ lb}$

37. a) 83%

   b) $65\% - 45\% = 20\%$

   c) 83% of 110,567

   $\approx 0.83 \times 110,567 = 91,770.61 \approx 91,771$ sq mi

   d) No, since we are not given the area of each state.

38. a) $2(410) + 4(545)$

   $\approx 2(400) + 4(550) = 800 + 2200 = 3000$ calories

   b) Running: $4(920) \approx 4(925) = 3700$ calories

   Casual bike riding: $4(300) = 1200$ calories,

   $3700 - 1200 = 2500$ calories

   c) $3(545) + 3(545) \approx 3(550) + 3(550)$

   $= 1650 + 1650 = 3300$ calories per week,

   3300 calories per week $(52$ weeks$)$

   $\approx 3000 \times 50 = 150,000$ calories

39. 25

40. 32

41. ≈ 90 berries

42. ≈ 160 leaves

43. 150°

44. 315°

45. 10%

46. 25%

47. 9 square units

48. 12 square units

49. 150 feet

50. $5(62) = 310$ in. or $\dfrac{310}{12} = 25.8\overline{3} \approx 25.8$ ft

51.-59. Answers will vary.

60. There are 118 ridges around the edge.

61. There are 336 dimples on a regulation golf ball.

62. a) Answers will vary.

   b) 60 seconds per minute $\times 60$ minutes per hour

   $\times 24$ hours per day $= 60 \times 60 \times 24$ seconds per day

   $= 86,400$ seconds per day,

   $\dfrac{1,000,000}{86,400} = 11.57407407 \approx 11.6$ days

63. Answers will vary. The U.S. government categorized the middle class as $32,000 - $50,000 in 2001.

Exercise Set 1.3

1. $\dfrac{1 \text{ in.}}{50 \text{ mi}} = \dfrac{3.75 \text{ in.}}{x \text{ mi}}$

   $1x = 50(3.75)$

   $x = 187.5$ mi

2. $\dfrac{1 \text{ in.}}{12 \text{ ft}} = \dfrac{x \text{ in.}}{82 \text{ ft}}$

   $12x = 1(82)$

   $\dfrac{12x}{12} = \dfrac{82}{12}$

   $x = \dfrac{82}{12} = 6\dfrac{10}{12} = 6\dfrac{5}{6}$ in. or $6.8\overline{3} \approx 6.83$ in.

3. $\dfrac{3 \text{ ft}}{1.2 \text{ ft}} = \dfrac{48.4 \text{ ft}}{x \text{ ft}}$

$3x = 1.2(48.4)$

$\dfrac{3x}{3} = \dfrac{58.08}{3}$

$x = \dfrac{58.08}{3} = 19.36 \text{ ft}$

4. $\dfrac{1 \text{ bag}}{6000 \text{ ft}^2} = \dfrac{x \text{ bags}}{26,000 \text{ ft}^2}$

$6000x = 1(26,000)$

$\dfrac{6000x}{6000} = \dfrac{26,000}{6000}$

$x = \dfrac{26,000}{6000} = 4.\overline{3} \approx 4.33 \text{ bags}$

5. 11.5% of $\$4222 = 0.115(\$4222) = \$485.53$

$\$4222 + \$485.53 = \$4707.53$

6. Cost for mileage:

$\$0.30\left(\dfrac{12}{\frac{1}{5}}\right) = \$0.30(12)(5) = \$18.00$

Cost for sitting still:
2 minutes $= 2(60) = 120$ seconds

$\$0.30\left(\dfrac{120}{30}\right) = \$0.30(4) = \$1.20$

Cost for ride: $\$2.00 + \$18.00 + \$1.20 = \$21.20$

7. $\dfrac{20,000 \text{ miles}}{20 \text{ miles per gallon}} = 1000 \text{ gallons}$

Hawaii: $1000(\$2.02) = \$2020$

South Carolina: $1000(\$1.22) = \$1220$

$\$2020 - \$1220 = \$800$

8. a) $\approx 1980 - 1900$ or 80 hours

b) $\approx \dfrac{2000 \text{ hours}}{52 \text{ weeks}} = 38.46153846 \approx 38.5 \text{ hr/wk}$

c) $\approx \dfrac{1500 \text{ hours}}{52 \text{ weeks}} = 28.84615385 \approx 28.8 \text{ hr/wk}$

9. Denise parks her car for eight hours per day.

$5\big[\$2.50 + \$1.00(7 \text{ hours per day})\big]$

$= 5[\$2.50 + \$7.00] = 5(\$9.50) = \$47.50$

Savings: $\$47.50 - \$35.00 = \$12.50$

10. $\$3.75 + (21 - 3)(\$0.50) = \$3.75 + 18(\$0.50)$

$= \$3.75 + \$9 = \$12.75$

11. $\$120 + \$80(15) = \$120 + \$1200 = \$1320$

Savings: $\$1320 - \$1250 = \$70$

12. $\$20,000$ down payment:

$\$20,000 + \$699.99(12)(30)$

$= \$20,000 + \$251,996.40 = \$271,996.40$

$\$40,000$ down payment:

$\$40,000 + \$559.20(12)(30)$

$= \$40,000 + \$201,312 = \$241,312$

Savings: $\$271,996.40 - \$241,312 = \$30,684.40$

13. 20 year mortgage: $\$752.40(12)(20) = \$180,576$

30 year mortgage: $\$660.60(12)(30) = \$237,816$

Savings: $\$237,816 - \$180,576 = \$57,240$

14. Points needed for 80 average: $80(5) = 400$ points

Wallace s points so far:
$77 + 93 + 90 + 76 = 336$ points

Grade needed on fifth exam: $400 - 336 = 64$

15. a) $\dfrac{86.5}{34} \approx 2.54; \dfrac{91.5}{36} \approx 2.54; \dfrac{96.5}{38} \approx 2.54;$

$\dfrac{101.5}{40} \approx 2.54; \dfrac{106.5}{42} \approx 2.54 \ldots$

So, $48(2.54) \approx 122.$

b) Answers will vary. A close approximation can be obtained by multiplying the U.S. sizes by 2.54.

16. a) $10 \cdot 10 \cdot 10 \cdot 10 = 10,000$

b) 1 in 10,000

17. a) $\dfrac{460}{50} = 9.2$ min

b) $\dfrac{1550}{25} = 62$ min

c) $\dfrac{1400}{35} = 40$ min

d) $\dfrac{1550}{25} + \dfrac{2200}{25} = \dfrac{3750}{25} = 150$ min

18. $38,687.0 \text{ mi} - 38,451.4 \text{ mi} = 235.6 \text{ mi}$

$\dfrac{235.6 \text{ mi}}{12.6 \text{ gal}} = 18.6984127 \approx 18.7 \text{ mpg}$

19. a) 11% of 273,300,000

$= 0.11(273,300,000) = 30,063,000$

b) 10% of 970,000 $= 0.10(970,000) = 97,000$

c) 3% of 970,000 $= 0.03(970,000) = 29,100$

20. a) $40 \times \$8.50 \times 52 = \$17,680$

b) Each week he makes $40 \times \$8.50 = \$340.$

$\dfrac{\$1275}{\$340} = 3.75$ weeks

21. By mail: $(\$52.80 + \$5.60 + \$8.56) \times 4$

$= \$66.96 \times 4 = \$267.84$

Tire store: $\$324 + 0.08 \times \$324$

$= \$324 + \$25.92 = \$349.92$

Savings: $\$349.92 - \$267.84 = \$82.08$

22. $(1 \text{ yd})^2 = (3 \text{ ft})^2 = 9 \text{ ft}^2$

$2400 \times 9 = 21,600 \text{ ft}^2$

$\dfrac{1 \text{ gal}}{350 \text{ ft}^2} = \dfrac{x \text{ gal}}{21,600 \text{ ft}^2}$

$350x = 1(21,600)$

$\dfrac{350x}{350} = \dfrac{21,600}{350}$

$x = \dfrac{21,600}{350} = 61.71428571 \approx 62 \text{ gal}$

23. a) $\$620(0.12) = \$74.40$

b) $\$1200(0.22) = \$264$

c) The store lost $\$1200 - \$1000 = \$200$ on the purchase.

Store's profit: $\$264 - \$200 = \$64$

24. a) $0.1 \text{ cm}^3 \times 60 \text{ sec} \times 60 \text{ min} \times 24 \text{ hr} \times 365 \text{ days}$

$= 3,153,600 \text{ cm}^3$

b) $30 \text{ cm} \times 20 \text{ cm} \times 20 \text{ cm} = 12,000 \text{ cm}^3$

$0.1 \text{ cm}^3 \times 60 \text{ sec} \times 60 \text{ min} \times 24 \text{ hr} = 8640$

$\dfrac{12,000}{8640} = 1.3\overline{8} \approx 1.4 \text{ days}$

25. Let $x =$ the amount above $12,000

    $4950 - $1200 = $3390

    $$\frac{0.15x}{0.15} = \frac{$3390}{0.15}$$

    $$x = $22,600$$

    $12,000 + $22.600 = $34,600$

26. a) $1 \text{ oz} \times 60 \text{ min} \times 24 \text{ hr} \times 365 \text{ days} = 525,600 \text{ oz}$

    $$\frac{525,600}{128} = 4106.25 \text{ gal}$$

    b) $\dfrac{4106.25}{1000} \times $11.20 = 4.10625 \times $11.20 = 45.99

27. $7(2) + 5(1) + 4(29) + 3(201) + 2(1408) + 1(10,352)$

    $= 14 + 5 + 116 + 603 + 2816 + 10,352$

    $= 13,906$ violations

28. a) $\dfrac{20,000}{20.8} - \dfrac{20,000}{21.6} = 961.5384615 - 925.9259259$

    $= 35.6125356 \approx 35.61$ gal

    b) $35.61 \times $1.60 = 56.976 \approx 56.98

    c) $140,000,000 \times 35.61 = 4,985,400,000$ gal

29. a) Yes, divide the total emissions by the emissions per capita.

    b) $\dfrac{6503.8}{24.3} = 267.6460905 \approx 267.65$ million

    c) $\dfrac{4964.8}{4.0} = 1241.2$ million or 1.2412 billion

30. Cost after 1 year: $450 + 0.06($450)$

    $= $450 + $27 = 477

    Cost after 2 years: $477 + 0.06($477)$

    $= $477 + $28.62 = 505.62

31. Value after first year: $1000 + 0.10($1000)$

    $= $1000 + $100 = 1100

    Value after second year: $1100 - 0.10($1100)$

    $= $1100 - $110 = 990

    $990 is less than the intial investment of $1000.

32. After paying the $100 deductible, Yungchen must pay 20% of the cost of x-rays.

    First x-ray:

    $100 + 0.20($540) = $100 + $108 = 208

    Second x-ray: $0.20($920) = 184

    Total: $208 + $184 = 392

33. a) $\dfrac{$200}{$41} \approx 4.87804878$ The maximum number of 10 packs is 4.

    $200 - (4 \times $41) = $200 - $164 = 36 , $\dfrac{$36}{$17} = 2.117647059$ Deirdre can also buy two 4 packs.

10 packs	4 packs	Number of rolls	Cost
4	2	$4(10) + 2(4) = 48$	$4($41) + 2($17) = 198
3	4	46	$191
2	6	44	$184
1	9	46	$194
0	11	44	$187

    Maximum number of rolls of film is 48.

    b) The cost is $198 when she purchases four 10 packs and two 4 packs.

34. a) $\dfrac{\$50}{\$5.76} \approx 8.680\overline{5}$ The maximum number of 4 packs of 36 exposures is 8.

$\$50 - (8 \times \$5.76) = \$50 - \$46.08 = \$3.92$, Erika cannot buy any 24 exposures.

4 packs of 36 exp.	4 packs of 24 exp.	Number of exposures	Cost
8	0	$8(36) + 0(24) = 288$	$8(\$5.76) + 0(\$4.08) = \$46.08$
7	2	300	$48.48
6	3	288	$46.80
5	5	300	$49.20
4	6	288	$47.52
3	8	300	$49.92
2	9	288	$48.24
1	10	276	$46.56
0	12	288	$48.96

2 packs of 24 exposures and 7 packs of 36 exposures, or 5 packs of 24 exposures and 5 packs of 36 exposures, or 8 packs of 24 exposures and 3 packs of 36 exposures

b) 300 exposures in each case

c) The minimum cost is $48.48 when she purchases 2 packs of 24 exposures and 7 packs of 36 exposures.

35. a) water/milk: $3(1) = 3$ cups   salt: $3\left(\dfrac{1}{8}\right) = \dfrac{3}{8}$ tsp

cream: $3(3) = 9$ tbsp $= \dfrac{9}{16}$ cup (because 16 tbsp = 1 cup)

b) water/milk: $\dfrac{2 + 3.75}{2} = \dfrac{5.75}{2} = 2.875$ cups $= 2\dfrac{7}{8}$ cups

salt: $\dfrac{0.25 + 0.5}{2} = \dfrac{0.75}{2} = 0.375$ tsp $= \dfrac{3}{8}$ tsp   cream: $\dfrac{0.5 + 0.75}{2} = \dfrac{1.25}{2} = 0.625$ cups $= \dfrac{5}{8}$ cup

$$= \dfrac{5}{8}(16 \text{ tbsp}) = 10 \text{ tbsp}$$

c) water/milk: $3\dfrac{3}{4} - 1 = \dfrac{15}{4} - \dfrac{4}{4} = \dfrac{11}{4} = 2\dfrac{3}{4}$ cups

salt: $\dfrac{1}{2} - \dfrac{1}{8} = \dfrac{4}{8} - \dfrac{1}{8} = \dfrac{3}{8}$ tsp   cream: $\dfrac{3}{4} - \dfrac{3}{16} = \dfrac{12}{16} - \dfrac{3}{16} = \dfrac{9}{16}$ cup $= 9$ tbsp

d) Differences exist in water/milk because the amount for 4 servings is not twice that for 2 servings. Differences also exist in Cream of Wheat because $\dfrac{1}{2}$ cup is not twice 3 tbsp.

36. a) rice: $\frac{1}{2}(4) = 2$ cups b) rice: $1(2) = 2$ cups

 water: $1\frac{1}{3}(4) = \frac{4}{3}(4) = \frac{16}{3} = 5\frac{1}{3}$ cups water: $2\frac{1}{4}(2) = \frac{9}{4}(2) = \frac{18}{4} = 4\frac{2}{4} = 4\frac{1}{2}$ cups

 salt: $\frac{1}{4}(4) = 1$ tsp salt: $\frac{1}{2}(2) = 1$ tsp

 butter/margarine: $1(4) = 4$ tsp butter/margarine: $2(2) = 4$ tsp

 c) rice: $\frac{1}{2} + 1\frac{1}{2} = \frac{1}{2} + \frac{3}{2} = \frac{4}{2} = 2$ cups

 water: $1\frac{1}{3} + 3\frac{1}{3} = \frac{4}{3} + \frac{10}{3} = \frac{14}{3} = 4\frac{2}{3}$ cups

 salt: $\frac{1}{4} + \frac{3}{4} = \frac{4}{4} = 1$ tsp

 butter/margarine: 1 tsp + 1 tbsp = 1 tsp + 3 tsp = 4 tsp

 d) rice: $3 - 1 = 2$ cups

 water: $6 - 2\frac{1}{4} = \frac{24}{4} - \frac{9}{4} = \frac{15}{4} = 3\frac{3}{4}$ cups

 salt: $1\frac{1}{2} - \frac{1}{2} = 1$ tsp

 butter/margarine: 2 tbsp $= 2(3\text{tsp}) = 6$ tsp

 6 tsp $- 2$ tsp $= 4$ tsp

 e) Differences exist in water because the amount for 4 servings is not twice that for 2 servings.

37. 1 ft^2 would be 12 in. by 12 in.

 Thus, $1 \text{ ft}^2 = 12 \text{ in.} \times 12 \text{ in.} = 144 \text{ in.}^2$

38. $1 \text{ ft}^3 = 12 \text{ in.} \times 12 \text{ in.} \times 12 \text{ in.} = 1728 \text{ in.}^3$

39. Area of original rectangle $= lw$

 Area of new rectangle $= (2l)(2w) = 4lw$

 Thus, if the length and width of a rectangle are doubled, the area is 4 times as large.

40. Volume of original cube $= lwh$

 Volume of new cube $= (2l)(2w)(2h) = 8lwh$ Thus, if the length, width, and height of a cube are doubled, the volume is 8 times as large or increases eight-fold.

41. 1 and 9

 $1 \times 9 = 9$

 $1 + 9 = 10$

42. $\dfrac{10 \text{ pieces}}{\$x} = \dfrac{1000 \text{ pieces}}{\$10}$

 $1000x = 10(10)$

 $\dfrac{1000x}{1000} = \dfrac{100}{1000}$

 $x = \dfrac{100}{1000} = \$0.10 = 10\text{¢}$

43. Left side: $1(-6) = -6$ Right side: $1(2) = 2$

 $2(-2) = -4$ $1(3) = 3$

 $-6 + -4 = -10$ $1(6) = 6$

  $2 + 3 + 6 = 11$

 Place it at -1 so the left side would total $-10 + -1 = -11$.

44. 3

45.

Birds	Lizards	Number of Heads	Number of Feet
8	14	22	$8(2)+14(4)=72$
9	13	22	$9(2)+13(4)=70$
10	12	22	$10(2)+12(4)=68$

Therefore, there are 10 birds and 12 lizards.

46. $10; 2002, 2112, 2222, 2332, 2442, 2552,$
$2662, 2772, 2882, 2992$

47. a) $(4 \times 4)+(3 \times 3)+(2 \times 2)+(1 \times 1)$
$=16+9+4+1=30$
b) $(7 \times 7)+(6 \times 6)+(5 \times 5)+30$
$=49+36+25+30=140$

48. a) Place the object, 1 g, and 3 g on one side and
9 g on the other side.
b) Place the object, 9 g, and 3 g on
one side and 27 g and 1 g on the other side.

49.

50. Eight pieces

51.

8	6	16
18	10	2
4	14	12

52.

15	1	11
5	9	13
7	17	3

53. $6+10+8+4=28; 3+7+5+1=16;$
$10+14+12+8=44$
The sum of the four corner entries is
4 times the number in the center of the middle row.

54. $21, 12, 33$
Multiply the number in the center of the middle
row by 3.

55. $63, 36, 99$
Multiply the number in the center of the middle
row by 9.

56. $35-15=20$ cubes

57. $3 \times 2 \times 1 = 6$ ways

58. Each shakes with four people.

59.

	7	
3	1	4
5	8	6
	2	

Other answers are possible, but 1 and
8 must appear in the center.

60.

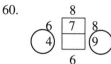

61.

1	2	3	4	5
2	3	4	5	1
3	4	5	1	2
4	5	1	2	3
5	1	2	3	4

Other answers are possible.

62. With umbrella policy:

Mustang reduced premium: $1648 - $90 = $1558

Focus reduced premium:

$1530 - 0.12($1530)$

$= $1530 - $183.60 = 1346.40

Total for umbrella policy:

$1558 + $1346.40 + $450 = 3354.40

Without umbrella policy: $1648 + $1530 = 3178

Net amount for umbrella policy:

$3354.40 - $3178 = 176.40

63. Mary is the skier.

64. $16 + 16 + 4 + 4 + 4 = 44$

65. Areas of the colored regions are:

1×1, 1×1, 2×2, 3×3, 5×5, 8×8, 13×13,

21×21 ; $1 + 1 + 4 + 9 + 25 + 64 + 169 + 441$

$= 714$ square units

66. 1 giraffe = 2 frogs

1 giraffe = 3 lions

3 lions = 2 frogs

Therefore, $\dfrac{3}{3}$ lion = $\dfrac{2}{3}$ frog.

Therefore, 1 lion = $\dfrac{2}{3}$ frog.

1 lion = 2 ostriches

Therefore, $\dfrac{2}{3}$ frog = 2 ostriches.

$\dfrac{2}{3}\left(\dfrac{3}{2}\right)$ frog = $2\left(\dfrac{3}{2}\right)$ ostriches

Therefore, 1 frog = 3 ostriches.

Review Exercises

1. 23, 28, 33 (Add 5 to previous number.)

2. 25, 36, 49 (next three perfect squares)

3. -48, 96, -192 (Multiply previous number by -2.)

4. 25, 32, 40 (19 + 6 = 25, 25 + 7 = 32, 32 + 8 = 40)

5. 15, 9, 2 $(20 - 5 = 15, 15 - 6 = 9, 9 - 7 = 2)$

6. $\dfrac{3}{8}$, $\dfrac{3}{16}$, $\dfrac{3}{32}$ (Multiply previous number by $\dfrac{1}{2}$.)

7.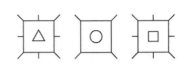

8.

9. c

10. a) The original number and the final number are the same.

    b) The original number and the final number are the same.

    c) Conjecture: The final number is the same as the original number.

    d) $n, 2n, 2n+10, \dfrac{2n+10}{2} = \dfrac{2n}{2} + \dfrac{10}{2} = n+5, n+5-5 = n$

11. This process will always result in an answer of 3. $n, n+5, 6(n+5) = 6n+30, 6n+30-12$

$$= 6n+18, \dfrac{6n+18}{2} = \dfrac{6n}{2} + \dfrac{18}{2} = 3n+9, \dfrac{3n+9}{3} = \dfrac{3n}{3} + \dfrac{9}{3} = n+3, n+3-n = 3$$

12. $1^2 + 2^2 = 5$, 5 is an odd number.

(Note: Answers for Ex. 13 - 25 will vary depending on how you round your numbers. The answers may differ from the answers in the back of the textbook. However, your answers should be something near the answers given. All answers are approximate.)

13. $210,302 \times 1992 \approx 210,000 \times 2000 = 420,000,000$

14. $346.2 + 96.402 + 1.04 + 897 + 821$
$\approx 350 + 100 + 0 + 900 + 800 = 2150$

15. 21% of $1012 \approx$ 20% of 1000
$= 0.20 \times 1000 = 200$

16. Answers will vary.

17. $82 \times \$1.09 \approx 80 \times \$1.10 = \$88$

18. 6% of $\$202 \approx$ 6% of $200 = 0.06 \times 200 = \$12$

19. $\dfrac{1.1 \text{ mi}}{22 \text{ min}} \approx \dfrac{1 \text{ mi}}{20 \text{ min}} = \dfrac{3 \text{ mi}}{60 \text{ min}} = 3$ mph

20. $\$2.49 + \$0.79 + \$1.89 + \$0.10 + \$2.19 + \$6.75$
$\approx \$2 + \$1 + \$2 + \$0 + \$2 + \$7 = \$14.00$

21. $5 \text{ in.} = \dfrac{20}{4} \text{ in.} = 20\left(\dfrac{1}{4}\right) \text{ in.} = 20(0.1) \text{ mi} = 2 \text{ mi}$

22. 70%

23. 5%

24. 13 square units

25. Length $= 1.75$ in., $1.75(12.5) = 21.875 \approx 22$ ft

Height $= 0.625$ in., $0.625(12.5) = 7.8125 \approx 8$ ft

26. $\$2.00 + 7(\$1.50) = \$2.00 + \$10.50 = \$12.50$

Change: $\$20.00 - \$12.50 = \$7.50$

27. $4(\$2.69) = \$10.76$ for four six-packs

Savings: $\$10.76 - \$9.60 = \$1.16$

28. Akala's: 2 hr = 120 min, $\dfrac{120}{15} = 8$, $8 \times \$15 = \$120$

Berkman's: 2 hr = 120 min, $\dfrac{120}{30} = 4$,

$4 \times \$25 = \$100$

Berkman's is the better deal by
$\$120 - \$100 = \$20.00$.

29. To produce the 52 Oscars he found:

$52 \times \$327 = \$17,004$

He was awarded

$\$50,000 - \$17,004 = \$32,996$ more.

30. $\$1.50 + \left[\left(10 - \dfrac{1}{5}\right)(5)\right]\$0.30$

$= \$1.50 + \left[\left(\dfrac{50}{5} - \dfrac{1}{5}\right)(5)\right]\$0.30$

$= \$1.50 + \left[\dfrac{49}{5}(5)\right]\$0.30$

$= \$1.50 + 49 \times \$0.30 = \$1.50 + \$14.70 = \$16.20$

31. 10% of $530 = 0.10 \times \$530 = \$53$
    $\$53 \times 7 = \$371$
    Savings: $371 - $60 = $311

32. $\dfrac{1.5 \text{ mg}}{10 \text{ lb}} = \dfrac{x \text{ mg}}{47 \text{ lb}}$
    $10x = 47(1.5)$
    $\dfrac{10x}{10} = \dfrac{70.5}{10}$
    $x = 7.05 \text{ mg}$

33. $\$3800 - 0.30(\$3800) = \$3800 - \$1140$
    $= \$2660$ take-home
    28% of $2660 = 0.28 \times \$2660 = \$744.80$

34. 9 A.M. Eastern is 6 A.M. Pacific,
    from 6 A.M. Pacific to 1:35 P.M. Pacific
    is 7 hr 35 min , 7 hr 35 min $-$ 50 min stop
    = 6 hr 45 min

35. 3 P.M. $- 4 \text{ hr} = 11$ A.M.
    July 26, 11:00 A.M.

36. a) 1 in.$\times$1 in. = 2.54 cm$\times$2.54 cm
    $= 6.4516 \text{ cm}^2 \approx 6.45 \text{ cm}^2$
    b) 1 in.$\times$1 in.$\times$1 in.
    $= 2.54 \text{ cm} \times 2.54 \text{ cm} \times 2.54 \text{ cm}$
    $= 16.387064 \text{ cm}^3 \approx 16.39 \text{ cm}^3$
    c) $\dfrac{1 \text{ in.}}{2.54 \text{ cm}} = \dfrac{x \text{ in.}}{1 \text{ cm}}$
    $2.54x = 1(1)$
    $\dfrac{2.54x}{2.54} = \dfrac{1}{2.54}$
    $x = 0.393700787 \approx 0.39 \text{ in.}$

37. Each figure has an additional two dots. To get the hundredth figure, 97 more figures must be drawn, $97(2) = 194$ dots added to the third figure. Thus, $194 + 7 = 201$.

38.

21	7	8	18
10	16	15	13
14	12	11	17
9	19	20	6

39.

23	25	15
13	21	29
27	17	19

40. 59 min 59 sec   Since it doubles every second, the jar was half full 1 second earlier than 1 hour.

41. 6

42. Nothing. Each friend paid $9 for a total of $27; $25 to the hotel, $2 to the clerk.
    $25 for the room + $3 for each friend + $2 for the clerk = $30

43. Let $x =$ the total weight of the four women
    $\dfrac{x}{4} = 130, \quad x = 520, \quad \dfrac{520 + 180}{5} = \dfrac{700}{5} = 140$ lb

44. Yes; 3 quarters and 4 dimes, or 1 half dollar, 1 quarter and 4 dimes, or 1 quarter and 9 dimes.
    Other answers are possible.

45. 6 cm$\times$6 cm$\times$6 cm = 216 cm^3

46. Place six coins in each pan with one coin off to the side. If it balances, the heavier coin is the one on the side. If the pan does not balance, take the six coins on the heavier side and split them into two groups of three. Select the three heavier coins and weigh two coins. If the pan balances, it is the third coin. If the pan does not balance, you can identify the heavier coin.

47. $\dfrac{n(n+1)}{2} = \dfrac{500(501)}{2} = \dfrac{250,500}{2} = 125,250$

48. 16 blue: 4 green $\rightarrow$ 8 blue, 2 yellow $\rightarrow$ 5 blue, 2 white $\rightarrow$ 3 blue

49. 90: 101, 111, 121, 131, 141, 151, 161, 171, 181, 191,

50. The fifth figure will be an octagon with sides of equal length. Inside the octagon will be a seven sided figure with each side of equal length. The figure will have one antenna.

51. 61: The sixth figure will have 6 rows of 6 tiles and 5 rows of 5 tiles ($6\times 6 + 5\times 5 = 36 + 25 = 61$).

52. Some possible answers are given below. There are other possibilities.

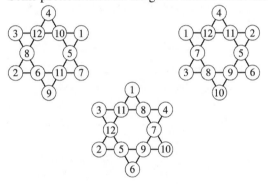

53. a) 2

    b) There are 3 choices for the first spot. Once that person is standing, there are 2 choices for the second spot and 1 for the third. Thus, $3\times 2\times 1 = 6$.

    c) $4\times 3\times 2\times 1 = 24$

    d) $5\times 4\times 3\times 2\times 1 = 120$

    e) $n(n-1)(n-2)\cdots 1$, (or $n!$), where $n = $ the number of people in line

Chapter Test

1. 18, 21, 24 (Add 3 to previous number.)

2. $\dfrac{1}{81}, \dfrac{1}{243}, \dfrac{1}{729}$ (Multiply previous number by $\dfrac{1}{3}$.)

3. a) The result is the original number plus 1.

   b) The result is the original number plus 1.

   c) Conjecture: The result will always be the original number plus 1.

   d) $n, 5n, 5n+10, \dfrac{5n+10}{5} = \dfrac{5n}{5} + \dfrac{10}{5} = n+2, n+2-1 = n+1$

(Note: Answers for #4 - #6 will vary depending on how you round your numbers. The answers may differ from the answers in the back of the textbook. However, your answers should be something near the answers given. All answers are approximate.)

4. $0.06 \times 98,000 \approx 0.06 \times 100,000 = 6000$

5. $\dfrac{102,000}{0.00302} \approx \dfrac{100,000}{0.003} = 33,333,333.\overline{3} \approx 33,000,000$

6. 7 square units

7. a) $\dfrac{130 \text{ lb}}{63 \text{ in.}} = 2.063492063$

   $\dfrac{2.063492063}{63 \text{ in.}} = 0.032753842$

   $0.032753842 \times 703 = 23.02595093 \approx 23.03$

   b) He is in the at risk range.

8. $\$122.13 - \$9.63 = \$112.50$

   $\dfrac{\$112.50}{\$0.72} = 156.25 \text{ therms}$

   $156.25 \text{ therms} + \text{ first 3 therms} = 159.25 \text{ therms}$

9. $\dfrac{\$15}{\$2.59} = 5.791505792$

   The maximum number of 6 packs is 5.

   $\$15.00 - (5 \times \$2.59) = \$15.00 - \$12.95 = \$2.05$

   $\dfrac{\$2.05}{\$0.80} = 2.5625$

   Thus, two individual cans can be purchased.

6 packs	Indiv. cans	Number of cans
5	2	32
4	5	29
3	9	27
2	12	24
1	15	21
0	18	18

   The maximum number of cans is 32.

10. 1 cut yields 2 equal pieces. Cut each of these 2 equal pieces to get 4 equal pieces.

    3 cuts $\rightarrow$ 3(2.5 min) = 7.5 min

11. 2.5 in. by 1.875 in.

    $\approx 2.5 \times 15.8$ by $1.875 \times 15.8 = 39.5$ in. by 29.625 in.

    ≈ 39.5 in. by 29.6 in.

    (The actual dimensions are 100.5 cm by 76.5 cm.)

12. $\$12.75 \times 40 = \$510$

    $\$12.75 \times 1.5 \times 10 = \$191.25$

    $\$510 + \$191.25 = \$701.25$

    $\$701.25 - \$652.25 = \$49.00$

13.

40	15	20
5	25	45
30	35	10

14. Christine drove the first 15 miles at 60 mph which took $\dfrac{15}{60} = \dfrac{1}{4}$ hr, and the second 15 miles at 30 mph which took $\dfrac{15}{30} = \dfrac{1}{2}$ hr for a total time of $\dfrac{3}{4}$ hr. If she drove the entire 30 miles at 45 mph, the trip would take $\dfrac{30}{45} = \dfrac{2}{3}$ hr (40 min) which is less than $\dfrac{3}{4}$ hr (45 min).

15. $2\times6\times8\times9\times13 = 11,232$; 11 does not divide 11,232.

16. 243 jelly beans; $260-17 = 243, 234+9 = 243, 274-31 = 243$

17. a) $3 \times \$3.99 = \$11.97$

b) $9(\$1.75\times0.75) = 11.8125 \approx \$11.81$

c) $\$11.97 - \$11.81 = \$0.16$ Using the coupon is least expensive by $0.16.

18. 8: $\$ \rightarrow$ on $* \rightarrow$ off

$\$\$\$\$$, $\$\$\$*$, $\$\$*\$$, $\$*\$\$$, $*\$\$\$$, $*\$*\$$, $*\$\$*$, $\$*\$*$

Group Projects

1. a) $\dfrac{\$325}{3} \approx \$108.33$

b) Let $x = $ the amount before tax

$x+0.07x = 325$

$\dfrac{1.07x}{1.07} = \dfrac{325}{1.07}$

$x = 303.7383178 \approx \$303.74$

$\dfrac{\$303.74}{3} = 101.24\overline{6} \approx \$101.25$

c) Inductive reasoning - arriving at a general conclusion from specific cases

d) Combination set: $\$62.00-(\$62.00\times0.10) = \$62.00-\$6.20 = \$55.80$

Individual sets: $2\times\$36.00 = \$72.00, \$72.00-(\$72.00\times0.20) = \$72.00-\$14.40 = \$57.60$

Therefore, the combination set is cheaper.

e) Combintion with tax: $\$55.80 \times 1.07 \approx \$59.71$

Individual set with tax: $\$57.60 \times 1.07 \approx \$61.63$

$\$61.63 - \$59.71 = \$1.92$

2. a)   d) Answers will vary.

e)   400 mi $\div$ 50 mi/hr $= 8$ hrs, 9 A.M. $+ 8$ hrs $= 5$ P.M.

f)   h) Answers will vary.

3.

Order	Name	Apparel
1	Ernie	holster
2	Zeke	vest
3	Jed	chaps
4	Tex	stetson

CHAPTER TWO

SETS

Exercise Set 2.1

1. A **set** is a collection of objects.
2. An **ellipsis** is three dots in a set indicating the elements continue in the same manner.
3. Description: the set of counting numbers less than 7

   Roster form: $\{1, 2, 3, 4, 5, 6\}$

   Set-builder notation: $\{x \mid x \in N \text{ and } x < 7\}$

4. A set is **finite** if it either contains no elements or the number of elements in the set is a natural number.
5. An **infinite** set is a set that is not finite.
6. Set A is **equal** to set B, symbolized by $A = B$, if and only if they contain exactly the same elements.
7. Two sets are **equivalent** if they contain the same number of elements.
8. The **cardinal number** of a set A, symbolized by $n(A)$, is the number of elements in set A.
9. A set that contains no elements is called the **empty set or null set**.
10. $\{\ \}, \varnothing$
11. Set A and set B can be placed in **one-to-one correspondence** if every element of set A can be matched with exactly one element of set B and every element of set B can be matched with exactly one element of set A.
12. A **universal set**, symbolized by U, is a set that contains all the elements for any specific discussion.
13. Not well defined, "large" is interpreted differently by different people.
14. Not well defined, "best" is interpreted differently by different people.
15. Well defined, the contents can be clearly determined.
16. Well defined, the contents can be clearly determined.
17. Well defined, the contents can be clearly determined.
18. Not well defined, "nicest" is interpreted differently by different people.
19. Infinite, the number of elements in the set is not a natural number.
20. Finite, the number of elements in the set is a natural number.
21. Infinite, the number of elements in the set is not a natural number.
22. Infinite, the number of elements in the set is not a natural number.
23. Infinite, the number of elements in the set is not a natural number.
24. Finite, the number of elements in the set is a natural number.
25. $\{$Atlantic, Pacific, Arctic, Indian$\}$
26. $\{$Idaho, Illinois, Indiana, Iowa$\}$
27. $\{11, 12, 13, 14, \ldots, 177\}$
28. $C = \{4\}$
29. $B = \{2, 4, 6, 8, \ldots\}$
30. $\{\ \}$ or $\varnothing$
31. $\{\ \}$ or $\varnothing$
32. $\{$Hawaii, Alaska$\}$
33. $E = \{6, 7, 8, 9, \ldots, 71\}$
34. $\{$Mark McGwire$\}$

35. {Sony DSC-S50, Sony DSC-S70, Sony Mavica FD-90}

36. {Olympus D-360L}

37. {Sony Mavica FD-73, Olympus D-360L, Sony DSC-S50, Kodak DC215, H-P Photo Smart C315}

38. {Sony DSC-S50, Sony DSC-S70, Sony Mavica FD-90}

39. $\{2002, 2003, 2004, 2005, 2006, 2007, 2008\}$

40. $\{2005, 2006, 2007, 2008\}$

41. $\{2005, 2006, 2007, 2008\}$

42. $\{2002, 2003, 2004\}$

43. $B = \{x | x \in N \text{ and } 3 < x < 11\}$ or

    $B = \{x | x \in N \text{ and } 4 \le x \le 10\}$

44. $A = \{x | x \in N \text{ and } x < 8\}$ or

    $A = \{x | x \in N \text{ and } x \le 7\}$

45. $C = \{x | x \in N \text{ and } x \text{ is a multiple of } 3\}$

46. $D = \{x | x \in N \text{ and } x \text{ is a multiple of } 5\}$

47. $E = \{x | x \in N \text{ and } x \text{ is odd}\}$

48. $A = \{x | x \text{ is Labor Day}\}$

49. $C = \{x | x \text{ is February}\}$

50. $F = \{x | x \in N \text{ and } 14 < x < 101\}$ or $F = \{x | x \in N \text{ and } 15 \le x \le 100\}$

51. Set A is the set of natural numbers less than or equal to 7.
52. Set D is the set of natural numbers that are multiples of 4.
53. Set V is the set of vowels in the English alphabet.
54. Set S is the set of the seven dwarfs in *Snow White and the Seven Dwarfs*.
55. Set C is the set of companies that make calculators.
56. Set B is the set of the five longest rivers in the United States.
57. Set B is the set of members of the Beatles.
58. Set E is the set of natural numbers greater than 5 and less than or equal to 12.

59. {St. Louis}

60. {Scranton}

61. { } or $\varnothing$

62. {Spokane, Detroit}

63. $\{1999, 2000, 2001, 2002\}$

64. $\{1998\}$

65. $\{1999, 2001, 2002\}$

66. { } or $\varnothing$

67. False; $\{b\}$ is a set, and not an element of the set.

68. True; b is an element of the set.

69. False; h is not an element of the set.

70. True; Cat in the Hat is an element of the set.

71. False; 3 is an element of the set.

72. False; the capital of Hawaii is Honolulu, not Maui.

73. True; *Titanic* is an element of the set.

74. False; 2 is an even natural number.

75. $n(A) = 4$

76. $n(B) = 6$

77. $n(C) = 0$

78. $n(D) = 5$

79. Both; A and B contain exactly the same elements.
80. Equivalent; both sets contain the same number of elements, 3.
81. Neither; the sets have a different number of elements.
82. Neither; not all dogs are collies.
83. Equivalent; both sets contain the same number of elements, 3.
84. Equivalent; both sets contain the same number of elements, 50.

85. a) Set A is the set of natural numbers greater than 2. Set B is the set of all numbers greater than 2.

b) Set A contains only natural numbers. Set B contains other types of numbers, including fractions and decimal numbers.

c) $A = \{3, 4, 5, 6, \ldots\}$

d) No; set B cannot be written in roster form since we cannot list all the elements in set B.

86. a) Set A is the set of natural numbers greater than 2 and less than or equal to 5. Set B is the set of numbers greater than 2 and less than or equal to 5.

b) Set A contains only natural numbers. Set B contains other types of numbers, including fractions and decimal numbers.

c) $A = \{3, 4, 5\}$

d) No, set B cannot be written in roster form since there is no smallest number that is greater than 2.

87. Cardinal; 12 tells how many.

88. Ordinal; 25 tells the relative position of the chart.

89. Ordinal; sixteenth tells Lincoln s relative position.

90. Cardinal; 35 tells how many dollars she spent.

91. Answers will vary.

92. Answers will vary. Examples: the set of people in the class who were born on the moon, the set of automobiles that get 400 miles on a gallon of gas, the set of fish that can talk

93. Answers will vary.

94. Answers will vary. Here are some examples.

a) The set of men. The set of actors. The set of people over 12 years old. The set of people with two legs. The set of people who have been in a movie.

b) The set of all the people in the world.

95.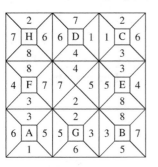

Exercise Set 2.2

1. Set A is a **subset** of set B, symbolized by $A \subseteq B$, if and only if all the elements of set A are also elements of set B.

2. Set A is a **proper subset** of set B, symbolized by $A \subset B$, if and only if all the elements of set A are also elements of set B and set $A \neq$ set B.

3. If $A \subseteq B$, then every element of set A is also an element of set B. If $A \subset B$, then every element of set A is also an element of set B and set $A \neq$ set B.

4. 2^n, where n is the number of elements in the set.

5. $2^n - 1$, where n is the number of elements in the set.

6. No, if two sets are equal one cannot be a proper subset of the other.

7. False; gold is an element of the set, not a subset.

8. False; the empty set is a subset of $\{knee, ankle, shoulder, hip\}$.

9. True; the empty set is a subset of every set.

10. False; red is an element of the set, not a proper subset.

11. True; 5 is not an element of $\{2,4,6\}$.

12. False; Pete and Mike are not in the second set.

13. False; the set $\{\varnothing\}$ contains the element $\varnothing$.

14. True; $\{engineer\}$ is a subset of $\{architect, physician, attorney, engineer\}$.

15. True; $\{\ \}$ and $\varnothing$ each represent the empty set.

16. False; 0 is a number and $\{\ \}$ is a set.

17. False; the set $\{0\}$ contains the element 0.

18. True; $\{3,8,11\}$ is a subset of $\{3,8,11\}$.

19. False; $\{swimming\}$ is a subset, not an element.

20. True; $\{3,5,9\} = \{3,9,5\}$.

21. True; the empty set is a subset of every set, including itself.

22. True; the elements of the set are themselves sets.

23. False; no set is a proper subset of itself.

24. True; $\{b,a,t\}$ is a subset of $\{t,a,b\}$.

25. $B \subseteq A, B \subset A$

26. $A = B, A \subseteq B, B \subseteq A$

27. $A \subseteq B, A \subset B$

28. None

29. $B \subseteq A, B \subset A$

30. $B \subseteq A, B \subset A$

31. $A = B, A \subseteq B, B \subseteq A$

32. $B \subseteq A, B \subset A$

33. $\{\ \}$ is the only subset.

34. $\{\ \},\{O\}$

35. $\{\ \},\{pen\},\{pencil\},\{pen, pencil\}$

36. $\{\ \},\{apple\},\{peach\},\{banana\},\{apple, peach\},$
$\{apple, banana\},\{peach, banana\},$
$\{apple, peach, banana\}$

37. a) $\{\ \},\{a\},\{b\},\{c\},\{d\},\{a,b\},\{a,c\},\{a,d\},$
$\{b,c\},\{b,d\},\{c,d\},\{a,b,c\},\{a,b,d\},$
$\{a,c,d\},\{b,c,d\},\{a,b,c,d\}$

b) All the sets in part a) are proper subsets of A except $\{a,b,c,d\}$.

38. a) $2^9 = 2 \times 2 \times 2 \times 2 \times 2 \times 2 \times 2 \times 2 \times 2 = 512$ subsets
b) $2^9 - 1 = 512 - 1 = 511$ proper subsets

39. False; A could be equal to B.

40. True; every proper subset is a subset.

41. True; every set is a subset of itself.

42. False; no set is a proper subset of itself.

43. True; $\varnothing$ is a proper subset of every set except itself.

44. True; $\varnothing$ is a subset of every set.

45. True; every set is a subset of the universal set.

46. False; a set cannot be a proper subset of itself.

47. True; $\varnothing$ is a proper subset of every set except itself and $U = \varnothing$.

48. False; the only subset of $\varnothing$ is itself and $U = \varnothing$.

49. True; $\varnothing$ is a subset of every set.

50. False; U is not a subset of $\varnothing$.
(See answer for #48.)

51. The number of different variations of the house is equal to the number of subsets of $\{deck, jacuzzi, security system, hardwood flooring\}$, which is $2^4 = 2 \times 2 \times 2 \times 2 = 16$.

52. The number of options is equal to the number of subsets of $\{RAM, modem, video card, hard drive, processor, sound card\}$, which is $2^6 = 2 \times 2 \times 2 \times 2 \times 2 \times 2 = 64$.

53. The number of different variations is equal to the number of subsets of
    {call waiting, call forwarding, caller identification, three way calling, voice mail, fax line},

    which is $2^6 = 2 \times 2 \times 2 \times 2 \times 2 \times 2 = 64$.

54. The number of variations is equal to the number of subsets of
    {ketchup, mustard, relish, hot sauce, onions, lettuce, tomato}, which is $2^7 = 2 \times 2 \times 2 \times 2 \times 2 \times 2 \times 2 = 128$.

55. $E = F$ since they are both subsets of each other.

56. Count the number of boys then count the number of girls. If the number is the same, then they are equivalent.

57. a) Yes, because a is a member of set D.

    b) No, c is an element of set D.

    c) Yes, each element of $\{a, b\}$ is an element of set D.

58. a) Each person has 2 choices, namely yes or no. $2 \times 2 \times 2 \times 2 = 16$

    b) YYYY, YYYN, YYNY, YNYY, NYYY, YYNN, YNYN, YNNY, NYNY, NNYY, NYYN, YNNN, NYNN, NNYN, NNNY, NNNN

    c) 5 out of 16

59. A one element set has one proper subset, namely the empty set. A one element set has two subsets, namely itself and the empty set. One is one-half of two. Thus, the set must have one element.

60. Yes

61. Yes

62. No

Section 2.3

1.

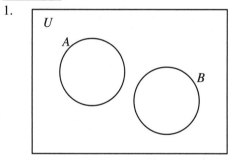

2.

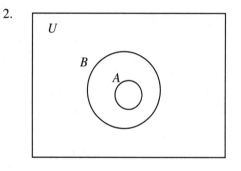

3.

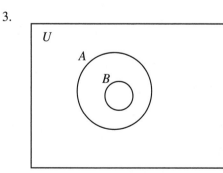

4.

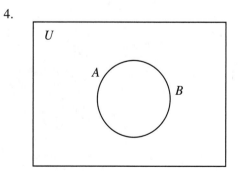

5.

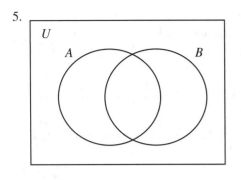

6. Determine the elements that are in the universal set that are not in set A.
7. Combine the elements from set A and set B into one set. List any element that is contained in both sets only once.
8. I, II, III
9. Take the elements common to both set A and set B.
10. II
11. a) *Or* is generally interpreted to mean *union*.
    b) *And* is generally interpreted to mean *intersection*.
12. $n(A \cup B) = n(A) + n(B) - n(A \cap B)$
13. Region II, the intersection of the two sets.
14. Region IV which contains any element not belonging to either set.

15.

16.

17.

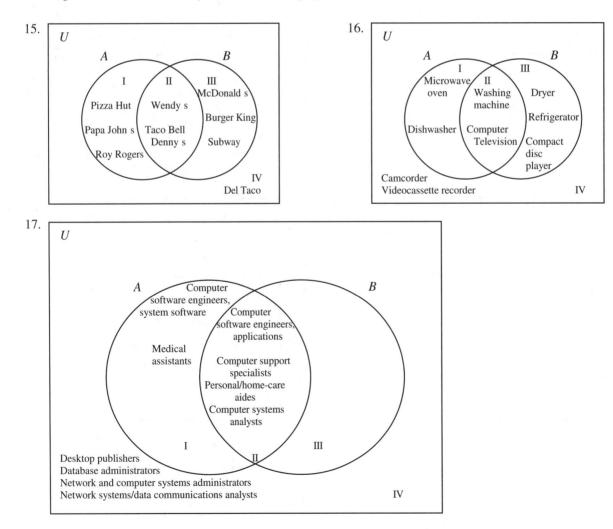

18.

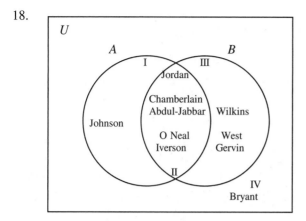

19. The set of U.S. colleges and universities that are not in the state of North Dakota

20. The set of marbles in the box that contain no blue coloring

21. The set of insurance companies in the U.S. that do not offer life insurance

22. The set of insurance companies in the U.S. that do no offer car insurance

23. The set of insurance companies in the U.S. that offer life insurance or car insurance

24. The set of insurance companies in the U.S. that offer life insurance and car insurance

25. The set of insurance companies in the U.S. that offer life insurance and do not offer car insurance

26. The set of insurance companies in the U.S. that offer life insurance or do not offer car insurance

27. The set of U.S. corporations whose headquarters are in New York State and whose chief executive officer is a woman

28. The set of U.S. corporations whose headquarters are in New York State or that employ at least 100 people

29. The set of U.S. corporations whose chief executive officer is not a woman and who employ at least 100 people

30. The set of U.S. corporations whose headquarters are in New York State and whose chief executive officer is a woman and who employ at least 100 people

31. The set of U.S. corporations whose headquarters are in New York State or whose chief executive officer is a woman or that employ at least 100 people

32. The set of U.S. corporations whose headquarters are not in New York State or that do not employ at least 100 people

33. $A = \{a, b, c, h, t, w\}$

34. $B = \{a, f, g, h, r\}$

35. $A \cap B = \{a, b, c, h, t, w\} \cap \{a, f, g, h, r\} = \{a, h\}$

36. $U = \{c, w, b, t, a, h, f, g, r, p, m, z\}$

37. $A \cup B = \{a, b, c, h, t, w\} \cup \{a, f, g, h, r\} = \{a, b, c, f, g, h, r, t, w\}$

38. $(A \cup B)'$ From #37, $A \cup B = \{a, b, c, f, g, h, r, t, w\}$. $(A \cup B)' = \{a, b, c, f, g, h, r, t, w\}' = \{m, p, z\}$

39. $A' \cap B' = \{a, b, c, h, t, w\}' \cap \{a, f, g, h, r\}' = \{f, g, r, p, m, z\} \cap \{c, w, b, t, p, m, z\} = \{p, m, z\}$

40. $(A \cap B)'$ From #35, $A \cap B = \{a, h\}$. $(A \cap B)' = \{a, h\}' = \{b, c, f, g, m, p, r, t, w, z\}$

41. $A = \{L, \Delta, @, *, \$\}$

42. $B = \{*, \$, R, \square, \alpha\}$

43. $U = \{L, \Delta, @, *, \$, R, \square, \alpha, \infty, \Sigma, Z\}$

44. $A \cup B = \{L, \Delta, @, *, \$\} \cup \{*, \$, R, \square, \alpha\} = \{L, \Delta, @, *, \$, R, \square, \alpha\}$

45. $A \cap B = \{L, \Delta, @, *, \$\} \cap \{*, \$, R, \square, \alpha\} = \{*, \$\}$

46. $A \cup B' = \{L, \Delta, @, *, \$\} \cup \{*, \$, R, \square, \alpha\}' = \{L, \Delta, @, *, \$\} \cup \{L, \Delta, @, \infty, \Sigma, Z\} = \{L, \Delta, @, *, \$, \infty, \Sigma, Z\}$

47. $A' \cap B = \{L, \Delta, @, *, \$\}' \cap \{*, \$, R, \square, \alpha\} = \{R, \square, \alpha, \infty, \Sigma, Z\} \cap \{*, \$, R, \square, \alpha\} = \{R, \square, \alpha\}$

48. $(A \cup B)'$ From #44, $A \cup B = \{L, \Delta, @, *, \$, R, \square, \alpha\}$. $(A \cup B)' = \{L, \Delta, @, *, \$, R, \square, \alpha\}' = \{\infty, \Sigma, Z\}$

49. $A \cup B = \{1, 2, 4, 5, 8\} \cup \{2, 3, 4, 6\} = \{1, 2, 3, 4, 5, 6, 8\}$

50. $A \cap B = \{1, 2, 4, 5, 8\} \cap \{2, 3, 4, 6\} = \{2, 4\}$

51. $B' = \{2, 3, 4, 6\}' = \{1, 5, 7, 8\}$

52. $A \cup B' = \{1, 2, 4, 5, 8\} \cup \{2, 3, 4, 6\}' = \{1, 2, 4, 5, 8\} \cup \{1, 5, 7, 8\} = \{1, 2, 4, 5, 7, 8\}$

53. $(A \cup B)'$ From #49, $A \cup B = \{1, 2, 3, 4, 5, 6, 8\}$. $(A \cup B)' = \{1, 2, 3, 4, 5, 6, 8\}' = \{7\}$

54. $A' \cap B' = \{1, 2, 4, 5, 8\}' \cap \{2, 3, 4, 6\}' = \{3, 6, 7\} \cap \{1, 5, 7, 8\} = \{7\}$

55. $(A \cup B)' \cap B$ From #53, $(A \cup B)' = \{7\}$. $(A \cup B)' \cap B = \{7\} \cap \{2, 3, 4, 6\} = \{\ \}$

56. $(A \cup B) \cap (A \cup B)'$ From #49, $A \cup B = \{1, 2, 3, 4, 5, 6, 8\}$ and from #53, $(A \cup B)' = \{7\}$.

    $(A \cup B) \cap (A \cup B)' = \{1, 2, 3, 4, 5, 6, 8\} \cap \{7\} = \{\ \}$

57. $(B \cup A)' \cap (B' \cup A')$ From #53, $(A \cup B)' = (B \cup A)' = \{7\}$.

    $(B \cup A)' \cap (B' \cup A') = \{7\} \cap \left(\{2, 3, 4, 6\}' \cup \{1, 2, 4, 5, 8\}'\right) = \{7\} \cap \left(\{1, 5, 7, 8\} \cup \{3, 6, 7\}\right)$

    $= \{7\} \cap \{1, 3, 5, 6, 7, 8\} = \{7\}$

58. $A' \cup (A \cap B)$ From #50, $A \cap B = \{2, 4\}$. $A' \cup (A \cap B) = \{1, 2, 4, 5, 8\}' \cup \{2, 4\} = \{3, 6, 7\} \cup \{2, 4\} = \{2, 3, 4, 6, 7\}$

59. $B' = \{b, c, d, f, g\}' = \{a, e, h, i, j, k\}$

60. $B \cup C = \{b, c, d, f, g\} \cup \{a, b, f, i, j\} = \{a, b, c, d, f, g, i, j\}$

61. $A \cap C = \{a, c, d, f, g, i\} \cap \{a, b, f, i, j\} = \{a, f, i\}$

62. $A \cup B'$ From #59, $B' = \{a, e, h, i, j, k\}$. $A \cup B' = \{a, c, d, f, g, i\} \cup \{a, e, h, i, j, k\} = \{a, c, d, e, f, g, h, i, j, k\}$

63. $(A \cap C)'$ From #61, $A \cap C = \{a, f, i\}$. $(A \cap C)' = \{a, f, i\}' = \{b, c, d, e, g, h, j, k\}$

64. $(A \cap B) \cup C = \left(\{a, c, d, f, g, i\} \cap \{b, c, d, f, g\}\right) \cup \{a, b, f, i, j\} = \{c, d, f, g\} \cup \{a, b, f, i, j\} = \{a, b, c, d, f, g, i, j\}$

65. $A \cup (C \cap B)' = \{a, c, d, f, g, i\} \cup \left(\{a, b, f, i, j\} \cap \{b, c, d, f, g\}\right)' = \{a, c, d, f, g, i\} \cup \{b, f\}'$

    $= \{a, c, d, f, g, i\} \cup \{a, c, d, e, g, h, i, j, k\} = \{a, c, d, e, f, g, h, i, j, k\}$

66. $A \cup (C' \cup B') = \{a, c, d, f, g, i\} \cup \left(\{a, b, f, i, j\}' \cup \{b, c, d, f, g\}'\right)$

    $= \{a, c, d, f, g, i\} \cup \left(\{c, d, e, g, h, k\} \cup \{a, e, h, i, j, k\}\right) = \{a, c, d, f, g, i\} \cup \{a, c, d, e, g, h, i, j, k\}$

    $= \{a, c, d, e, f, g, h, i, j, k\}$

67. $(A' \cup C) \cup (A \cap B) = \left(\{a, c, d, f, g, i\}' \cup \{a, b, f, i, j\}\right) \cup \left(\{a, c, d, f, g, i\} \cap \{b, c, d, f, g\}\right)$

    $= \left(\{b, e, h, j, k\} \cup \{a, b, f, i, j\}\right) \cup \{c, d, f, g\} = \{a, b, e, f, h, i, j, k\} \cup \{c, d, f, g\}$

    $= \{a, b, c, d, e, f, g, h, i, j, k\}$, or U

68. $(C \cap B) \cap (A' \cap B)$ From #65, $C \cap B = \{b, f\}$.

$(C \cap B) \cap (A' \cap B) = \{b, f\} \cap \left(\{a, c, d, f, g, i\}' \cap \{b, c, d, f, g\}\right) = \{b, f\} \cap \left(\{b, e, h, j, k\} \cap \{b, c, d, f, g\}\right)$

$= \{b, f\} \cap \{b\} = \{b\}$

For exercises 69-82: $U = \{1, 2, 3, 4, 5, 6, 7, 8, 9\}$, $A = \{1, 3, 5, 7, 9\}$, $B = \{2, 4, 6, 8\}$, $C = \{1, 2, 3, 4, 5\}$

69. $A \cap B = \{1, 3, 5, 7, 9\} \cap \{2, 4, 6, 8\} = \{\ \}$

70. $A \cup B = \{1, 3, 5, 7, 9\} \cup \{2, 4, 6, 8\} = \{1, 2, 3, 4, 5, 6, 7, 8, 9\}$, or U

71. $A' \cup B = \{1, 3, 5, 7, 9\}' \cup \{2, 4, 6, 8\} = \{2, 4, 6, 8\} \cup \{2, 4, 6, 8\} = \{2, 4, 6, 8\}$, or B

72. $(B \cup C)' = (\{2, 4, 6, 8\} \cup \{1, 2, 3, 4, 5\})' = \{1, 2, 3, 4, 5, 6, 8\}' = \{7, 9\}$

73. $A \cap C' = \{1, 3, 5, 7, 9\} \cap \{1, 2, 3, 4, 5\}' = \{1, 3, 5, 7, 9\} \cap \{6, 7, 8, 9\} = \{7, 9\}$

74. $A \cap B' = \{1, 3, 5, 7, 9\} \cap \{2, 4, 6, 8\}' = \{1, 3, 5, 7, 9\} \cap \{1, 3, 5, 7, 9\} = \{1, 3, 5, 7, 9\}$, or A

75. $(B \cap C)' = (\{2, 4, 6, 8\} \cap \{1, 2, 3, 4, 5\})' = \{2, 4\}' = \{1, 3, 5, 6, 7, 8, 9\}$

76. $(A \cup C) \cap B = (\{1, 3, 5, 7, 9\} \cup \{1, 2, 3, 4, 5\}) \cap \{2, 4, 6, 8\} = \{1, 2, 3, 4, 5, 7, 9\} \cap \{2, 4, 6, 8\} = \{2, 4\}$

77. $(C \cap B) \cup A$ From #75, $C \cap B = \{2, 4\}$. $(C \cap B) \cup A = \{2, 4\} \cup \{1, 3, 5, 7, 9\} = \{1, 2, 3, 4, 5, 7, 9\}$

78. $(C' \cup A) \cap B = \left(\{1, 2, 3, 4, 5\}' \cup \{1, 3, 5, 7, 9\}\right) \cap \{2, 4, 6, 8\} = (\{6, 7, 8, 9\} \cup \{1, 3, 5, 7, 9\}) \cap \{2, 4, 6, 8\}$

$= \{1, 3, 5, 6, 7, 8, 9\} \cap \{2, 4, 6, 8\} = \{6, 8\}$

79. $(A' \cup C) \cap B = \left(\{1, 3, 5, 7, 9\}' \cup \{1, 2, 3, 4, 5\}\right) \cap \{2, 4, 6, 8\} = (\{2, 4, 6, 8\} \cup \{1, 2, 3, 4, 5\}) \cap \{2, 4, 6, 8\}$

$= \{1, 2, 3, 4, 5, 6, 8\} \cap \{2, 4, 6, 8\} = \{2, 4, 6, 8\}$, or B

80. $(A \cap B)' \cup C$ From #69, $A \cap B = \{\ \}$.

$(A \cap B)' \cup C = \{\ \}' \cup \{1, 2, 3, 4, 5\} = \{1, 2, 3, 4, 5, 6, 7, 8, 9\} \cup \{1, 2, 3, 4, 5\} = \{1, 2, 3, 4, 5, 6, 7, 8, 9\}$, or U

81. $(A' \cup B') \cap C = \left(\{1, 3, 5, 7, 9\}' \cup \{2, 4, 6, 8\}'\right) \cap \{1, 2, 3, 4, 5\}$

$= (\{2, 4, 6, 8\} \cup \{1, 3, 5, 7, 9\}) \cap \{1, 2, 3, 4, 5\} = \{1, 2, 3, 4, 5, 6, 7, 8, 9\} \cap \{1, 2, 3, 4, 5\} = \{1, 2, 3, 4, 5\}$, or C

82. $(A' \cap C) \cup (A \cap B)$ From #69, $A \cap B = \{\ \}$.

$(A' \cap C) \cup (A \cap B) = \left(\{1, 3, 5, 7, 9\}' \cap \{1, 2, 3, 4, 5\}\right) \cup \{\ \} = (\{2, 4, 6, 8\} \cap \{1, 2, 3, 4, 5\}) \cup \{\ \} = \{2, 4\} \cup \{\ \} = \{2, 4\}$

83. A set and its complement will always be disjoint since the complement of a set is all of the elements in the universal set that are not in the set. Therefore, a set and its complement will have no elements in common.

    For example, if $U = \{1, 2, 3\}$, $A = \{1, 2\}$, and $A' = \{3\}$, then $A \cap A' = \{\ \}$.

84. $n(A \cap B) = 0$ when A and B are disjoint sets. For example, if $U = \{1, 2, 3, 4, 5, 6\}$, $A = \{1, 3\}$, $B = \{2, 4\}$,

    then $A \cap B = \{\ \}$. $n(A \cap B) = 0$

85. Let $A = \{\text{visitors who visited the Hollywood Bowl}\}$ and $B = \{\text{visitors who visited Disneyland}\}$.

$$n(A \cup B) = n(A) + n(B) - n(A \cap B) = 27 + 38 - 16 = 49$$

86. Let $A = \{\text{students who sang in the chorus}\}$ and $B = \{\text{students who played in the stage band}\}$.

$$n(A \cup B) = n(A) + n(B) - n(A \cap B)$$
$$46 = n(A) + 30 - 4$$
$$46 = n(A) + 26$$
$$46 - 26 = n(A) + 26 - 26$$
$$20 = n(A)$$

87. a) $A \cup B = \{a,b,c,d\} \cup \{b,d,e,f,g,h\} = \{a,b,c,d,e,f,g,h\}$, $n(A \cup B) = 8$,

$A \cap B = \{a,b,c,d\} \cap \{b,d,e,f,g,h\} = \{b,d\}$, $n(A \cap B) = 2$.

$n(A) + n(B) - n(A \cap B) = 4 + 6 - 2 = 8$

Therefore, $n(A \cup B) = n(A) + n(B) - n(A \cap B)$.

b) Answers will vary.

c) Elements in the intersection of A and B are counted twice in $n(A) + n(B)$.

88. $A \cap B'$ defines Region I. $A \cap B$ defines Region II. $A' \cap B$ defines Region III.

$A' \cap B'$ or $(A \cup B)'$ defines Region IV.

89. $A \cup B = \{1,2,3,4,...\} \cup \{4,8,12,16,...\} = \{1,2,3,4,...\}$, or A

90. $A \cap B = \{1,2,3,4,...\} \cap \{4,8,12,16,...\} = \{4,8,12,16,...\}$, or B

91. $B \cap C = \{4,8,12,16,...\} \cap \{2,4,6,8,...\} = \{4,8,12,16,...\}$, or B

92. $B \cup C = \{4,8,12,16,...\} \cup \{2,4,6,8,...\} = \{2,4,6,8,...\}$, or C

93. $A \cap C = \{1,2,3,4,...\} \cap \{2,4,6,8,...\} = \{2,4,6,8,...\}$, or C

94. $A' \cap C = \{1,2,3,4,...\}' \cap \{2,4,6,8,...\} = \{0\} \cap \{2,4,6,8,...\} = \{\ \}$

95. $B' \cap C = \{4,8,12,16,...\}' \cap \{2,4,6,8,...\} = \{0,1,2,3,5,6,7,9,10,11,13,14,15,...\} \cap \{2,4,6,8,...\}$

$= \{2,6,10,14,18,...\}$

96. $(B \cup C)' \cup C$ From #92, $B \cup C = C$. $(B \cup C)' \cup C = C' \cup C = \{2,4,6,8,...\}' \cup \{2,4,6,8,...\}$

$= \{0,1,2,3,4,...\}$, or U

97. $(A \cap C) \cap B'$ From #93, $A \cap C = C$. $(A \cap C) \cap B' = C \cap B'$.

From #95, $B' \cap C = C \cap B' = \{2,6,10,14,18,...\}$

98. $U' \cap (A \cup B)$ From #89, $A \cup B = A$. $U' \cap (A \cup B) = U' \cap A = \{\ \} \cap \{1,2,3,4,...\} = \{\ \}$

99. $A \cup A' = U$

100. $A \cap A' = \{\ \}$

101. $A \cup \varnothing = A$

102. $A \cap \varnothing = \varnothing$

103. $A' \cup U = U$

104. $A \cap U = A$

105. $A \cup U = U$

106. $A \cup U' = A \cup \{\ \} = A$

107. If $A \cap B = B$, then $B \subseteq A$.

108. If $A \cup B = B$, then $A \subseteq B$.

109. If $A \cap B = \varnothing$, then A and B are disjoint sets.

110. If $A \cup B = A$, then $B \subseteq A$.

111. If $A \cap B = A$, then A $\subseteq$ B.

112. If $A \cup B = \emptyset$, then $A = \emptyset$ and $B = \emptyset$. Therefore, they are equal sets.

113. $A - B = \{b,c,e,f,g,h\} - \{a,b,c,g,i\} = \{e,f,h\}$

114. $B - A = \{a,b,c,g,i\} - \{b,c,e,f,g,h\} = \{a,i\}$

115. $A' - B = \{b,c,e,f,g,h\}' - \{a,b,c,g,i\}$

$= \{a,d,i,j,k\} - \{a,b,c,g,i\} = \{d,j,k\}$

116. $A - B' = \{b,c,e,f,g,h\} - \{a,b,c,g,i\}'$

$= \{b,c,e,f,g,h\} - \{d,e,f,h,j,k\} = \{b,c,g\}$

117. $A - B = \{2,4,5,7,9,11,13\} - \{1,2,4,5,6,7,8,9,11\}$

$= \{13\}$

118. $B - A = \{1,2,4,5,6,7,8,9,11\} - \{2,4,5,7,9,11,13\}$

$= \{1,6,8\}$

119. $(A - B)'$ From #117, $A - B = \{13\}$.

$(A - B)' = \{13\}'$

$= \{1,2,3,4,5,6,7,8,9,10,11,12,14,15\}$

120. $A - B'$

$= \{2,4,5,7,9,11,13\} - \{1,2,4,5,6,7,8,9,11\}'$

$= \{2,4,5,7,9,11,13\} - \{3,10,12,13,14,15\}$

$= \{2,4,5,7,9,11\}$

121. $(B - A)'$ From #118, $B - A = \{1,6,8\}$.

$(B - A)' = \{1,6,8\}'$

$= \{2,3,4,5,7,9,10,11,12,13,14,15\}$

122. $A \cap (A - B)$ From #117, $A - B = \{13\}$.

$A \cap (A - B) = \{2,4,5,7,9,11,13\} \cap \{13\} = \{13\}$

123. Complement

124.

Exercise Set 2.4

1. 8

2. Region V, the intersection of all three sets

3. Regions II, IV, VI

4. $A \cap B$ is represented by regions II and V. If $A \cap B$ contains 10 elements and region V contains 6 elements, then region II contains $10 - 6 = 4$ elements.

5. $B \cap C$ is represented by regions V and VI. If $B \cap C$ contains 12 elements and region V contains 4 elements, then region VI contains $12 - 4 = 8$ elements.

6. $(A \cup B)' = A' \cap B'; (A \cap B)' = A' \cup B'$

7. a) Yes

$A \cup B = \{1,4,5\} \cup \{1,4,5\} = \{1,4,5\}$

$A \cap B = \{1,4,5\} \cap \{1,4,5\} = \{1,4,5\}$

b) No, one specific case cannot be used as proof.

c)

$A \cup B$		$A \cap B$	
Set	Regions	Set	Regions
A	I, II	A	I, II
B	II, III	B	II, III
$A \cup B$	I, II, III	$A \cap B$	II

Since the two statements are not represented by the same regions, $A \cup B \neq A \cap B$ for all sets A and B.

8. Deductive reasoning – the process of reasoning to a specific conclusion from a general statement.

9.

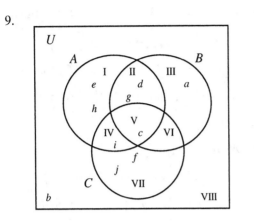

10.

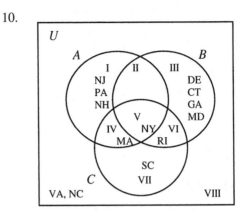

11.

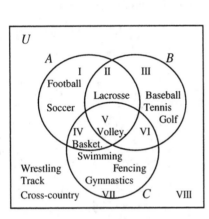

12.

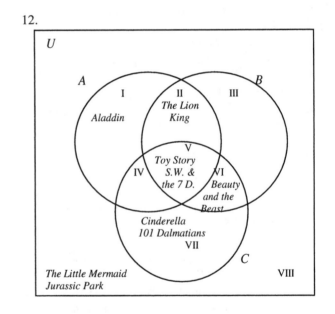

13.

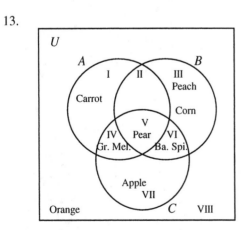

14.

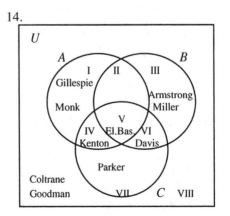

15.

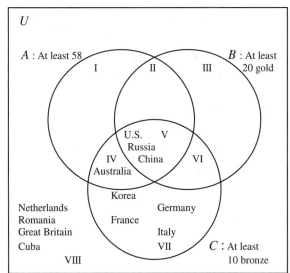

16.

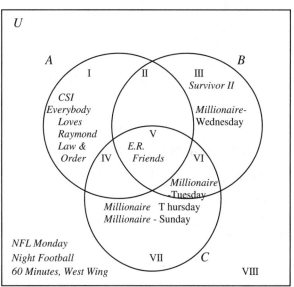

17. Harvard, V
18. Yale, I
19. Boston College, VIII
20. University of California   Berk eley, VI
21. Northwestern, VI
22. Duke, IV
23. Washington D.C., IV
24. Pittsburgh, III
25. Denver, II
26. Houston, I
27. Rochester, NY, VII
28. Chicago, VI

29. VI
30. VIII
31. III
32. IV
33. III
34. I
35. V
36. III
37. II
38. VIII
39. VII
40. VI
41. I
42. VII
43. VIII
44. V
45. VI
46. III

47. $A = \{1, 2, 3, 4, 5, 6\}$
48. $U = \{1, 2, 3, 4, 5, 6, 7, 8, 9, 10, 11, 12\}$

49. $B = \{3, 4, 5, 7, 8, 9, 12\}$
50. $C = \{4, 5, 6, 7, 8, 10\}$

51. $A \cap B = \{3, 4, 5\}$
52. $A \cap C = \{4, 5, 6\}$

53. $(B \cap C)' = \{1, 2, 3, 6, 9, 10, 11, 12\}$
54. $A \cap B \cap C = \{4, 5\}$

55. $A \cup B = \{1, 2, 3, 4, 5, 6, 7, 8, 9, 12\}$
56. $B \cup C = \{3, 4, 5, 6, 7, 8, 9, 10, 12\}$

57. $(A \cup C)' = \{9, 11, 12\}$
58. $A \cup B \cup C = \{1, 2, 3, 4, 5, 6, 7, 8, 9, 10, 12\}$

59. $A' = \{7, 8, 9, 10, 11, 12\}$
60. $(A \cup B \cup C)' = \{11\}$

61. $(A \cup B)'$ $A' \cap B'$

Set	Regions
A	I, II
B	II, III
$A \cup B$	I, II, III
$(A \cup B)'$	IV

Set	Regions
A	I, II
A'	III, IV
B	II, III
B'	I, IV
$A' \cap B'$	IV

Both statements are represented by the same region, IV, of the Venn diagram. Therefore, $(A \cup B)' = A' \cap B'$ for all sets A and B.

62. $(A \cap B)'$ $A' \cup B$

Set	Regions
A	I, II
B	II, III
$A \cap B$	II
$(A \cap B)'$	I, III, IV

Set	Regions
A	I, II
A'	III, IV
B	II, III
$A' \cup B$	II, III, IV

Since the two statements are not represented by the same regions, $(A \cap B)' \neq A' \cup B$ for all sets A and B.

63. $A' \cup B'$ $A \cap B$

Set	Regions
A	I, II
A'	III, IV
B	II, III
B'	I, IV
$A' \cup B'$	I, III, IV

Set	Regions
A	I, II
B	II, III
$A \cap B$	II

Since the two statements are not represented by the same regions, $A' \cup B' \neq A \cap B$ for all sets A and B.

64. $(A \cup B)'$ $(A \cap B)'$

Set	Regions
A	I, II
B	II, III
$A \cup B$	I, II, III
$(A \cup B)'$	IV

Set	Regions
A	I, II
B	II, III
$A \cap B$	II
$(A \cap B)'$	I, III, IV

Since the two statements are not represented by the same regions, $(A \cup B)' \neq (A \cap B)'$ for all sets A and B.

65. $A' \cup B'$ $(A \cup B)'$

Set	Regions
A	I, II
A'	III, IV
B	II, III
B'	I, IV
$A' \cup B'$	I, III, IV

Set	Regions
A	I, II
B	II, III
$A \cup B$	I, II, III
$(A \cup B)'$	IV

Since the two statements are not represented by the same regions, $A' \cup B' \neq (A \cup B)'$ for all sets A and B.

66. $A' \cap B'$ $A \cup B'$

Set	Regions
A	I, II
A'	III, IV
B	II, III
B'	I, IV
$A' \cap B'$	IV

Set	Regions
A	I, II
B	II, III
B'	I, IV
$A \cup B'$	I, II, IV

Since the two statements are not represented by the same regions, $A' \cap B' \neq A \cup B'$ for all sets A and B.

67. $(A' \cap B)'$ $A \cup B'$

Set	Regions
A	I, II
A'	III, IV
B	II, III
$A' \cap B$	III
$(A' \cap B)'$	I, II, IV

Set	Regions
A	I, II
B	II, III
B'	I, IV
$A \cup B'$	I, II, IV

Both statements are represented by the same regions, I, II, IV, of the Venn diagram. Therefore,

$(A' \cap B)' = A \cup B'$ for all sets A and B.

68. $A' \cap B'$ $(A' \cap B')'$

Set	Regions
A	I, II
A'	III, IV
B	II, III
B'	I, IV
$A' \cap B'$	IV

Set	Regions
A	I, II
A'	III, IV
B	II, III
B'	I, IV
$A' \cap B'$	IV
$(A' \cap B')'$	I, II, III

Since the two statements are not represented by the same regions, $A' \cap B' \neq (A' \cap B')'$ for all sets A and B.

69. $A \cap (B \cup C)$ $(A \cap B) \cup C$

Set	Regions	Set	Regions
B	II, III, V, VI	A	I, II, IV, V
C	IV , V, VI, VII	B	II, III, V, VI
$B \cup C$	II, III, IV, V, VI, VII	$A \cap B$	II, V
A	I, II, IV, V	C	IV, V, VI, VII
$A \cap (B \cup C)$	II, IV, V	$(A \cap B) \cup C$	II, IV, V, VI, VII

Since the two statements are not represented by the same regions, $A \cap (B \cup C) \neq (A \cap B) \cup C$

for all sets $A, B,$ and $C.$

70. $A \cup (B \cap C)$ $(B \cap C) \cup A$

Set	Regions	Set	Regions
B	II, III, V, VI	B	II, III, V, VI
C	IV , V, VI, VII	C	IV, V, VI, VII
$B \cap C$	V, VI	$B \cap C$	V, VI
A	I, II, IV, V	A	I, II, IV, V
$A \cup (B \cap C)$	I, II, IV, V, VI	$(B \cap C) \cup A$	I, II, IV, V, VI

Both statements are represented by the same regions, I, II, IV, V, VI, of the Venn diagram.

Therefore, $A \cup (B \cap C) = (B \cap C) \cup A$ for all sets $A, B,$ and $C.$

71. $A \cap (B \cup C)$ $(B \cup C) \cap A$

Set	Regions	Set	Regions
B	II, III, V, VI	B	II, III, V, VI
C	IV , V, VI, VII	C	IV, V, VI, VII
$B \cup C$	II, III, IV, V, VI, VII	$B \cup C$	II, III, IV, V, VI, VII
A	I, II, IV, V	A	I, II, IV, V
$A \cap (B \cup C)$	II, IV, V	$(B \cup C) \cap A$	II, IV, V

Both statements are represented by the same regions, II, IV, V, of the Venn diagram.

Therefore, $A \cap (B \cup C) = (B \cup C) \cap A$ for all sets $A, B,$ and $C.$

72. $A \cup (B \cap C)'$ $A' \cap (B \cup C)$

Set	Regions	Set	Regions
B	II, III, V, VI	B	II, III, V, VI
C	IV , V, VI, VII	C	IV, V, VI, VII
$B \cap C$	V, VI	$B \cup C$	II, III, IV, V, VI, VII
$(B \cap C)'$	I, II, III, IV, VII, VIII	A	I, II, IV, V
A	I, II, IV, V	A'	III, VI, VII, VIII
$A \cup (B \cap C)'$	I, II, III, IV, V, VII, VIII	$A' \cap (B \cup C)$	III, VI, VII

Since the two statements are not represented by the same regions, $A \cup (B \cap C)' \neq A' \cap (B \cup C)$

for all sets $A, B,$ and $C.$

73. $A \cap (B \cup C)$ $(A \cap B) \cup (A \cap C)$

Set	Regions	Set	Regions
B	II, III, V, VI	A	I, II, IV, V
C	IV, V, VI, VII	B	II, III, V, VI
$B \cup C$	II, III, IV, V, VI, VII	$A \cap B$	II, V
A	I, II, IV, V	C	IV, V, VI, VII
$A \cap (B \cup C)$	II, IV, V	$A \cap C$	IV, V
		$(A \cap B) \cup (A \cap C)$	II, IV, V

Both statements are represented by the same regions, II, IV, V, of the Venn diagram.

Therefore, $A \cap (B \cup C) = (A \cap B) \cup (A \cap C)$ for all sets $A, B,$ and C.

74. $A \cup (B \cap C)$ $(A \cup B) \cap (A \cup C)$

Set	Regions	Set	Regions
B	II, III, V, VI	A	I, II, IV, V
C	IV, V, VI, VII	B	II, III, V, VI
$B \cap C$	V, VI	$A \cup B$	I, II, III, IV, V, VI
A	I, II, IV, V	C	IV, V, VI, VII
$A \cup (B \cap C)$	I, II, IV, V, VI	$A \cup C$	I, II, IV, V, VI, VII
		$(A \cup B) \cap (A \cup C)$	I, II, IV, V, VI

Both statements are represented by the same regions, I, II, IV, V, VI, of the Venn diagram.

Therefore, $A \cup (B \cap C) = (A \cup B) \cap (A \cup C)$ for all sets $A, B,$ and C.

75. $A \cap (B \cup C)'$ $A \cap (B' \cap C')$

Set	Regions	Set	Regions
B	II, III, V, VI	B	II, III, V, VI
C	IV, V, VI, VII	B'	I, IV, VII, VIII
$B \cup C$	II, III, IV, V, VI, VII	C	IV, V, VI, VII
$(B \cup C)'$	I, VIII	C'	I, II, III, VIII
A	I, II, IV, V	$B' \cap C'$	I, VIII
$A \cap (B \cup C)'$	I	A	I, II, IV, V
		$A \cap (B' \cap C')$	I

Both statements are represented by the same region, I, of the Venn diagram.

Therefore, $A \cap (B \cup C)' = A \cap (B' \cap C')$ for all sets $A, B,$ and C.

76. $(A \cup B) \cap (B \cup C)$ $B \cup (A \cap C)$

Set	Regions	Set	Regions
A	I, II, IV, V	A	I, II, IV, V
B	II, III, V, VI	C	IV, V, VI, VII
$A \cup B$	I, II, III, IV, V, VI	$A \cap C$	IV, V
C	IV, V, VI, VII	B	II, III, V, VI
$B \cup C$	II, III, IV, V, VI, VII	$B \cup (A \cap C)$	II, III, IV, V, VI
$(A \cup B) \cap (B \cup C)$	II, III, IV, V, VI		

Both statements are represented by the same regions, II, III, IV, V, VI, of the Venn diagram.

Therefore, $(A \cup B) \cap (B \cup C) = B \cup (A \cap C)$ for all sets $A, B,$ and C.

77. $(A \cup B)' \cap C$ $(A' \cup C) \cap (B' \cup C)$

Set	Regions	Set	Regions
A	I, II, IV, V	A	I, II, IV, V
B	II, III, V, VI	A'	III, VI, VII, VIII
$A \cup B$	I, II, III, IV, V, VI	C	IV, V, VI, VII
$(A \cup B)'$	VII, VIII	$A' \cup C$	III, IV, V, VI, VII, VIII
C	IV, V, VI, VII	B	II, III, V, VI
$(A \cup B)' \cap C$	VII	B'	I, IV, VII, VIII
		$B' \cup C$	I, IV, V, VI, VII, VIII
		$(A' \cup C) \cap (B' \cup C)$	IV, V, VI, VII, VIII

Since the two statements are not represented by the same regions, $(A \cup B)' \cap C \neq (A' \cup C) \cap (B' \cup C)$

for all sets $A, B,$ and C.

78. $(C \cap B)' \cup (A \cap B)'$ $A \cap (B \cap C)$

Set	Regions	Set	Regions
C	IV, V, VI, VII	B	II, III, V, VI
B	II, III, V, VI	C	IV, V, VI, VII
$C \cap B$	V, VI	$B \cap C$	V, VI
$(C \cap B)'$	I, II, III, IV, VII, VIII	A	I, II, IV, V
A	I, II, IV, V	$A \cap (B \cap C)$	V
$A \cap B$	II, V		
$(A \cap B)'$	I, III, IV, VI, VII, VIII		
$(C \cap B)' \cup (A \cap B)'$	I, II, III, IV, VI, VII, VIII		

Since the two statements are not represented by the same regions, $(C \cap B)' \cup (A \cap B)' \neq A \cap (B \cap C)$

for all sets $A, B,$ and C.

79. $(A \cup B)'$

80. $A \cap B'$

81. $(A \cup B) \cap C'$

82. $A' \cap B \cap C$

83. a) $(A \cup B) \cap C = (\{1,2,3,4\} \cup \{3,6,7\}) \cap \{6,7,9\} = \{1,2,3,4,6,7\} \cap \{6,7,9\} = \{6,7\}$

 $(A \cap C) \cup (B \cap C) = (\{1,2,3,4\} \cap \{6,7,9\}) \cup (\{3,6,7\} \cap \{6,7,9\}) = \varnothing \cup \{6,7\} = \{6,7\}$

 Therefore, for the specific sets, $(A \cup B) \cap C = (A \cap C) \cup (B \cap C)$.

 b) Answers will vary.

 c) $\quad (A \cup B) \cap C \qquad\qquad\qquad\qquad (A \cap C) \cup (B \cap C)$

Set	Regions	Set	Regions
A	I, II, IV, V	A	I, II, IV, V
B	II, III, V, VI	C	IV, V, VI, VII
$A \cup B$	I, II, III, IV, V, VI	$A \cap C$	IV, V
C	IV, V, VI, VII	B	II, III, V, VI
$(A \cup B) \cap C$	IV, V, VI	$B \cap C$	V, VI
		$(A \cap C) \cup (B \cap C)$	IV, V, VI

Both statements are represented by the same regions, IV, V, VI, of the Venn diagram.

Therefore, $(A \cup B) \cap C = (A \cap C) \cup (B \cap C)$ for all sets $A, B,$ and C.

84. a) $(A \cup C)' \cap B = (\{a, c, d, e, f\} \cup \{a, b, c, d, e\})' \cap \{c, d\} = \{a, b, c, d, e, f\}' \cap \{c, d\}$

 $= \{g, h, i\} \cap \{c, d\} = \varnothing$

 $(A \cap C)' \cap B = (\{a, c, d, e, f\} \cap \{a, b, c, d, e\})' \cap \{c, d\} = \{a, c, d, e\}' \cap \{c, d\} = \{b, f, g, h, i\} \cap \{c, d\} = \varnothing$

 Therefore, for the specific sets, $(A \cup C)' \cap B = (A \cap C)' \cap B$.

 b) Answers will vary.

 c) $\quad (A \cup C)' \cap B \qquad\qquad\qquad\qquad (A \cap C)' \cap B$

Set	Regions	Set	Regions
A	I, II, IV, V	A	I, II, IV, V
C	IV, V, VI, VII	C	IV, V, VI, VII
$A \cup C$	I, II, IV, V, VI, VII	$A \cap C$	IV, V
$(A \cup C)'$	III, VIII	$(A \cap C)'$	I, II, III, VI, VII, VIII
B	II, III, V, VI	B	II, III, V, VI
$(A \cup C)' \cap B$	III	$(A \cap C)' \cap B$	II, III, VI

Since the two statements are not represented by the same regions, $(A \cup C)' \cap B \neq (A \cap C)' \cap B$

for all sets $A, B,$ and C.

85.

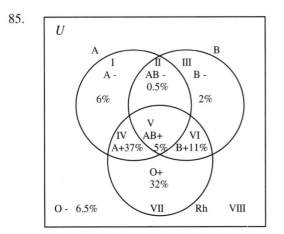

86.

Region	Set	Region	Set
I	$A \cap B' \cap C'$	V	$A \cap B \cap C$
II	$A \cap B \cap C'$	VI	$A' \cap B \cap C$
III	$A' \cap B \cap C'$	VII	$A' \cap B' \cap C$
IV	$A \cap B' \cap C$	VIII	$A' \cap B' \cap C'$

87.   a) A : Office Building Construction Projects, B : Plumbing Projects, C : Budget Greater Than \$300,000

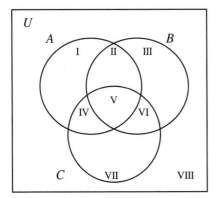

b) Region V; $A \cap B \cap C$

c) Region VI; $A' \cap B \cap C$

d) Region I; $A \cap B' \cap C'$

88. $n(A \cup B \cup C) = n(A) + n(B) + n(C) - 2n(A \cap B \cap C) - n(A \cap B \cap C') - n(A \cap B' \cap C) - n(A' \cap B \cap C)$

89. a)

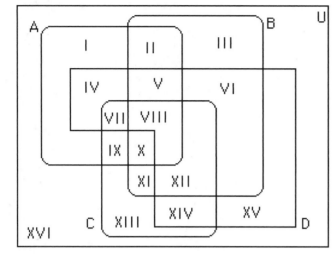

b)

Region	Set	Region	Set
I	$A \cap B' \cap C' \cap D'$	IX	$A \cap B' \cap C \cap D'$
II	$A \cap B \cap C' \cap D'$	X	$A \cap B \cap C \cap D'$
III	$A' \cap B \cap C' \cap D'$	XI	$A' \cap B \cap C \cap D'$
IV	$A \cap B' \cap C' \cap D$	XII	$A' \cap B \cap C \cap D$
V	$A \cap B \cap C' \cap D$	XIII	$A' \cap B' \cap C \cap D'$
VI	$A' \cap B \cap C' \cap D$	XIV	$A' \cap B' \cap C \cap D$
VII	$A \cap B' \cap C \cap D$	XV	$A' \cap B' \cap C' \cap D$
VIII	$A \cap B \cap C \cap D$	XVI	$A' \cap B' \cap C' \cap D'$

90.

```
S U B T R H M D F R T S
T S J O N B U I E L E M
W E E S N E R N I T V O
P R S A K C P A R N E P
O R P O N R R O Y I M T
I R J I E E S O C L U S
F M U N G R T R N E P P
```

Exercise Set 2.5

1. a) 33, Region I
    b) 29, Region III
    c) 27, Region IV

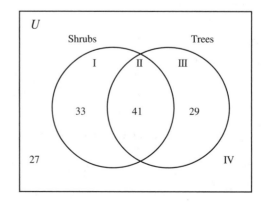

2. a) 36, Region I
    b) 22, Region III
    c) 59, Region IV

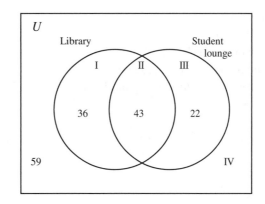

3. a) 17, Region I
    b) 12, Region III
    c) 59, the sum of the numbers in Regions I, II, III

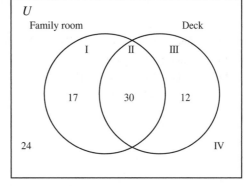

4. a) 39, Region I
    b) 27, Region III
    c) 101, the sum of the numbers in Regions I, II, III

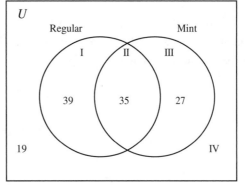

5. a) 27, Region VIII
    b) 80, Region VII
    c) 340, the sum of the numbers in Regions I, III, VII
    d) 55, the sum of the numbers in Regions II, IV, VI
    e) 337, the sum of the numbers in Regions I, II, III, IV, V, VI

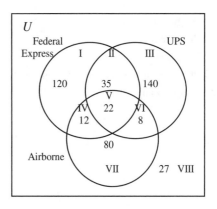

6. a) 2, Region III
   b) 6, Region II
   c) 22, the sum of the numbers in Regions I, II, III, IV, V, VI
   d) 11, the sum of the numbers in Regions I, II, III
   e) 12, the sum of the numbers in Regions II, IV, VI

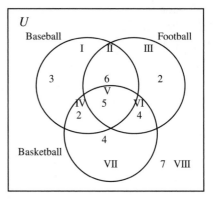

7. a) 22, Region I
   b) 11, Region II
   c) 64, the sum of the numbers in Regions I, II, III, IV, V, VI
   d) 50, the sum of the numbers in Regions I, II, III
   e) 23, the sum of the numbers in Regions II, IV, VI

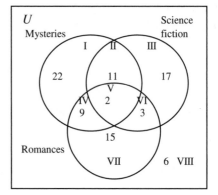

8. a) 9, Region I
   b) 20, the sum of the numbers in Regions I, III, VII
   c) 57, the sum of the numbers in Regions I, II, III, IV, V, VI, VII
   d) 30, the sum of the numbers in Regions II, IV, VI
   e) 8, Region VIII

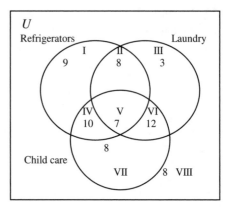

9. a) 17, Region I
   b) 27, Region VII
   c) 2, Region II
   d) 31, the sum of the numbers in Regions I, II, III
   e) 2, Region VIII

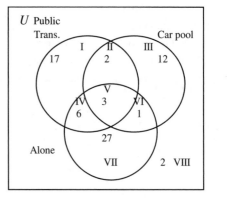

10. a) 496, the sum of the numbers in
    Regions I, II, III, IV, V, VI, VII, VIII
    b) 132, Region IV
    c) 29, Region III
    d) 328, the sum of the numbers in Regions II, IV, VI
    e) 470, the sum of the numbers in Regions I, II, III, IV, V, VI, VII

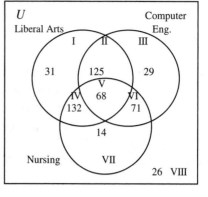

11. a) 10, the sum of the numbers in Regions III and VI
    b) 15, the sum of the numbers in Regions I, II, III, IV, V, VI
    c) 0, Region II
    d) 6, Region VIII

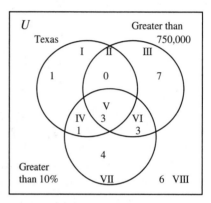

12. No. The sum of the numbers in the Venn diagram
    is 99. Dennis claims he surveyed 100 people.

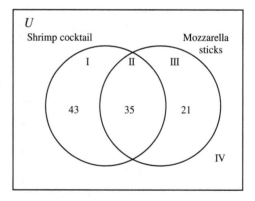

13. The Venn diagram shows the number of cars
    driven by women is 37, the sum of the numbers
    in Regions II, IV, V. This exceeds the 35 women
    the agent claims to have surveyed.

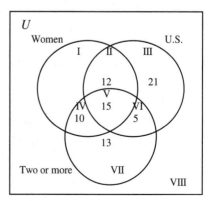

14. a) 290, the sum of the numbers in
Regions I, II, III, IV, V, VI, VII
b) 95, Region V
c) 10, Region VIII
d) 125, the sum of the numbers in Regions II, IV, VI
The number of parks that had only camping, Region I,
is 15. The number of parks that had only hiking trails,
Region III, is 20. The number of parks that had only
picnicking, Region VII, is 35. 140 parks had camping
and hiking trails, Regions II and V. 185 parks had camping.
Therefore, the sum of the numbers in Regions I, II, IV, V
must equal 185. 15 + 140 + number in Region IV = 185.

Thus, the number in Region IV is 30.

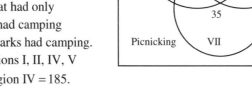

15. a) 410, the sum of the numbers in Regions I through VII
b) 35, Region V
c) 90, Region VIII
d) 50, the sum of the numbers in Regions II, IV, VI
The number of farmers growing wheat only, Region I,
is 125. The number growing corn only, Region III,
is 110. The number growing oats only, Region VII,
is 90. 60 farmers grew wheat and corn, Regions II and V.
200 farmers grew wheat. Therefore, the sum of the numbers
in Regions I, II, IV, V must equal 200.

125 + 60 + number in Region IV = 200.

Thus, the number in Region IV is 15.

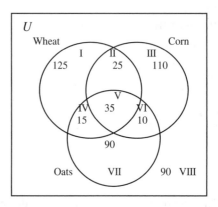

16. From the given information, we get the following Venn diagram:

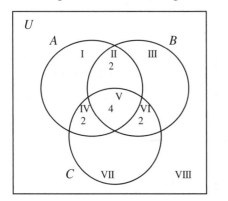

Since $n(A \cup B \cup C) = 10$ and $n(U) = 12$, the remaining 2 elements in the universal set must be in Region VIII.

a) 10, the sum of the numbers in Regions II, IV, V, VI

b) 10, the sum of the numbers in Regions IV, V, VI, VIII

c) 6, the sum of the numbers in Regions IV, VI, VIII

Exercise Set 2.6

1. An **infinite set** is a set that can be placed in a one-to-one correspondence with a proper subset of itself.

2. a) A set is **countable** if it is finite or if it can be placed in a one-to-one correspondence with the set of counting numbers.

   b) Any set that can be placed in a one-to-one correspondence with the set of counting numbers has cardinality $\aleph_0$.

3. $\{7, 8, 9, \ 10, 11, \ ..., n+6, ...\}$
   $\quad \downarrow \downarrow \downarrow \downarrow \ \downarrow \qquad \downarrow$
   $\{8, 9, 10, 11, 12, \ ..., n+7, ...\}$

4. $\{12, 13, 14, 15, 16, \ ..., n+11, ...\}$
   $\quad \downarrow \ \downarrow \ \downarrow \ \downarrow \ \downarrow \qquad \downarrow$
   $\{13, 14, 15, 16, 17, \ ..., n+12, ...\}$

5. $\{3, 5, \ 7, \ 9, 11, \ ..., 2n+1, ...\}$
   $\quad \downarrow \downarrow \downarrow \downarrow \ \downarrow \qquad \downarrow$
   $\{5, 7, \ 9, 11, 13, \ ..., 2n+3, ...\}$

6. $\{20, 22, 24, 26, 28, \ ..., 2n+18, ...\}$
   $\quad \downarrow \ \downarrow \ \downarrow \ \downarrow \ \downarrow \qquad \downarrow$
   $\{22, 24, 26, 28, 30, \ ..., 2n+20, ...\}$

7. $\{4, \ 7, \ 10, 13, 16, \ ..., 3n+1, ...\}$
   $\quad \downarrow \downarrow \ \downarrow \ \downarrow \ \downarrow \qquad \downarrow$
   $\{7, 10, 13, 16, 19, \ ..., 3n+4, ...\}$

8. $\{4, \ 8, \ 12, 16, 20, \ ..., \ 4n, ...\}$
   $\quad \downarrow \ \downarrow \ \downarrow \ \downarrow \ \downarrow \qquad \downarrow$
   $\{8, 12, 16, 20, 24, \ ..., 4n+4, ...\}$

9. $\{6, \ \ 11, 16, 21, 26, \ ..., 5n+1, ...\}$
   $\quad \downarrow \ \ \downarrow \ \downarrow \ \downarrow \ \downarrow \qquad \downarrow$

   $\{11, 16, 21, 26, 31, \ ..., 5n+6, ...\}$

10. $\left\{1, \dfrac{1}{2}, \dfrac{1}{3}, \dfrac{1}{4}, \dfrac{1}{5}, ..., \dfrac{1}{n}, ...\right\}$
    $\quad \downarrow \downarrow \downarrow \downarrow \downarrow \qquad \downarrow$
    $\left\{\dfrac{1}{2}, \dfrac{1}{3}, \dfrac{1}{4}, \dfrac{1}{5}, \dfrac{1}{6}, ..., \dfrac{1}{n+1}, ...\right\}$

11. $\left\{\dfrac{1}{2}, \dfrac{1}{4}, \dfrac{1}{6}, \dfrac{1}{8}, ..., \ \dfrac{1}{2n}, ...\right\}$
    $\quad \downarrow \downarrow \downarrow \downarrow \qquad \downarrow$
    $\left\{\dfrac{1}{4}, \dfrac{1}{6}, \dfrac{1}{8}, \dfrac{1}{10}, ..., \dfrac{1}{2n+2}, ...\right\}$

12. $\left\{\dfrac{6}{11}, \dfrac{7}{11}, \dfrac{8}{11}, \dfrac{9}{11}, \dfrac{10}{11}, ..., \dfrac{n+5}{11}, ...\right\}$
    $\quad \downarrow \ \downarrow \ \downarrow \ \downarrow \ \downarrow \qquad \downarrow$
    $\left\{\dfrac{7}{11}, \dfrac{8}{11}, \dfrac{9}{11}, \dfrac{10}{11}, \dfrac{11}{11}, ..., \dfrac{n+6}{11}, ...\right\}$

13. $\{1, \ 2, \ \ 3, \ 4, \ ..., \ n, ...\}$
    $\quad \downarrow \downarrow \ \downarrow \ \downarrow \qquad \downarrow$
    $\{6, 12, 18, 24, \ ..., 6n, ...\}$

14. $\{1, \ \ 2, \ 3, \ 4, \ ..., \ \ n, ...\}$
    $\quad \downarrow \ \ \downarrow \ \downarrow \ \downarrow \qquad \downarrow$
    $\{50, 51, 52, 53, \ ..., n+49, ...\}$

15. $\{1, 2, 3, 4, \ 5, \ ..., \ \ n, ...\}$
    $\quad \downarrow \downarrow \downarrow \downarrow \ \downarrow \qquad \downarrow$
    $\{4, 6, 8, 10, 12, \ ..., 2n+2, ...\}$

16. $\{1, 2, 3, 4, 5, \ ..., \ \ n, ...\}$
    $\quad \downarrow \downarrow \downarrow \downarrow \qquad \downarrow$
    $\{0, 2, 4, 6, 8, \ ..., 2n-2, ...\}$

17. $\{1, 2, 3, \ 4, \ \ 5, \ ..., \ \ \ n, ...\}$
    $\quad \downarrow \downarrow \downarrow \downarrow \ \downarrow \qquad \downarrow$
    $\{2, 5, 8, 11, 14, \ ..., 3n-1, ...\}$

18. $\{1, \ 2, \ 3, \ \ 4, \ \ 5, \ ..., \ \ \ n, ...\}$
    $\quad \downarrow \downarrow \downarrow \ \downarrow \ \downarrow \qquad \downarrow$
    $\{4, 9, 14, 19, 24, \ ..., 5n-1, ...\}$

19. $\{1, \ 2, \ 3, \ 4, \ \ 5, \ ..., \ \ \ n , ...\}$
    $\quad \downarrow \downarrow \downarrow \downarrow \ \downarrow \qquad \downarrow$
    $\{5, 8, 11, 14, 17, \ ..., 3n+2, ...\}$

20. $\{ \ 1, \ 2, 3, 4, \ \ 5, \ ..., n, ...\}$
    $\quad \downarrow \downarrow \downarrow \downarrow \ \downarrow \qquad \downarrow$
    $\left\{\dfrac{1}{2}, \dfrac{1}{4}, \dfrac{1}{6}, \dfrac{1}{8}, \dfrac{1}{10}, ..., \dfrac{1}{2n}, ...\right\}$

21. $\{ \ 1, \ 2, \ 3, 4, 5, \ ..., \ \ n, ...\}$
    $\quad \downarrow \downarrow \downarrow \downarrow \ \downarrow \qquad \downarrow$
    $\left\{\dfrac{1}{3}, \dfrac{1}{4}, \dfrac{1}{5}, \dfrac{1}{6}, \dfrac{1}{7}, ..., \dfrac{1}{n+2}, ...\right\}$

22. $\{1, \ 2, 3, 4, \ \ 5, \ ..., \ \ n, ...\}$
    $\quad \downarrow \downarrow \downarrow \downarrow \ \downarrow \qquad \downarrow$
    $\left\{\dfrac{1}{2}, \dfrac{2}{3}, \dfrac{3}{4}, \dfrac{4}{5}, \dfrac{5}{6}, ..., \dfrac{n}{n+1}, ...\right\}$

23. $\{1, 2, 3, 4, 5, ..., n, ...\}$
    ↓ ↓ ↓ ↓ ↓   ↓
    $\{1, 4, 9, 16, 25, ..., n^2, ...\}$

24. $\{1, 2, 3, 4, 5, ..., n, ...\}$
    ↓ ↓ ↓ ↓ ↓   ↓
    $\{2, 4, 8, 16, 32, ..., 2^n, ...\}$

25. $\{1, 2, 3, 4, 5, ..., n, ...\}$
    ↓ ↓ ↓ ↓ ↓   ↓
    $\{3, 9, 27, 81, 243, ..., 3^n, ...\}$

26. $\{1, 2, 3, 4, 5, ..., n, ...\}$
    ↓ ↓ ↓ ↓ ↓   ↓
    $\left\{\dfrac{1}{3}, \dfrac{1}{6}, \dfrac{1}{12}, \dfrac{1}{24}, \dfrac{1}{48}, ..., \dfrac{1}{3 \times 2^{n-1}}, ...\right\}$

27. =

28. =

29. =

30. =

31. =

32. a) Answers will vary.
    b) No

Review Exercises

1. True

2. False; the word *best* makes the statement not well defined.

3. True

4. False; no set is a proper subset of itself.

5. False; the elements 6, 12, 18, 24, ... are members of both sets.

6. True

7. False; both sets do not contain exactly the same elements.

8. True

9. True

10. True

11. True

12. True

13. True

14. True

15. $A = \{7, 9, 11, 13, 15\}$

16. $B = \{\text{California, Oregon, Idaho, Utah, Arizona}\}$

17. $C = \{1, 2, 3, 4, ..., 296\}$

18. $D = \{9, 10, 11, 12, ..., 96\}$

19. $A = \{x \mid x \in N \text{ and } 52 < x < 100\}$

20. $B = \{x \mid x \in N \text{ and } x > 63\}$

21. $C = \{x \mid x \in N \text{ and } x < 3\}$

22. $D = \{x \mid x \in N \text{ and } 23 \le x \le 41\}$

23. A is the set of capital letters in the English alphabet from E through M, inclusive.

24. B is the set of U.S. coins with a value of less than one dollar.

25. C is the set of the last three lowercase letters in the English alphabet.

26. D is the set of numbers greater than or equal to 3 and less than 9.

27. $A \cap B = \{1, 3, 5, 6\} \cap \{5, 6, 9, 10\} = \{5, 6\}$

28. $A \cup B' = \{1, 3, 5, 6\} \cup \{5, 6, 9, 10\}' = \{1, 3, 5, 6\} \cup \{1, 2, 3, 4, 7, 8\} = \{1, 2, 3, 4, 5, 6, 7, 8\}$

29. $A' \cap B = \{1, 3, 5, 6\}' \cap \{5, 6, 9, 10\} = \{2, 4, 7, 8, 9, 10\} \cap \{5, 6, 9, 10\} = \{9, 10\}$

30. $(A \cup B)' \cup C = (\{1, 3, 5, 6\} \cup \{5, 6, 9, 10\})' \cup \{1, 6, 10\} = \{1, 3, 5, 6, 9, 10\}' \cup \{1, 6, 10\}$

    $= \{2, 4, 7, 8\} \cup \{1, 6, 10\} = \{1, 2, 4, 6, 7, 8, 10\}$

31. $2^4 = 2 \times 2 \times 2 \times 2 = 16$

32. $2^4 - 1 = (2 \times 2 \times 2 \times 2) - 1 = 16 - 1 = 15$

33.

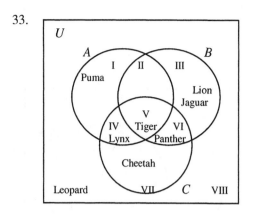

34. $A \cup B = \{b, e, g, k, c, d, f, a\}$

35. $A \cap B' = \{b, d\}$

37. $A \cap B \cap C = \{f\}$

39. $(A \cap B) \cup C = \{e, g, f, d, a, i\}$

36. $A \cup B \cup C = \{b, e, g, k, c, d, f, a, i\}$

38. $(A \cup B) \cap C = \{d, f, a\}$

40. $(A' \cup B')'$ $\qquad$ $A \cap B$

Set	Regions	Set	Regions
A	I, II	A	I, II
A'	III, IV	B	II, III
B	II, III	$A \cap B$	II
B'	I, IV		
$A' \cup B'$	I, III, IV		
$(A' \cup B')'$	II		

Both statements are represented by the same region, II, of the Venn diagram. Therefore, $(A' \cup B')' = A \cap B$ for all sets A and B.

41. $(A \cup B') \cup (A \cup C')$ $\qquad\qquad$ $A \cup (B \cap C)'$

Set	Regions	Set	Regions
A	I, II, IV, V	B	II, III, V, VI
B	II, III, V, VI	C	IV, V, VI, VII
B'	I, IV, VII, VIII	$B \cap C$	V, VI
$A \cup B'$	I, II, IV, V, VII, VIII	$(B \cap C)'$	I, II, III, IV, VII, VIII
C	IV, V, VI, VII	A	I, II, IV, V
C'	I, II, III, VIII	$A \cup (B \cap C)'$	I, II, III, IV, V, VII, VIII
$A \cup C'$	I, II, III, IV, V, VIII		
$(A \cup B') \cup (A \cup C')$	I, II, III, IV, V, VII, VIII		

Both statements are represented by the same regions, I, II, III, IV, V, VII, VIII, of the Venn diagram.

Therefore, $(A \cup B') \cup (A \cup C') = A \cup (B \cap C)'$ for all sets $A, B,$ and C.

42. II

43. V

44. VIII

45. IV

46. IV

47. VII

48. The company paid $450 since the sum of the numbers in Regions I through IV is 450.

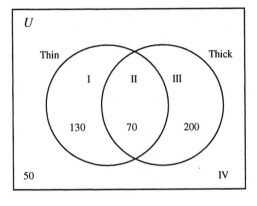

49. a) 315, the sum of the numbers in Regions I through VIII
    b) 10, Region III
    c) 30, Region II
    d) 110, the sum of the numbers in Regions III, VI, VII

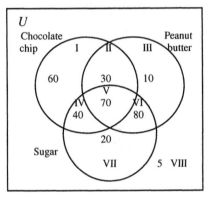

50. a) 38, Region I
    b) 298, the sum of the numbers in Regions I, III, VII
    c) 28, Region VI
    d) 236, the sum of the numbers in Regions I, IV, VII
    e) 106, the sum of the numbers in Regions II, IV, VI

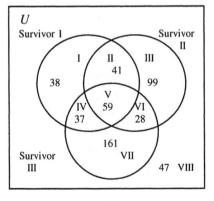

51.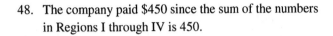
    $\{2, 4, 6,\ 8,\ 10, \ldots, 2n, \ldots\}$
    $\downarrow \downarrow \downarrow \downarrow \ \downarrow \qquad \downarrow$
    $\{4, 6,\ 8, 10, 12, \ldots, 2n + 2, \ldots\}$

52. $\{3, 5, 7,\ 9,\ 11, \ldots, 2n + 1, \ldots\}$
    $\downarrow \downarrow \downarrow \downarrow \ \downarrow \qquad \downarrow$
    $\{5, 7, 9, 11, 13, \ldots, 2n + 3, \ldots\}$

53. $\{1, 2,\ 3,\ 4,\ 5, \ldots,\ \ n, \ldots\}$
    $\downarrow \downarrow \downarrow \downarrow \ \downarrow \qquad \downarrow$
    $\{5, 8,\ 11, 14, 17, \ldots, 3n + 2, \ldots\}$

54. $\{1, 2, 3, 4,\ 5, \ldots,\ \ n, \ldots\}$
    $\downarrow \downarrow \downarrow \downarrow \ \downarrow \qquad \downarrow$
    $\{4, 9, 14, 19, 24, \ldots, 5n - 1, \ldots\}$

Chapter Test

1. True

2. False; the sets do not contain exactly the same elements.

3. True

4. False; the second set has no subset that contains the element 7.

5. False; the empty set is a proper subset of every set except itself.

6. False; the set has $2^3 = 2 \times 2 \times 2 = 8$ subsets.

7. True

8. False; for any set A, $A \cup A' = U$, not $\{\ \}$.

9. True

10. $A = \{1, 2, 3, 4, 5, 6, 7, 8\}$

11. Set A is the set of natural numbers less than 9.

12. $A \cap B = \{3, 5, 7, 9\} \cap \{7, 9, 11, 13\} = \{7, 9\}$

13. $A \cup C' = \{3, 5, 7, 9\} \cup \{3, 11, 15\}' = \{3, 5, 7, 9\} \cup \{5, 7, 9, 13\} = \{3, 5, 7, 9, 13\}$

14. $A \cap (B \cap C)' = \{3, 5, 7, 9\} \cap (\{7, 9, 11, 13\} \cap \{3, 11, 15\})' = \{3, 5, 7, 9\} \cap \{11\}' = \{3, 5, 7, 9\} \cap \{3, 5, 7, 9, 13, 15\}$

    $= \{3, 5, 7, 9\}$, or A.

15. $n(A \cap B') = n\left(\{3, 5, 7, 9\} \cap \{7, 9, 11, 13\}'\right) = n\left(\{3, 5, 7, 9\} \cap \{3, 5, 15\}\right) = n(\{3, 5\}) = 2$

16.

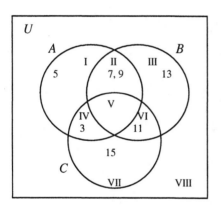

17. $A \cap (B \cup C')$ $(A \cap B) \cup (A \cap C')$

Set	Regions	Set	Regions
B	II, III, V, VI	A	I, II, IV, V
C	IV, V, VI, VII	B	II, III, V, VI
C'	I, II, III, VIII	$A \cap B$	II, V
$B \cup C'$	I, II, III, V, VI, VIII	C	IV, V, VI, VII
A	I, II, IV, V	C'	I, II, III, VIII
$A \cap (B \cup C')$	I, II, V	$A \cap C'$	I, II
		$(A \cap B) \cup (A \cap C')$	I, II, V

Both statements are represented by the same regions, I, II, V, of the Venn diagram.

Therefore, $A \cap (B \cup C') = (A \cap B) \cup (A \cap C')$ for all sets A, B, and C.

18.

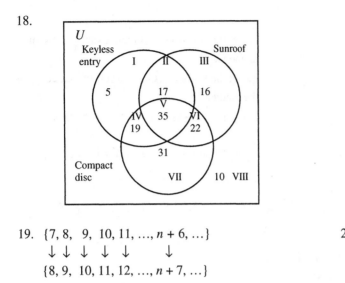

a) 52, the sum of the numbers in Regions I, III, VII
b) 10, Region VIII
c) 93, the sum of the numbers in Regions II, IV, V, VI
d) 17, Region II
e) 38, the sum of the numbers in Regions I, II, III
f) 31, Region VII

19. $\{7, 8, \ 9, \ 10, 11, \ldots, n + 6, \ldots\}$
    $\downarrow \downarrow \downarrow \ \downarrow \ \downarrow \qquad \downarrow$
    $\{8, 9, \ 10, 11, 12, \ldots, n + 7, \ldots\}$

20. $\{1, 2, 3, 4, \ 5, \ldots, \ \ n, \ldots\}$
    $\downarrow \downarrow \downarrow \downarrow \downarrow \qquad \downarrow$
    $\{1, 3, 5, \ 7, 9, \ldots, 2n - 1, \ldots\}$

Group Projects

1. a) A : Does not shed, B : Less than 16 in. tall, C : Good with kids

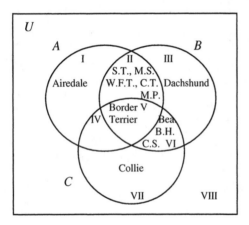

   b) Border terrier, Region V

2. a) Animal   b) Chordate   c) Mammalia   d) Carnivore
   e) Felidae   f) Felis   g) Catus

3.

	First	Second	Third	Fourth	Fifth
a) Color	yellow	blue	red	ivory	green
b) Nationality	Norwegian	Afghan.	Senegalese	Spanish	Japanese
c) Food	apple	cheese	banana	peach	fish
d) Drink	vodka	tea	milk	whiskey	ale
e) Pet	fox	horse	snail	dog	zebra
f) Ale					

CHAPTER THREE

LOGIC

Exercise Set 3.1

1.a. A simple statement is a sentence that conveys one idea and can be identified as either true or false.

  b. Statements consisting of two or more simple statements are called compound statements

2. All, none (no), some

3. a) Some are   b) All are
   c) Some are not   d) None are

4. Let p: The ink is purple.
   The symbolic form is ~ p. The negation symbol, ~ , represents the word <u>not</u>.

5. a) →   b) ∨   c) ∧
   d) ~ e) ↔

6. The exclusive OR means that one or the other event can can occur, but not both. b. Yes; the inclusive OR means that one or more events can occur simultaneously. c. The inclusive OR is used in this chapter, unless otherwise stated.

7. When a compound statement contains more than one connective a comma can be used to indicate which simple statements are to be grouped together. When writing a statement symbolically, the simple statements on the same side of the comma are to be grouped together within parentheses.

8. 1st Biconditional ↔   2nd Conditional → 3rd Conjunction ∧   Disjunction ∨   4th Negation ~

9. compound; conjunction, ∧
11. compound; biconditional ↔
13. compound; disjunction, ∨
15. simple statement
17. compound; negation, ~
19. compound; conjunction, ∧
21. compound; conditional; negation, ~
23. No picnic tables are portable.
25. Some chicken do not fly.
27. All turtles have claws.
29. Some bicycles have three wheels.
31. All pine trees produce pine cones.
33. No pedestrians are in the crosswalk.
35. ~ p
38. ~ q ↔ ~ p
41. ~ q → ~ p
44. ~ p ∧ q
47. Firemen do not work hard.
49. Firemen wear red suspenders or firemen work hard.

10. compound; negation, ~
12. compound; conditional, →
14. compound; conjunction, ∧
16. compound; biconditional, ↔
18. compound; conditional, →
20. compound; conjunction, ∧
22. compound; conditional, →
24. Some stock mutual funds have guaranteed yields.
26. Some plants do not create (contain) chlorphyll.
28. Some teachers made the roster.
30. Some horses do not have manes.
32. Someone likes asparagus.
34. All dogs with long hair get cold.

36. p ∧ q
39. ~ p → ~ q
42. ~ q ↔ ~ p
45. ~ (q → ~ p)

37. ~ q ∨ ~ p
40. ~ q ∧ p
43. ~ p ∧ ~ q
46. ~ (p ∧ q)

48. Firemen do not wear red suspenders.
50. Firemen work hard and wear red suspenders.

51. Firemen do not work hard if and only if firemen do not wear red suspenders.

52. If firemen do not work hard, then firemen wear red suspenders.

53. It is false that firemen wear red suspenders or firemen work hard.

54. Firemen do not work hard or firemen do not wear red suspenders.

55. Firemen do not work hard and firemen do not wear red suspenders.

56. It is false that firemen work hard and firemen wear red suspenders.

57. $(p \lor \sim q) \to r$

58. $(r \leftrightarrow \sim p) \lor \sim q$

59. $(p \land q) \lor r$

60. $(r \land q) \to p$

61. $p \to (q \lor \sim r)$

62. $(\sim p \leftrightarrow \sim q) \lor \sim r$

63. $(r \leftrightarrow q) \land p$

64. $\sim (r \to \sim q)$

65. $q \to (p \leftrightarrow r)$

66. $(r \lor \sim q) \leftrightarrow p$

67. The water is $70°$ or the sun is shining, and we do not go swimming.

68. The water is $70°$ and the sun is shining, or we go swimming.

69. If water is not , and the sun is shining or we do go swimming.

70. If the sun is shining then the water is $70°$, or we go swimming.

71. If we do not go swimming, then the sun is shining and the water is $70°$.

72. If the sun is shining and we go swimming, then the water is $70°$.

73. If the sun is shining then we go swimming, and the water is $70°$.

74. If the water is not $70°$, then the sun is shining or we will go swimming.

75. The sun is shinning if and only if the water is $70°$, and we go swimming.

76. If the sun is shining, then the water is $70°$ if and only if we go swimming.

77. Not permissible. In the list of choices, the connective or is the exclusive or, thus one can order either the soup or the salad but not both items.

78. Permissible.

79. Not permissible. Potatoes and pasta cannot be ordered together.

80. Not permissible. Potatoes and pasta cannot be ordered together.

81. a) $(\sim p) \to q$ b) conditional

82. a) $(\sim p \land r) \leftrightarrow (\sim q)$ b) biconditional

83. a) $(\sim q) \land (\sim r)$ b) conjunction

84. a) $(\sim p) \lor q$ b) disjunction

85. a) $(p \lor q) \to r$ b) conditional

86. a) $q \to (p \land \sim r)$ b) conditional

87. a) $r \to (p \lor q)$ b) conditional

88. a) $(q \to p) \leftrightarrow (p \to q)$ b) biconditional

89. a) $(\sim p) \leftrightarrow (\sim q \to r)$ b) biconditional

90. a) $(\sim q) \to (r \land p)$ b) conditional

91. a) $(r \land \sim q) \to (q \land \sim p)$ b) conditional

92. a) $\sim [p \to (q \lor r)]$ b) negation

93. a) $\sim [(p \land q) \leftrightarrow (p \lor r)]$ b) negation

94. a) $\sim [r \land \sim q) \to (q \land r)]$ b) negation

95. a) r: retired; c: concrete business; $r \land c$
    b) conjunction

96. a) w: water level up; c: go canoeing; r: go rafting; $w \to (c \lor r)$ b) conditional

97. a) b: below speed limit; p: pulled over
    $\sim (b \to \sim p)$ b) conditional, negation

98. a) d: dinner is ready; e: can eat; r: can go to restaurant; $(d \to e) \lor \sim r$ b) disjunction

99. a) f: food has fiber; v: food has vitamins   h: be healthy; $(f \lor v) \to h$ b). conditional

100. a) c: Corliss is teaching.; f: Faye in Math.lab. w: a weekend; $(c \to f) \leftrightarrow \sim w$ b) biconditional

101. a) c: may take course; f: fail previous exam; p: passed placement test;
    $c \leftrightarrow (\sim f \lor p)$ b) biconditional

102. a) g: car has gas; b: battery charged; s: car will start; $(g \land b) \to s$ b) conditional

103. a) c: classroom is empty; w: is the
   weekend s: is 7:00 a.m.; (c ↔ w) ∨ s
   b) disjunction

104. This statement/question is a paradox. Therefore
   it is false.

105. [(~ q) → (r ∨ p)] ↔ [(~ r) ∧ q],
   biconditional

106. ~[[(~r) → (p ∧ q)] ↔[(~p) ∨ r]], negation

107. a) The conjunction and disjunction have
   the same dominance.
   b) Answers will vary.

107. c) If we evaluate the truth table for p ∨ q ∧ r
   using the order (p ∨ q) ∧ r we get a different
   solution than if we used the order p ∨ (q ∧ r).
   Therefore, unless we are told where the
   parentheses belong, we do not know which
   solution is correct.

Exercise Set 3.2

1. a) $2^2 = 2 \times 2 = 4$ distinct cases
   b)

	p	q
Case 1:	T	T
Case 2:	T	F
Case 3:	F	T
Case 4:	F	F

2. a) $2^3 = 2 \times 2 \times 2 = 8$ distinct cases
   b)

	p	q	r
case 1:	T	T	T
case 2:	T	T	F
case 3:	T	F	T
case 4:	T	F	F
case 5:	F	T	T
case 6:	F	T	F
case 7:	F	F	T
case 8:	F	F	F

3. a)

p	q	p	∨	q
T	T	T	T	T
T	F	T	T	F
F	T	F	T	T
F	F	F	F	F
		1	3	2

   b) Only in Case 4, in which both simple
   statements are false.

4. a)

p	q	p	∧	q
T	T	T	T	T
T	F	T	F	F
F	T	F	F	T
F	F	F	F	F
		1	3	2

   b) Only in case 1, when both simple
   statements are true.

5.

p	p	∨	~ p
T	T	T	F
F	F	T	T

6.

p	p	∧	~ p
T	T	F	F
F	F	F	T
	1	3	2

7.

p	q	p	∧	~ q
T	T	T	F	F
T	F	T	T	T
F	T	F	F	F
F	F	F	F	T
		1	3	2

8.

P	q	q	∨	~ p
T	T	T	T	F
T	F	F	F	F
F	T	T	T	T
F	F	F	T	T
		1	3	2

9.

P	q	~ (p ∨ ~ q)
T	T	F T T F
T	F	F T T T
F	T	T F F F
F	F	F F T T
		4 1 3 2

10.

p	q	~ p ∨ ~ q
T	T	F F F
T	F	F T T
F	T	T T F
F	F	T T T
		1 3 2

11.

p	q	~(p ∧ ~ q)
T	T	T T F F
T	F	F T T T
F	T	T F F F
F	F	T F F T
		4 1 3 2

12.

p	q	~(~ p ∧ ~ q)
T	T	T F F F
T	F	T F F T
F	T	T T F F
F	F	F T T T
		4 1 3 2

13.

p	q	r	~ q ∨ (p ∧ r)
T	T	T	F T T T T
T	T	F	F F T F F
T	F	T	T T T T T
T	F	F	T F T F F
F	T	T	F F F F T
F	T	F	F F F F F
F	F	T	T T F F T
F	F	F	T T F F F
			1 5 2 4 3

14.

p	q	r	(p ∨ ~ q) ∧ r)
T	T	T	T T F T T
T	T	F	T T F F F
T	F	T	T T T T T
T	F	F	T T T F F
F	T	T	F F F F T
F	T	F	F F F F F
F	F	T	F T T T T
F	F	F	F T T F F
			1 3 2 5 4

15.

p	q	r	r ∨ (p ∧ ~ q)
T	T	T	T T T F F
T	T	F	F F T F F
T	F	T	T T T T T
T	F	F	F T T T T
F	T	T	T T F F F
F	T	F	F F F F F
F	F	T	T T F F T
F	F	F	F F F F T
			1 5 2 4 3

16.

p	q	r	(r ∧ q) ∧ ~ p
T	T	T	T F F
T	T	F	F F F
T	F	T	F F F
T	F	F	F F F
F	T	T	T T T
F	T	F	F F T
F	F	T	F F T
F	F	F	F F T
			1 3 2

17.

p	q	r	~ q ∧ (r ∨ ~ p)
T	T	T	F F T T F
T	T	F	F F F F F
T	F	T	T T T T F
T	F	F	T F F F F
F	T	T	F F T T T
F	T	F	F F F T T
F	F	T	T T T T T
F	F	F	T T F T T
			1 5 2 4 3

18.

p	q	r	~ p ∧ (q ∨ r)
T	T	T	F F T T T
T	T	F	F F T T F
T	F	T	F F F T T
T	F	F	F F F F F
F	T	T	T T T T T
F	T	F	T T T T F
F	F	T	T T F T T
F	F	F	T F F F F
			1 5 2 4 3

19.

p	q	r	(~q ∧ r) ∨ p
T	T	T	F F T T T
T	T	F	F F F T T
T	F	T	T T T T T
T	F	F	T F F T T
F	T	T	F F T F F
F	T	F	F F F F F
F	F	T	T T T T F
F	F	F	T F F F F
			1 3 2 5 4

20.

p	q	r	~r ∨ (~p ∧ q)
T	T	T	F F F F T
T	T	F	T T F F T
T	F	T	F F F F F
T	F	F	T T F F F
F	T	T	F T T T T
F	T	F	T T T T T
F	F	T	F F T F F
F	F	F	T T T F F
			4 5 1 3 2

21. p: Meetings are dull.
q: Teaching is fun.
In symbolic form the statement is p ∧ q.

p	q	p ∧ q
T	T	T
T	F	F
F	T	F
F	F	F
		1

22. p: The stadium is enclosed.
q: The stadium is air-conditioned.
In symbolic form the statement is p ∧ ~ q.

p	q	p	∧	~ q
T	T	T	F	F
T	F	T	T	T
F	T	F	F	F
F	F	F	F	T
		1	3	2

23. p: Bob will get a haircut.
q: Bob will shave his beard.
In symbolic form the statement is p ∧ ~ q.

p	q	p	∧	~ q
T	T	T	F	F
T	F	T	T	T
F	T	F	F	F
F	F	F	F	T
		1	3	2

24. p: The class has 15 minutes.
q: The class is cancelled.
In symbolic form the statement is ~ (p ∨ q).

p	q	~	(p	∨	q)
T	T	F	T	T	T
T	F	F	T	T	F
F	T	F	F	T	T
F	F	T	F	F	F
		4	1	3	2

25. p: Jasper Adams is the tutor.
q: Mark Russo is a secretary.
In symbolic form the statement is ~ (p ∧ q).

p	q	~ (p ∧ q)
T	T	F T T T
T	F	T T F F
F	T	T F F T
F	F	T F F F
		4 1 3 2

26. p: Mike made pizza.
q: Dennis made a chef salad.
r : Gil burned the lemon squares.
In symbolic form the statement is (p ∧ q) ∧ r.

p	q	r	(p ∧ q) ∧ r
T	T	T	T T T T T
T	T	F	T T T F F
T	F	T	T F F F T
T	F	F	T F F F F
F	T	T	T F T F T
F	T	F	T F T F F
F	F	T	T F F F T
F	F	F	T F F F F
			1 3 2 5 4

27. p: The copier is out of toner.
    q: The lens is dirty.
    r : The corona wires are broken.
    The statement is p ∨ (q ∨ r).

p	q	r	p ∨ (q ∨ r)
T	T	T	T T T
T	T	F	T T T
T	F	T	T T T
T	F	F	T T F
F	T	T	F T T
F	T	F	F T T
F	F	T	F T T
F	F	F	F F F
			2 3 1

28. p: I am hungry.
    q: I want to eat a healthy lunch.
    r : I want to eat in a hurry.
    The statement is p ∧ (q ∧ r).

p	q	r	p ∧ (q ∧ r)
T	T	T	T T T
T	T	F	T F F
T	F	T	T F F
T	F	F	T F F
F	T	T	F F T
F	T	F	F F F
F	F	T	F F F
F	F	F	F F F
			2 3 1

29. p: Congress acts on the bill.
    q: The President signs the bill.
    In symbolic form, the statement is
    p ∧ (q ∨ ~ q).

p	q	p ∧(q∨ ~q)
T	T	T T T T F
T	F	T T F T T
F	T	F F T T F
F	F	F F F T T
		1 5 2 4 3

30. p: Gordon likes the PowerMac G4 Cube.
    q: Gordon likes the iBook.
    r : Gordon likes the Pentium IV.
    In symbolic form, the statement is
    (p ∧ q) ∧ ~ q.

p	q	(p ∧ q)∧~q
T	T	T T T F F
T	F	T F F F T
F	T	F F T F F
F	F	F F F F T
		1 3 2 5 4

31. (a) ~ p ∨ (q ∨ r)
        F ∨ (F ∧ T)
        F ∨   F
             F
    Therefore the statement is false.
    (b) ~ p ∨ (q ∧ r)
        T ∨ (T ∧ T)
        T ∨   T
             T
    Therefore the statement is true.

32. (a) (~ p ∧ r) ∧ q
        (F ∧ T) ∧ F
         F   ∧ F
             F
    Therefore the statement is false.
    (b) (~ p ∧ r) ∧ q
        (T ∧ T) ∧ T
         T   ∧ T
             T
    Therefore the statement is true.

33. (a) (~ q ∧ ~p) ∨ ~ r
        (T ∧ F) ∨ F
           F   ∨ F
             F
    Therefore the statement is false.
    (b) (~ q ∧ ~p) ∨ ~ r
        (F ∧ T) ∨ F
           F   ∨ F
             F
    Therefore the statement is false.

34. (a) (~ p ∨ ~q) ∨ ~ r
        (F ∨ T) ∨ F
           T   ∨ F
             T
    Therefore the statement is true.
    (b) (~ p ∨ ~q) ∨ ~ r
        (T ∨ F) ∨ F
           T   ∨ F
             T
    Therefore the statement is true.

35. (a) (p ∧ ~q) ∨ r
        (T ∧ T) ∨ T
          T   ∨ T
            T
    Therefore the statement is true.
    (b) (p ∧ ~q) ∨ r
        (F ∧ F) ∨ T
          F   ∨ T
            T
    Therefore the statement is true.

36. (a) (p ∨ ~q) ∧ ~ (p ∧ ~ r)
        (T ∨ T) ∧ ~ (T ∧ F)
          T   ∧   ~F
            T
    Therefore the statement is true.
    (b) (p ∨ ~q) ∧ ~(p ∧ ~ r)
        (F ∨ F) ∧ (F ∧ F)
          F   ∧   ~F
            F
    Therefore the statement is false.

37. (a) (~ r ∧ p) ∨ q
        (F ∧ T) ∨ F
          F   ∨ F
            F
    Therefore the statement is false.
    (b) (~ r ∧ p) ∨ q
        (F ∧ F) ∨ T
          F   ∨ T
            T
    Therefore the statement is true.

38. (a) ~q ∨ (r ∧ p)
        T ∨ (T ∧ T
        T ∨   T
            T
    Therefore the statement is true.
    (b) ~q ∨ (r ∧ p)
        F ∨ (T ∧ F
        F ∨   F
            F
    Therefore the statement is false.

39. (a) (~q ∨ ~p) ∧ r
        (T ∨ F) ∧ T
          T   ∧ T
            T
    Therefore the statement is true.
    (b) (~q ∨ ~p) ∧ r
        (F ∨ T) ∧ T
          T   ∧ T
            T
    Therefore the statement is true.

40. (a) (~r ∨ ~p) ∨ ~q
        (F ∨ F) ∨ T
          F   ∨ T
            T
    Therefore the statement is true.
    (b) (~r ∨ ~p) ∨ ~q
        (F ∨ T) ∨ F
          T   ∨ F
            T
    Therefore the statement is true.

41. (a) (~p ∨ ~q) ∨ (~r ∨ q)
        (F ∨ T) ∨ (F ∨ F)
          T   ∨   F
            T
    Therefore the statement is true.
    (b) (~p ∨ ~q) ∨ (~r ∨ q)
        (T ∨ F) ∨ (F ∨ T)
          T   ∨   T
            T
    Therefore the statement is true.

42. (a) (~r ∧ ~q) ∧ (~r ∨ ~p)
        (F ∧ T) ∧ (F ∨ F)
          F   ∧   F
            F
    Therefore the statement is false.
    (b) (~r ∧ ~q) ∧ (~r ∨ ~p)
        (F ∧ F) ∧ (F ∨ T)
          F   ∧   T
            F
    Therefore the statement is true.

43. $3 + 5 = 4 + 47$ or $10 \quad 9 = 9 \quad 1 \, 0$

    $8 = 8 \qquad \lor \qquad 1 \neq -1$

      T $\lor$ F

            T

Therefore the statement is true.

44. $5 < 4$ and $4 < 5$

    F $\land$ T

         F

Therefore the statement is false.

45. E: Elvis was a singer.

    C: Chickens can swim.

      E $\lor$ C

      T $\lor$ F

          T

Therefore the statement is true.

46. AL: Alaska is the 50th state.

    HI: Hawaii is a group of islands.

    AT: Atlanta is the capitol of Alabama.

      (AL $\lor$ HI) $\land$ AT

      (F $\lor$ T) $\land$ F

         T $\land$ F

Therefore the statement is false.

47. U2: U2 is a rock band.

    DW: Denzel Washington is an actor.

    JS: Jerry Seinfeld is a comedian.

      (U2 $\land$ DW) $\land$ ~JS

      (T $\land$ T) $\land$ F

          T $\land$ F

           F

Therefore the statement is false.

48. T: Toronto is a city in Minnesota.

    M: Mexico City is in Texas.

    C: Cairo is in Egypt.

      (T $\lor$ M) $\land$ C

      (F $\lor$ F) $\land$ T

        F $\land$ T

          F

Therefore the statement is false.

49. CR: Cal Ripken played football.

    GB: Bush was prime minister of England.

    CP: Colon Powell was in the Army.

      (CR $\lor$ GB) $\land$ CP

      (F $\lor$ F) $\land$ T

         F $\land$ T

          F

Therefore the statement is false.

50. H: Holstein is a breed of cattle.

    C: Collie is a breed of dog.

    B: Beagle is a breed of cat.

      (H $\land$ C) $\lor$ ~B

      (T $\land$ T) $\lor$ T

        T $\lor$ T

          T

Therefore the statement is true.

51. p: 30 pounds of cheese was consumed by the average American in 1909.

    q: The average American consumed 154 pounds of sweetners in 2001.

    p $\land$ ~ q

    F $\land$ ~ T

    F $\land$ F

    False

52. p: Per capita consumption of red meat was less for average American in 2001 than in 1909.

    q: The average American consumed 154 pounds of sweetners in 2001.

    p $\lor$ q

    F $\lor$ T

    True

53. p: In 1909, average American ate approximately the same amount of fish and poultry.

    q: Between 1909 and 2001, average American consumed more poultry.

    $p \wedge q$

    $T \wedge T$

    True

54. p: In 1909, average American ate approximately 9 times as much red meat as fish.

    q: By 2001, average American only ate 8 times as much red meat as fish.

    $p \wedge q$

    $T \wedge T$

    True

55. p: 30% of Americans get 6 hours of sleep.

    q: 9% get 5 hours of sleep.

    $\sim (p \wedge q)$

    $\sim (F \wedge T)$

    $\sim F$

    True

56. p: 25% of Americans get 6 hours of sleep.

    q: 30% of Americans get 7 hours of sleep.

    r: 9% of Americans get 5 hours of sleep.

    $p \wedge (q \vee \sim r)$

    $T \wedge (T \vee \sim T)$

    $T \wedge (T \vee F)$

    $T \wedge \quad T$

    True

57. p: 13% of Americans get $\leq$ 5 hrs. of sleep.

    q: 32% of Americans get $\geq$ 6 hrs. of sleep.

    r: 30% of Americans get $\geq$ 8 hrs. of sleep.

    $(p \vee q) \wedge r$

    $(T \vee F) \wedge F$

    $T \quad \wedge F)$

    False

58. p: > 50% of Americans get $\leq$ 7 hrs. of sleep.

    q: > 25% of Americans get $\leq$ 6 hrs. of sleep.

    $p \wedge q)$

    $T \wedge T$

    True

59. $p \wedge \sim q$

60. $\sim p \wedge q$

61. $p \vee \sim q$

62. $\sim p \vee \sim q$

63. $(r \vee q) \wedge p$

64. $(p \wedge q) \vee r$

65. $q \vee (p \wedge \sim r)$

66. $\sim p \wedge (\sim r \wedge q)$

67. (a) Mr. Duncan qualifies for the loan. Mrs. Tuttle qualifies for the loan.

    (b) The Rusineks do not qualify because their gross income is too low.

68. (a) The Argentos qualify for the loan. Mrs. Tuttle qualifies for the loan.

    (b) Ms. McVey does not qualify because her gross income is too low. Ms. Fox and Mr. Siewert do not qualify because their assets are too low.

69. (a) Wing Park qualifies for the special fare.

    (b) The other 4 do not qualify: Gina V. returns after 04/01; Kara S. returns on Monday; Christos G. does not stay at least one Saturday; and Alex C. returns on Monday.

70.

p	q	r	$\sim$	[($\sim$	(p $\wedge$ q))	$\wedge$	(q $\vee$ r)]
T	T	T	F	F	T	T	T
T	T	F	T	F	T	F	F
T	F	T	T	F	T	F	F
T	F	F	T	F	T	F	F
F	T	T	F	F	T	T	T
F	T	F	T	F	T	F	F
F	F	T	F	T	F	T	F
F	F	F	F	T	F	T	F
			5	2	1	4	3

71.

p	q	r	[(q ∧ ~ r) ∧ (~ p ∨ ~ q)] ∨ ~ (p ∨ ~ r)
T	T	T	F F F T T
T	T	F	T F F T T
T	F	T	F F T T T
T	F	F	F F T T T
F	T	T	F F T F F
F	T	F	T T T T T
F	F	T	F F T F F
F	F	F	F F T T T
			1 4 2 5 3

72. (a) $2^4 = 16$ distinct cases

74. Answers will vary (out of order).

72. (b)

p	q	r	s
T	T	T	T
T	T	T	F
T	T	F	T
T	T	F	F
T	F	T	T
T	F	T	F
T	F	F	T
T	F	F	F
F	T	T	T
F	T	T	F
F	T	F	T
F	T	F	F
F	F	T	T
F	F	T	F
F	F	F	T
F	F	F	F

72. (c)

p	q	r	s	(q ∧ p) ∨ (~r ∧ s)
T	T	T	T	T T F
T	T	T	F	T T F
T	T	F	T	T T T
T	T	F	F	T T F
T	F	T	T	F F F
T	F	T	F	F F F
T	F	F	T	F T T
T	F	F	F	F F F
F	T	T	T	F F F
F	T	T	F	F F F
F	T	F	T	F T T
F	T	F	F	F F F
F	F	T	T	F F F
F	F	T	F	F F F
F	F	F	T	F T T
F	F	F	F	F F F

72 (d)

p	q	r	s	(~r ∧ ~s)∧(~p ∨ q)
T	T	T	T	F F T
T	T	T	F	F F T
T	T	F	T	F F T
T	T	F	F	T T T
T	F	T	T	F F F
T	F	T	F	F F F
T	F	F	T	F F F
T	F	F	F	T F F
F	T	T	T	F F T
F	T	T	F	F F T
F	T	F	T	F F T
F	T	F	F	T T T
F	F	T	T	F F T
F	F	T	F	F F T
F	F	F	T	F F T
F	F	F	F	T T T

73.

p	q	r	s	(p ∧~q) ∨ r	(q ∧ ~r) ∨ p
T	T	T	T	F T T	F T T
T	T	T	F	F T T	F T T
T	T	F	T	F F F	T T T
T	T	F	F	F F F	T T T
T	F	T	T	T T T	F T T
T	F	T	F	T T T	F T T
T	F	F	T	T T F	F T T
T	F	F	F	T T F	F T T
F	T	T	T	F T T	F F F
F	T	T	F	F T T	F F F
F	T	F	T	F F F	T T F
F	T	F	F	F F F	T T F
F	F	T	T	F T T	F F F
F	F	T	F	F T T	F F F
F	F	F	T	F F F	F F F
F	F	F	F	F F F	F F F

Exercise Set 3.3

1.a)

p	q	p	→	q
T	T	T	T	T
T	F	T	F	F
F	T	F	T	T
F	F	F	T	F
		1	3	2

b) The conditional statement is false only in the case when antecedent is true and the consequent is false, otherwise it is true.

3.a) Substitute the truth values for the simple statement. Then evaluate the compound statement for that specific case.

b) $[(p \leftrightarrow q) \vee (\sim r \rightarrow q)] \rightarrow \sim r$

$[(T \rightarrow T) \vee (\sim T \rightarrow T)] \rightarrow \sim T$

$[\quad T \quad \vee (T \rightarrow T)] \rightarrow T$

$[\quad T \quad \vee \quad T \quad] \rightarrow T$

$\qquad\qquad T \qquad\qquad \rightarrow T$

$\qquad\qquad\qquad\qquad T$

In this specific case the statement is true.

2.a)

p	q	p ↔ q
T	T	T T T
T	F	T F F
F	T	F F T
F	F	F T F
		1 3 2

b) The biconditional statement is true when the statements to the left and right of the biconditional symbol match, otherwise, false.

4. A tautology is a compound statement that is true in every case.

5. A self-contradiction is a compound statement that is false in every case.

6. An implication is a <u>conditional</u> statement that is a tautology.

7.

p	q	~q → p
T	T	F T F
T	F	T F F
F	T	F T T
F	F	T T T
		1 3 2

8.

p	q	p → ~q
T	T	T F F
T	F	T T T
F	T	F T F
F	F	F T T
		1 3 2

9.

p	q	~ (q→p)
T	T	F T
T	F	F T
F	T	T F
F	F	F T
		2 1

10.

p	q	~ (p↔q)
T	T	F T
T	F	T F
F	T	T F
F	F	F T
		2 1

11.

p	q	~q	↔	p
T	T	F	F	T
T	F	T	T	T
F	T	F	T	F
F	F	T	F	F
		1	3	2

12.

p	q	(p	↔	q)	→	p
T	T		T		T	T
T	F		F		T	T
F	T		F		T	F
F	F		T		F	F
			1		3	2

13.

p	q	p	↔	(q ∨ p)
T	T	T	T	T
T	F	T	T	T
F	T	F	F	T
F	F	F	T	F
		1	3	2

14.

p	q	(~ q ∧ p) → ~ q
T	T	F F T T F
T	F	T T T T T
F	T	F F F T F
F	F	T F F T T
		1 3 2 5 4

15.

p	q	q	→	(p	→	~q)
T	T	T	F	T	F	F
T	F	F	T	T	T	T
F	T	T	T	F	T	F
F	F	F	T	F	T	T
		4	5	1	3	2

16.

p	q	(p ∨ q)	↔	(p ∧ q)
T	T	T	T	T
T	F	T	F	F
F	T	T	F	F
F	F	F	T	F
		1	3	2

17.

p	q	r	r	∧	(~q	→	p)
T	T	T	T	T	F	T	T
T	T	F	F	F	F	T	T
T	F	T	T	T	T	T	T
T	F	F	F	F	T	T	T
F	T	T	T	T	F	T	F
F	T	F	F	F	F	T	F
F	F	T	T	F	T	F	F
F	F	F	F	F	T	F	F
			4	5	1	3	2

18.

p	q	r	p	→	(q ∨ r)
T	T	T	T	T	T
T	T	F	T	T	T
T	F	T	T	T	T
T	F	F	T	F	F
F	T	T	F	T	T
F	T	F	F	T	T
F	F	T	F	T	T
F	F	F	F	T	F
			2	3	1

19.

p	q	r	(q	↔	p)	∧	~r
T	T	T		T		F	F
T	T	F		T		T	T
T	F	T		F		F	F
T	F	F		F		F	T
F	T	T		F		F	F
F	T	F		F		F	T
F	F	T		T		F	F
F	F	F		T		T	T
				1		3	2

20.

p	q	r	q	↔	(r	∧	p)
T	T	T		T		T	
T	T	F		F		F	
T	F	T		F		T	
T	F	F		T		F	
F	T	T		F		F	
F	T	F		F		F	
F	F	T		T		F	
F	F	F		T		F	
				2		3	1

21.

p	q	r	(q	∨	~r)	↔	~p
T	T	T	T	T	F	F	F
T	T	F	T	T	T	F	F
T	F	T	F	F	F	T	F
T	F	F	F	T	T	F	F
F	T	T	T	T	F	T	T
F	T	F	T	T	T	T	T
F	F	T	F	F	F	F	T
F	F	F	F	T	T	T	T
			1	3	2	5	4

22.

P	q	r	(p	∧	r)	→	(q∨r)
T	T	T		T		T	T
T	T	F		F		T	T
T	F	T		T		T	T
T	F	F		F		T	F
F	T	T		F		T	T
F	T	F		F		T	T
F	F	T		F		T	T
F	F	F		F		T	F
				1		3	2

23.

p	q	r	(~r	∨	~q)	→	p
T	T	T	F	F	F	T	T
T	T	F	T	T	F	T	T
T	F	T	F	T	T	T	T
T	F	F	T	T	T	T	T
F	T	T	F	F	F	T	F
F	T	F	T	T	F	F	F
F	F	T	F	T	T	F	F
F	F	F	T	T	T	F	F
			1	3	2	5	4

24.

p	q	r	[r	∧	(q	∨	~p)]	↔	~p
T	T	T	T	T	T	T	F	F	F
T	T	F	F	F	T	T	F	T	F
T	F	T	T	F	F	F	F	T	F
T	F	F	F	F	F	F	F	T	F
F	T	T	T	T	T	T	T	T	T
F	T	F	F	F	T	T	T	F	T
F	F	T	T	T	F	T	T	T	T
F	F	F	F	F	F	T	T	F	T
			4	5	1	3	2	7	6

25.

p	q	r	(p	→	q)	↔	(~q	→	~r)
T	T	T		T		T	F	T	F
T	T	F		T		T	F	T	T
T	F	T		F		T	T	F	F
T	F	F		F		F	T	T	T
F	T	T		T		T	F	T	F
F	T	F		T		T	F	T	T
F	F	T		T		F	T	F	F
F	F	F		T		T	T	T	T
				1		5	2	4	3

26.

p	q	r	(~p	↔	~q)	→	(~q	↔	r)
T	T	T	F	T	F	F	F	F	T
T	T	F	F	T	F	T	F	T	F
T	F	T	F	F	T	T	T	T	T
T	F	F	F	F	T	T	T	F	F
F	T	T	T	F	F	T	F	F	T
F	T	F	T	F	F	T	F	T	F
F	F	T	T	T	T	T	T	T	T
F	F	F	T	T	T	F	T	F	F
			1	3	2	7	4	6	5

Exercise Set 3.4

1. Two statements are equivalent if both statements have exactly the same truth values in the answer column of the truth table.

2. Construct a truth table for each statement and then compare the columns. If they are identical, then the statements are equivalent. If the answer columns are not identical, then the statements are not equivalent.

3. The two statements must be equivalent. A biconditional is a tautology only when the statements on each side of the biconditional are equivalent.

4. $\sim (p \wedge q) \Leftrightarrow \sim p \vee \sim q$
   $\sim (p \vee q) \Leftrightarrow \sim p \wedge \sim q$

5. a) $q \rightarrow p$ b) $\sim p \rightarrow \sim q$ c) $\sim q \rightarrow \sim p$

6. converse $\Leftrightarrow$ inverse;   conditional $\Leftrightarrow$ contrapositive

7. $\sim p \vee q$

8. p: T $p \leftrightarrow q \Leftrightarrow p \rightarrow q) \wedge (q \rightarrow p)$
   q: T $p \leftrightarrow q \Leftrightarrow p \rightarrow q) \wedge (\sim q \rightarrow p)$

9. Using DeMorgan's Laws on the statement $\sim p \vee \sim q$, we get the following: (1) $\sim (\sim p \vee \sim q)$ (2) $\sim (p \vee q)$,   (3) $\sim (p \wedge q)$. Therefore $\sim p \vee \sim q \Leftrightarrow \sim (p \wedge q)$.

10. Using DeMorgan's Laws on the statement $\sim (p \vee q)$, we get the following: (1) $p \vee q$, (2) $\sim p \vee \sim q$, (3) $\sim p \wedge \sim q$. Therefore $\sim (p \vee q) \Leftrightarrow \sim p \wedge \sim q$.

11. Using DeMorgan's Laws on the statement $\sim (p \wedge q)$, we get the following: (1) $(p \wedge q)$, (2) $\sim p \wedge \sim q$, (3) $\sim p \vee \sim q$. Therefore $\sim (p \wedge q)$ is not equivalent to $p \vee \sim q$.

12. Using DeMorgan's Laws on the statement $\sim (p \wedge q)$, we get the following: (1) $p \wedge q$, (2) $\sim p \wedge \sim q$, (3) $\sim p \vee \sim q$. Therefore $\sim (p \wedge q)$ is not equivalent to $\sim p \wedge q$.

13. Yes, $\sim (p \vee \sim q) \Leftrightarrow \sim p \wedge q$
14. No, $\sim (p \wedge \sim q) \Leftrightarrow \sim p \vee \sim q$
15. Yes, $(\sim p \wedge \sim q) \rightarrow r$
16. Yes, $q \rightarrow \sim (p \wedge \sim r)$; $q \rightarrow \sim p \vee r$
17. Yes, $\sim (p \rightarrow \sim q) \Leftrightarrow \sim (\sim p \vee q) \Leftrightarrow p \wedge q$
18. Yes, $\sim (\sim p \rightarrow q) \Leftrightarrow \sim (p \vee q) \Leftrightarrow \sim p \wedge \sim q$

19.

p	q	p → q	~p ∨ q
T	T	T	F T T
T	F	F	F F F
F	T	T	T T T
F	F	T	T T F
		1	1 3 2

The statements are equivalent.

20.

p	q	~p → q	p ∧ q
T	T	F T T	T
T	F	F T F	F
F	T	T T T	F
F	F	T F F	F
		1 3 2	1

The statements are not equivalent.

21.

p	q	r	(p ∧q) ∧r	p ∧ (q ∧r)
T	T	T	T TT	T T T
T	T	F	T FF	T F F
T	F	T	F FT	T F F
T	F	F	F FF	T F F
F	T	T	F FT	F F T
F	T	F	F FF	F F F
F	F	T	F FT	F F F
F	F	F	F FF	F F F
			1 3 2	2 3 1

The statements are equivalent.

22.

p	q	p → q	~ q → ~ p
T	T	T	F T F
T	F	F	T F F
F	T	T	F T T
F	F	T	T T T
		1	3 2

The statements are equivalent.

23.

p	q	r	(p ∨q) ∨ r	p ∨ (q ∨r)
T	T	T	T TT	T T T
T	T	F	T TF	T T T
T	F	T	T TT	T T T
T	F	F	T TF	T T F
F	T	T	T TT	F T T
F	T	F	T TF	F T T
F	F	T	F TT	F T T
F	F	F	F FF	F F F
			1 32	2 3 1

The statements are equivalent.

24.

p	q	r	p ∨ (q ∧r)	~ p → (q ∧r)
T	T	T	T T T	F T T
T	T	F	T T F	F T F
T	F	T	T T F	F T F
T	F	F	T T F	F T F
F	T	T	F T T	T T T
F	T	F	F F F	T F F
F	F	T	F F F	T F F
F	F	F	F F F	T F F
			2 3 1	1 3 2

The statements are equivalent.

25.

p	q	r	p ∧(q ∨ r)	(p ∧q) ∨ r
T	T	T	TT TTT	TTTTT
T	T	F	TT TTF	TTTTT
T	F	T	TT FTT	TFF TT
T	F	F	TF FF F	TFF FF
F	T	T	FF TTT	FFT TT
F	T	F	FF TT F	FFT FF
F	F	T	FF FT T	FFF TT
F	F	F	FF FF F	FFF FF
			1 5 2 4 3	1 3 2 5 4

The statements are not equivalent.

26.

p	q	r	~ (q → p) ∨ r	(p ∨q) ∧~ r
T	T	T	F T TT	T F F
T	T	F	F T FF	T T T
T	F	T	F T TT	T F F
T	F	F	F T FF	T T T
F	T	T	T F TT	T F F
F	T	F	T F TF	T T T
F	F	T	F T TT	F F F
F	F	F	F T FF	F F T
			2 1 4 3	1 3 2

The statements are not equivalent.

27.

p	q	r	(p → q) ∧(q → r)	(p → q) → r
T	T	T	T T T	T T T
T	T	F	T F F	T F F
T	F	T	F F T	F T T
T	F	F	F F T	F T F
F	T	T	T T T	T T T
F	T	F	T F F	T F F
F	F	T	T T T	T T T
F	F	F	T T T	T F F
			1 3 2	1 3 2

The statements are not equivalent.

28.

p	q	r	~ q → (p ∧r)	~ (p ∨r) → q
T	T	T	F T T	F TT T
T	T	F	F T F	F TT T
T	F	T	T T T	F TT F
T	F	F	T F F	F TT F
F	T	T	F T F	F TT T
F	T	F	F T F	T FT T
F	F	T	T F F	F TT F
F	F	F	T F F	T FF F
			2 3 1	2 1 4 3

The statements are not equivalent.

29.

p	q	(p → q) ∧ (q → p)				p ↔ q
T	T	T	T	T		T
T	F	F	F	T		F
F	T	T	F	F		F
F	F	T	T	T		T
		1	3	2		1

The statements are equivalent.

30.

p	q	[~ (p → q)] ∧ [~ (q → p)]						~ (p ↔ q)	
T	T	F	T	F	F	T		F	T
T	F	T	F	F	F	T		T	F
F	T	F	T	F	T	F		T	F
F	F	F	T	F	F	T		F	T
		2	1	5	4	3		2	1

The statements are not equivalent.

31. p: The Mississippi River runs through Ohio.
q: The Ohio River runs through Mississippi.
In symbolic form, the statement is ~ (p ∨ q).
Applying DeMorgan s Laws we get: ~ p ∧ ~ q.
The Mississippi River does not run through Ohio
And the Ohio River does not run through Miss.

32. p: The printer is out of toner.
q: The fax machine is out of paper.
In symbolic form, the statement is ~ (p ∧ q).
Applying DeMorgan s Laws we get:
~ p ∨ ~ q. The printer is not out of toner
or the fax machine is not out of paper.

33. p: The snowmobile was not an Arctic Cat.
q: The snowmobile was not a Ski-Do.
In symbolic form, the statement is ~ p ∨ ~ q.
Applying DeMorgan s Laws we get: ~ (p ∧ q).
It is false that the snowmobile was an Arctic
Cat and was a Ski-Do.

34. p: The pot roast is hot.
q: The pot roast is not well done.
In symbolic form, the statement is p ∧ ~ q .
Applying DeMorgan s Laws we get: ~ (~ p ∨ q). It
is false that the pot roast is not hot or it is well done.

35. p: The hotel does not have a weight room.
q: The conference center does not have an
auditorium.
In symbolic form, the statement is ~ p ∧ ~ q.
Applying DeMorgan s Laws we get: ~ (p ∨ q).
It is false that the hotel has a weight room and
the conference center has an auditorium.

36. p: Robert Farinelli is authorized WedgCor dealer.
q: He is not going to work for Prism Constr. Co.
In symbolic form, the statement is p ∨ ~ q.
Applying DeMorgan s Laws we get: ~ (~ p ∧ q). It is
False that Robert Farinelli isn t an authorized WedgCor
Dealer and he is going to work for Prism Constr. Co.

37. p: We go to Cozemel.
q: We will go snorkeling.
r: We will go to Senior Frogs.
In symbolic form, the statement is
p → (q ∨ ~ r). Applying DeMorgan s Laws
we get: p → ~ (~ q ∧ r). If we go to Cozemel,
then it is false that we will not go snorkeling
and we will go to Senior Frogs.

38. p: Phil Murphy buys us dinner.
q: We will not go to the top of the CN Tower.
r: We will be able to walk to the Bistro Restaurant.
In symbolic form the statement is p → (~ q ∧ r).
Applying DeMorgan s Laws we get: p → ~ (q ∨ ~ r).
If Phil Murphy buy us dinner, then it is false that
we will go to the top of the CN Tower and that
we will not be able to walk to the Bistro Restaurant.

39. p: You drink a glass of orange juice.
q: You ll get a full day s supply of folic acid.
In symbolic form, the statement is p → q.
Since p → q ⇔ ~ p ∨ q, an equivalent
Statement is: You do not drink a glass of OJ
or you will get a full day s supply of folic acid.

40. p: Nick-at-Nite is showing *Family Ties*.
q: Nick-at-Nite is showing *The Facts of Life*.
In symbolic form, the statement is p ∨ q.
Since p → q ⇔ ~ p ∨ q, an equivalent statement
is: If Nick-at-Nite is not showing *Family Ties*,
then they are showing *The Facts of Life*.

41. p: Bob the Tomato visited the nursing home.

    q: Bob the Tomato visited the Cub Scout meeting.

    In symbolic form, the statement is $p \vee \sim q$.

    Since $p \rightarrow q \Leftrightarrow \sim p \vee q$, an equivalent Statement is: If Bob the Tomato did not visit the nursing home, then he did not visit the Cub Scout meeting.

42. p: John Peden will buy a Harley Davidson.

    q: John Peden will buy a Honda.

    In symbolic form, the statement is $p \rightarrow \sim q$.

    Since $p \rightarrow q \Leftrightarrow \sim p \vee q$, an equivalent statement is: John Peden will not buy a H-D or he will not buy a Honda.

43. p: The plumbers meet in Kansas City.

    q: The *Rainmakers* will provide the entertainment.

    In symbolic form, the statement is $\sim (p \rightarrow q)$.

    $\sim (p \rightarrow q) \Leftrightarrow \sim (\sim p \vee q) \Leftrightarrow p \wedge \sim q$. The plumbers meet in KC and the *Rainmakers* did not provide the entertainment.

44. p: Mary Beth Headlee organized the conference.

    q: John Waters works at Sinclair Community College.

    In symbolic form, the statement is $p \vee \sim q$.

    $\sim p \rightarrow \sim q \Leftrightarrow p \vee \sim q$. If Mary Beth Headless did not organize the conference, then John Waters does not work for SCC.

45. p: It is cloudy.

    q: The front is coming through.

    In symbolic form, the statement is $(p \rightarrow q) \wedge (q \rightarrow p)$. $(p \rightarrow q) \wedge (q \rightarrow p) \Leftrightarrow p \leftrightarrow q$. It is cloudy if and only if the front is coming through.

46. p: Model Road is closed.

    q: Use Kirkwood Road.

    In symbolic form, the statement is $(p \rightarrow q) \wedge (q \rightarrow p)$. $(p \rightarrow q) \wedge (q \rightarrow p) \Leftrightarrow p \leftrightarrow q$. Model Rd. is closed if and only if we use Kirkwood Rd.

47. p: The chemistry teacher uses mathematics.

    q: There is a shortage of math. teachers.

    In symbolic form, the statement is $p \leftrightarrow q$.

    $(p \rightarrow q) \wedge (q \rightarrow p) \Leftrightarrow p \leftrightarrow q$.

    If the chemistry teachers uses math., then there Is a shortage of math. teachers and if there is a shortage of math. teachers, then the chemistry teacher uses math.

48. p: John Deere will hire new workers.

    q: The City of Dubuque will retrain the workers.

    In symbolic form, the statement is $p \leftrightarrow q$.

    $(p \rightarrow q) \wedge (q \rightarrow p) \Leftrightarrow p \leftrightarrow q$.

    If John Deere hires new workers, then the City of Dubuque will retrain the workers and if the City of Dubuque retrains the workers, then John Deere will hire new workers.

49. Converse: If I finish the book in 1 week, then it is interesting.

    Inverse: If the book is not interesting, then I do not finish it in 1 week.

    Contrapositive: If I do not finish the book in One week, then it is not interesting.

50. Converse: If you need to replace the blower fan, then the dryer is making a loud noise.

    Inverse: If the dryer is not making a loud noise, Then you do not need to replace the blower fan.

    Contrapositive: If you do not need to replace the blower fan, then the dryer is not making a loud noise.

51. Converse: If you can watch TV, then you finish your HW.

    Inverse: If you do not finish your HW, then you cannot watch TV.

    Contrapositive: If you do not watch TV, then You did not finish your HW.

52. Converse: If Bob Dylan goes on tour, then he releases a new CD.

    Inverse: If Bob Dylan does not release a new CD, then he does not go on tour.

    Contrapositive: If Bob Dylan does not go on tour, then he does not release a new CD.

53. Converse: If I scream, then that annoying paper clip (Clippie) shows up on my screen.
    Inverse: If Clippie does not show up on my screen, then I will not scream.
    Contrapositive: If I do not scream, then Clippie does not show up on my screen.

54. Converse: If I watch the same channel all night, then the remote control is not within my reach
    Inverse: If the remote control is within my reach, then I will not watch the dame channel all.night.
    Contrapositive: If I do not watch the same channel all night, then the remote control is within my reach.

55. Converse: If we go down to the marina and take out a sailboat, then the sun is shining.
    Inverse: If the sun is not shining, then we do not go down to the marina and take out a sailboat.
    Contrapositive: If we do not go down to the marina and do not take out a sailboat, then the sun is not shining.

56. Converse: If we eat a piece of apple and save some for later, then the apple pie is baked.
    Inverse: If the apple pie is not baked, then we do not eat a piece of pie nor do we save some for later.
    Contrapositive: If we do not eat a piece of pie nor do we save some for later, then the apple pie is not baked.

57. If a natural number is divisible by 10, then it is divisible by 5. True

58. If a quadrilateral is a parallelogram, then the opposite sides are parallel. True

59. If a natural number is not divisible by 6, then it is not divisible by 3. False

60. If n is a natural number, then 1/n is a natural number. False

61. If two lines are not parallel, then the two lines intersect in at least one point. True

62. If m is a counting number, then $\frac{ma}{mb} = \frac{a}{b}$. True

63. If the polygon is a quadrilateral, then the sum of the interior angles is 360 degrees. True

64. If the product of a and b is an even counting number then a and b are both even counting numbers. False

65. p: Maria has retired.
    q: Maria is still working.
    In symbolic form, the statements are:
    a) $\sim p \vee q$, b) $q \rightarrow \sim p$, c) $p \rightarrow \sim q$
    Statement (c) is the contrapositive of statement. (b). Therefore, statements (b) and (c) are equivalent.

P	q	$\sim p \vee q$	$q \rightarrow \sim p$
T	T	F T T	T F F
T	F	F F F	F T F
F	T	T T T	T T T
F	F	T T F	F T T
		1 3 2	1 3 2

Since the truth tables for (a) and (b) are different we conclude that only statements (b) and (c) are equivalent.

66. p: Today is Monday.
    q: Tomorrow is Wednesday.
    In symbolic form, the statements are:
    a) $p \rightarrow \sim q$, b) $\sim (p \wedge \sim q)$, c) $\sim p \vee q$.
    If we use DeMorgan s Laws on statement (b) we get statement (c).
    Therefore, statements (b) and (c) are equivalent. If we look at the truth tables for all three statements we can see that only statements (b) and (c) are equivalent.

p	q	a) $p \rightarrow \sim q$	b) $\sim (p \wedge \sim q)$	c) $\sim p \vee q$
T	T	T F F	T T T F	F T T
T	F	T T T	F T T T	F F F
F	T	F T F	T F F F	T T T
F	F	F T T	T F F T	T T F
		1 3 2	4 1 3 2	1 3 2

67. p: The car is reliable.

q: The car is noisy.

In symbolic form, the statements are: a) ~ p ∧ q, b) ~ p → ~ q, c) ~ (p ∨ ~ q. If we use DeMorgan s Laws on statement (a), we get statement (c). Therefore, statements (a) and (c) are equivalent. If we look at the truth tables for statements (a), (b), and (c), we see that only statements (a) and (c) are equivalent.

		a)	b)	c)
p	q	~ p ∧ q	~ p → ~ q	~ (p ∨ ~ q)
T	T	F F T	F T F	F T T F
T	F	F F F	F T T	F T T T
F	T	T T T	T F F	T F F F
F	F	T F F	T T T	F F T T
		1 3 2	1 3 2	4 1 3 2

68. p: The house is made of wood.

q: The shed is made of wood.

In symbolic form, the statements are: a) ~ p ∨ ~ q, b) p → ~ q, and c) ~ (q ∧ ~ p). Using the fact that p → q ⇔ ~ p ∨ q to rewrite statement (b), we get ~ p ∨ ~ q. Therefore, statements (a) and (b) are equivalent. Looking at the truth tables for all three, it can be determined that only statements (a) and (b) are equivalent.

		a)	b)	c)
p	q	~ p ∨ ~ q	p → ~ q	~ (q ∧ ~ p)
T	T	F F F	T F F	T T F F
T	F	F T T	T T T	T F F F
F	T	T T F	F T F	F T T T
F	F	T T T	F T T	T F F T
		1 3 2	1 3 2	4 1 3 2

69. p: Today is Sunday.

q: The library is open.

In symbolic form, the statements are: a) ~ p ∨ q, b) p → ~ q, c) q → ~ p. Looking at the truth table for all three statements, we can determine that only statements (b) and (c) are equivalent.

		a)	b)	c)
p	q	~ p ∨ q	p → ~ q	q → ~ p
T	T	F T T	T F F	T F F
T	F	F F F	T T T	F T F
F	T	T T T	F T F	T T T
F	F	T T F	F T T	F T T
		1 3 2	1 3 2	1 3 2

70. p: You are fishing at 1 PM.

q: You are driving a car at 1 PM.

In symbolic form, the statements are: a) p → q, b) ~ p ∨ q, c) ~ (p ∧ ~ q). Using the fact that p → q ⇔ ~ p ∨ q, we see that (a) and (b) are equivalent statements. If we use DeMorgan s Laws on statement (b), we get statement (c). Therefore all three statements are equivalent.

71. p: The grass grows.

q: The trees are blooming.

In symbolic form, the statements are: a) p ∧ q, b) q → ~ p, c) ~ q ∨ ~ p. Using the fact that p → q ⇔ ~ p ∨ q, on statement (b) we get ~ q ∨ ~ p. Therefore, statements (b) and (c) are equivalent. Looking at the truth table for statements (a) and (b) we can conclude that only statements (b) and (c) are equivalent.

		p ∧ q	q → ~ p
p	q		
T	T	T	T F F
T	F	F	F T F
F	T	F	T T T
F	F	F	F T T
		1	1 3 2

72. p: Johnny Patrick is chosen as department chair.

q: Johnny Patrick is the only candidate.

In symbolic form, the statements are: a) p ↔ q, b) (p → q) ∧ (q → p), and c) ~ p ∧ ~ q .

p	q	p ↔ q	(p → q) ∧ (q → p)	~ p ∧ ~ q
T	T	T T T	T T T T T T T	F F F
T	F	T F F	T F F F F T T	F F T
F	T	F F T	F T T F T F F	T F F
F	F	F T F	F T F T F T F	T T T
		1 3 2	4 6 5 10 7 8 9	11 13 12

Therefore, p is equivalent to q.

73. p: Johnny Patrick is chosen as department chair.
q: Johnny Patrick is the only candidate.
In symbolic form, the statements are: a) p ↔ q,
b) (p → q) ∧ (q→ p), and c) ~ p ∧~q .

p	q	p ↔ q	(p → q) ∧ (q→p)	~ p ∧~q
T	T	T T T	T T T T T T T	F F F
T	F	T F F	T F F F F T T	F F T
F	T	F F T	F T T F T F F	T F F
F	F	F T F	F T F T F T F	T T T
		1 3 2	4 6 5 10 7 8 9	11 13 12

Therefore, p is equivalent to q.

75. p: The pay is good.
q: Today is Monday.
r : I will take the job.
Looking at the truth tables for statements (a), (b), and
(c), we can determine that none of these statements are
equivalent.

			a)	b)	c)
p	q	r	(p ∧q) → r	~ r → ~(p ∨q)	(p ∧q) ∨ r
T	T	T	T T T	F T F T	T T T
T	T	F	T F F	T F F T	T T F
T	F	T	F T T	F T F T	F T T
T	F	F	F T F	T F F T	F F F
F	T	T	F T T	F T F T	F T T
F	T	F	F T F	T F F T	F F F
F	F	T	F T T	F T T F	F T T
F	F	F	F T F	T T T F	F F F
			1 3 2	1 4 3 2	1 3 2

74. p: You drink milk.
q: Your cholesterol count will be lower.
In symbolic form, the statements are:
a) ~ (~ p → q), b) q ↔ p, and c) ~ (p → ~ q) .

P	q	~ (~ p → q)	q ↔ p	~ (p → ~ q)
T	T	F F T T	T T T	T T F F
T	F	F F T F	F F T	F T T T
F	T	F T T T	T F F	F F T F
F	F	T T F F	F T F	F F T T
		4 1 3 2	5 7 6	11 8 10 9

Therefore, none of the statements are equivalent.

76. p: You are 18 years old.
q: You are a citizen of the United States.
r : You can vote in a presidential election.
Looking at the truth tables for statements
(a), (b), and (c), we can determine that
none of these statements are equivalent.

			a)	b)	c)
p	q	r	(p ∧ q) → r	r → (q ∧ p)	~ r ∨ (p ∧ ~ q)
T	T	T	T T T	T T T	F F T F F
T	T	F	T F F	F F T	T T T F F
T	F	T	F T T	T F F	F T T T T
T	F	F	F T F	F T F	T T T T T
F	T	T	F T T	T F F	F F F F F
F	T	F	F T F	F T F	T T F F F
F	F	T	F T T	T F F	F F F F T
F	F	F	F T F	F T F	T T F F T
			1 3 2	1 3 2	1 5 2 4 3

77. p: The package was sent by Federal Express.
q: The package was sent by United Parcel Service.
r : The package arrived on time.
Using the fact that p → q ⇔ ~ p ∨q to rewrite
statement (c), we get p ∨ (~ q ∧r). Therefore,
statements (a) and (c) are equivalent. Looking at the
truth table for statements (a) and (b), we can conclude
that only statements (a) and (c) are equivalent.

			a)	b)
p	q	r	p ∨(~q ∧ r)	r → (p ∨ ~ q)
T	T	T	T T F F T	T T T T F
T	T	F	T T F F F	F F T T F
T	F	T	T T T T T	T T T T T
T	F	F	T T T F F	F F T T T
F	T	T	F F F F T	T F F F F
F	T	F	F F F F F	F T F F F
F	F	T	F T T T T	T T F T T
F	F	F	F F T F F	F F F T T
			1 5 2 4 3	1 5 2 4 3

78. p: We will put the dog outside.
q: We feed the dog.
r : The dog will bark.
In symbolic form, the statements are:
a) (p ∨q) → ~ r, b) r → (~ p ∧~ q),
and c) r ⇔ ~ (p ∨q). Statement (c) is the
contrapositive of statement (b) and if we
use DeMorgan s Laws on statement (b)
we obtain statement (c). Therefore,
statements (a), (b), and (c) are
equivalent.

79. p: The car needs oil.
    q: The car needs gas.
    r : The car is new.
    In symbolic form, the statements are: a) p ∧(q ∨r),
    b) p ∧~ (~ q ∧~ r), and c) p → (q ∨~ r). If we use
    DeMorgan s Laws on the disjunction in statement (a),
    we obtain p ∧~ (~ q ∧~ r). Therefore, statements (a)
    and (b) are equivalent. If we compare the truth tables
    for (a) and (c) we see that they are not equivalent.
    Therefore, only statements (a) and (b) are equivalent.

p	q	r	p ∧ (q ∨ r)	p → (q ∨ ~ r)
T	T	T	TT T	TT TT F
T	T	F	TT T	TT TT T
T	F	T	TT T	TF FF F
T	F	F	TF F	TT FT T
F	T	T	FF T	FT TT F
F	T	F	FF T	FT TT T
F	F	T	FF T	FT FF F
F	F	F	FF F	FT FT T
			1 3 2	1 5 2 4 3

80. p: The mortgage rate went down.
    q: Tim purchased the house.
    r : The down payment was 10%.
    Looking at the truth tables for statements
    (a), (b), and (c), we can determine that
    none of these statements are equivalent.

			a)	b)	c)
p	q	r	p ↔ (q ∧r)	r ∧(q → p)	q → (p ∧ ~ r)
T	T	T	TT T	TT T	TF TF F
T	T	F	TF F	FF T	TT TT T
T	F	T	TF F	TT T	FT TF F
T	F	F	TF F	FF T	FT TT T
F	T	T	FF T	TF F	TF FF F
F	T	F	FT F	FF F	TF FF T
F	F	T	FT F	TT T	FT FF F
F	F	F	FT F	FF T	FT FF T
			1 3 2	1 3 2	1 5 2 4 3

81. Yes conditional: If it is a bird, then it can fly.
(False); converse: If it can fly, then it is a bird. (F)

82. Yes conditional: If 5 + 1 = 9, then 2 + 5 = 7. (T)
converse: If 2 + 5 = 7, then 5 + 1 = 9. (F)

83. Yes conditional: If 2 + 5 = 7, then 5 + 1 = 4. (F)
contrapositive: If 5 + 1 ≠ 4, then 2 + 5 ≠ 7. (False)

84. No. A conditional statement and its contrapositive
are equivalent statements.

85. If we use DeMorgan s Laws to rewrite ~ p ∨q, we
get ~ (p ∧~ q). Since ~ p ∨q ⇔ ~ (p ∧~ q) and
p → q ⇔ ~ p ∨q, we can conclude that p → q ⇔
~ (p ∧~ q). Other answers are possible.

86. ~ [~ (p ∨~ q)] ⇔ p ∨~ q. Make use of the fact that
~ (~ p) ⇔ p, then use DeMorgan s Law twice.
~ [~ (p ∨~ q)] = ~[~ p ∧q] = p ∨~ q

87. Research problem -- Answers will vary.

88. a) ~ p = 1   p = 1   0 .25 = 0.75
    b) ~ q = 1   q = 1   0.20 = 0.80
    c) p ∧q has a truth value equal to the lesser of
    p = 0.75 and q = 0.20. Thus p ∧q = 0.20
    d) p ∨q has truth value equal to the greater of
    p = 0.25 and q = 0.20. Thus p ∨ q = 0.25
    e) p → q has truth value equal to the lesser of 1
    and 1 (p + q) = 1   0. 25 + 0.20 = 0.95. Thus
    p → q = 0.95
    f) p ↔ q has a truth value equal to 1 |p   q| =
    1   (0.25   0.20) = 1   0.05 = 0.95. Thus p ↔ q
    = 0.95

89. (a) conditional; (b) biconditional; (c) inverse;
    (d) converse; (e) contrapositive

Exercise Set 3.5

1. An argument is valid when its conclusion necessarily follows from the given set of premises.

2. An argument is invalid or a fallacy when the conclusion is false.

3. Yes. It is not necessary for the premises or the conclusion to be true statements for the argument to be valid.

4. Yes. If the conclusion does not follow from the set of premises, then the argument is invalid.

5. Yes. If the conclusion follows from the set of premises, then the argument is valid, even if the premises are false.

6. If the truth table is a tautology, then the argument is valid. If the truth table is not a tautology, then the argument is invalid.

7. a) $p \rightarrow q$ b) Pizza is served on time or is free.
   $\underline{\sim p}$ The pizza was not served on time.
   q The pizza is free.

8. a) $p \rightarrow q$ b) If soil is dry, then grass needs water.
   $\underline{\sim q}$ The grass does not need water.
   $\sim p$ The soil is not dry.

9. a) $p \rightarrow q$ b) If the sky is clear, then it will be hot.
   $\underline{q \rightarrow r}$ If it is hot, then will wear shorts.
   $p \rightarrow r$ If sky is clear, then wear shorts.

10. a) $p \rightarrow q$ b) If sky is clear, then I ll go to game.
    $\underline{p}$ The sky is clear.
    q I will go to the game.

11. a) $p \rightarrow q$ b) If you wash my car, then I pay $5.
    $\underline{\sim p}$ You did not wash my car.
    $\sim q$ I will not give you $5.

12. a) $p \rightarrow q$ b) If you wash my car, then pay $5.
    $\underline{q}$ I will give you $5.
    . p You washed my car.

13. This argument is the law of detachment and therefore it is valid.

14.

p	q	$[(p \rightarrow q) \wedge \sim p] \rightarrow q$
T	T	T F F T T
T	F	F F F T F
F	T	T T T T T
F	F	T T T F F
		1 3 2 5 4

The argument is invalid.

15.

p	q	$[(p \wedge \sim q) \wedge q] \rightarrow \sim p$
T	T	T F F F T T F
T	F	T T T F F T F
F	T	F F F F T T T
F	F	F F T F F T T
		1 3 2 5 4 7 6

The argument is valid.

16.

p	q	$[(\sim p \vee q) \wedge q] \rightarrow p$
T	T	F T T T T T T
T	F	F F F F F T T
F	T	T T T T T F F
F	F	T T F F F T F
		1 3 2 5 4 7 6

The argument is not valid.

17.

p	q	$[\sim p \wedge (p \vee q)] \rightarrow \sim q$
T	T	F F T T F
T	F	F F T T T
F	T	T T T F F
F	F	T F F T T
		1 3 2 5 4

The argument is a fallacy.

20.

p	q	$[(p \vee q) \wedge \sim q] \rightarrow p$
T	T	T F F T T
T	F	T T T T T
F	T	T F F T F
F	F	F F T T F
		1 3 2 5 4

The argument is valid.

18. This argument is the law of contraposition and therefore it is valid.

19. This argument is the fallacy of the inverse. Therefore, it is not valid.

21.

p	q	[(~ p → q) ∧ ~ q] → ~ p						
T	T	F	T	T	F	F	T	F
T	F	F	T	F	T	T	F	F
F	T	T	T	T	F	F	T	T
F	F	T	F	F	F	T	T	T
		1	3	2	5	4	7	6

The argument is invalid.

22.

p	q	[(q ∧ ~ p) ∧ ~ p] → q						
T	T	T	F	F	F	F	T	T
T	F	F	F	F	F	F	T	F
F	T	T	T	T	T	T	T	T
F	F	F	F	T	F	T	T	F
		1	3	2	5	4	7	6

The argument is valid.

23. This argument is the law of syllogism and therefore it is valid.

24.

p	q	[(q ∧ p) ∧ q] → ~ p				
T	T	T	T	T	F	F
T	F	F	F	F	T	F
F	T	F	F	T	T	T
F	F	F	F	F	T	T
		1	3	2	5	4

The argument is invalid.

25.

p	q	r	[(p ↔ q) ∧ (q ∧ r)] → (p ∨ r)				
T	T	T	T	T	T	T	T
T	T	F	T	F	F	T	T
T	F	T	F	F	F	T	T
T	F	F	F	F	F	T	T
F	T	T	F	F	T	T	T
F	T	F	F	F	F	T	F
F	F	T	T	F	F	T	T
F	F	F	T	F	F	T	F
			1	3	2	5	4

The argument is valid.

26.

p	q	r	[(p ↔ q) ∧ (q → r)] → (~ r → ~ p)						
T	T	T	T	T	T	T	F	T	F
T	T	F	T	F	F	T	T	F	F
T	F	T	F	F	T	T	F	T	F
T	F	F	F	F	T	T	T	F	F
F	T	T	F	F	T	T	F	T	T
F	T	F	F	F	F	T	T	T	T
F	F	T	T	T	T	T	F	T	T
F	F	F	T	T	T	T	T	T	T
			1	3	2	7	4	6	5

The argument is valid.

27.

p	q	r	[(r ↔ p) ∧ (~p ∧ q)] → (p ∧ r)							
T	T	T	T	F	F	F	T	T	T
T	T	F	F	F	F	F	T	T	F
T	F	T	T	F	F	F	F	T	T
T	F	F	F	F	F	F	F	T	F
F	T	T	F	F	T	T	T	T	F
F	T	F	T	T	T	T	T	F	F
F	F	T	F	F	T	F	F	T	F
F	F	F	T	F	T	F	F	T	F
			1	5	2	4	3	7	6

The argument is invalid.

28.

p	q	r	[(p ∨ q) ∧ (r ∧ p)] → q				
T	T	T	T	T	T	T	T
T	T	F	T	F	F	T	T
T	F	T	T	T	T	F	F
T	F	F	T	F	F	T	F
F	T	T	T	F	F	T	T
F	T	F	T	F	F	T	T
F	F	T	F	F	F	T	F
F	F	F	F	F	F	T	F
			1	3	2	5	4

The argument is invalid.

29.

p	q	r	[(p→ q) ∧ (q ∨ r) ∧ (r ∨ p)] → p
T	T	T	T T T T T T T
T	T	F	T T T T T T T
T	F	T	F F T F T T T
T	F	F	F F F F T T T
F	T	T	T T T T T F F
F	T	F	T T T F F T F
F	F	T	T T T T T F F
F	F	F	T F F F F T F
			1 3 2 5 4 7 6

The argument is invalid.

30. This argument is the law of syllogism and therefore it is valid.

31.

p	q	r	[(p → q) ∧ (r → ~ p) ∧ (p ∨ r)] → (q ∨ ~ p)
T	T	T	T F TF F F T T TT F
T	T	F	T T FT F T T T TT F
T	F	T	F F TF F F T T FF F
T	F	F	F F FT F F T T FF F
F	T	T	T T TT T T T T TT T
F	T	F	T T FT T F F T TT T
F	F	T	T T TT T T T T FT T
F	F	F	T T FT T F F T FT T
			1 5 24 3 7 6 11 9 10 8

The argument is valid.

32.

p	q	r	[(p ↔ q) ∧ (p ∨ q) ∧ (q → r)] → (q ∨ r)
T	T	T	T T T T T T T
T	T	F	T T T F F T T
T	F	T	F F T F T T T
T	F	F	F F T F T T F
F	T	T	F F T F T T T
F	T	F	F F T F F T T
F	F	T	T F F F T T T
F	F	F	T F F F T T F
			1 4 2 5 3 7 6

The argument is valid.

33. p: Will Smith wins an Academy Award.
q: Will Smith retires from acting.

p	q	[(p → q) ∧ ~ p] → ~ q
T	T	T F FT F
T	F	F F FT T
F	T	T T TF F
F	F	T T TT T
		1 3 25 4

The argument is invalid.

34. p: The president resigns.
q: The vice president becomes president.

p	q	[(p → q) ∧ q] → p
T	T	T T T T T
T	F	F F F T T
F	T	T T T F F
F	F	T F F T F
		1 3 2 5 4

The argument is invalid.

35. p: The baby is a boy.
    q: The baby will be named Alexander Martin.

p	q	[(p → q) ∧ q] → p			
T	T	T	T T T		
T	F	F	F F T		T
F	T	T	T T F		F
F	F	T	F F T		F
		1	3 2 5		4

The argument is valid.

36. p: I get my child to preschool by 8:45 a.m.
    q: I take the 9:00 a.m. class.
    r: I am done by 2:00 p.m.
    [(p → q) ∧ (q → r)] → (p → r)

This argument is valid because of the Law of Syllogism.

37. p: Monkeys can fly.
    q: Scarecrows can dance.
    [(p → q) ∧ ~q] → ~ p

This argument valid because of the Law of Contraposition.

38. p: Rob Calcatera will go on sabbatical.
    q: Frank Cheek will teach Logic.

p	q	[(p ∨ q) ∧ ~ q] → p				
T	T	T	F F	T	T	
T	F	T	T T	T	T	
F	T	T	F F	F	F	
F	F	T	T T	T	F	
		1	3 2	5	4	

The argument is valid.

39. p: The orange was left on the tree for one year.
    q: The orange is ripe.

p	q	[(p → q) ∧ q] → p			
T	T	T	T T T		
T	F	F	F F T		T
F	T	T	T T F		F
F	F	T	F F T		F
		1	3 2 5		4

This is the Fallacy of the Converse; thus the argument is valid.

40. p: You pass General Chemistry.
    q: You take Organic Chemistry.

p	q	[(p → q) ∧ p] → q			
T	T	T	T T T		
T	F	F	F T T		F
F	T	T	F F T		T
F	F	T	F F T		F
		1	3 2 5		4

This is the Law of Detachment, thus the argument is valid.

41. p: The X-Games will be in San Diego.
    q: The X-Games will be in Corpus Christi.

P	q	[(p ∨ q) ∧ ~p] → q				
T	T	T	F F	T	T	
T	F	T	F F	T	F	
F	T	T	F T	T	T	
F	F	F	T F	T	F	
		1	3 2	5	4	

The argument is valid.

42. p: Nicholas Thompson teaches this course.
    q: I will get a passing grade.

p	q	[(p → q) ∧ ~ q] → ~ p				
T	T	T	F F	T	F	
T	F	F	F T	T	F	
F	T	T	F F	T	T	
F	F	T	T T	T	T	
		1	3 2	5	4	

This argument is valid - Contraposition.

43. p: It is cold.
    q: The graduation will be held indoors.
    r: The fireworks will be postponed.
    [(p → q) ∧ (q → r)] → (p → r)

This argument is valid because of the Chain Rule.

44. p: Miles Davis played with Louis Armstrong.
    q: Charlie Parker played with Dizzy Gillespie.

p	q	[(p → q) ∧ ~ p] → ~ q				
T	T	T	F F	T	F	
T	F	F	F F	T	T	
F	T	T	T T	F	F	
F	F	T	T T	T	T	
		1	3 2	5	4	

This argument is invalid.

45. f: The canteen is full
    w: We can go for a walk.
    t: We will get thirsty.

f	w	t	[(f → w) ∧ (w ∧ ~ t)] → (w → ~ f)
T	T	T	T F T F F T T F F
T	T	F	T T T T T F T F F
T	F	T	F F F F F T F T F
T	F	F	F F F F T T F T F
F	T	T	T F T F F T T T T
F	T	F	T T T T T T T T T
F	F	T	T F F F F T F T T
F	F	F	T F F F T T F T T
			1 5 2 4 3 9 6 8 7

The argument is not valid.

46. p: Bryce Canyon National Park is in Utah.
    q: Bryce Canyon National Park is in Arizona.

p	q	[(p ∨ q) ∧ (q → ~ p)] → ~ q
T	T	T F T F F T F
T	F	T T T F T F T T
F	T	T T T T T F F
F	F	F F F T T T T
		1 5 2 4 3 7 6

The argument is invalid.

47. s: It is snowing.
    g: I am going skiing.
    c: I will wear a coat.

s	g	c	[(s ∧ g) ∧ (g → c)] → (s → c)
T	T	T	T T T T T
T	T	F	T F F T F
T	F	T	F F T T T
T	F	F	F F T T F
F	T	T	F F T T T
F	T	F	F F F T T
F	F	T	F F T T T
F	F	F	F F T T T
			1 3 2 5 4

The argument is valid.

48. g: The garden has vegetables.
    f: The garden has flowers.

g	f	[(g ∨ f) ∧ (~ f → g)] → (f ∨ g)
T	T	T T F T T T T
T	F	T T T T T T T
F	T	T T F T F T T
F	F	F F T F F T F
		1 5 2 4 3 7 6

The argument is valid.

49. h: The house has electric heat.
    b: The Flynns will buy the house.
    p: The price is less than $100,000.

h	b	p	[(h → b) ∧ (~p → ~b)] → (h → p)
T	T	T	T T F T F T T
T	T	F	T F T F F T F
T	F	T	F F F T T T T
T	F	F	F F T T T T F
F	T	T	T T F T F T T
F	T	F	T F T F F T T
F	F	T	T T F T T T T
F	F	F	T T T T T T T
			1 5 2 4 3 7 6

The argument is valid.

50. a: There is an atmosphere.
    g: There is gravity.
    w: An object has weight.

a	g	w	[(a → g) ∧ (w → g)] → (a → w)
T	T	T	T T T T T
T	T	F	T T T F F
T	F	T	F F F T T
T	F	F	F F T T F
F	T	T	T T T T T
F	T	F	T T T T T
F	F	T	T F F T T
F	F	F	T T T T T
			1 3 2 5 4

The argument is invalid.

51. p: The prescription is called in to Walgreen s.
    q. You pick up the prescription at 4:00 p.m.

p	q	[(p	→	q)	∧	~ q]	→	~ p
T	T	T	T	T	F	F	T	F
T	F	T	F	F	F	T	T	F
F	T	F	T	T	F	F	T	T
F	F	F	T	F	T	T	T	T
		1	3	2	5	4	7	6

The argument is valid.

52. p: The printer has a clogged nozzle.
    q. The printer has no toner.

p	q	[(p	∨	q)	∧	~ q]	→	p
T	T	T	T	T	F	F	T	T
T	F	T	T	F	T	T	T	F
F	T	F	T	T	F	F	T	T
F	F	F	F	F	F	T	T	F
		1	3	2	5	4	7	6

The argument is valid.

53. t: The television is on.
    p: The plug is plugged in.

t	p	[(t	∨	~ p)	∧	(p)]	→	t
T	T	T	F	T	T	T
T	F	T	T	T	T	T
F	T	F	F	F	T	F
F	F	T	T	F	T	F
		2	1	3	5	4

The argument is valid.

54. c: The cat is in the room.   c → m
    m: The mice are hiding.   ~ m
                                      ~ c

This argument is the law of
contraposition and is valid.

55. t: The test was easy.
    g: I received a good grade.

t	g	[(t ∧ g)	∧	(~ t	∨	~ g)]	→	~ t
T	T	T	F	F	F	F	T	F
T	F	F	F	F	T	T	T	F
F	T	F	F	T	T	F	T	T
F	F	F	F	T	T	T	T	T
		1	5	2	4	3	7	6

The argument is valid.

56. b: Bonnie passed the bar exam.   b → p
    p: Bonnie will practice law.   ~ p
                                    . ~ b

This argument is the law of
contraposition and is valid.

57. c: The baby is crying.
    h: The baby is hungry.

c	h	[(c ∧	~ h)	∧	(h	→	c)]	→	h
T	T	T	F	F	F	T	T	T	T
T	F	T	T	T	T	T	F	F	F
F	T	F	F	F	F	T	T	T	T
F	F	F	F	T	F	T	T	T	F
		1	3	2	5	4	7	6	

The argument is invalid.

58. n: The car is new.
    a: The car has air conditioning.

n	a	[(n → a)	∧	(~ n	∧	a)]	→	~ n
T	T	T	F	F	F	T	T	F
T	F	F	F	F	F	F	T	F
F	T	T	T	T	T	T	T	T
F	F	T	F	T	F	F	T	T
		1	5	2	4	3	7	6

The argument is valid.

59. f: The football team wins the game.   f → d
    d: Dave played quarterback.   d → ~ s
    s: The team is in second place.   f → s

    Using the law of syllogism, this argument is
    invalid.

52. e: The engineering courses are difficult.
    c: The chemistry labs are long.
    A: The art tests are easy.

e	c	a	[(e ∧ c) ∧ (c → a)] → (e ∧ ~ a)
T	T	T	T T T F T F F
T	T	F	T F F T T T T
T	F	T	F F T T T F F
T	F	F	F F T T T T T
F	T	T	F F T T F F F
F	T	F	F F F T F F T
F	F	T	F F T T F F F
F	F	F	F F T T F F T
			1 3 2 7 4 6 5

The argument is invalid.

61. p: You eat an entire bag of M&Ms.
    q: Your face will break out.

p	q	[(p → q) ∧ p] → q
T	T	T T T T T
T	F	F F T T F
F	T	T F F T T
F	F	T F F T F
		1 3 2 5 4

The argument is valid.

62. p: The temperature hits 100 degrees.
    q: We go swimming.

p	q	[(p → q) ∧ ~ q] → ~ p
T	T	T F F T F
T	F	F F T T F
F	T	T F F T T
F	F	T T T T T
		1 3 2 5 4

The argument is valid.

63. p: A tick is an insect.
    Q: A tick is an arachnid.

p	q	[(p ∨ q) ∧ ~ p] → q
T	T	T F F T T
T	F	T F F T F
F	T	T T T T T
F	F	F F T T F
		1 3 2 5 4

The argument is valid.

64. p: Margaret Chang arranged the conference.
    q: Many people attend the conference.
    r: Our picture will be in the paper.

p	q	r	[(p → q) ∧ (q → r)] → (p → r)
T	T	T	T T T T T
T	T	F	T F F T F
T	F	T	F F T T T
T	F	F	F F T T F
F	T	T	T T T T T
F	T	F	T F F T T
F	F	T	T T T T T
F	F	F	T T T T T
			1 3 2 5 4

Using the Chain Rule, the argument is valid.

65. d: You close the deal.   d → c
    c: You get a commission.   ~ c
                                   ~ d

    Using the Law of Contraposition, you did not
    close the deal.

66. r: You read a lot.   ~ r → ~ k
    k: You gain knowledge.   ~ r
                                    ~ k

    Using the Law of Detachment, you will
    not gain knowledge.

67. c: You pay off your credit card bill.   ~ c → p
    p: You will have to pay interest.   p → m
    m: The bank makes money.   ~ c → m
    [(~ c → p) ∧ (p → m) → (~ c → m)]
    Using the Law of Syllogism, the bank makes money.

68. p: Lynn wins the contest.
    q: Lynn strikes oil.
    r : Lynn will be rich.
    s: Lynn will stop working.

p	q	r	s	[((p ∨ q) → r) ∧ (r → s)]	(~ s → ~ p)
T	T	T	T	T T T T T	T T
T	T	T	F	T T T F F	T F
T	T	F	T	T F F F T	T T
T	T	F	F	T F F F T	T F
T	F	T	T	T T T T T	T T
T	F	T	F	T T T F F	T F
T	F	F	T	T F F F T	T T
T	F	F	F	T F F F T	T F
F	T	T	T	T T T T T	T T
F	T	T	F	T T T F F	T T
F	T	F	T	T F F F T	T T
F	T	F	F	T F F F T	T T
F	F	T	T	F T T T T	T T
F	F	T	F	F T T F F	T T
F	F	F	T	F T F T T	T T
F	F	F	F	F T F T T	T T
				1 3 2 5 4	7 6

The argument is valid.

69. No. An argument is <u>invalid</u> only when the conjunction of the premises are true and the conclusion is false.

63. p: I think.
    q: I am.

p	q	[(p → q) ∧ ~p] → ~ q
T	T	T F F T F
T	F	F F F T T
F	T	T T T F F
F	F	T T T T T
		1 3 2 5 4

By the Fallacy of the Inverse, the argument is invalid.

Exercise Set 3.6

1. The argument is invalid.

2. The argument is valid.

3. The conclusion necessarily follows from the set of premises.

4. Symbolic arguments use the connectives "and," "or," "not," "but," "if-then," and "if and only if", while syllogistic arguments use the quantifiers "all," "some," and "none."

5. Yes. If the conjunction of the premises is false in all cases, then the argument is valid regardless of the truth value of the conclusion.

6. Yes. An argument in which the conclusion does not necessarily follow from the given set of premises is invalid, even if the conclusion is a true statement.

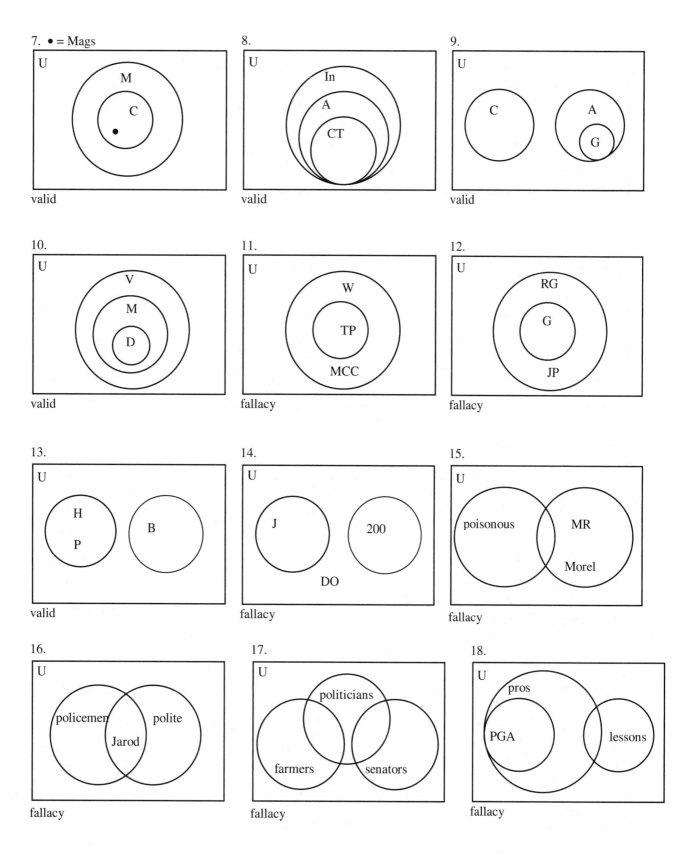

7. • = Mags

U
M
C
•

valid

8.

U
In
A
CT

valid

9.

U
C
A
G

valid

10.

U
V
M
D

valid

11.

U
W
TP
MCC

fallacy

12.

U
RG
G
JP

fallacy

13.

U
H
P
B

valid

14.

U
J
200
DO

fallacy

15.

U
poisonous
MR
Morel

fallacy

16.

U
policemen
polite
Jarod

fallacy

17.

U
politicians
farmers
senators

fallacy

18.

U
pros
PGA
lessons

fallacy

19.

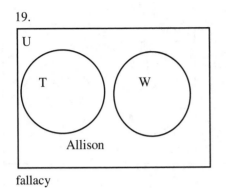

fallacy

20.

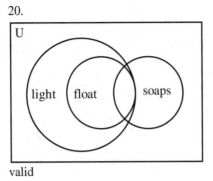

valid

21.

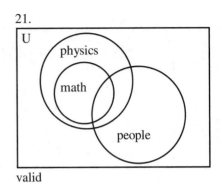

valid

22.

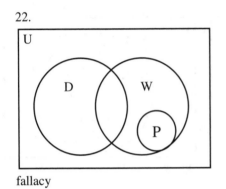

fallacy

23.

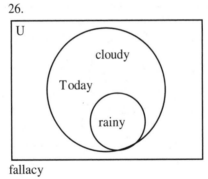

fallacy

24.

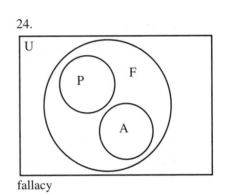

fallacy

25.

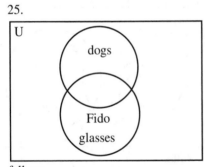

fallacy

26.

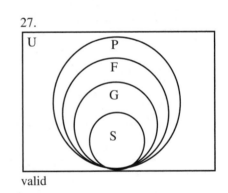

fallacy

27.

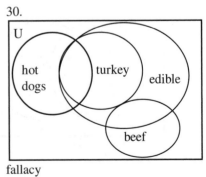

valid

28.

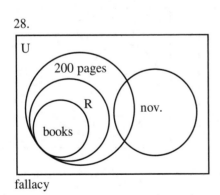

fallacy

29.

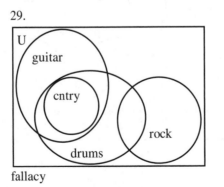

fallacy

30.

fallacy

31. [(P → Q) ∧ (P ∨ Q)] → ~ P can be expressed as a set
statement by [(P'∪ Q) ∩ (P ∪ Q)] ∩ P'. If this
statement is true, then the argument is valid; otherwise,
the argument is invalid.

Set	Regions
P'∪ Q	II, III, IV
P ∪ Q	I, II, III
(P'∪ Q) ∩ (P ∪ Q)	II, III
P'	III, IV

Since (P'∪ Q) ∩ (P ∪ Q) is not a subset of P', the
argument is invalid.

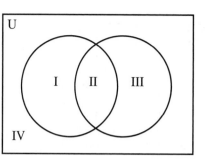

Review Exercises

1. No rock bands play ballads.

2. All bananas are ripe.

3. Some chickens have lips.

4. Some panthers are not endangered.

5. Some pens do not use ink.

6. Some rabbits wear glasses.

7. The coffee is Maxwell House or the coffee is hot.

8. The coffee is not hot and the coffee is strong.

9. If the coffee is hot, then the coffee is strong and it is not Maxwell House.

10. The coffee is Maxwell House if and only if the coffee is not strong.

12. The coffee is not Maxwell House, if and only if the coffee is strong and the coffee is not hot.

11. The coffee is Maxwell House or the coffee is not hot, and the coffee is not strong.

13. r ∧ q

14. p → r

15. (r → q) ∨ ~ p

16. (q ↔ p) ∧ ~ r

17. (r ∧ q) ∨ ~ p

18. ~ (r ∧ q)

19.

p	q	(p	∨	q)	∧	~ p
T	T		T		F	F
T	F		T		F	F
F	T		T		T	T
F	F		F		F	T
			1		3	2

20.

p	q	r	p	∧	(~ q	∨	r)
T	T	T	T	T	F	T	T
T	T	F	T	F	F	F	F
T	F	T	T	T	T	T	T
T	F	F	T	T	T	T	F
F	T	T	F	F	F	T	T
F	T	F	F	F	F	F	F
F	F	T	F	F	T	T	T
F	F	F	F	F	T	T	F
			4	5	1	3	2

21.

p	q	r	(p ∨ q)	↔	(p ∨ r)
T	T	T	T	T	T
T	T	F	T	T	T
T	F	T	T	T	T
T	F	F	T	T	T
F	T	T	T	T	T
F	T	F	T	F	F
F	F	T	F	F	T
F	F	F	F	T	F
			1	3	2

22.

p	q	q	↔	(p	∨	~ q)
T	T	T	T	T	T	F
T	F	F	F	T	T	T
F	T	T	F	F	F	F
F	F	F	F	F	T	T
		1	5	2	4	3

23.

p	q	r	P	→	(q	∧	~r)
T	T	T	T	F	T	F	F
T	T	F	T	T	T	T	T
T	F	T	T	F	F	F	F
T	F	F	T	F	F	F	T
F	T	T	F	T	T	F	F
F	T	F	F	T	T	T	T
F	F	T	F	T	F	F	F
F	F	F	F	T	F	F	T
			4	5	1	3	2

24.

p	q	r	(p	∧ q)	→	~ r
T	T	T	T		F	F
T	T	F	T		T	T
T	F	T	F		T	F
T	F	F	F		T	T
F	T	T	F		T	F
F	T	F	F		T	T
F	F	T	F		T	F
F	F	F	F		T	T
			1		3	2

25. p: 7 is odd. $p \to q$
    q: 11 is even. $T \to F$
                             F

26. p: The St. Louis arch is in St. Louis. $p \lor q$
    q: Abraham Lincoln is buried in $T \lor F$
       in Grant s Tomb.   T

27. p: Oregon borders the Pacific Ocean. $p \lor q$
    q: California borders the Atlantic Ocean. $T \lor F$
                                   T

28. p: 15 7 = 22 $(p \lor q) \land r$
    q: 4 + 9 = 13 $(F \lor T) \land T$
    r : 9 8 = 1   T $\land T$
                             T

29. p: 32% of OR s electricity - coal $(p \leftrightarrow q) \lor r$
    q: 54% of OR s electricity - hydro $(T \leftrightarrow T) \lor F$
    r: 38% of OR s electricity - nuclear   T $\lor F$
                               T

30. p: 3% of OR s electricity - gas/oil $p \to (q \land r)$
    q: 45% of OR s electricity - coal $F \to (F \land T)$
    r: 3% of OR s electricity - nuclear $F \to$ F
                               T

31. $(p \to \, \sim r) \lor (p \land q)$
   $(T \to \, T \,) \lor (T \land F)$
     T $\lor$ F
           T

32. $(p \lor q) \leftrightarrow (\sim r \land p)$
   $(T \lor F) \leftrightarrow (T \land T)$
     T $\leftrightarrow$ T
         T

33. $\sim r \leftrightarrow [(p \lor q) \leftrightarrow \, \sim p]$
   $T \leftrightarrow [(T \lor F) \leftrightarrow F]$
   $T \leftrightarrow [\quad T \quad \leftrightarrow F]$
   $T \leftrightarrow$ F
      F

34. $\sim [(q \land r) \to (\sim p \lor r)]$
   $\sim [(F \land F) \to (F \lor F)]$
   $\sim [\quad F \quad \to \quad\quad F]$
         $\sim$ T
          F

35.

p	q	~p	∨	~q	~p	↔	q
T	T	F	F	F	F	F	T
T	F	F	T	T	F	T	F
F	T	T	T	F	T	T	T
F	F	T	T	T	T	F	F
		1	3	2	1	3	2

The statements are not equivalent.

36. Using the fact that $(p \to q) \Leftrightarrow (\sim p \lor q)$,
    we can conclude that $\sim p \to \, \sim q \Leftrightarrow p \lor \, \sim q$.

37.

p	q	r	~p ∨ (q ∧ r)	(~p ∨ q) ∧ (~p ∨ r)
T	T	T	F T T	F T T T F T T
T	T	F	F F F	F T T F F F F
T	F	T	F F F	F F F F F T T
T	F	F	F F F	F F F F F F F
F	T	T	T T T	T T T T T T T
F	T	F	T T F	T T T T T T F
F	F	T	T T F	T T F T T T T
F	F	F	T T F	T T F T T T F
			2 3 1	1 3 2 7 4 6 5

The statements are equivalent.

38.

p	q	(~q → p) ∧ p	~ (~p ↔ q) ∨ p
T	T	F T T T T	T F F T T T
T	F	T T T T T	F F T F T T
F	T	F T F F F	F T T T F F
F	F	T F F F F	T T F F T F
		1 3 2 5 4	4 1 3 2 6 5

The statements are not equivalent.

39. p: Johnny Cash is in the Rock and Roll (R&R) Hall of Fame.

q: India Arie recorded *Acoustic Soul*.

In symbolic form, the statement is p ∧ q. Using DeMorgan's Laws, we get p ∧ q ⇔ ~ (~ p ∨ ~ q). It is false that Johnny Cash is not in the R&R Hall of Fame or India Arie did not sing *Acoustic Soul*.

40. p: Her foot fell asleep.

q: She has injured her ankle.

In symbolic form, the statement is p ∨ q. Using the fact that p → q ⇔ ~ p ∨ q, we can rewrite the given statement as ~ p → q. If her foot did not fall asleep, then she has injured it.

41. p: Altec Lansing only produces speakers.

q: Harmon Kardon only produces stereo receivers.

The symbolic form is ~ (p ∨ q).

Using DeMorgan's Laws, we get ~ (p ∨ q) ⇔ ~ p ∧ ~ q. Altec Lansing does not produce only speakers and Harmon Karson does not produce only stereo receivers.

42. p: Travis Tritt won an Academy Award.

q: Randy Jackson does commercials for Milk Bone dog biscuits.

The symbolic form is ~ p ∧ ~ q. Using DeMorgan's Laws, we get ~ p ∧ ~ q ⇔ ~ (p ∨ q). It is false that Travis Tritt won an Academy Award or Randy Jackson does commercials for Milk Bone dog biscuits.

43. p: The temperature is above 32 degrees Fahrenheit.

q: We will go ice fishing at O'Leary's Lake.

The symbolic form is ~ p → q.

Using DeMorgan's Laws, we get ~ p → q ⇔ p ∨ q. The temperature is above 32 degrees Fahrenheit or we will go ice fishing at O'Leary's Lake.

44. Converse: If you enjoy life, then you will hear a beautiful songbird today.

Inverse: If you do not hear a beautiful songbird today, then you will not enjoy life.

Contrapositive: If you will not enjoy life, then you will not hear beautiful songbird today.

45. Converse: If the quilt has a uniform design, then you followed the correct pattern.

Inverse: If you do not follow the correct pattern, then the quilt will not have a uniform design.

Contrapositive: If the quilt does not have a uniform design, then you did not follow the correct pattern.

46. Converse: If Maureen Gerald is helping at school, then she is not in attendance.

Inverse: If Maureen Gerald is in attendance, then she is not helping at school.

Contrapositive: If Maureen Gerald is not helping at school, then she is in attendance.

47. Converse: If we do not buy a desk at Miller's
    Furniture, then the desk is made by Winner's Only
    and is in the Rose catalog.
    Inverse: If we did not buy the desk at Miller's
    Furniture, then it is not made by Winner's Only
    and is not in the Rose catalog.
    Contrapositive: If the desk is not made by
    Winner's Only and is not in the Rose catalog.,
    then we did not buy it at Miller's Furniture.

48. Converse: If I let you attend the prom,
    then you get straight A's on your report
    card.
    Inverse: If you do not get straight A's on
    your report card, then I will not let you
    attend the prom.
    Contrapositive: If I do not let you attend
    the prom, then you do not get straight
    A's on your report card.

49. p: The temperature is over 80^0.
    q: The air conditioner will come on.
    In symbolic form, the statements are: a) $p \rightarrow q$,
    b) $\sim p \vee q$, and c) $\sim (p \wedge \sim q)$. Using the fact that
    $p \rightarrow q$ is equivalent to $\sim p \vee q$, statements (a) and (b)
    are equivalent. Using DeMorgan's Laws on
    statement (b) we get $\sim (p \wedge \sim q)$.

    Therefore all 3 statements are equivalent.

50. p: The screwdriver is on the workbench.
    q: The screwdriver is on the counter.
    In symbolic form, the statements are: a) $p \leftrightarrow \sim q$,
    b) $\sim q \rightarrow \sim p$, and c) $\sim (q \wedge \sim p)$. Looking at the truth
    tables for statements (a), (b), and (c) we can conclude
    that none of the statements are equivalent.

p	q	a) $p \leftrightarrow \sim q$	b) $\sim q \rightarrow \sim p$	c) $\sim (q \wedge \sim p)$
T	T	T F F	F T F	T T F F
T	F	T T T	T F F	T F F F
F	T	F T F	F T T	F T T T
F	F	F F T	T T T	T F F T
		1 3 2	1 3 2	4 1 3 2

51. p: $2 + 3 = 6$.
    q: $3 + 1 = 5$.
    In symbolic form, the statements are: a) $p \rightarrow q$,
    b) $p \leftrightarrow \sim q$, and c) $\sim q \rightarrow \sim p$.
    Statement (c) is the contrapositive of statement
    (a). Therefore statements (a) and (c) are equivalent.
    (a) $F \rightarrow F$ (b) $F \leftrightarrow T$ (c) $T \rightarrow T$
         T   F   T

52. p: The sale is on Tuesday.
    q: I have money.
    r : I will go to the sale.
    In symbolic form the statements are: a) $(p \wedge q) \rightarrow r$,
    b) $r \rightarrow (p \wedge q)$, and c) $r \vee (p \wedge q)$. The truth table for
    statements (a), (b), and (c) shows that none of the
    statements are equivalent.

p	q	r	$(p \wedge q) \rightarrow r$	$r \rightarrow (p \wedge q)$	$r \vee (p \wedge q)$
T	T	T	T T T	T T T	T T T
T	T	F	T F F	F T T	F T T
T	F	T	F T T	T F F	T T F
T	F	F	F T F	F T F	F F F
F	T	T	F T T	T F F	T T F
F	T	F	F T F	F T F	F F F
F	F	T	F T T	T F F	T T F
F	F	F	F T F	F T F	F F F
			1 3 2	1 3 2	1 3 2

53.

p	q	[(p → q)	∧	~ p]	→	q
T	T	T	F	F	T	T
T	F	F	F	F	T	F
F	T	T	T	T	T	T
F	F	T	T	T	F	F
		1	3	2	5	4

The argument is invalid.

54.

p	q	r	[(p ∧ q)	∧	(q → r)]	→	(p → r)
T	T	T	T	T	T	T	T
T	T	F	T	F	F	T	F
T	F	T	F	F	T	T	T
T	F	F	F	F	T	T	F
F	T	T	F	F	T	T	T
F	T	F	F	F	F	T	T
F	F	T	F	F	T	T	T
F	F	F	F	F	T	T	T
			1	3	2	5	4

The argument is valid.

55. p: Nicole is in the hot tub. p ∨ q
    q: Nicole is in the shower. _____p_
                                  ~ q

p	q	[(p ∨ q)	∧	p]	→	~ q
T	T	T	T	T	F	F
T	F	T	T	T	T	T
F	T	T	F	F	T	F
F	F	F	F	F	T	T
		1	3	2	5	4

The argument is invalid.

56. p: The car has a sound system.   p → q
    p: Rick will buy the car.   ~ r → ~ q
    r : The price is less than $18,000. p → r

p	q	r	[(p → q)	∧	(~ r → ~q)]	→	(p → r)
T	T	T	T	T	F T F	T	T
T	T	F	T	F	T F F	T	F
T	F	T	F	F	F T T	T	F
T	F	F	F	F	T T T	T	F
F	T	T	T	T	F T F	T	T
F	T	F	T	F	T F F	T	T
F	F	T	T	T	F T T	T	T
F	F	F	T	T	T T T	T	T
			1	5	2 4 3	7	6

The argument is valid.

57.

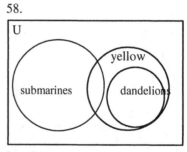

invalid

58.

invalid

Chapter Test

1. (p ∧ r) ∨ ~ q

2. (r → q) ∨ ~ p

3. ~ (r ↔ ~ q)

4. Ann Veneman is the Secretray of Agriculture, if and only if Dick Cheney is the Vice President and Elaine Chao is the Secretary of Labor.

5. It is false that if Ann Veneman is the Secretary of Agriculture, then Elaine Chao is not the Secretary the Secretary of Labor.

6.

p	q	r	[~ (p → r)] ∧ q			
T	T	T	F	T	F	T
T	T	F	T	F	T	T
T	F	T	F	T	F	F
T	F	F	T	F	F	F
F	T	T	F	T	F	T
F	T	F	F	T	F	T
F	F	T	F	T	F	F
F	F	F	F	T	F	F
			2	1	4	3

7.

p	q	r	(q ↔ ~ r) ∨ p				
T	T	T	T	F	F	T	T
T	T	F	T	T	T	T	T
T	F	T	F	T	F	T	T
T	F	F	F	F	T	T	T
F	T	T	T	F	F	F	F
F	T	F	T	T	T	T	F
F	F	T	F	T	F	T	F
F	F	F	F	F	T	F	F
			1	3	2	5	4

8. p: $2 + 6 = 8$
  q: $7 - 12 = 5$
  p ∨ q
  T ∨ F
    T

9. p: A scissors can cut paper.
  q: A dime equals 2 nickels.
  r : Louisville is a city in Kentucky.
    (p ∨ q) ↔ r
    (T ∨ T) ↔ T
     T ↔ T
       T

10. (r ∨ q) ↔ (p ∧ ~ q)
  (T ∨ F) ↔ (T ∧ T)
   T ↔ T
     T

11. [~(r → ~ p)] ∧ (q → p)
  [~(T → F)] ∧ (F → T)
  [~ (F)] ∧   T
      T   ∧   T
          T

12. Applying DeMorgan's Law to
statement (a), we get:
  (1) ~(~p ∨ q), (2) ~(p ∨ ~ q), and
  (3) ~(p ∧ ~ q).
Therefore, ~ p ∨ q ⇔ ~(p ∧ ~ q).

13. p: The bird is red.
  q: It is a cardinal.
In symbolic form the statements
are: a) p → q, b) ~ p ∨ q,
and c) ~ p → ~ q.
Statement (c) is the inverse of
statement (a) and thus they cannot
be equivalent. Using the fact that
p → q ⇔ ~ p ∨ q, to rewrite
statement (a) we get ~ p ∨ q.
Therefore statements (a) and (b)
are equivalent.

14. p: The test is today.   q: The concert is tonight.   In symbolic form the
statements are: a) ~ (p ∨ q), b) ~ p ∧ ~ q, and ~ p → ~ q.
Applying DeMorgan's Law to statement (a) we get: ~ p ∧ ~ q.
Therefore statements (a) and (b) are equivalent. When we compare the
Truth tables for statements (a), (b), and (c) we see that only statements
(a) and (b) are equivalent.

p	q	~ (p ∨ q)		~ p ∧ ~ q			~ p → ~ q		
T	T	F	T	F	F	F	F	T	F
T	F	F	T	F	F	T	F	T	T
F	T	F	T	T	F	F	T	F	F
F	F	T	F	T	T	T	T	T	T
		2	1	1	3	2	1	3	2

15. s: The soccer team won the game.
  f: Sue played fullback.
  p: The team is in second place.
This argument is the law of
syllogism and therefore it is
valid.   s → f
             f → p
             s → p
This argument is the law of
syllogism and therefore it is valid.

16.

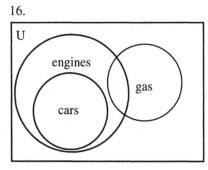

Fallacy

17. Some leopards are not spotted.

18. No Jacks-in-the-box are electronic.

19. Converse: If today is Saturday, then the
garbage truck comes.
Inverse: If the garbage truck does not come
today, then today is not Saturday.
Contrapositive: If today is not Saturday, then
the garbage truck does not come.

20. Yes. An argument is valid when its conclusion
necessarily follows from the given set of premises. It
doesn't matter whether the conclusion is a true or
false statement.

Group Projects

1. a) 4,   p closed, q closed   p closed, q open
p open, q closed   p open, q open

1. b)

p	q	p∧q
1	1	1
1	0	0
0	1	0
0	0	0

1. c) If a closed switch is represented as T and an open
switch is represented as F, and the bulb lighting as T, and
the bulb not lighting as F, then the table would be identical
to the truth table for p ∧q.

1. d)

p	q	p∨q
1	1	1
1	0	1
0	1	1
0	0	0

1. f) (p ∧ q) ∨ r

1. g)

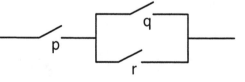

2. a) The tables have the same truth values as the *not, and*
and *or* tables respectively.
b) 0   c) 1   d) 0
e) $I_a = 0$, $I_b = 1$ or $I_a = 1$, $I_b = 0$

2.f)

I_a	I_b	0
1	1	1
1	0	1
0	1	1
0	0	0

CHAPTER FOUR

SYSTEMS OF NUMERATION

Exercise Set 4.1

1. A **number** is a quantity, and it answers the question, How many? A **numeral** is a symbol used to represent the number.

2. ∩, x, ✝, ɩ, ◁, 10

3. A **system of numeration** consists of a set of numerals and a scheme or rule for combining the numerals to represent numbers.

4. ꒐, C, 百, ρ, 100

5. The Hindu-Arabic numeration system

6. In an **additive system**, the sum of the values of the numerals equals the number.

7. In a **multiplicative system**, there are numerals for each number less than the base and for powers of the base. Each numeral less than the base is multiplied by a numeral for the power of the base, and these products are added to obtain the number.

8. In a **ciphered system**, the number represented by a particular set of numerals is the sum of the values of the numerals.

9. $100 + 10 + 10 + 10 + 10 + 1 + 1 = 142$

10. $100 + 100 + 10 + 10 + 1 + 1 = 222$

11. $1000 + 1000 + 100 + 100 + 100 + 100 + 10 + 10$
    $+ 1 + 1 + 1 = 2423$

12. $10,000 + 10,000 + 10,000 + 10,000 + 1000$
    $+ 100 + 100 + 10 = 41,210$

13. $100,000 + 100,000 + 100,000 + 10,000$
    $+ 10,000 + 10,000 + 1000 + 1000 + 1000 + 1000$
    $+ 100 + 100 + 10 + 1 + 1 + 1 + 1 = 334,214$

14. $1,000,000 + 1,000,000 + 1,000,000 + 100,000$
    $+ 100,000 + 100,000 + 100 + 100 + 100 + 100$
    $+ 10 + 10 + 10 + 1 = 3,300,431$

15. 99999∩∩∩||||

16. 9999999∩∩∩||

17. ⌇⌇∩∩∩|||||

18. ⌇99999999∩||

19. ◁|||||||⌇⌇⌇99999999∩∩|||||

20. 𓀀𓀀𓀀◁◁|||⌇⌇⌇⌇⌇999999∩|||||

21. $10 + (10 - 1) = 19$

22. $10 + 5 + 1 = 16$

23. $500 + (50 - 10) + 5 + 1 + 1 = 547$

24. $500 + 50 + 10 + 10 + 5 = 575$

25. $1000 + (500 - 100) + (100 - 10) + 1 + 1 = 1492$

26. $1000 + (1000 - 100) + 10 + 5 + 1 + 1 + 1 = 1918$

27. $1000 + 1000 + (1000 - 100) + (50 - 10) + 5 + 1$
    $= 2946$

28. $1000 + 500 + 100 + 100 + (50 - 10) + 5 + 1$
    $= 1746$

29. $10(1000)+1000+1000+500+100+50+10$
$+5+1=12{,}666$

30. $50(1000)+1000+(1000-100)+(50-10)$
$+(5-1)=51{,}944$

31. $9(1000)+(500-100)+50+10+(5-1)=9464$

32. $5(1000)+1000+100+100+100+10+10$
$+10+1+1+1=6333$

33. LIX
34. XCIV
35. CXXXIV
36. CCLXIX
37. MMV
38. $\overline{\text{IV}}$CCLXXXV

39. $\overline{\text{IV}}$DCCXCIII
40. $\overline{\text{VI}}$CCLXXIV
41. $\overline{\text{IX}}$CMXCIX
42. $\overline{\text{XIV}}$CCCXV
43. $\overline{\text{XX}}$DCXLIV
44. $\overline{\text{XCIX}}$CMXCIX

45. $7(10)+4=74$
46. $6(10)+2=62$
47. $4(1000)+8(10)+1=4081$
48. $3(1000)+2(10)+9=3029$
49. $8(1000)+5(100)+5(10)=8550$
50. $3(1000)+4(100)+8(10)+7=3487$
51. $4(1000)+3=4003$
52. $5(1000)+6(100)+2=5602$

53. 五十三

54. 一百七十八

55. 三百七十八

56. 二千零一

57. 四千二百六十

58. 六千九百零五

59. 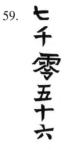　七千零五十六

60. 三千零九

61. $300 + 40 + 1 = 341$
62. $700 + 30 + 6 = 736$
63. $20(1000) + 2(1000) + 500 + 5 = 22,505$
64. $100(1000) + 50(1000) + 800 + 10 + 3 = 150,813$
65. $9(1000) + 600 + 7 = 9607$
66. $4(1000) + 900 + 90 + 9 = 4999$
67. $\nu\ \theta$
68. $\rho\ o\ \eta$
69. $\psi\ \kappa\ \zeta$
70. $\beta'\ \alpha$

71. $\pi'\ \beta'\ \psi\ \delta$
72. $\chi'\ Q'\ \phi\ \mu$

73. Advantage: You can write some numbers more compactly.
    Disadvantage: There are more numerals to memorize.
74. Advantage: Numbers are written in a more compact form.
    Disadvantage: There are more symbols to remember.
75. Advantage: You can write some numbers more compactly.
    Disadvantage: There are more numerals to memorize.
    The Hindu-Arabic system has fewer symbols, more compact notation, the inclusion of zero, and the capability of expressing decimal numbers and fractions.

76. $1000 + 10 + 10 + 1 = 1021$, MXXI, $\alpha'\kappa\alpha$
    　一千零二十一

77. $1000 + (1000\ 1\,00) + 10 + 10 + 10 + 5 + 1 = 1936$,
    99999999ΛΛΛΛΙΙΙΙΙΙ,
    $\alpha'\,\pi\,\lambda\,\zeta$,
    一千九百三十六

78. $5(100) + 2(10) + 7 = 527$,
    99999ΛΙΙΙΙΙΙΙ, DXXVII, $\phi\ \kappa\ \zeta$

79. $400 + 20 + 2 = 422$, 9999ΛΙΙ, CDXXII,
    四百二十二

80. $\overline{\text{CMXCIX}}$CMXCIX

81. $\text{T}'\text{Q}'\,\theta'\,\text{T}\,\text{Q}\,\theta$

82. a) c) Answers will vary.

83. Turn the book upside down.

84. MM

85. 1888, MDCCCLXXXVIII

Exercise Set 4.2

1. A base 10 place-value system
2. Positional value system
3. $40 \to$ four tens, $400 \to$ four hundreds
4. Base 10, because we have 10 fingers.
5. A true positional-value system requires a base and a set of symbols, including a symbol for zero and one for each counting number less than the base.
6. a) 10
   b) 0, 1, 2, 3, 4, 5, 6, 7, 8, 9
7. Write each digit times its corresponding positional value.
8. It lacked a symbol for zero.

9. a) There may be confusion because numbers could be interpreted in different ways. For example, $\mathbf{V}$ could be interpreted to be either 1 or 60.

   b) $\mathbf{VV}$ $\mathbf{\langle VVV}$ for both numbers; $133 = 2(60) + 13(1)$ and $7980 = 133(60)$

10. $(10+1)(1) = 11$ and $(10+1)(60) = 660$

11. $1, 20, 18 \times 20, 18 \times (20)^2, 18 \times (20)^3$

12. The Mayan system has a different base and the numbers are written vertically.

13. $(6 \times 10) + (3 \times 1)$

14. $(7 \times 10) + (5 \times 1)$

15. $(3 \times 100) + (5 \times 10) + (9 \times 1)$

16. $(5 \times 100) + (6 \times 10) + (2 \times 1)$

17. $(8 \times 100) + (9 \times 10) + (7 \times 1)$

18. $(3 \times 1000) + (7 \times 100) + (6 \times 10) + (9 \times 1)$

19. $(4 \times 1000) + (3 \times 100) + (8 \times 10) + (7 \times 1)$

20. $(2 \times 10,000) + (3 \times 1000) + (4 \times 100) + (6 \times 10) + (8 \times 1)$

21. $(1 \times 10,000) + (6 \times 1000) + (4 \times 100) + (0 \times 10) + (2 \times 1)$

22. $(1 \times 100,000) + (2 \times 10,000) + (5 \times 1000) + (6 \times 100) + (7 \times 10) + (8 \times 1)$

23. $(3 \times 100,000) + (4 \times 10,000) + (6 \times 1000) + (8 \times 100) + (6 \times 10) + (1 \times 1)$

24. $(3 \times 1,000,000) + (7 \times 100,000) + (6 \times 10,000) + (5 \times 1000) + (9 \times 100) + (3 \times 10) + (4 \times 1)$

25. $(10 + 10 + 10 + 10 + 1 + 1)(1) = 42$

26. $(10 + 10 + 10)(1) - (1 + 1 + 1 + 1)(1) = 30 - 4 = 26$

27. $(10 + 1 + 1 + 1)(60) + (1 + 1 + 1 + 1)(1) = 13(60) + 4(1) = 780 + 4 = 784$

28. $(10 + 1)(60) + ((10 + 10) - (1 + 1 + 1))(1) = 11(60) + (20 - 3)(1) = 660 + 17 = 677$

29. $1(60^2) + (10 + 10 + 1)(60) + (10 - (1 + 1))(1) = 3600 + 21(60) + (10 - 2)(1) = 3600 + 1260 + 8 = 4868$

30. $10(60^2) + ((10 + 10) - (1 + 1 + 1))(60) + (1 + 1)(1) = 10(3600) + (20 - 3)(60) + 2 = 36,000 + 17(60) + 2$
    $= 36,000 + 1020 + 2 = 37,022$

31. 88 is 1 group of 60 and 28 units remaining. $\mathbf{V}$ $\mathbf{\langle\langle\langle\hat{V}VV}$

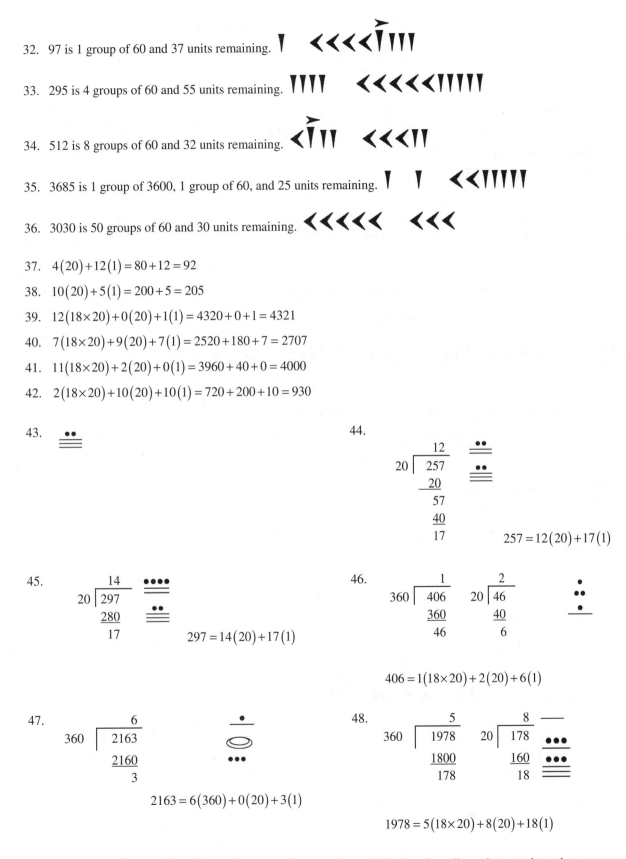

32. 97 is 1 group of 60 and 37 units remaining.

33. 295 is 4 groups of 60 and 55 units remaining.

34. 512 is 8 groups of 60 and 32 units remaining.

35. 3685 is 1 group of 3600, 1 group of 60, and 25 units remaining.

36. 3030 is 50 groups of 60 and 30 units remaining.

37. $4(20)+12(1)=80+12=92$

38. $10(20)+5(1)=200+5=205$

39. $12(18\times20)+0(20)+1(1)=4320+0+1=4321$

40. $7(18\times20)+9(20)+7(1)=2520+180+7=2707$

41. $11(18\times20)+2(20)+0(1)=3960+40+0=4000$

42. $2(18\times20)+10(20)+10(1)=720+200+10=930$

43.

44.

$$257=12(20)+17(1)$$

45.

$$297=14(20)+17(1)$$

46.

$$406=1(18\times20)+2(20)+6(1)$$

47.

$$2163=6(360)+0(20)+3(1)$$

48.

$$1978=5(18\times20)+8(20)+18(1)$$

49. Advantages: In general, a place-value system is more compact; large and small numbers can be written more easily; there are fewer symbols to memorize.
Disadvantage: If many of the symbols in the numeral represent zero, then a place-value system may be less compact.

50. Answers will vary.

51. Hindu-Arabic: $10+10+10+1+1+1=33$

Mayan: $33=1(20)+13(1)$

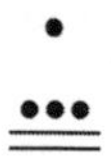

52. Hindu-Arabic:

$5(18\times20)+7(20)+4(1)=1800+140+4=1944$

Babylonian: $1944=32(60)+24(1)$

53. 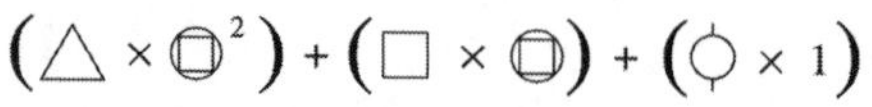 $\left(\triangle\times\bigcirc^{2}\right)+\left(\square\times\bigcirc\right)+\left(\Diamond\times1\right)$

54. 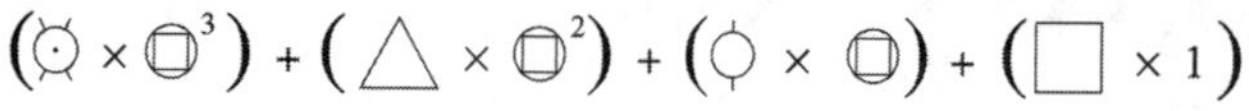 $\left(\dot{\bigcirc}\times\bigcirc^{3}\right)+\left(\triangle\times\bigcirc^{2}\right)+\left(\Diamond\times\bigcirc\right)+\left(\square\times1\right)$

55. a) No largest number; The positional values are $\dots,(60)^{3},(60)^{2},60,1$.

b) $999{,}999=4(60)^{3}+37(60)^{2}+46(60)+39(1)$

56. a) No largest number; The positional values above 18×20 are $18\times20^{2},18\times20^{3},\dots$

b) $999{,}999=6\left(18\times20^{3}\right)+18\left(18\times20^{2}\right)+17(18\times20)+13(20)+19(1)$

57. $2(60)+23(1)=120+23=143$

23

$143+23=166$

$166=2(60)+46(1)$

58. $3(60)+33(1)=180+33=213$

32

$213-32=181$

$181=3(60)+1(1)$

59. $7(18 \times 20) + 6(20) + 15(1) = 2520 + 120 + 15 = 2655$

$6(18 \times 20) + 7(20) + 13(1) = 2160 + 140 + 13 = 2313$

$2655 + 2313 = 4968$

$4968 = 13(18 \times 20) + 14(20) + 8(1)$

60. $7(18 \times 20) + 6(20) + 15(1) = 2520 + 120 + 15 = 2655$

$6(18 \times 20) + 7(20) + 13(1) = 2160 + 140 + 13 = 2313$

$2655 - 2313 = 342$

$342 = 17(20) + 2(1)$

61.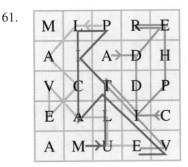

Exercise Set 4.3

1. Answers will vary.

2. Answers will vary.

3. $5_6 = 5(1) = 5$

4. $60_7 = 6(7) + 0(1) = 42 + 0 = 42$

5. $42_5 = 4(5) + 2(1) = 20 + 2 = 22$

6. $101_2 = 1(2^2) + 0(2) + 1(1) = 4 + 0 + 1 = 5$

7. $1011_2 = 1(2^3) + 0(2^2) + 1(2) + 1(1) = 8 + 0 + 2 + 1 = 11$

8. $1101_2 = 1(2^3) + 1(2^2) + 0(2) + 1(1) = 8 + 4 + 0 + 1 = 13$

9. $84_{12} = 8(12) + 4(1) = 96 + 4 = 100$

10. $21021_3 = 2(3^4) + 1(3^3) + 0(3^2) + 2(3) + 1(1) = 2(81) + 27 + 0(9) + 6 + 1 = 162 + 27 + 0 + 6 + 1 = 196$

11. $565_8 = 5(8^2) + 6(8) + 5(1) = 5(64) + 48 + 5 = 320 + 48 + 5 = 373$

12. $654_7 = 6(7^2) + 5(7) + 4(1) = 6(49) + 35 + 4 = 294 + 35 + 4 = 333$

13. $20432_5 = 2(5^4) + 0(5^3) + 4(5^2) + 3(5) + 2(1) = 2(625) + 0 + 4(25) + 15 + 2 = 1250 + 0 + 100 + 15 + 2 = 1367$

14. $101111_2 = 1(2^5) + 0(2^4) + 1(2^3) + 1(2^2) + 1(2) + 1(1) = 32 + 0 + 8 + 4 + 2 + 1 = 47$

15. $4003_6 = 4(6^3) + 0(6^2) + 0(6) + 3(1) = 4(216) + 0 + 0 + 3 = 864 + 0 + 0 + 3 = 867$

16. $123E_{12} = 1(12^3) + 2(12^2) + 3(12) + 11(1) = 1728 + 2(144) + 36 + 11 = 1728 + 288 + 36 + 11 = 2063$

17. $123_8 = 1(8^2) + 2(8) + 3(1) = 64 + 16 + 3 = 83$

18. $2043_8 = 2(8^3) + 0(8^2) + 4(8) + 3(1) = 2(512) + 0 + 32 + 3 = 1024 + 0 + 32 + 3 = 1059$

19. $14705_8 = 1(8^4) + 4(8^3) + 7(8^2) + 0(8) + 5(1) = 4096 + 4(512) + 7(64) + 0 + 5 = 4096 + 2048 + 448 + 0 + 5 = 6597$

20. $67342_9 = 6(9^4) + 7(9^3) + 3(9^2) + 4(9) + 2(1) = 6(6561) + 7(729) + 3(81) + 36 + 2$

    $= 39,366 + 5103 + 243 + 36 + 2 = 44,750$

21. To convert 8 to base 2   16 8 4 2 1

$$
\begin{array}{cccc}
\ \ 1 & \ \ 0 & \ \ 0 & \ \ 0 \\
8\,\overline{)\,8} & 4\,\overline{)\,0} & 2\,\overline{)\,0} & 1\,\overline{)\,0} \\
\ \ \underline{8} & \ \ \underline{0} & \ \ \underline{0} & \ \ \underline{0} \\
\ \ 0 & \ \ 0 & \ \ 0 & \ \ 0
\end{array}
$$

$8 = 1000_2$

22. To convert 16 to base 2   32 16 8 4 2 1

$$
\begin{array}{ccccc}
\ \ 1 & \ \ 0 & \ \ 0 & \ \ 0 & \ \ 0 \\
16\,\overline{)\,16} & 8\,\overline{)\,0} & 4\,\overline{)\,0} & 2\,\overline{)\,0} & 1\,\overline{)\,0} \\
\ \ \underline{16} & \ \ \underline{0} & \ \ \underline{0} & \ \ \underline{0} & \ \ \underline{0} \\
\ \ 0 & \ \ 0 & \ \ 0 & \ \ 0 & \ \ 0
\end{array}
$$

$16 = 10000_2$

23. To convert 23 to base 2   32 16 8 4 2 1

$$
\begin{array}{ccccc}
\ \ 1 & \ \ 0 & \ \ 1 & \ \ 1 & \ \ 1 \\
16\,\overline{)\,23} & 8\,\overline{)\,7} & 4\,\overline{)\,7} & 2\,\overline{)\,3} & 1\,\overline{)\,1} \\
\ \ \underline{16} & \ \ \underline{0} & \ \ \underline{4} & \ \ \underline{2} & \ \ \underline{1} \\
\ \ 7 & \ \ 7 & \ \ 3 & \ \ 1 & \ \ 0
\end{array}
$$

$23 = 10111_2$

24. To convert 243 to base 6   1296 216 36 6 1

$$
\begin{array}{cccc}
\ \ 1 & \ \ 0 & \ \ 4 & \ \ 3 \\
216\,\overline{)\,243} & 36\,\overline{)\,27} & 6\,\overline{)\,27} & 1\,\overline{)\,3} \\
\ \ \underline{216} & \ \ \underline{0} & \ \ \underline{24} & \ \ \underline{3} \\
\ \ 27 & \ \ 27 & \ \ 3 & \ \ 0
\end{array}
$$

$243 = 1043_6$

25. To convert 635 to base 6   1296 216 36 6 1

$$
\begin{array}{cccc}
\ \ 2 & \ \ 5 & \ \ 3 & \ \ 5 \\
216\,\overline{)\,635} & 36\,\overline{)\,203} & 6\,\overline{)\,23} & 1\,\overline{)\,5} \\
\ \ \underline{432} & \ \ \underline{180} & \ \ \underline{18} & \ \ \underline{5} \\
\ \ 203 & \ \ 23 & \ \ 5 & \ \ 0
\end{array}
$$

$635 = 2535_6$

26. To convert 908 to base 4   1024 256 64 16 4 1

$$
\begin{array}{ccccc}
\ \ 3 & \ \ 2 & \ \ 0 & \ \ 3 & \ \ 0 \\
256\,\overline{)\,908} & 64\,\overline{)\,140} & 16\,\overline{)\,12} & 4\,\overline{)\,12} & 1\,\overline{)\,0} \\
\ \ \underline{768} & \ \ \underline{128} & \ \ \underline{0} & \ \ \underline{12} & \ \ \underline{0} \\
\ \ 140 & \ \ 12 & \ \ 12 & \ \ 0 & \ \ 0
\end{array}
$$

$908 = 32030_4$

27. To convert 2061 to base 12   20,736 1728 144 12 1

$$
\begin{array}{cccc}
\ \ 1 & \ \ 2 & \ \ 3 & \ \ 9 \\
1728\,\overline{)\,2061} & 144\,\overline{)\,333} & 12\,\overline{)\,45} & 1\,\overline{)\,9} \\
\ \ \underline{1728} & \ \ \underline{288} & \ \ \underline{36} & \ \ \underline{9} \\
\ \ 333 & \ \ 45 & \ \ 9 & \ \ 0
\end{array}
$$

$2061 = 1239_{12}$

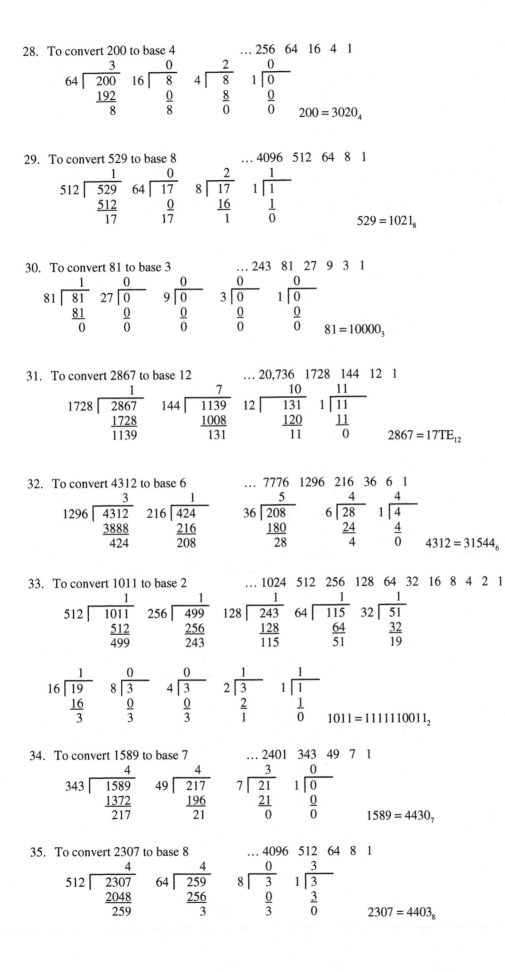

28. To convert 200 to base 4   ... 256 64 16 4 1

$$64\overline{)200} \quad 16\overline{)8} \quad 4\overline{)8} \quad 1\overline{)0}$$

with quotients 3, 0, 2, 0

$$\begin{array}{cccc} 3 & 0 & 2 & 0 \\ 64\overline{)200} & 16\overline{)8} & 4\overline{)8} & 1\overline{)0} \\ \underline{192} & \underline{0} & \underline{8} & \underline{0} \\ 8 & 8 & 0 & 0 \end{array}$$

$200 = 3020_4$

29. To convert 529 to base 8   ... 4096 512 64 8 1

$$\begin{array}{cccc} 1 & 0 & 2 & 1 \\ 512\overline{)529} & 64\overline{)17} & 8\overline{)17} & 1\overline{)1} \\ \underline{512} & \underline{0} & \underline{16} & \underline{1} \\ 17 & 17 & 1 & 0 \end{array}$$

$529 = 1021_8$

30. To convert 81 to base 3   ... 243 81 27 9 3 1

$$\begin{array}{ccccc} 1 & 0 & 0 & 0 & 0 \\ 81\overline{)81} & 27\overline{)0} & 9\overline{)0} & 3\overline{)0} & 1\overline{)0} \\ \underline{81} & \underline{0} & \underline{0} & \underline{0} & \underline{0} \\ 0 & 0 & 0 & 0 & 0 \end{array}$$

$81 = 10000_3$

31. To convert 2867 to base 12   ... 20,736 1728 144 12 1

$$\begin{array}{cccc} 1 & 7 & 10 & 11 \\ 1728\overline{)2867} & 144\overline{)1139} & 12\overline{)131} & 1\overline{)11} \\ \underline{1728} & \underline{1008} & \underline{120} & \underline{11} \\ 1139 & 131 & 11 & 0 \end{array}$$

$2867 = 17TE_{12}$

32. To convert 4312 to base 6   ... 7776 1296 216 36 6 1

$$\begin{array}{ccccc} 3 & 1 & 5 & 4 & 4 \\ 1296\overline{)4312} & 216\overline{)424} & 36\overline{)208} & 6\overline{)28} & 1\overline{)4} \\ \underline{3888} & \underline{216} & \underline{180} & \underline{24} & \underline{4} \\ 424 & 208 & 28 & 4 & 0 \end{array}$$

$4312 = 31544_6$

33. To convert 1011 to base 2   ... 1024 512 256 128 64 32 16 8 4 2 1

$$\begin{array}{ccccc} 1 & 1 & 1 & 1 & 1 \\ 512\overline{)1011} & 256\overline{)499} & 128\overline{)243} & 64\overline{)115} & 32\overline{)51} \\ \underline{512} & \underline{256} & \underline{128} & \underline{64} & \underline{32} \\ 499 & 243 & 115 & 51 & 19 \end{array}$$

$$\begin{array}{ccccc} 1 & 0 & 0 & 1 & 1 \\ 16\overline{)19} & 8\overline{)3} & 4\overline{)3} & 2\overline{)3} & 1\overline{)1} \\ \underline{16} & \underline{0} & \underline{0} & \underline{2} & \underline{1} \\ 3 & 3 & 3 & 1 & 0 \end{array}$$

$1011 = 1111110011_2$

34. To convert 1589 to base 7   ... 2401 343 49 7 1

$$\begin{array}{cccc} 4 & 4 & 3 & 0 \\ 343\overline{)1589} & 49\overline{)217} & 7\overline{)21} & 1\overline{)0} \\ \underline{1372} & \underline{196} & \underline{21} & \underline{0} \\ 217 & 21 & 0 & 0 \end{array}$$

$1589 = 4430_7$

35. To convert 2307 to base 8   ... 4096 512 64 8 1

$$\begin{array}{cccc} 4 & 4 & 0 & 3 \\ 512\overline{)2307} & 64\overline{)259} & 8\overline{)3} & 1\overline{)3} \\ \underline{2048} & \underline{256} & \underline{0} & \underline{3} \\ 259 & 3 & 3 & 0 \end{array}$$

$2307 = 4403_8$

36. To convert 13,469 to base 8 ... 32,768 4096 512 64 8 1

$$
\begin{array}{r} 3 \\ 4096\overline{)13{,}469} \\ \underline{12{,}288} \\ 1181 \end{array}
\quad
\begin{array}{r} 2 \\ 512\overline{)1181} \\ \underline{1024} \\ 157 \end{array}
\quad
\begin{array}{r} 2 \\ 64\overline{)157} \\ \underline{128} \\ 29 \end{array}
\quad
\begin{array}{r} 3 \\ 8\overline{)29} \\ \underline{24} \\ 5 \end{array}
\quad
\begin{array}{r} 5 \\ 1\overline{)5} \\ \underline{5} \\ 0 \end{array}
$$

$13{,}469 = 32235_8$

37. $735_{16} = 7\left(16^2\right) + 3(16) + 5(1) = 7(256) + 48 + 5 = 1792 + 48 + 5 = 1845$

38. $581_{16} = 5\left(16^2\right) + 8(16) + 1(1) = 5(256) + 128 + 1 = 1280 + 128 + 1 = 1409$

39. $6D3B7_{16} = 6\left(16^4\right) + 13\left(16^3\right) + 3\left(16^2\right) + 11(16) + 7(1) = 6(65{,}536) + 13(4096) + 3(256) + 176 + 7$

    $= 393{,}216 + 53{,}248 + 768 + 176 + 7 = 447{,}415$

40. $24FEA_{16} = 2\left(16^4\right) + 4\left(16^3\right) + 15\left(16^2\right) + 14(16) + 10(1) = 2(65{,}536) + 4(4096) + 15(256) + 224 + 10$

    $= 131{,}072 + 16{,}384 + 3840 + 224 + 10 = 151{,}530$

41. To convert 573 to base 16 ... 4096 256 16 1

$$
\begin{array}{r} 2 \\ 256\overline{)573} \\ \underline{512} \\ 61 \end{array}
\quad
\begin{array}{r} 3 \\ 16\overline{)61} \\ \underline{48} \\ 13 \end{array}
\quad
\begin{array}{r} 13 = \text{D} \\ 1\overline{)13} \\ \underline{13} \\ 0 \end{array}
$$

$573 = 23D_{16}$

42. To convert 349 to base 16 ... 4096 256 16 1

$$
\begin{array}{r} 1 \\ 256\overline{)349} \\ \underline{256} \\ 93 \end{array}
\quad
\begin{array}{r} 5 \\ 16\overline{)93} \\ \underline{80} \\ 13 \end{array}
\quad
\begin{array}{r} 13 = \text{D} \\ 1\overline{)13} \\ \underline{13} \\ 0 \end{array}
$$

$349 = 15D_{16}$

43. To convert 5478 to base 16 ... 65,536 4096 256 16 1

$$
\begin{array}{r} 1 \\ 4096\overline{)5478} \\ \underline{4096} \\ 1382 \end{array}
\quad
\begin{array}{r} 5 \\ 256\overline{)1382} \\ \underline{1280} \\ 102 \end{array}
\quad
\begin{array}{r} 6 \\ 16\overline{)102} \\ \underline{96} \\ 6 \end{array}
\quad
\begin{array}{r} 6 \\ 1\overline{)6} \\ \underline{6} \\ 0 \end{array}
$$

$5478 = 1566_{16}$

44. To convert 34,721 to base 16 ... 65,536 4096 256 16 1

$$
\begin{array}{r} 8 \\ 4096\overline{)34{,}721} \\ \underline{32{,}768} \\ 1953 \end{array}
\quad
\begin{array}{r} 7 \\ 256\overline{)1953} \\ \underline{1792} \\ 161 \end{array}
\quad
\begin{array}{r} 10 = \text{A} \\ 16\overline{)161} \\ \underline{160} \\ 1 \end{array}
\quad
\begin{array}{r} 1 \\ 1\overline{)1} \\ \underline{1} \\ 0 \end{array}
$$

$34{,}721 = 87A1_{16}$

45. To convert 2005 to base 2 ... 2048 1024 512 256 128 64 32 16 8 4 2 1

$$
\begin{array}{r} 1 \\ 1024\overline{)2005} \\ \underline{1024} \\ 981 \end{array}
\quad
\begin{array}{r} 1 \\ 512\overline{)981} \\ \underline{512} \\ 469 \end{array}
\quad
\begin{array}{r} 1 \\ 256\overline{)469} \\ \underline{256} \\ 213 \end{array}
\quad
\begin{array}{r} 1 \\ 128\overline{)213} \\ \underline{128} \\ 85 \end{array}
\quad
\begin{array}{r} 1 \\ 64\overline{)85} \\ \underline{64} \\ 21 \end{array}
$$

$$
\begin{array}{r} 0 \\ 32\overline{)21} \\ \underline{0} \\ 21 \end{array}
\quad
\begin{array}{r} 1 \\ 16\overline{)21} \\ \underline{16} \\ 5 \end{array}
\quad
\begin{array}{r} 0 \\ 8\overline{)5} \\ \underline{0} \\ 5 \end{array}
\quad
\begin{array}{r} 1 \\ 4\overline{)5} \\ \underline{4} \\ 1 \end{array}
\quad
\begin{array}{r} 0 \\ 2\overline{)1} \\ \underline{0} \\ 1 \end{array}
\quad
\begin{array}{r} 1 \\ 1\overline{)1} \\ \underline{1} \\ 0 \end{array}
$$

$2005 = 11111010101_2$

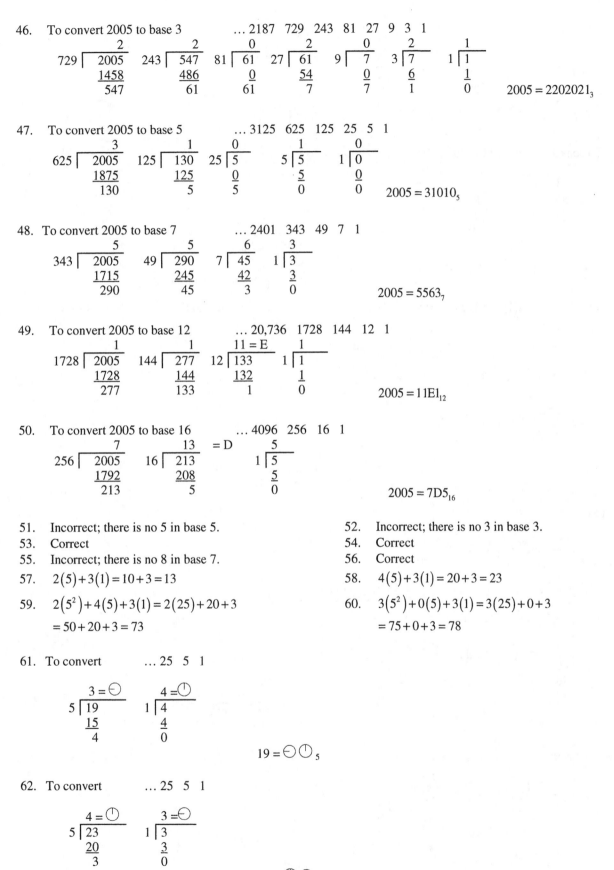

46. To convert 2005 to base 3   ... 2187 729 243 81 27 9 3 1

$2005 = 2202021_3$

47. To convert 2005 to base 5   ... 3125 625 125 25 5 1

$2005 = 31010_5$

48. To convert 2005 to base 7   ... 2401 343 49 7 1

$2005 = 5563_7$

49. To convert 2005 to base 12   ... 20,736 1728 144 12 1

$2005 = 11E1_{12}$

50. To convert 2005 to base 16   ... 4096 256 16 1

$2005 = 7D5_{16}$

51. Incorrect; there is no 5 in base 5.

52. Incorrect; there is no 3 in base 3.

53. Correct

54. Correct

55. Incorrect; there is no 8 in base 7.

56. Correct

57. $2(5) + 3(1) = 10 + 3 = 13$

58. $4(5) + 3(1) = 20 + 3 = 23$

59. $2(5^2) + 4(5) + 3(1) = 2(25) + 20 + 3$
    $= 50 + 20 + 3 = 73$

60. $3(5^2) + 0(5) + 3(1) = 3(25) + 0 + 3$
    $= 75 + 0 + 3 = 78$

61. To convert   ... 25 5 1

$19 = \ominus\, \bigcirc\,_5$

62. To convert   ... 25 5 1

$23 = \bigcirc\, \ominus\,_5$

63. To convert   ... 125 25 5 1

$$25\overline{)74} \quad \begin{array}{r} 2 \\ \underline{50} \\ 24 \end{array} = ⊕ \qquad 5\overline{)24} \quad \begin{array}{r} 4 \\ \underline{20} \\ 4 \end{array} = ⊖ \qquad 1\overline{)4} \quad \begin{array}{r} 4 \\ \underline{4} \\ 0 \end{array} = ⊖$$

$$74 = ⊕ ⊖ ⊖ {}_5$$

64. To convert   ... 125 25 5 1

$$25\overline{)85} \quad \begin{array}{r} 3 \\ \underline{75} \\ 10 \end{array} = ⊝ \qquad 5\overline{)10} \quad \begin{array}{r} 2 \\ \underline{10} \\ 0 \end{array} = ⊕ \qquad 1\overline{)0} \quad \begin{array}{r} 0 \\ \underline{0} \\ 0 \end{array} = ◯$$

$$85 = ⊝ ⊕ ◯ {}_5$$

65. $1(4) + 3(1) = 4 + 3 = 7$ 66. $3(4) + 2(1) = 12 + 2 = 14$

67. $2(4^2) + 1(4) + 0(1) = 2(16) + 4 + 0 = 32 + 4 + 0 = 36$ 68. $3(4^2) + 2(4) + 1(1) = 3(16) + 8 + 1 = 48 + 8 + 1 = 57$

For #69-72, blue = 0 = b, red = 1 = r, gold = 2 = go, green = 3 = gr

69. To convert   ... 16 4 1

$$4\overline{)10} \quad \begin{array}{r} 2 \\ \underline{8} \\ 2 \end{array} = \boxed{go} \qquad 1\overline{)2} \quad \begin{array}{r} 2 \\ \underline{2} \\ 0 \end{array} = \boxed{go}$$

$$10 = \boxed{go} \; \boxed{go} {}_4$$

70. To convert   ... 16 4 1

$$4\overline{)15} \quad \begin{array}{r} 3 \\ \underline{12} \\ 3 \end{array} = \boxed{gr} \qquad 1\overline{)3} \quad \begin{array}{r} 3 \\ \underline{3} \\ 0 \end{array} = \boxed{gr}$$

$$15 = \boxed{gr} \; \boxed{gr} {}_4$$

71. To convert   ... 64 16 4 1

$$16\overline{)60} \quad \begin{array}{r} 3 \\ \underline{48} \\ 12 \end{array} = \boxed{gr} \qquad 4\overline{)12} \quad \begin{array}{r} 3 \\ \underline{12} \\ 0 \end{array} = \boxed{gr} \qquad 1\overline{)0} \quad \begin{array}{r} 0 \\ \underline{0} \\ 0 \end{array} = \boxed{b}$$

$$60 = \boxed{gr} \; \boxed{gr} \; \boxed{b} {}_4$$

72. To convert   ... 64 16 4 1

$$16\overline{)56} \quad \begin{array}{r} 3 \\ \underline{48} \\ 8 \end{array} = \boxed{gr} \qquad 4\overline{)8} \quad \begin{array}{r} 2 \\ \underline{8} \\ 0 \end{array} = \boxed{go} \qquad 1\overline{)0} \quad \begin{array}{r} 0 \\ \underline{0} \\ 0 \end{array} = \boxed{b}$$

$$56 = \boxed{gr} \; \boxed{go} \; \boxed{b} {}_4$$

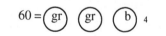

73. a) Each remainder is multiplied by the proper power of 5.

b)

5	683		
5	136	3	↑
5	27	1	↑
5	5	2	↑
5	1	0	↑
	0	1	↑

$$683 = 10213_5$$

c)

8	763		
8	95	3	↑
8	11	7	↑
8	1	3	↑
	0	1	↑

$$763 = 1373_8$$

74. a) $1_3, 2_3, 10_3, 11_3, 12_3, 20_3, 21_3, 22_3, 100_3, 101_3, 102_3, 110_3, 111_3, 112_3, 120_3, 121_3, 122_3, 200_3, 201_3, 202_3$

b) 1000_3

75. Answers will vary.

76. $2^7 = 2 \times 2 \times 2 \times 2 \times 2 \times 2 \times 2 = 128$

77. $1(b^2) + 1(b) + 1 = 43$

$b^2 + b + 1 = 43$

$b^2 + b - 42 = 0$

$(b+7)(b-6) = 0$

$b + 7 = 0$ or $b - 6 = 0$

$b = -7$ or $b = 6$

Since the base cannot be negative, $b = 6$.

78. $d(5^2) + d(5) + d(1) = 124$

$25d + 5d + d = 124$

$\dfrac{31d}{31} = \dfrac{124}{31}$

$d = 4$

79. a) $3(4^4) + 1(4^3) + 2(4^2) + 3(4) + 0(1) = 3(256) + 64 + 2(16) + 12 + 0 = 768 + 64 + 32 + 12 + 0 = 876$

b) To convert   … 256 64 16 4 1

$$64\overline{)177} \begin{array}{c} 2 \\ \underline{128} \\ 49 \end{array} = \text{(go)} \qquad 16\overline{)49} \begin{array}{c} 3 \\ \underline{48} \\ 1 \end{array} = \text{(gr)} \qquad 4\overline{)1} \begin{array}{c} 0 \\ \underline{0} \\ 1 \end{array} = \text{(b)} \qquad 1\overline{)1} \begin{array}{c} 1 \\ \underline{1} \\ 0 \end{array} = \text{(r)}$$

$$177 = \text{(go)} \ \text{(gr)} \ \text{(b)} \ \text{(r)}_4$$

Exercise Set 4.4

1. a) $b^0 = 1, b^1 = b, b^2, b^3, b^4$

   b) $6^0 = 1, 6^1 = 6, 6^2, 6^3, 6^4$

2. $8^0 = 1, 8^1 = 8, 8^2 = 64$ using base 8.

3. No; there is no 6 in base 5.

4. No; there is no 3 in base 3.

5. Answers will vary.

6. Answers will vary.

7.
$$\begin{array}{r} 43_5 \\ \underline{41_5} \\ 134_5 \end{array}$$

8.
$$\begin{array}{r} 33_8 \\ \underline{65_8} \\ 120_8 \end{array}$$

9.
$$\begin{array}{r} 2303_4 \\ \underline{232_4} \\ 3201_4 \end{array}$$

10.
$$\begin{array}{r} 101_2 \\ \underline{11_2} \\ 1000_2 \end{array}$$

11.
$$\begin{array}{r} 799_{12} \\ \underline{218_{12}} \\ 9E5_{12} \end{array}$$

12.
$$\begin{array}{r} 222_3 \\ \underline{22_3} \\ 1021_3 \end{array}$$

13.
$$\begin{array}{r} 1112_3 \\ \underline{1011_3} \\ 2200_3 \end{array}$$

14.
$$\begin{array}{r} 470_{12} \\ \underline{347_{12}} \\ 7E7_{12} \end{array}$$

15.
$$\begin{array}{r} 14631_7 \\ \underline{6040_7} \\ 24001_7 \end{array}$$

16.
$$\begin{array}{r} 1341_8 \\ \underline{341_8} \\ 1702_8 \end{array}$$

17.
$$\begin{array}{r} 1110_2 \\ \underline{110_2} \\ 10100_2 \end{array}$$

18.
$$\begin{array}{r} 43A_{16} \\ \underline{496_{16}} \\ 8D0_{16} \end{array}$$

19.
$$\begin{array}{r} 322_4 \\ \underline{-\,103_4} \\ 213_4 \end{array}$$

20.
$$\begin{array}{r} 526_7 \\ \underline{-\,145_7} \\ 351_7 \end{array}$$

21.
$$\begin{array}{r} 2342_5 \\ \underline{-\,1442_5} \\ 400_5 \end{array}$$

22.
$$\begin{array}{r} 1011_2 \\ \underline{-\,101_2} \\ 110_2 \end{array}$$

23.
$$\begin{array}{r} 782_{12} \\ \underline{-\,13T_{12}} \\ 644_{12} \end{array}$$

24.
$$\begin{array}{r} 1221_3 \\ \underline{-\,202_3} \\ 1012_3 \end{array}$$

25.
$$\begin{array}{r} 1001_2 \\ \underline{-\,110_2} \\ 11_2 \end{array}$$

26.
$$\begin{array}{r} 2T34_{12} \\ \underline{-\,345_{12}} \\ 26TE_{12} \end{array}$$

27.
$$\begin{array}{r} 4223_7 \\ \underline{-\,304_7} \\ 3616_7 \end{array}$$

28.
$$\begin{array}{r} 4232_5 \\ \underline{-\,2341_5} \\ 1341_5 \end{array}$$

29.
$$\begin{array}{r} 2100_3 \\ \underline{-\,1012_3} \\ 1011_3 \end{array}$$

30.
$$\begin{array}{r} 4E7_{16} \\ \underline{-\,189_{16}} \\ 35E_{16} \end{array}$$

31.
$$\begin{array}{r} 33_5 \\ \underline{\times\,2_5} \\ 121_5 \end{array}$$

32.
$$\begin{array}{r} 323_6 \\ \underline{\times\,4_6} \\ 2140_6 \end{array}$$

33.
$$\begin{array}{r} 342_7 \\ \underline{\times\,5_7} \\ 2403_7 \end{array}$$

34.
$$\begin{array}{r} 101_2 \\ \underline{\times\,11_2} \\ 101 \\ \underline{101} \\ 1111_2 \end{array}$$

35.
$$\begin{array}{r} 512_6 \\ \underline{\times\,23_6} \\ 2340 \\ \underline{1424} \\ 21020_6 \end{array}$$

36.
$$\begin{array}{r} 124_{12} \\ \underline{\times\,6_{12}} \\ 720_{12} \end{array}$$

37.
$$\begin{array}{r} 436_9 \\ \underline{\times\,25_9} \\ 2403 \\ \underline{873} \\ 12233_9 \end{array}$$

38.
$$\begin{array}{r} 6T3_{12} \\ \underline{\times\,24_{12}} \\ 2350 \\ \underline{1186} \\ 13EE0_{12} \end{array}$$

39.
$$\begin{array}{r} 111_2 \\ \underline{\times\,101_2} \\ 111 \\ 000 \\ \underline{111} \\ 100011_2 \end{array}$$

40.
$$\begin{array}{r} 584_9 \\ \underline{\times\,24_9} \\ 2567 \\ \underline{1278} \\ 15457_9 \end{array}$$

41.
$$\begin{array}{r} 316_7 \\ \underline{\times\,16_7} \\ 2541 \\ \underline{316} \\ 6031_7 \end{array}$$

42.
$$\begin{array}{r} 8T_{12} \\ \underline{\times\,2T_{12}} \\ 744 \\ \underline{158} \\ 2104_{12} \end{array}$$

43. $1_2 \times 1_2 = 1_2$

$$\begin{array}{r} \,110_2 \\ 1_2\,\overline{)\,110_2} \\ \underline{1} \\ 01 \\ \underline{1} \\ 0 \\ \underline{0} \\ 0 \end{array}$$

44. $4_6 \times 1_6 = 4_6$ 34_6 $R3_6$

$4_6 \times 2_6 = 12_6$ $4_6\,\overline{)\,231_6}$

$4_6 \times 3_6 = 20_6$ $\underline{20}$

$4_6 \times 4_6 = 24_6$ 31

$4_6 \times 5_6 = 32_6$ $\underline{24}$

 3

45. $3_5 \times 1_5 = 3_5$ 31_5

$3_5 \times 2_5 = 11_5$ $3_5\,\overline{)\,143_5}$

$3_5 \times 3_5 = 14_5$ $\underline{14}$

$3_5 \times 4_5 = 22_5$ 03

 $\underline{3}$

 0

46.
$7_8 \times 1_8 = 7_8$
$7_8 \times 2_8 = 16_8$
$7_8 \times 3_8 = 25_8$
$7_8 \times 4_8 = 34_8$
$7_8 \times 5_8 = 43_8$
$7_8 \times 6_8 = 52_8$
$7_8 \times 7_8 = 61_8$

$$
\begin{array}{r}
37_8 \quad \text{R4}_8 \\
7_8 \overline{)335_8} \\
\underline{25} \\
65 \\
\underline{61} \\
4
\end{array}
$$

47.
$2_4 \times 1_4 = 2_4$
$2_4 \times 2_4 = 10_4$
$2_4 \times 3_4 = 12_4$

$$
\begin{array}{r}
123_4 \\
2_4 \overline{)312_4} \\
\underline{2} \\
11 \\
\underline{10} \\
12 \\
\underline{12} \\
0
\end{array}
$$

48.
$6_{12} \times 1_{12} = 6_{12}$
$6_{12} \times 2_{12} = 10_{12}$
$6_{12} \times 3_{12} = 16_{12}$
$6_{12} \times 4_{12} = 20_{12}$
$6_{12} \times 5_{12} = 26_{12}$
$6_{12} \times 6_{12} = 30_{12}$
$6_{12} \times 7_{12} = 36_{12}$
$6_{12} \times 8_{12} = 40_{12}$

$$
\begin{array}{r}
86_{12} \quad \text{R1}_{12} \\
6_{12} \overline{)431_{12}} \\
\underline{40} \\
31 \\
\underline{30} \\
1
\end{array}
$$

49.
$2_4 \times 1_4 = 2_4$
$2_4 \times 2_4 = 10_4$
$2_4 \times 3_4 = 12_4$

$$
\begin{array}{r}
103_4 \quad \text{R1}_4 \\
2_4 \overline{)213_4} \\
\underline{2} \\
01 \\
\underline{00} \\
13 \\
\underline{12} \\
1
\end{array}
$$

50.
$5_6 \times 1_6 = 5_6$
$5_6 \times 2_6 = 14_6$
$5_6 \times 3_6 = 23_6$
$5_6 \times 4_6 = 32_6$
$5_6 \times 5_6 = 41_6$

$$
\begin{array}{r}
24_6 \quad \text{R2}_6 \\
5_6 \overline{)214_6} \\
\underline{14} \\
34 \\
\underline{32} \\
2
\end{array}
$$

51.
$3_5 \times 1_5 = 3_5$
$3_5 \times 2_5 = 11_5$
$3_5 \times 3_5 = 14_5$
$3_5 \times 4_5 = 22_5$

$$
\begin{array}{r}
41_5 \quad \text{R1}_5 \\
3_5 \overline{)224_5} \\
\underline{22} \\
04 \\
\underline{3} \\
1
\end{array}
$$

52.
$4_6 \times 1_6 = 4_6$
$4_6 \times 2_6 = 12_6$
$4_6 \times 3_6 = 20_6$
$4_6 \times 4_6 = 24_6$
$4_6 \times 5_6 = 32_6$

$$
\begin{array}{r}
31_6 \quad \text{R2}_6 \\
4_6 \overline{)210_6} \\
\underline{20} \\
10 \\
\underline{4} \\
2
\end{array}
$$

53.
$6_7 \times 1_7 = 6_7$
$6_7 \times 2_7 = 15_7$
$6_7 \times 3_7 = 24_7$
$6_7 \times 4_7 = 33_7$
$6_7 \times 5_7 = 42_7$
$6_7 \times 6_7 = 51_7$

$$
\begin{array}{r}
45_7 \quad \text{R2}_7 \\
6_7 \overline{)404_7} \\
\underline{33} \\
44 \\
\underline{42} \\
2
\end{array}
$$

54.
$3_7 \times 1_7 = 3_7$
$3_7 \times 2_7 = 6_7$
$3_7 \times 3_7 = 12_7$
$3_7 \times 4_7 = 15_7$
$3_7 \times 5_7 = 21_7$
$3_7 \times 6_7 = 24_7$

$$
\begin{array}{r}
500_7 \quad \text{R1}_7 \\
3_7 \overline{)2101_7} \\
\underline{21} \\
00 \\
\underline{00} \\
01 \\
\underline{00} \\
1
\end{array}
$$

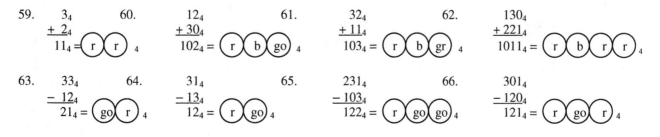

55.
$\begin{array}{r} 2_5 \\ + \ 3_5 \\ \hline 10_5 \end{array} = \ominus \bigcirc _5$

56.
$\begin{array}{r} 3_5 \\ + \ 3_5 \\ \hline 11_5 \end{array} = \ominus \ominus _5$

57.
$\begin{array}{r} 21_5 \\ + \ 43_5 \\ \hline 114_5 \end{array} = \ominus \ominus \bigcirc\!\!\!| _5$

58.
$\begin{array}{r} 23_5 \\ + \ 13_5 \\ \hline 41_5 \end{array} = \bigcirc\!\!\!| \ominus _5$

For #59-66, blue = 0 = b, red = 1 = r, gold = 2 = go, green = 3 = gr

59.
$\begin{array}{r} 3_4 \\ + \ 2_4 \\ \hline 11_4 \end{array} = \boxed{r}\,\boxed{r} _4$

60.
$\begin{array}{r} 12_4 \\ + \ 30_4 \\ \hline 102_4 \end{array} = \boxed{r}\,\boxed{b}\,\boxed{go} _4$

61.
$\begin{array}{r} 32_4 \\ + \ 11_4 \\ \hline 103_4 \end{array} = \boxed{r}\,\boxed{b}\,\boxed{gr} _4$

62.
$\begin{array}{r} 130_4 \\ + \ 221_4 \\ \hline 1011_4 \end{array} = \boxed{r}\,\boxed{b}\,\boxed{r}\,\boxed{r} _4$

63.
$\begin{array}{r} 33_4 \\ - \ 12_4 \\ \hline 21_4 \end{array} = \boxed{go}\,\boxed{r} _4$

64.
$\begin{array}{r} 31_4 \\ - \ 13_4 \\ \hline 12_4 \end{array} = \boxed{r}\,\boxed{go} _4$

65.
$\begin{array}{r} 231_4 \\ - \ 103_4 \\ \hline 122_4 \end{array} = \boxed{r}\,\boxed{go}\,\boxed{go} _4$

66.
$\begin{array}{r} 301_4 \\ - \ 120_4 \\ \hline 121_4 \end{array} = \boxed{r}\,\boxed{go}\,\boxed{r} _4$

67. $2302_5 = 2(5^3) + 3(5^2) + 0(5) + 2(1) = 2(125) + 3(25) + 0 + 2 = 250 + 75 + 0 + 2 = 327$

68. To convert 327 to base 9   ... 729   81   9   1

$$
\begin{array}{r} 4 \\ 81\overline{\smash{\big)}\,327} \\ \underline{324} \\ 3 \end{array}
\qquad
\begin{array}{r} 0 \\ 9\overline{\smash{\big)}\,3} \\ \underline{0} \\ 3 \end{array}
\qquad
\begin{array}{r} 3 \\ 1\overline{\smash{\big)}\,3} \\ \underline{3} \\ 0 \end{array}
$$

$327 = 403_9$

$9^2 = \bullet\bullet\bullet\bullet$
$9^1 = \text{none}$
$9^0 = \bullet\bullet\bullet$

69. $14_5 \times 1_5 = 14_5$ 13_5 70. $20_4 \times 1_4 = 20_4$ 11_4 R3$_4$

$14_5 \times 2_5 = 33_5$ $14_5\overline{\smash{\big)}\,242_5}$ $20_4 \times 2_4 = 100_4$ $20_4\overline{\smash{\big)}\,223_4}$

$14_5 \times 3_5 = 102_5$ $\underline{14}$ $20_4 \times 3_4 = 120_4$ $\underline{20}$

$14_5 \times 4_5 = 121_5$ 102   23

                         $\underline{102}$ $\underline{20}$

                           0   3

71. a) 462_8

     $\underline{\times 35_8}$

     2772

     $\underline{1626}$

     21252_8

b) $462_8 = 4(8^2) + 6(8) + 2(1) = 4(64) + 48 + 2 = 256 + 48 + 2 = 306$

    $35_8 = 3(8) + 5(1) = 24 + 5 = 29$

c) $306 \times 29 = 8874$

d) $21252_8 = 2(8^4) + 1(8^3) + 2(8^2) + 5(8) + 2(1)$

     $= 2(4096) + 512 + 2(64) + 40 + 2$

     $= 8192 + 512 + 128 + 40 + 2 = 8874$

e) Yes, in part a), the numbers were multiplied in base 8 and then converted to base 10 in part d).
In part b), the numbers were converted to base 10 first, then multiplied in part c).

72. $b = 5$

73. Orange = 0; purple = 1; turquoise = 2; brown = 3

Exercise Set 4.5

1. Duplation and mediation, the galley method and Napier rods

2. a) Answers will vary.

   b) 267 –   193

      133 –   386

      ~~66~~   ~~772~~

      33 –   1544

      ~~16~~   ~~3088~~

      8 –   6176

      ~~4~~ ~~12,352~~

      ~~2~~ ~~24,704~~

      1 – 49,408

         51,531

3. a) Answers will vary.

   b)

$362 \times 29 = 10,498$

4. a)   Answers will vary.

b)

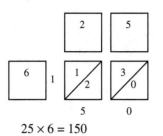

$$25 \times 6 = 150$$

5.
```
23 –  31
11 –  62
 5 – 124
 2 – 248
 1 – 496
      713
```

6.
```
35 –  23
17 –  46
 8 –  92
 4 – 184
 2 – 368
 1 – 736
      805
```

7.
```
9 –  162
4 –  324
2 –  648
1 – 1296
     1458
```

8.
```
175 –     86
 87 –    172
 43 –    344
 21 –    688
 10 –   1376
  5 –   2752
  2 –   5504
  1 – 11,008
       15,050
```

9.
```
35 –  236
17 –  472
 8 –  944
 4 – 1888
 2 – 3776
 1 – 7552
      8260
```

10.
```
96 –   53
48 –  106
24 –  212
12 –  424
 6 –  848
 3 – 1696
 1 – 3392
      5088
```

11.
```
93 –   93
46 –  186
23 –  372
11 –  744
 5 – 1488
 2 – 2976
 1 – 5952
      8649
```

12.
```
49 –  124
24 –  248
12 –  496
 6 –  992
 3 – 1984
 1 – 3968
      6076
```

13.

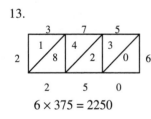

$$6 \times 375 = 2250$$

14.

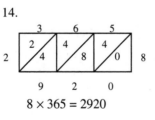

$$8 \times 365 = 2920$$

15.

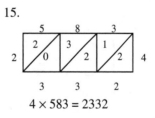

$$4 \times 583 = 2332$$

16.

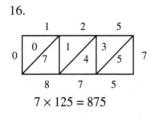

$$7 \times 125 = 875$$

17.

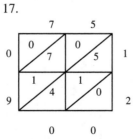

$$75 \times 12 = 900$$

18.

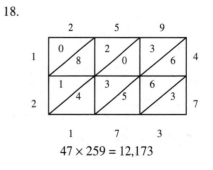

$$47 \times 259 = 12,173$$

19.

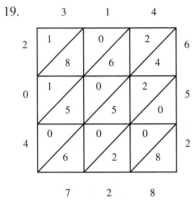

$314 \times 652 = 204,728$

20.

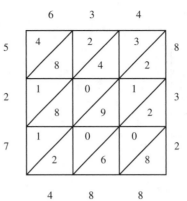

$634 \times 832 = 527,488$

21.

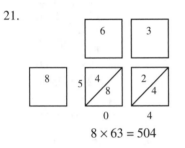

$8 \times 63 = 504$

22.

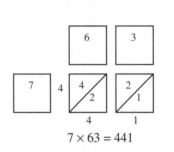

$7 \times 63 = 441$

23.

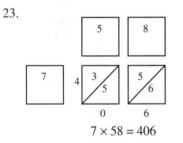

$7 \times 58 = 406$

24.

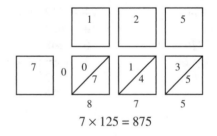

$7 \times 125 = 875$

25.

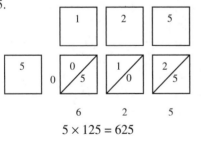

$5 \times 125 = 625$

26. $75 \times 125 = (70 + 5)125 = 70(125) + 5(125)$

From # 24, $70 \times 125 = 8750$

From # 25, $5 \times 125 = \underline{625}$

9375

$75 \times 125 = 9375$

27.

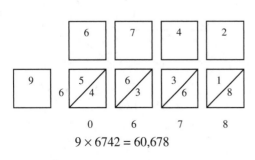

$9 \times 6742 = 60,678$

28.

$7 \times 3456 = 24,192$

29. a) 253 × 46; Place the factors of 8 until the
    correct factors and placements are found
    so the rest of the rectangle can be completed.

    b)

    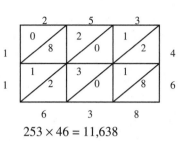

    253 × 46 = 11,638

30. a) 475 × 263; Place the factors of 8 until the
    correct factors and placements are found
    so the rest of the rectangle can be
    completed.

    b)

    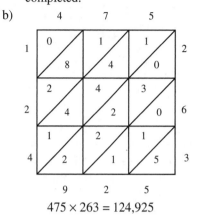

    475 × 263 = 124,925

31. a) 4 × 382; Place the factors of 12 until the correct
    factors and placements are found so the rest
    can be completed.

    b)

    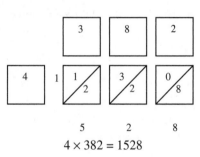

    4 × 382 = 1528

32. a) 7 × 685; Place the factors of 42 until the
    correct factors and placements are found
    so the rest can be completed.

    b)

    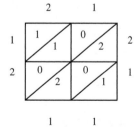

    7 × 685 = 4795

33. 13 – 22
     ~~6 — 44~~
     3 – 88
     1 – 176
     286 = ⌒⌒∩∩∩∩∩∩∩|||||

34. ~~26 —   67~~
     13 –   134
     ~~6 —   268~~
     3 –   536
     1 – 1072
     1742 = MDCCXLII

35. Answers will vary.

36.

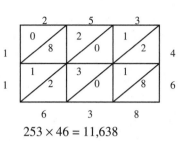

    $21_3 \times 21_3 = 1211_3$

37.

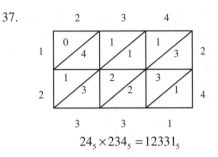

$$24_5 \times 234_5 = 12331_5$$

38. a) $1000+500+100+100+50+10+10+5+1=1776$
     b) Answers will vary.

Review Exercises

1. $1000 + 1000 + 1000 + 100 + 1 + 1 + 1 = 3103$

2. $100 + 100 + 10 + 1000 + 1 = 1211$

3. $10 + 100 + 100 + 100 + 1 + 1000 = 1311$

4. $100 + 10 + 1000 + 1 + 1000 + 1 + 1 + 1 = 2114$

5. $1000 + 1000 + 100 + 100 + 100 + 10$
    $+ 1 + 1 + 1 + 1 = 2314$

6. $100 + 100 + 10 + 1 + 1000 + 1000 + 1 + 100 = 2312$

7. *bbbbbaaaaaa*

8. *cbbaaaaa*

9. *ccbbbbbbbbbaaa*

10. *ddaaaaa*

11. *ddddddcccccccbbbbba*

12. *ddcccbaaaa*

13. $4 (10) + 3 = 40 + 3 = 43$

14. $2 (10) + 7 = 20 + 7 = 27$

15. $7 (100) + 4 (10) + 9 = 700 + 40 + 9 = 749$

16. $4 (1000) + 6 (10) + 8 = 4000 + 60 + 8 = 4068$

17. $5 (1000) + 6 (100) + 4 (10) + 8$
    $= 5000 + 600 + 40 + 8 = 5648$

18. $6 (1000) + 9 (100) + 5 = 6000 + 900 + 5 = 6905$

19. *hxb*

20. *byixe*

21. *hyfxb*

22. *czixd*

23. *fzd*

24. *bza*

25. $4 (10) + 5 (1) = 40 + 5 = 45$

26. $3 (100) + 8 (1) = 300 + 8 = 308$

27. $5 (100) + 6 (10) + 8 (1) = 500 + 60 + 8 = 568$

28. $4 (10,000) + 6 (1000) + 8 (100) + 8 (10) + 3(1)$
    $= 40,000 + 6000 + 800 + 80 + 3 = 46,883$

29. $6 (10,000) + 4 (1000) + 4 (100) + 8 (10) + 1$
    $= 60,000 + 4000 + 400 + 80 + 1 = 64,481$

30. $6 (10,000) + 5 (100) + 2 (10) + 9 (1)$
    $= 60,000 + 500 + 20 + 9 = 60,529$

31. qe

32. upb

33. vrc

34. BArg

35. ODvog

36. QFvrf

37.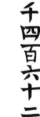

38. MCDLXII

39.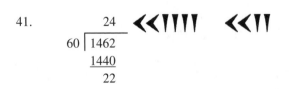

40. $\alpha' \upsilon \xi \beta$

41. $\begin{array}{r} 24 \\ 60\overline{)1462} \\ 1440 \\ \hline 22 \end{array}$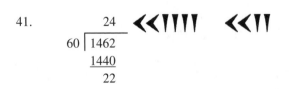

$$1462 = 24 (60) + 22$$

42. $\begin{array}{r} 4 \\ 360\overline{)1462} \\ 1440 \\ \hline 22 \end{array}$ $\begin{array}{r} 1 \\ 20\overline{)22} \\ 20 \\ \hline 2 \end{array}$

$$1462 = 4 (18 \quad 20) + 1 (20) + 2 (1)$$

43. $100,000 + 100,000 + 10,000 + 10,000 + 1000 + 1000 + 10 + 10 + 10 + 1 + 1 + 1 + 1 + 1 = 222,035$

44. $8(1000) + 2(100) + 5(10) + 4 = 8000 + 200 + 50 + 4 = 8254$

45. $600 + 80 + 5 = 685$

46. $1000 + (1000 - 100) + (100 - 10) + 1 = 1000 + 900 + 90 + 1 = 1991$

47. $21(60) + (20 - 3) = 1260 + 17 = 1277$

48. $7(18 \times 20) + 8(20) + 10(1) = 7(360) + 160 + 10 = 2520 + 160 + 10 = 2690$

49. $47_8 = 4(8) + 7(1) = 32 + 7 = 39$

50. $101_2 = 1(2^2) + 0(2) + 1(1) = 4 + 0 + 1 = 5$

51. $130_4 = 1(4^2) + 3(4) + 0(1) = 16 + 12 + 0 = 28$

52. $3425_7 = 3(7^3) + 4(7^2) + 2(7) + 5(1) = 3(343) + 4(49) + 14 + 5 = 1029 + 196 + 14 + 5 = 1244$

53. $T0E_{12} = 10(12^2) + 0(12) + 11(1) = 10(144) + 0 + 11 = 1440 + 0 + 11 = 1451$

54. $20220_3 = 2(3^4) + 0(3^3) + 2(3^2) + 2(3) + 0(1) = 2(81) + 0 + 2(9) + 6 + 0 = 162 + 0 + 18 + 6 + 0 = 186$

55. To convert 463 to base 4 ... 1024 256 64 16 4 1

$463 = 13033_4$

56. To convert 463 to base 3 ... 729 243 81 27 9 3 1

$463 = 122011_3$

57. To convert 463 to base 2 ... 512 256 128 64 32 16 8 4 2 1

$463 = 111001111_2$

58. To convert 463 to base 5 ... 625 125 25 5 1

$463 = 3323_5$

59. To convert 463 to base 12 ... 1728 144 12 1

$463 = 327_{12}$

60. To convert 463 to base 8 ... 512 64 8 1

$463 = 717_8$

61.
$$\begin{array}{r} 52_7 \\ \underline{55_7} \\ 140_7 \end{array}$$

62.
$$\begin{array}{r} 10110_2 \\ \underline{11001_2} \\ 101111_2 \end{array}$$

63.
$$\begin{array}{r} TE_{12} \\ \underline{87_{12}} \\ 176_{12} \end{array}$$

64.
$$\begin{array}{r} 234_7 \\ \underline{456_7} \\ 1023_7 \end{array}$$

65.
$$\begin{array}{r} 3024_5 \\ \underline{4023_5} \\ 12102_5 \end{array}$$

66.
$$\begin{array}{r} 3407_8 \\ \underline{7014_8} \\ 12423_8 \end{array}$$

67.
$$\begin{array}{r} 4032_7 \\ \underline{-321_7} \\ 3411_7 \end{array}$$

68.
$$\begin{array}{r} 1001_2 \\ \underline{-101_2} \\ 100_2 \end{array}$$

69.
$$\begin{array}{r} 3TT_{12} \\ \underline{-E7_{12}} \\ 2E3_{12} \end{array}$$

70.
$$\begin{array}{r} 4321_5 \\ \underline{-442_5} \\ 3324_5 \end{array}$$

71.
$$\begin{array}{r} 1713_8 \\ \underline{-1243_8} \\ 450_8 \end{array}$$

72.
$$\begin{array}{r} 2021_3 \\ \underline{-212_3} \\ 1102_3 \end{array}$$

73.
$$\begin{array}{r} 32_6 \\ \underline{\times\ 4_6} \\ 212_6 \end{array}$$

74.
$$\begin{array}{r} 34_5 \\ \underline{\times\ 21_5} \\ 34 \\ \underline{123} \\ 1314_5 \end{array}$$

75.
$$\begin{array}{r} 126_{12} \\ \underline{\times\ 47_{12}} \\ 856 \\ \underline{4T0} \\ 5656_{12} \end{array}$$

76.
$$\begin{array}{r} 221_3 \\ \underline{\times\ 22_3} \\ 1212 \\ \underline{1212} \\ 21102_3 \end{array}$$

77.
$$\begin{array}{r} 1011_2 \\ \underline{\times\ 101_2} \\ 1011 \\ 0000 \\ \underline{1011} \\ 110111_2 \end{array}$$

78.
$$\begin{array}{r} 476_8 \\ \underline{\times\ 23_8} \\ 1672 \\ \underline{1174} \\ 13632_8 \end{array}$$

79. $1_2 \times 1_2 = 1_2$

$$\begin{array}{r} 1011_2 \\ 1_2\ \overline{)\ 1011_2} \\ \underline{1} \\ 00 \\ \underline{00} \\ 01 \\ \underline{1} \\ 01 \\ \underline{1} \\ 0 \end{array}$$

80.
$2_4 \times 1_4 = 2_4$
$2_4 \times 2_4 = 10_4$
$2_4 \times 3_4 = 12_4$

$$\begin{array}{r} 130_4 \\ 2_4\ \overline{)\ 320_4} \\ \underline{2} \\ 12 \\ \underline{12} \\ 0 \\ \underline{0} \\ 0 \end{array}$$

81.
$3_5 \times 1_5 = 3_5$
$3_5 \times 2_5 = 11_5$
$3_5 \times 3_5 = 14_5$
$3_5 \times 4_5 = 22_5$

$$\begin{array}{r} 23_5 \quad R1_5 \\ 3_5\ \overline{)\ 130_5} \\ \underline{11} \\ 20 \\ \underline{14} \\ 1 \end{array}$$

82.
$4_6 \times 1_6 = 4_6$
$4_6 \times 2_6 = 12_6$
$4_6 \times 3_6 = 20_6$
$4_6 \times 4_6 = 24_6$
$4_6 \times 5_6 = 32_6$

$$\begin{array}{r} 433_6 \\ 4_6\ \overline{)\ 3020_6} \\ \underline{24} \\ 22 \\ \underline{20} \\ 20 \\ \underline{20} \\ 0 \end{array}$$

83.
$3_6 \times 1_6 = 3_6$
$3_6 \times 2_6 = 10_6$
$3_6 \times 3_6 = 13_6$
$3_6 \times 4_6 = 20_6$
$3_6 \times 5_6 = 23_6$

$$\begin{array}{r} 411_6 \quad R1_6 \\ 3_6\ \overline{)\ 2034_6} \\ \underline{20} \\ 03 \\ \underline{3} \\ 04 \\ \underline{3} \\ 1 \end{array}$$

84.
$6_8 \times 1_8 = 6_8$
$6_8 \times 2_8 = 14_8$
$6_8 \times 3_8 = 22_8$
$6_8 \times 4_8 = 30_8$
$6_8 \times 5_8 = 36_8$
$6_8 \times 6_8 = 44_8$
$6_8 \times 7_8 = 52_8$

$$\begin{array}{r} 664_8 \quad R2_8 \\ 6_8\ \overline{)\ 5072_8} \\ \underline{44} \\ 47 \\ \underline{44} \\ 32 \\ \underline{30} \\ 2 \end{array}$$

85.

~~142~~   ~~24~~
71 -   48
35 -   96
17 -   192
~~8~~   ~~384~~
~~4~~   ~~768~~
1 -   <u>3072</u>
       3408

86.

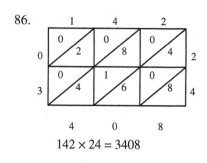

$$142 \times 24 = 3408$$

87.

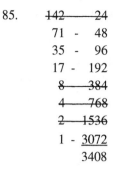

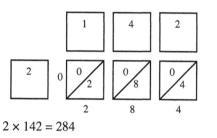

$2 \times 142 = 284$ $4 \times 142 = 568$

$2 \times 142 = 284$, therefore $20 \times 142 = 2840$
Therefore, $142 \times 24 = 2840 + 568 = 3408$.

Chapter Test

1. A **number** is a quantity and answers the question "How many?" A **numeral** is a symbol used to represent the number.

2. $1000 + 1000 + 1000 + 500 + 100 + (50 - 10) + 5 + 1 = 3646$

3. $21(60) + 15(1) = 1260 + 15 = 1275$

4. $8(1000) + 0 + 9(10) = 8000 + 0 + 90 = 8090$

5. $2(18 \times 20) + 12(20) + 9(1) = 2(360) + 240 + 9$
   $= 720 + 240 + 9 = 969$

6. $100,000 + 10,000 + 10,000 + 1000 + 1000 + 100 + 10 + 10 + 10 + 10 + 1 + 1 = 122,142$

7. $9(1000) + 900 + 90 + 9 = 9000 + 900 + 90 + 9$
   $= 9999$

8.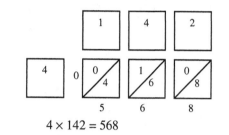

9. $\beta' \upsilon \, \delta \zeta$

10.

$$\begin{array}{r} 3 \\ 360\overline{)1434} \\ \underline{1080} \\ 354 \end{array} \qquad \begin{array}{r} 17 \\ 20\overline{)354} \\ \underline{340} \\ 14 \end{array}$$

$1434 = 3(18 \times 20) + 17(20) + 14(1)$

11.

$$\begin{array}{r} 26 \\ 60\overline{)1596} \\ \underline{1560} \\ 36 \end{array}$$

$1596 = 26(60) + 36(1)$

12. MMCCCLXXVIII

13. In an additive system, the number represented by a particular set of numerals is the sum of the values of the numerals.

14. In a multiplicative system, there are numerals for each number less than the base and for powers of the base. Each numeral less than the base is multiplied by a numeral for the power of the base, and these products are added to obtain the number.

15. In a ciphered system, the number represented by a particular set of numerals is the sum of the values of the numerals. There are numerals for each number up to and including the base and multiples of the base.

16. In a place-value system, each number is multiplied by a power of the base. The position of the numeral indicates the power of the base by which it is multiplied.

17. $56_7 = 5\,(7) + 6\,(1) = 35 + 6 = 41$

18. $403_5 = 4(5^2) + 0(5) + 3(1) = 4(25) + 0 + 3$
$= 100 + 0 + 3 = 103$

19. $101101_2 = 1(2^5) + 0(2^4) + 1(2^3) + 1(2^2) + 0(2) + 1(1) = 32 + 0 + 8 + 4 + 0 + 1 = 45$

20. $368_9 = 3(9^2) + 6(9) + 8(1) = 3(81) + 54 + 8 = 243 + 54 + 8 = 305$

21. To convert 36 to base 2   … 64 32 16 8 4 2 1

$$
32\overline{\smash{\big)}\,36} \quad 16\overline{\smash{\big)}\,4} \quad 8\overline{\smash{\big)}\,4} \quad 4\overline{\smash{\big)}\,4} \quad 2\overline{\smash{\big)}\,0} \quad 1\overline{\smash{\big)}\,0}
$$

1	0	0	1	0	0
32	0	0	4	0	0
4	4	4	0	0	0

$36 = 100100_2$

22. To convert 93 to base 5   … 125 25 5 1

3	3	3
25⟌93	5⟌18	1⟌3
75	15	3
18	3	0

$93 = 333_5$

23. To convert 2356 to base 12   … 20,736 1728 144 12 1

1	4	4	4
1728⟌2356	144⟌628	12⟌52	1⟌4
1728	576	48	4
628	52	4	0

$2356 = 1444_{12}$

24. To convert 2938 to base 7   … 16,807 2401 343 49 7 1

1	1	3	6	5
2401⟌2938	343⟌537	49⟌194	7⟌47	1⟌5
2401	343	147	42	5
537	194	47	5	0

$2938 = 11365_7$

25.
$$
\begin{array}{r}
133_5 \\
434_5 \\
\hline
1122_5
\end{array}
$$

26.
$$
\begin{array}{r}
324_6 \\
-\,142_6 \\
\hline
142_6
\end{array}
$$

27.
$$
\begin{array}{r}
45_6 \\
\times\,23_6 \\
\hline
223 \\
134 \\
\hline
2003_6
\end{array}
$$

28. $3_5 \times 1_5 = 3_5$
$3_5 \times 2_5 = 11_5$
$3_5 \times 3_5 = 14_5$
$3_5 \times 4_5 = 22_5$

$$
\begin{array}{r}
220_5 \\
3_5\,\overline{\smash{\big)}\,1210_5} \\
11 \\
\hline
11 \\
11 \\
\hline
00 \\
00 \\
\hline
0
\end{array}
$$

29. 35 - 28
    17 - 56
    ~~8 - 112~~
    ~~4 - 224~~
    ~~2 - 448~~
    1 - 896
       980

30.

$43 \times 196 = 8428$

Group Projects

1.   a) 06470-9869-1

     b) i) 51593-4837-7   ii) 14527-8924-75-6

   c) i)

      ii)

   d) Answers will vary.

CHAPTER FIVE

NUMBER THEORY AND THE REAL NUMBER SYSTEM

Exercise Set 5.1

1. Number theory is the study of numbers and their properties.
2. If a and b are factors of c, then c ÷ a is an integer and c ÷ b is an integer.
3. a) *a* divides *b* means that *b* divided by *a* has a remainder of zero.
   b) *b* is divisible by *b* means that *a* divided by *b* has a remainder of zero.
4. A prime number is natural number greater than 1 that has exactly two factors (or divisors), itself and one.
5. A composite number is a natural number that is divisible by a number other than itself and 1. Any natural number that is not prime is composite.
6. Every composite number can be expressed as a unique product of prime numbers.

7. a) The least common multiple (LCM) of a set of natural numbers is the smallest natural number that is divisible (without remainder) by each element of the set.
   b) Determine the prime factorization of each number. Then find the product of the prime factors with the largest exponent in each of the prime factorizations.

8. a) The greatest common divisor (GCD) of a set of natural numbers is the largest natural number that divides (without remainder) every number in that set.
   b) Determine the prime factorization of each number. Then find the product of the prime factors with the smallest exponent that appears in each of the prime factorizations.
   c)

2	16
2	8
2	4
	2

$16 = 2^4$

5	40
2	8
2	4
	2

$40 = 2^3 \cdot 5$

   The prime factors with the smallest exponents that appear in each of the factorizations are 2^3.
   The GCD of 16 and 40 is $2^3 = 8$.

9. Mersenne Primes are prime numbers of the form $2^n - 1$ where n is a prime number.

10. A conjecture is a supposition that has not been proved nor disproved.

11. Goldbach's conjecture states that every even number greater than or equal to 4 can be represented as the sum of two (not necessarily distinct) prime numbers.

12. Twin primes are of the form p, p+2, where p is a prime number. An example is 5 & 7.

13. The prime numbers between 1 and 100 are: 2, 3, 5, 7, 11, 13, 17, 19, 23, 29, 31, 37, 41, 43, 47, 53, 59, 61, 67, 71, 73, 79, 83, 89, 91, 97.

14.

1	2	3	4	5	6	7	8	9	10
11	12	13	14	15	16	17	18	19	20
21	22	23	24	25	26	27	28	29	30
31	32	33	34	35	36	37	38	39	40
41	42	43	44	45	46	47	48	49	50
51	52	53	54	55	56	57	58	59	60
61	62	63	64	65	66	67	68	69	70
71	72	73	74	75	76	77	78	79	80
81	82	83	84	85	86	87	88	89	90
91	92	93	94	95	96	97	98	99	100
101	102	103	104	105	106	107	108	109	110
111	112	113	114	115	116	117	118	119	120
121	122	123	124	125	126	127	128	129	130
131	132	133	134	135	136	137	138	139	140
141	142	143	144	145	146	147	148	149	150

The prime numbers between 1 and 150 are: 2, 3, 5, 7, 11, 13, 17, 19, 23, 29, 31, 37, 41, 43, 47, 53, 59, 61, 67, 71, 73, 79, 83, 87, 89, 91, 97,

15. True; since $54 \div 9 = 6$

16. True; since $36 \div 4 = 9$

17. False; since 21 is divisible by 7.

18. False; since 35 is a multiple of 5.

19. False; since 56 is divisible by 8.

20. True; since $45 \div 15 = 3$.

21. True; If a number is divisible by 10, then it is also divisible by 5.

22. False; If a number is divisible by 10, then it is also divisible by 5.

23. False; If a number is divisible by 3, then the sum of the number's digits is divisible by 3.

24. True.

25. True; since $2 \cdot 3 = 6$.

26. True; since $3 \cdot 4 = 12$.

27. Divisible by 2, 3, 4, 6, 8 and 9.

28. Divisible by 2, 3, 4, 5, 6, 8, and 10.

29. Divisible by 3 and 5.

30. Divisible by 2, 3, 4, 5, 6, 8, and 10.

31. Divisible by 2, 3, 4, 5, 6, 8, and 10.

32. Divisible by none of the numbers.

33. $2 \cdot 3 \cdot 4 \cdot 5 \cdot 6 = 720$. (other answers are possible)

34. $3 \cdot 4 \cdot 5 \cdot 9 \cdot 10 = 5400$. (other answers are possible

35.
```
5 | 45
3 | 9
    3
```
$45 = 3^2 \cdot 5$

36.
```
2 | 52
2 | 26
   13
```
$52 = 2^2 \cdot 13$

37.
```
2 | 196
2 | 98
7 | 49
    7
```
$196 = 2^2 \cdot 7^2$

38.
```
2 | 198
3 | 99
3 | 33
   11
```
$198 = 2 \cdot 3^2 \cdot 11$

39.
```
3 | 303
   101
```
$303 = 3 \cdot 101$

40.
```
2 | 400
2 | 200
2 | 100
2 | 50
5 | 25
    5
```
$400 = 2^4 \cdot 5^2$

41.
$$\begin{array}{r|l} 3 & 513 \\ 3 & 171 \\ 3 & 57 \\ & 19 \end{array}$$
$513 = 3^3 \cdot 19$

42.
$$\begin{array}{r|l} 3 & 663 \\ 13 & 221 \\ & 17 \end{array}$$
$663 = 3 \cdot 13 \cdot 17$

43.
$$\begin{array}{r|l} 2 & 1336 \\ 2 & 668 \\ 2 & 334 \\ & 167 \end{array}$$
$1336 = 2^3 \cdot 167$

44.
$$\begin{array}{r|l} 13 & 1313 \\ & 101 \end{array}$$
$1313 = 13 \cdot 101$

45.
$$\begin{array}{r|l} 3 & 2001 \\ 23 & 667 \\ & 29 \end{array}$$
$2001 = 3 \cdot 23 \cdot 29$

46.
$$\begin{array}{r|l} 2 & 3190 \\ 5 & 1595 \\ 11 & 319 \\ & 29 \end{array}$$
$3190 = 2 \cdot 5 \cdot 11 \cdot 29$

47. The prime factors of 15 and 18 are: $6 = 3 \cdot 2$, $\quad 15 = 3 \cdot 5$
a) The common factor is 3, thus, the GCD = 3.
b) The factors with the greatest exponent that appear in either are 2, 3, 5. Thus, the LCM = $2 \cdot 3 \cdot 5 = 30$.

48. The prime factors of 20 and 36 are: $20 = 2^2 \cdot 5$ and $36 = 2^2 \cdot 3^2$
a) The common factor is 2^2; thus, the GCD = 4.
b) The factors with the greatest exponent that appear in either is $2^2, 3^2$; the LCM = $2^2 \cdot 3^2 \cdot 5 = 180$.

49. The prime factors of 48 and 54 are: $48 = 2^4 \cdot 3$, $54 = 2 \cdot 3^3$
a) The common factors are: 2,3; thus, the GCD = $2 \cdot 3 = 6$.
b) The factors with the greatest exponent that appear in either are: $2^4, 3^3$; thus, the LCM = $2^4 \cdot 3^3 = 432$

50. The prime factors of 22 and 231 are: $22 = 2 \cdot 11$, $231 = 3 \cdot 7 \cdot 11$
a) The common factor is: 11; thus, the GCD = 11.
b) The factors with the greatest exponent that appear in either are: 2, 3, 7, 11; thus, the LCM = $2 \cdot 3 \cdot 7 \cdot 11 = 462$

51. The prime factors of 40 and 900 are: $40 = 2^3 \cdot 5$, $900 = 2^2 \cdot 3^2 \cdot 5^2$
a) The common factors are: 2^2, 5; thus, the GCD = $2^2 \cdot 5 = 20$.
b) The factors with the greatest exponent that appear in either are: $2^3, 3^2, 5^2$; thus, the LCM = $2^2 \cdot 3^2 \cdot 5^2 = 1800$

52. The prime factors of 120 and 240 are: $120 = 2^3 \cdot 3 \cdot 5$, $240 = 2^4 \cdot 3 \cdot 5$
a) The common factors are: 2^3, 3, 5; thus, the GCD = $2^3 \cdot 3 \cdot 5 = 120$.
b) The factors with the greatest exponent that appear in either are: 2^4, 3, 5; thus, the LCM = $2^4 \cdot 3 \cdot 5 = 240$

53. The prime factors of 96 and 212 are: $96 = 2^5 \cdot 3$, $212 = 2^2 \cdot 53$
a) The common factors are: 2^2; thus, the GCD = $2^2 = 4$.
b) The factors with the greatest exponent that appear in either are: 2^5, 3, 53; thus, the LCM = $2^5 \cdot 3 \cdot 53 = 5088$

55. The prime factors of 24, 48, and 128 are: $24 = 2^3 \cdot 3$, $48 = 2^4 \cdot 3$, $128 = 2^7$
a) The common factors are: 2^3; thus, the GCD = $2^3 = 8$.
b) The factors with the greatest exponent that appear in either are: 2^7, 3; thus, LCM = $2^7 \cdot 3 = 384$

56. The prime factors of 18, 78, and 198 are: $18 = 2 \cdot 3^2$, $78 = 2 \cdot 3 \cdot 13$, $198 = 2 \cdot 3^2 \cdot 11$
a) The common factors are: 2, 3; thus, the GCD = $2 \cdot 3 = 6$.
b) The factors with the greatest exponent that appear in either are: 2, 3^2, 11, 13; thus, the LCM = $2 \cdot 3^2 \cdot 11 \cdot 13 = 2574$

57. Use the list of primes generated in exercise 13. The next two sets of twin primes are: 17, 19, 29, 31.

58. No. Any other two consecutive natural numbers will include an even number, and even numbers greater than two are composite.

59. (a) 14, 15 Yes; (b) 21, 30 No; (c) 24, 25 Yes; (d) 119, 143 Yes

60. Fermat number $= 2^{2^n} + 1$, where n is a natural number. $2^{2^1} + 1 = 5$, $2^{2^2} + 1 = 2^4 + 1 = 17$,
    $2^{2^3} + 1 = 2^8 + 1 = 257$. These numbers are prime.

61. $4 = 2 + 2, 6 = 3 + 3, 8 = 3 + 5, 10 = 3 + 7, 12 = 5 + 7, 14 = 7 + 7, 16 = 3 + 13, 18 = 5 + 13, 20 = 3 + 17$

62. Use the formula $2^n - 1$, where n is a prime number. $2^2 - 1 = 3, 2^3 - 1 = 7, 2^5 - 1 = 31, 2^7 - 1 = 127$,
    $2^{13} - 1 = 8191$.

63. The gcd of 350 and 140 is 70 dolls.   64. The gcd of 288 and 192 is 96 cars.
65. The gcd of 432 and 360 is 72 cards.   66 The gcd of 150 and 180 is 30 trees.
67. The lcm of 45 and 60 is 180 mins.   68. The lcm of 3500 and 6000 is 42000 miles.

69. The least common multiple of 5 and 6 is 30. Thus, it will be 30 days before they both have the same
    night off again.

70. The least common multiple of 15 and 18 is 90. Thus, it will be 90 days before he visits both on the
    same day again.

71. a) The possible committee sizes are: 4, 5, 10, 20, or 25.   b) The number of committees possible are:
    25 committees of 4, 20 committees of 5, 10 committees of 10, 5 committees of 20, or
    4 committees of 25.

72. a) $5 = 6 - 1$ $7 = 6 + 1$ b) Conjecture: Every prime number greater than $11 = 12 - 1$ $13 = 12 + 1$
    3 differs by 1 from a multiple of the number 6. $17 = 18 - 1$ $19 = 18 + 1$ $23 = 24 - 1$
    $29 = 30 - 1$ c) The conjecture appears to be correct.

73. A number is divisible by 15 if both 3 and 5 divide the number.
74. A number is divisible by 22 if both 2 and 11 divide the number.

75. $35 \div 15 = 2$ with rem. $= 5$.   76. $28 \div 16 = 1$ with rem. $= 12$.   77. $108 \div 36 = 3$ with rem. $= 0$.
    $15 \div 5 = 3$ with rem. $= 0$. $16 \div 12 = 1$ with rem. $= 4$. $36 \div 3 = 12$ with rem. $= 0$.
    Thus, gcd of 35 and 15 is 5. $12 \div 4 = 3$ with rem. $= 0$.   Thus, gcd of 108 and 36
                                          Thus, gcd of 28 and 16 is 4.   is 36.

78. $240 \div 76 = 3$ with rem. $= 12$.   79. $180 \div 150 = 1$ with rem. $= 30$.   80. $560 \div 210 = 2$ w/rem. $= 140$.
    $76 \div 12 = 6$ with rem. $= 4$. $150 \div 30 = 5$ with rem. $= 0$. $210 \div 140 = 1$ w/rem. $= 70$.
    $12 \div 4 = 3$ with rem. $= 0$.   Thus, the gcd of 150 and 180 $140 \div 70 = 2$ w/rem. $= 0$.
    Thus, gcd of 240 and 76   is 30.   Thus, gcd of 210 and 560
    is 4.   is 70.

81. The proper factors of 12 are: 82. The proper factors of 28 are: 83. The proper factors of 496
    1, 2, 3, 4, and 6.   1, 2, 4, 7, and 14.   are: 1,2,4,8,16,31,62,124,
    $1 + 2 + 3 + 4 + 6 = 16 \neq 12$ $1 + 2 + 4 + 7 + 14 = 28$ and 248. $1+2+4+8+16$
    Thus, 12 is not a perfect #.   Thus, 28 is a perfect number. $+31+62+124+248 = 496$
                                                                                  Thus, 496 is a perfect #

84. The proper factors of 48 $1+2+3+4+6+8+12+16+24 = 76$
    are:1, 2, 3, 4, 6, 8, 12, 16,   Thus, 48 is not a perfect #.
    and 24.

85. a) $60 = 2^2 \cdot 3^1 \cdot 5^1$ Adding 1 to each exponent and then multiplying these numbers, we get
    $(2+1)(1+1)(1+1) = 3 \cdot 2 \cdot 2 = 12$ divisors of 60.

86. No, 2 and 4 are not unique prime factors since $4 = 2 \cdot 2$. Any number that 4 divides, 2 will also divide, but 8 does not divide all numbers that are divisible by 4. Some examples are: 4, 12, and 20.

87. The sum of the digits will be a number divisible by 3, thus the number is divisible by 6.

88. The sum of the groups which have the same three digits will always be divisible by three.
    (i.e. d + d + d = 3d and 3|3d)

89. $36,018 = (36,000 + 18)$; $36,000 \div 18 = 2,000$ and $18 \div 18 = 1$
    Thus, since 18 | 36000 and 18 | 18, 18 | 36018.

90. $2^2 - 1 = 3, 2^3 - 1 = 7, 2^5 - 1 = 31, 2^7 - 1 = 127$ are prime numbers, but $2^{11} - 1 = 2,048 - 1 = 2,047$;
    and since $23 \cdot 89 = 2,047$, 2047 is not prime.

91. 8 = 2+3+3, 9 = 3+3+3, 10 = 2+3+5, 11 = 2+2+7, 12 = 2+5+5, 13 = 3+3+7, 14 = 2+5+7, 15 = 3+5+7,
    16 = 2+7+7, 17 = 5+5+7, 18 = 2+5+11, 19 = 3+5+11, 20 = 2+7+11.

92. (a) 1000 = 3 + 997; (b) 2000 = 3 + 1997; (c) 3000 = 29 + 2971

Exercise Set 5.2

1. Begin at zero, draw an arrow to the value of the first number. From the tip of that arrow draw another arrow by moving a number of spaces equal to the value of the second number. Be sure to move left if the number is negative and move right if the number is positive. The sum of the two numbers is at the tip of the second arrow.
2. $-n$; Additive Inverse = that number when added to n yields the Additive Identity (= 0); n + (- n) = 0
3. To rewrite a subtraction problem as an addition problem, rewrite the subtraction sign as an addition sign and change the second number to its additive inverse.
4. The product of two numbers with like signs is a positive number, and the product of two numbers with unlike signs is a negative number.
5. The quotient of two numbers with like signs is a positive number, and the quotient of two numbers with unlike signs is a negative number.
6. If we set $5 \div 0 = x$ and we cross multiply, we get the equation $0x = 5$. Since $0 \cdot x = 0$, we get $5 = 0$, which is a false statement, which means that there is no such number x. Therefore, division by 0 is not allowed.

7. $-6 + 9 = 3$

8. $4 + (-5) = -1$

9. $(-7) + 9 = 2$

10. $(-3) + (-3) = -6$

11. $[6+(-11)]+0 = -5+0 = -5$

12. $(2+5)+(-4) = 7+(-4) = 3$

13. $[(-3)+(-4)]+9 = -7+9 = 2$

14. $[8+(-3)]+(-2) = [5]+(-2) = 3$

15. $[(-23)+(-9)]+11 = [-32]+11 = -21$

16. $[5+(-13)]+18 = [-8]+18 = 10$

17. $3 - 6 = -3$

18. $-3-7 = -10$

19. $-4-6 = -10$

20. $7-(-1) = 8$

21. $-5 - (-3) = -5+3 = -2$

22. $-4 - 4 = -4 + (-4) = -8$

23. $14 - 20 = 14 + (-20) = -6$

24. $8 - (-3) = 8+3 = 11$

25. $[5+(-3)] -4 = 2-4 =2+(-4) = -2$

26. $6-(8+6) = 6-14 = 6+(-14) =-8$

27. $-4 \cdot 5 = -20$

28. $4(-3) = -12$

29. $(-12)(-12) = 144$

30. $5(-5) = -25$

31. $[(-8)(-2)] \cdot 6 = 16 \cdot 6 = 96$

32. $(4)(-5)(-6) = (-20)(-6) = 120$

33. $(5 \cdot 6)(-2) = (30)(-2) = -60$

34. $(-9)(-1)(-2) = (9)(-2) = -18$

35. $[(-3)(-6)] \cdot [(-5)(8)] =$ $(18)(-40) = -720$

36. $[(-8)(4)(5)](-2) =$ $[(-32)(5)](-2) = [-160](-2) = 320$

37. $-26 \div (-13) = 2$

38. $-56 \div 8 = -7$

39. $23 \div (-23) = -1$

40. $-64 \div 16 = -4$

41. $\frac{56}{-8} = -7$

42. $\frac{-75}{15} = -5$

43. $\frac{-210}{14} = -15$

44. $\frac{186}{-6} = -31$

45. $144 \div (-3) = -48$

46. $(-900) \div (-4) = 225$

47. True; every whole number is an integer.

48. False; Negative numbers are not natural numbers.

49. False; the difference of two negative integers may be positive, negative, or zero.

50. True.

51. True; the product of two integers with like signs is a positive integer.

52. False; the difference of a positive integer and a neg. integer may be +, - or zero.

53. True; the quotient of two integers with unlike signs is a negative number.

54. False; the quotient of any two integers with like signs is a positive number.

55. False; the sum of a positive integer and a negative integer could be pos., neg., or zero.

56. False; the product of two integers with unlike signs is always a negative integer.

57. $(5 + 7) \div 2 = 12 \div 2 = 6$

58. $(-4) \div [14 \div (-7)] =$ $(-4) \div [-2] = 2$

59. $[6(-2)] - 5 = -12 + (-5) = -17$

60. $[(-5)(-6)] -3 = 30+(-3) = 27$

61. $(4 - 8)(3) = (-4)(3) = -12$

62. $[18 \div (-2)](-3) = (-9)(-3) = 27$

63. $[2+(-17)] \div 3 = [-15] \div 3 = -5$

64. $(5 - 9) \div (-4) = (-4) \div (-4) = 1$

65. $[(-22)(-3)] \div (2 - 13) =$ $66 \div (2+ (-13)) = 66 \div (-11)$ $= -6$

66. $[15(-4)] \div (-6) =$ $(-60) \div (-6) = 10$

67. $-15, -10, -5, 0, 5, 10$

68. $-40, -20, -10, 0, 10, 20, 40$

69. $-6, -5, -4, -3, -2, -1$

70. $-108, -76, -47, 33, 72, 106$

71. $134 - (-79.8) =$ $134 + 79.8 = 213.8°\,F.$

72. $1347 - 33 + 22 - 21 =$ $1314 + 22 - 21 = 1315$ pts.

73. $0 + 100 - 40 + 90 - 20 + 80 =$
    $60 + 90 - 20 + 80 = 210$ pts.

74. $14{,}495 - (-282) =$
    $14{,}495 + 282 = 14{,}777$ feet

75. $842 - (-927) = 842 + 927 =$
    $1{,}769$ feet

76. $8 - 5 + 3 + 4 = 3 + 3 + 4 =$
    $6 + 4 = 10$. The Texans did
    make a first down.

77. a) $+1 - (-8) = +1 + 8 = 9$.
       There is a 9 hr.time diff.
    b) $-5 - (-7) = -5 + 7 = 2$.
       There is a 2 hr. time diff.

78. $\dfrac{-a}{-b} = \dfrac{-1}{-1} \cdot \dfrac{a}{b} = \dfrac{a}{b}$

79.

$$\dfrac{-1+2-3+4-5+\ldots 99+100}{1-2+3-4+5\ldots +99-100} =$$

$$\dfrac{50}{-50} = -1$$

80. a) The next 3 pentagonal
       numbers are 35, 51, and
       70. The nth pentagonal.
    b) The number is obtained by
       adding the nth triangular #
       (see section 1.1) to the nth
       square number (see section
       1.1) and subtracting n. For
       example, if $n = 4$, the 4th
       triangular number is 10

80. b) continued: and the 4th
       square number is 16. The
       sum of 10 and 16 is 26 and
       $26 - n = 26 - 4 = 22$,
       which is the 4th
       pentagonal #. The next
       5 pentagonal numbers are
       92, 117, 145, 176, and
       210.
    c) Since 70 is the 7th
       pentagonal number and
       92 is the 8th pentagonal
       number, 72 cannot be a
       pentagonal number.

81. $0 + 1 - 2 + 3 + 4 - 5 + 6 - 7 - 8 + 9 = 1$ (other answers are possible)

82. (a) $\dfrac{4+4}{4+4} = 1$ (b) $4\left(4 - \dfrac{4}{4}\right) = 12$ $4 \cdot 4 - \dfrac{4}{4} = 15$ $\dfrac{4 \cdot 4 \cdot 4}{4} = 16$ $4 \cdot 4 - \dfrac{4}{4} = 17$ (c) $\dfrac{44 - 4}{4} = 10$

Exercise Set 5.3

1. Rational numbers is the set of all numbers of the form p/q, where p and q are integers, and $q \neq 0$.
2. a) Multiply and divide the number by the position value of the last nonzero digit to the right of the
      decimal point.

   b) $0.397 = \dfrac{1000(0.397)}{1000} = \dfrac{397}{1000}$

3. a) Divide both the numerator and the denominator by their greatest common divisor.

   b) $\dfrac{15}{27} = \dfrac{5 \div 3}{9 \div 3} = \dfrac{5}{9}$

4. Divide the numerator by the denominator. The quotient is the integer part of the mixed number.
   The fraction part of the mixed number is the remainder divided by the divisor.

5. For positive mixed numbers, multiply the denominator of the fraction by the integer preceding it.
   Add this product to the numerator. This sum is the numerator of the improper fraction; the denominator
   is the same as the denominator of the mixed number. For negative mixed numbers, you can temporarily
   ignore the negative sign, perform the conversion described above, and then reattach the negative sign.

6. a) The product of two fractions is found by multiplying the numerators and multiplying the
      denominators.

b) $\dfrac{15}{16} \cdot \dfrac{24}{25} = \dfrac{360}{400} = \dfrac{360 \div 40}{400 \div 40} = \dfrac{9}{10}$

7. a) The reciprocal of a number is 1 divided by the number.

   b) The reciprocal of -2 is $\dfrac{1}{-2} = -\dfrac{1}{2}$

8. a) To divide two fractions, multiply the first fraction by the reciprocal of the second fraction.

   b) $\dfrac{4}{15} \div \dfrac{16}{55} = \dfrac{4}{15} \cdot \dfrac{55}{16} = \dfrac{220}{240} = \dfrac{220 \div 20}{240 \div 20} = \dfrac{11}{12}$

9. a) To add or subtract two fractions with a common denominator, we add or subtract their numerators and keep the common denominator.

   b) $\dfrac{11}{36} + \dfrac{13}{36} = \dfrac{24}{36} = \dfrac{24 \div 12}{36 \div 12} = \dfrac{2}{3}$ c) $\dfrac{37}{48} - \dfrac{13}{48} = \dfrac{24}{48} = \dfrac{24 \div 24}{48 \div 24} = \dfrac{1}{2}$

10. a) First rewrite each fraction with a common denominator. Then add or subtract the fractions.

    b) $\dfrac{5}{12} + \dfrac{4}{9} = \dfrac{3}{3} \cdot \dfrac{5}{12} + \dfrac{4}{4} \cdot \dfrac{4}{9} = \dfrac{15}{36} + \dfrac{16}{36} = \dfrac{31}{36}$ c) $\dfrac{5}{6} - \dfrac{2}{15} = \dfrac{5}{5} \cdot \dfrac{5}{6} - \dfrac{2}{2} \cdot \dfrac{2}{15} = \dfrac{25}{30} - \dfrac{4}{30} = \dfrac{21}{30} = \dfrac{7}{10}$

11. We can multiply a fraction by the number one in the form of c/c (where c is a nonzero integer) and the number will maintain the same value.

12. Yes. $\dfrac{20}{35} = \dfrac{20 \div 5}{35 \div 5} = \dfrac{4}{7}$

13. GCD of 14 and 21 is 7.

    $\dfrac{14}{21} = \dfrac{14 \div 7}{21 \div 7} = \dfrac{2}{3}$

14. GCD of 21 and 35 is 7.

    $\dfrac{21}{35} = \dfrac{21 \div 7}{35 \div 7} = \dfrac{3}{5}$

15. GCD of 26 and 91 is 13.

    $\dfrac{26}{91} = \dfrac{26 \div 13}{91 \div 13} = \dfrac{2}{7}$

16. GCD of 36 and 56 is 4.

    $\dfrac{36}{56} = \dfrac{36 \div 4}{56 \div 4} = \dfrac{9}{14}$

17. GCD of 525 and 800 is 25.

    $\dfrac{525}{800} = \dfrac{525 \div 25}{800 \div 25} = \dfrac{21}{32}$

18. GCD of 13 and 221 is 13.

    $\dfrac{13}{221} = \dfrac{13 \div 13}{221 \div 13} = \dfrac{1}{17}$

19. GCD of 112 and 176 is 16.

    $\dfrac{112}{176} = \dfrac{112 \div 16}{176 \div 16} = \dfrac{7}{11}$

20. GCD of 120 and 135 is 15.

    $\dfrac{120}{135} = \dfrac{120 \div 15}{135 \div 15} = \dfrac{8}{9}$

21. GCD of 45 and 495 is 45.

    $\dfrac{45}{495} = \dfrac{45 \div 45}{495 \div 45} = \dfrac{1}{11}$

22. GCD of 124 and 148 is 4.

    $\dfrac{124}{148} = \dfrac{124 \div 4}{148 \div 4} = \dfrac{31}{37}$

23. $3\dfrac{4}{7} = \dfrac{(3)(7)+4}{7} = \dfrac{21+4}{7} = \dfrac{25}{7}$

24. $4\dfrac{5}{6} = \dfrac{(4)(6)+5}{6} = \dfrac{24+5}{6} = \dfrac{29}{6}$

25. $-1\dfrac{15}{16} = -\dfrac{((1)(16)+15)}{16}$

    $= -\dfrac{16+15}{16} = -\dfrac{31}{16}$

26. $-7\dfrac{1}{5} = -\dfrac{(7)(5)+1}{5}$

    $= -\dfrac{35+1}{5} = -\dfrac{36}{5}$

27. $-4\dfrac{15}{16} = -\dfrac{(4)(16)+15}{16}$

$= -\dfrac{64+15}{16} = -\dfrac{79}{16}$

28. $11\dfrac{9}{16} = \dfrac{(11)(16)+9}{16}$

$= \dfrac{176+9}{16} = \dfrac{185}{16}$

29.

$2\dfrac{1}{8} = \dfrac{(2)(8)+1}{8} = \dfrac{16+1}{8} = \dfrac{17}{8}$

30. $2\dfrac{3}{4} = \dfrac{(2)(4)+3}{4} = \dfrac{8+3}{4} = \dfrac{11}{4}$

31. $1\dfrac{7}{8} = \dfrac{(1)(8)+7}{8} = \dfrac{8+7}{8} = \dfrac{15}{8}$

32. $1\dfrac{1}{2} = \dfrac{(1)(2)+1}{2} = \dfrac{2+1}{2} = \dfrac{3}{2}$

33. $\dfrac{11}{8} = \dfrac{8+3}{8} = \dfrac{(1)(8)+3}{8} = 1\dfrac{3}{8}$

34. $\dfrac{23}{4} = \dfrac{20+3}{4} = \dfrac{(5)(4)+3}{4} = 5\dfrac{3}{4}$

35. $-\dfrac{73}{6} = \dfrac{-(72+1)}{6}$

$= \dfrac{-(12\square6+1)}{6} = -12\dfrac{1}{6}$

36. $-\dfrac{457}{11} = -\dfrac{451+6}{11}$

$= -\dfrac{(41)(11)+6}{11} =$

$-41\dfrac{6}{11}$

37. $-\dfrac{878}{15} = -\dfrac{870+8}{15}$

$= -\dfrac{(58)(15)+8}{15} = -58\dfrac{8}{15}$

38. $\dfrac{1028}{21} = \dfrac{1008+20}{21}$

$= \dfrac{(48)(21)+20}{21} = 48\dfrac{20}{21}$

39. $\dfrac{3}{5} = .60$

40. $\dfrac{15}{16} = .9375$

41. $\dfrac{2}{9} = .\overline{2}$

42. $5 \div 6 = 0.8\overline{3}$

43. $3 \div 8 = 0.375$

44. $23 \div 7 = 0.\overline{285714}$

45. $13 \div 3 = 4.\overline{3}$

46. $115 \div 15 = 7.\overline{6}$

47. $85 \div 15 = 5.\overline{6}$

48. $1002 \div 11 = 91.\overline{09}$

49. 0.6

$0.25 = \dfrac{25}{100} = \dfrac{25 \div 25}{100 \div 25} = \dfrac{1}{4}$

50. $0.29 = \dfrac{29}{100}$

51.

$0.045 = \dfrac{45}{1000} = \dfrac{45 \div 5}{1000 \div 5} = \dfrac{9}{200}$

52. $0.0125 = \dfrac{125}{10000} = \dfrac{1}{80}$

53. $0.2 = \dfrac{2}{10} = \dfrac{1}{5}$

54. $.251 = \dfrac{251}{1000}$

55. $.452 = \dfrac{452}{1000} = \dfrac{113}{250}$

56. $.2345 = \dfrac{2345}{10000} = \dfrac{469}{2000}$

57. $.0001 = \dfrac{1}{10000}$

58. $.2535 = \dfrac{2535}{10000} = \dfrac{507}{2000}$

59. Let $n = 0.\overline{3}$, $10n = 3.\overline{3}$

$10n = 6.\overline{6}$

$\underline{-n = 0.\overline{6}}$ $\dfrac{9n}{9} = \dfrac{6}{9} = \dfrac{2}{3} = n$

$9n = 6.0$

60. Let $n = 0.\overline{5}$, $10n = 5.\overline{5}$

$10n = 5.\overline{5}$

$\underline{-n = 0.\overline{5}}$ $\dfrac{9n}{9} = \dfrac{5}{9} = n$

$9n = 5.0$

61. Let $n = 1.\overline{9}$, $10n = 19.\overline{9}$

$10n = 19.\overline{9}$

$\underline{-n = \;1.\overline{9}}$

$9n = 18.0$

$\dfrac{9n}{9} = \dfrac{18}{9} = 2 = n$

62. Let $n = 0.\overline{51}$, $100n = 51.\overline{51}$

$100n = 51.\overline{51}$

$\underline{\quad -n = 0.\overline{51}}$

$99n = 51.0$

$\dfrac{99n}{99} = \dfrac{51}{99} = \dfrac{17}{33} = n$

63. Let $n = 1.\overline{36}$, $100n = 136.\overline{36}$

$100n = 136.\overline{36}$

$\underline{\quad -n = 1.\overline{36}}$

$99n = 135.0$

$\dfrac{99n}{99} = \dfrac{135}{99} = \dfrac{15}{11} = n$

64. Let $n = .\overline{135}$, $1000n = 135.\overline{135}$

$1000n = 135.\overline{135}$

$\underline{\quad -n = .\overline{135}}$

$999n = 135.0$

$\dfrac{999n}{999} = \dfrac{135}{999} = \dfrac{5}{37} = n$

65. Let $n = 1.0\overline{2}$, $100n = 102.\overline{2}$

$100n = 102.\overline{2}$

$\underline{\quad -10n = 10.\overline{2}}$

$90n = 92.0$

$\dfrac{90n}{90} = \dfrac{92}{90} = \dfrac{46}{45} = n$

66. Let $n = 2.4\overline{9}$, $100n = 249.\overline{9}$

$100n = 249.\overline{9}$

$\underline{\quad -10n = 24.\overline{9}}$

$90n = 225.0$

$\dfrac{90n}{90} = \dfrac{245}{90} = \dfrac{5}{2} = n$

67. Let $n = 3.4\overline{78}$,

$1000n = 3478.\overline{78}$

$1000n = 3478.\overline{78}$

$\underline{\quad -10n = 34.\overline{78}}$

$990n = 3444.0$

$\dfrac{990n}{990} = \dfrac{3444}{990} = \dfrac{574}{165} = n$

68. Let $n = 5.2\overline{39}$, $1000n = 5239.\overline{39}$

$1000n = 5239.\overline{39}$

$\underline{\quad -10n = 52.\overline{39}}$

$990n = 5187.0$

$\dfrac{990n}{990} = \dfrac{5187}{990} = \dfrac{1729}{330} = n$

69. $\dfrac{4}{11} \cdot \dfrac{3}{8} = \dfrac{4 \cdot 3}{11 \cdot 8} = \dfrac{12}{88} = \dfrac{12 \div 4}{88 \div 4} = \dfrac{3}{22}$

70. $\dfrac{3}{5} \div \dfrac{6}{7} = \dfrac{3}{5} \cdot \dfrac{7}{6} = \dfrac{21}{30} = \dfrac{21 \div 3}{30 \div 3} = \dfrac{7}{10}$

71. $\dfrac{-3}{8} \cdot \dfrac{-16}{15} = \dfrac{48}{120} = \dfrac{2}{5}$

72. $\left(-\dfrac{3}{5}\right) \div \dfrac{10}{21} = \left(-\dfrac{3}{5}\right) \cdot \dfrac{21}{10} = -\dfrac{63}{50}$

73. $\dfrac{7}{8} \div \dfrac{8}{7} = \dfrac{7}{8} \cdot \dfrac{7}{8} = \dfrac{49}{64}$

74. $\dfrac{3}{7} \div \dfrac{3}{7} = \dfrac{3}{7} \cdot \dfrac{7}{3} = \dfrac{21}{21} = 1$

75. $\left(\dfrac{3}{5} \cdot \dfrac{4}{7}\right) \div \dfrac{1}{3} = \dfrac{12}{35} \div \dfrac{1}{3} = \dfrac{12}{35} \cdot \dfrac{3}{1} = \dfrac{36}{35}$

76. $\left(\dfrac{4}{7} \div \dfrac{4}{5}\right) \cdot \dfrac{1}{7} = \left(\dfrac{4}{7} \cdot \dfrac{5}{4}\right) \cdot \dfrac{1}{7} = \dfrac{5}{7} \cdot \dfrac{1}{7} = \dfrac{5}{49}$

77. $\left[\left(\dfrac{-3}{4}\right)\left(\dfrac{-2}{7}\right)\right] \div \dfrac{3}{5} = \left(\dfrac{6}{28}\right) \div \dfrac{3}{5} = \dfrac{3}{14} \cdot \dfrac{5}{3} = \dfrac{15}{42} = \dfrac{5}{14}$

78. $\left(\dfrac{3}{8} \cdot \dfrac{5}{9}\right) \cdot \left(\dfrac{4}{7} \div \dfrac{5}{8}\right) = \left(\dfrac{15}{72}\right) \cdot \left(\dfrac{4}{7} \cdot \dfrac{8}{5}\right) = \dfrac{5}{24} \cdot \dfrac{32}{35} = \dfrac{160}{840} = \dfrac{4}{21}$

79. The lcm of 3 and 5 is 15.

$\dfrac{2}{3} + \dfrac{1}{5} = \left(\dfrac{2}{3} \cdot \dfrac{5}{5}\right) + \left(\dfrac{1}{5} \cdot \dfrac{3}{3}\right) = \dfrac{10}{15} + \dfrac{3}{15} = \dfrac{13}{15}$

80. The lcm of 6 and 8 is 24.

$\dfrac{5}{6} - \dfrac{1}{8} = \left(\dfrac{5}{6} \cdot \dfrac{4}{4}\right) - \left(\dfrac{1}{8} \cdot \dfrac{3}{3}\right) = \dfrac{20}{24} - \dfrac{3}{24} = \dfrac{17}{24}$

81. The lcm of 13 and 26 is 26.

$\dfrac{5}{13} + \dfrac{11}{26} = \left(\dfrac{5}{13} \cdot \dfrac{2}{2}\right) + \dfrac{11}{26} = \dfrac{10}{26} + \dfrac{11}{26} = \dfrac{21}{26}$

82. The lcm of 12 and 36 is 36.

$\dfrac{5}{12} + \dfrac{7}{36} = \left(\dfrac{5}{12} \cdot \dfrac{3}{3}\right) + \dfrac{7}{36} = \dfrac{15}{36} + \dfrac{7}{36} = \dfrac{22}{36} = \dfrac{22 \div 2}{36 \div 2} = \dfrac{11}{18}$

83. The lcm of 9 and 54 is 54.

$\dfrac{5}{9} - \dfrac{7}{54} = \left(\dfrac{5}{9} \cdot \dfrac{6}{6}\right) - \dfrac{7}{54} = \dfrac{30}{54} - \dfrac{7}{54} = \dfrac{23}{54}$

84. The lcm of 30 and 120 is 120.

$$\frac{13}{30} - \frac{17}{120} = \left(\frac{13}{30} \cdot \frac{4}{4}\right) - \frac{17}{120} = \frac{52}{120} - \frac{17}{120} = \frac{35}{120}$$

$$= \frac{35 \div 5}{120 \div 5} = \frac{7}{24}$$

85. The lcm of 12, 48, and 72 is 144.

$$\frac{1}{12} + \frac{1}{48} + \frac{1}{72} = \left(\frac{1}{12} \cdot \frac{12}{12}\right) + \left(\frac{1}{48} \cdot \frac{3}{3}\right) + \left(\frac{1}{72} \cdot \frac{2}{2}\right)$$

$$= \frac{12}{144} + \frac{3}{144} + \frac{2}{144} = \frac{17}{144}$$

86. The lcm of 5, 15, and 75 is 75.

$$\frac{3}{5} + \frac{7}{15} + \frac{9}{75} = \left(\frac{3}{5} \cdot \frac{15}{15}\right) + \left(\frac{7}{15} \cdot \frac{5}{5}\right) + \frac{9}{75}$$

$$= \frac{45}{75} + \frac{35}{75} + \frac{9}{75} = \frac{89}{75}$$

87. The lcm of 30, 40, and 50 is 600.

$$\frac{1}{30} - \frac{3}{40} - \frac{7}{50} = \left(\frac{1}{30} \cdot \frac{20}{20}\right) \left(\frac{3}{40} \cdot \frac{15}{15}\right) \left(\frac{7}{50} \cdot \frac{12}{12}\right)$$

$$= \frac{20}{600} - \frac{45}{600} - \frac{84}{600} = -\frac{109}{600}$$

88. The lcm of 25, 100, and 40 is 200.

$$\frac{4}{25} - \frac{9}{100} - \frac{7}{40} = \left(\frac{4}{25} \cdot \frac{8}{8}\right) \left(\frac{9}{100} \cdot \frac{2}{2}\right) \left(\frac{7}{40} \cdot \frac{5}{5}\right)$$

$$= \frac{32}{200} - \frac{18}{200} - \frac{35}{200} = -\frac{21}{200}$$

89. $$\frac{2}{5} + \frac{7}{8} = \frac{2 \cdot 8 + 7 \cdot 5}{8 \cdot 5} = \frac{16 + 35}{40} = \frac{51}{40}$$

90. $$\frac{3}{4} + \frac{2}{9} = \frac{3 \cdot 9 + 2 \cdot 4}{9 \cdot 4} = \frac{27 + 8}{36} = \frac{35}{36}$$

91. $$\frac{5}{6} - \frac{7}{8} = \frac{5 \cdot 4 - 7 \cdot 3}{24} = \frac{20 - 21}{24} = \frac{-1}{24}$$

92. $$\frac{7}{3} - \frac{5}{12} = \frac{7 \cdot 12 - 3 \cdot 5}{3 \cdot 12} = \frac{84 - 15}{36} = \frac{69}{36} = \frac{23}{12}$$

93. $$\frac{3}{8} + \frac{5}{12} = \frac{3 \cdot 12 + 8 \cdot 5}{8 \cdot 12} = \frac{36 + 40}{96} = \frac{76}{96} = \frac{19}{24}$$

94. $$\left(\frac{2}{3} + \frac{1}{4}\right) - \frac{3}{5} = \left(\frac{2 \cdot 4 + 3 \cdot 1}{3 \cdot 4}\right) - \frac{3}{5} = \frac{8 + 3}{12} - \frac{3}{5}$$

$$= \frac{11}{12} - \frac{3}{5} = \frac{11 \cdot 5 - 12 \cdot 3}{12 \cdot 5} = \frac{55 - 36}{60} = \frac{19}{60}$$

95. $$\left(\frac{2}{3} \cdot \frac{9}{10}\right) + \frac{2}{5} = \frac{18}{30} + \frac{2}{5} = \frac{18}{30} + \left(\frac{2}{5} \cdot \frac{6}{6}\right) =$$

$$= \frac{18}{30} + \frac{12}{30} = \frac{30}{30} = 1$$

96. $$\left(\frac{7}{6} \div \frac{4}{3}\right) - \frac{11}{12} = \left(\frac{7}{6} \cdot \frac{3}{4}\right) - \frac{11}{12} = \frac{21}{24} - \left(\frac{11}{12} \cdot \frac{2}{2}\right) =$$

$$= \frac{21}{24} - \frac{22}{24} = \frac{-1}{24}$$

97. $$\left(\frac{1}{2} + \frac{3}{10}\right) \div \left(\frac{1}{5} + 2\right) = \left(\frac{1}{2} \cdot \frac{5}{5} + \frac{3}{10}\right) \div \left(\frac{1}{5} + \frac{2}{1} \cdot \frac{5}{5}\right) = \left(\frac{5}{10} + \frac{3}{10}\right) \div \left(\frac{1}{5} + \frac{10}{5}\right) = \frac{8}{10} \div \frac{11}{5} = \frac{4}{5} \cdot \frac{5}{11} = \frac{20}{55} = \frac{4}{11}$$

98. $$\left(\frac{1}{9} \cdot \frac{3}{5}\right) + \left(\frac{2}{3} \cdot \frac{1}{5}\right) = \frac{3}{45} + \frac{2}{15} = \frac{1}{15} + \frac{2}{15} = \frac{3}{15} = \frac{1}{5}$$

99. $$\left(3\frac{4}{9}\right) \div \left(4 + \frac{2}{3}\right) = \left(\frac{3}{1} \cdot \frac{9}{9} - \frac{4}{9}\right) \div \left(\frac{4}{1} \cdot \frac{3}{3} + \frac{2}{3}\right) = \left(\frac{27}{9} - \frac{4}{9}\right) \div \left(\frac{12}{3} + \frac{2}{3}\right) = \frac{23}{9} \div \frac{14}{3} = \frac{23}{9} \cdot \frac{3}{14} = \frac{69}{126} = \frac{23}{42}$$

100. $$\left(\frac{2}{5} \div \frac{4}{9}\right) \left(\frac{3}{5} \cdot 6\right) = \left(\frac{2}{5} \cdot \frac{9}{4}\right) \left(\frac{3}{5} \cdot \frac{6}{1}\right) = \frac{18}{20} \cdot \frac{18}{5} = \frac{9}{10} \cdot \frac{18}{5} = \frac{162}{50} = \frac{81}{25}$$

101. The LCM of 2, 4, 6 is 12. $$\frac{1}{2} + \frac{1}{4} + \frac{1}{6} = \left(\frac{1}{2} \cdot \frac{6}{6}\right) + \left(\frac{1}{4} \cdot \frac{3}{3}\right) + \left(\frac{1}{6} \cdot \frac{2}{2}\right) = \frac{6}{12} + \frac{3}{12} + \frac{2}{12} = \frac{11}{12}$$ musk thistles

102.

$$73\frac{1}{4} \;\rightarrow\; 72\frac{5}{4}$$

$$-69\frac{3}{4} \;\rightarrow\; -69\frac{3}{4}$$

$$3\frac{2}{4} \rightarrow 3\frac{1}{2} \text{ inches}$$

103.

$$14\left(8\frac{5}{8}\right)=14\left(\frac{69}{8}\right)=\frac{966}{8}=\frac{966\div2}{8\div2}=\frac{483}{4}=120.75"$$

104.

$$67\left(\frac{5}{8}\right)\div6=\frac{67\cdot8+5}{8}\cdot\frac{1}{6}=\frac{541}{48}=11.27\,\text{oz.}=11\frac{13}{48}\,\text{oz}$$

105.

$$\left(1\frac{1}{2}\right)\left(\frac{1}{4}\right)=\left(\frac{3}{2}\right)\left(\frac{1}{4}\right)=\frac{3}{8} \text{ cups of snipped parsley}$$

$$\left(1\frac{1}{2}\right)\left(\frac{1}{8}\right)=\left(\frac{3}{2}\right)\left(\frac{1}{8}\right)=\frac{3}{16} \text{ tsp of pepper}$$

$$\left(1\frac{1}{2}\right)\left(\frac{1}{2}\right)=\left(\frac{3}{2}\right)\left(\frac{1}{2}\right)=\frac{3}{4} \text{ cups of sliced carrots}$$

106.

$$1-\left(\frac{1}{4}+\frac{1}{5}+\frac{1}{2}\right)2\frac{1}{4}+3\frac{7}{8}+4\frac{1}{4}=2\frac{4}{16}+3\frac{14}{16}+4\frac{4}{16}$$

$$=9\frac{22}{16}=10\frac{6}{16}$$

$$20\frac{5}{16}-10\frac{6}{16}=19\frac{21}{16}-10\frac{6}{16}=9\frac{15}{16}"$$

107. The LCM of 4, 5, 3 is 60.

$$\frac{1}{4}+\frac{2}{5}+\frac{1}{3}=\left(\frac{1}{4}\right)\left(\frac{15}{15}\right)+\left(\frac{2}{5}\right)\left(\frac{12}{12}\right)+\left(\frac{1}{3}\right)\left(\frac{20}{20}\right)$$

$$=\frac{15}{60}+\frac{24}{60}+\frac{20}{60}=\frac{59}{60}$$

108.

$$1-\left(\frac{1}{2}+\frac{2}{5}\right)=1-\left(\frac{5}{10}+\frac{4}{10}\right)=1-\frac{9}{10}=\frac{10}{10}-\frac{9}{10}=\frac{1}{10}$$

Student tutors represent 0.1 of the budget.

109.

$$1-\left(\frac{1}{4}+\frac{1}{5}+\frac{1}{2}\right)=1-\left(\frac{5}{20}+\frac{4}{20}+\frac{10}{20}\right)$$

$$=1-\frac{19}{20}=\frac{20}{20}-\frac{19}{20}=\frac{1}{20}$$

She must proofread .05 of the book or = 27 pages.

110. $\left(1\frac{1}{4}\right)(15)=\left(\frac{5}{4}\right)\left(\frac{15}{1}\right)=\frac{75}{4}=18\frac{3}{4}$ cups

111.

$$4\frac{1}{2}+30\frac{1}{4}+24\frac{1}{8}=4\frac{4}{8}+30\frac{2}{8}+24\frac{1}{8}=58\frac{7}{8}\text{ inches}$$

112.

a)
$$\left(15\frac{3}{8}\cdot\frac{1}{2}\right)=\left(\frac{15\cdot8+3}{8}\right)\cdot\left(\frac{1}{2}\right)=\left(\frac{123}{8}\cdot\frac{1}{2}\right)$$

$$=\frac{123}{16}=7\frac{11}{16}\text{ inches}$$

b) $7\frac{11}{16}-\frac{1}{16}=\frac{7\cdot16+11}{16}-\frac{1}{16}=\frac{123-1}{16}=\frac{122}{16}=7\frac{5}{8}"$

113. $\left(24\frac{7}{8}\right)\div2=\frac{199}{8}\cdot\frac{1}{2}=\frac{199}{16}=12\frac{7}{16}$ in.

114.

$$26\frac{1}{2}+105\frac{1}{4}+53\frac{1}{4}+106\frac{5}{16}=290+\frac{21}{16}=291\frac{5}{16}"$$

115. $8\frac{3}{4}\text{ ft}=\left(\frac{35}{4}\cdot\frac{12}{1}\right)\text{in.}=105\text{ in.}$

$$\left[105-(3)\left(\frac{1}{8}\right)\right]\div4=\left[\frac{840}{8}-\frac{3}{8}\right]\div4=\frac{837}{8}\cdot\frac{1}{4}=\frac{837}{32}=26\frac{5}{32}.\quad\text{The length of each piece is }26\frac{5}{32}\text{ in.}$$

116. original area $=8\frac{1}{2}\cdot9\frac{1}{4}=\frac{17}{2}\cdot\frac{37}{4}=\frac{629}{8}=78\frac{5}{8}$ sq. in.; new area $=8\frac{1}{2}\cdot10\frac{1}{4}=\frac{17}{2}\cdot\frac{41}{4}=\frac{697}{8}=87\frac{1}{8}$ sq. in.

area increase $=87\frac{1}{8}-78\frac{5}{8}=86\frac{9}{8}-78\frac{5}{8}=8\frac{4}{8}=8\frac{1}{2}$ sq. in.

117. width = 8 ft. 3 in. = 96 in. + 3 in. = 99 in.; length = 10 ft. 8 in. = 120 in. + 8 in. = 128 in.

    a) perimeter = 2L + 2W = 2(128) + 2(99) = 454 in $\dfrac{454''}{12''/\text{ft.}} = 37\dfrac{10}{12}\,\text{ft.} = 37\,\text{ft. } 10\,\text{in.}$

    b) width = 8ft. 3in. = $8\dfrac{3}{12}$ ft. = $8\dfrac{1}{4}$ ft. = $\dfrac{33}{4}$ ft .; length = 10ft. 8in. = $10\dfrac{8}{12}$ ft. = $10\dfrac{2}{3}$ ft. = $\dfrac{32}{3}$ ft

    Area = L × w = $\dfrac{32}{3} \times \dfrac{33}{4} = \dfrac{1056}{12} = 88$ sq.ft

117. c) Volume = L · W · H = $\dfrac{32}{3} \times \dfrac{33}{4} \times \dfrac{55}{6} = \dfrac{58080}{72} = 806.7$ cu. ft.

118. a) $20 + 18\dfrac{3}{8} \div 2 = 20 + 9\dfrac{3}{16} = 29\dfrac{3}{16}$ in.

    b) $26\dfrac{1}{4} + 6\dfrac{3}{4} = 33$ in.

    c) $26\dfrac{1}{4} + \left(6\dfrac{3}{4} - \dfrac{1}{4}\right) = 26\dfrac{1}{4} + 6\dfrac{2}{4} = 32\dfrac{3}{4}$ in.

119. $\dfrac{0.10 + 0.11}{2} = \dfrac{0.21}{2} = 0.105$

120. $\dfrac{5.03 + 5.003}{2} = \dfrac{10.033}{2} = 5.0165$

121. $\dfrac{-2.176 + (-2.175)}{2} = \dfrac{-4.351}{2} = -2.1755$

122. $\dfrac{1.3457 + 1.34571}{2} = \dfrac{2.69141}{2} = 1.345705$

123. $\dfrac{3.12345 + 3.123451}{2} = \dfrac{6.246901}{2} = 3.1234505$

124. $\dfrac{0.4105 + 0.4106}{2} = \dfrac{0.8211}{2} = 0.41055$

125. $\dfrac{4.872 + 4.873}{2} = \dfrac{9.745}{2} = 4.8725$

126. $\dfrac{3.7896 + (3.7895)}{2} = \dfrac{7.5791}{2} = 3.78955$

127. $\left(\dfrac{1}{3} + \dfrac{2}{3}\right) \div 2 = \dfrac{3}{3} \cdot \dfrac{1}{2} = \dfrac{3}{6} = \dfrac{1}{2}$

128. $\left(\dfrac{2}{7} + \dfrac{3}{7}\right) \div 2 = \dfrac{5}{7} \cdot \dfrac{1}{2} = \dfrac{5}{14}$

129. $\left(\dfrac{1}{100} + \dfrac{1}{10}\right) \div 2 = \dfrac{11}{100} \cdot \dfrac{1}{2} = \dfrac{11}{200}$

130. $\left(\dfrac{7}{13} + \dfrac{8}{13}\right) \div 2 = \dfrac{15}{13} \cdot \dfrac{1}{2} = \dfrac{15}{26}$

131. $\left(\dfrac{1}{4} + \dfrac{1}{5}\right) \div 2 = \left(\dfrac{5}{20} + \dfrac{4}{20}\right) \cdot \dfrac{1}{2} = \dfrac{9}{20} \cdot \dfrac{1}{2} = \dfrac{9}{40}$

132. $\left(\dfrac{1}{3} + \dfrac{2}{3}\right) \div 2 = \dfrac{3}{3} \cdot \dfrac{1}{2} = \dfrac{1}{1} \cdot \dfrac{1}{2} = \dfrac{1}{2}$

133. $\left(\dfrac{1}{10} + \dfrac{1}{100}\right) \div 2 = \left(\dfrac{10}{100} + \dfrac{1}{100}\right) \cdot \dfrac{1}{2} = \dfrac{11}{100} \cdot \dfrac{1}{2} = \dfrac{11}{200}$

134. $\left(\dfrac{1}{2} + \dfrac{2}{3}\right) \div 2 = \left(\dfrac{3}{6} + \dfrac{4}{6}\right) \cdot \dfrac{1}{2} = \dfrac{7}{6} \cdot \dfrac{1}{2} = \dfrac{7}{12}$

135. a) Water (or milk): $\left(1 + 1\dfrac{3}{4}\right) \div 2 = \left(\dfrac{4}{4} + \dfrac{7}{4}\right) \cdot \dfrac{1}{2} = \dfrac{11}{4} \cdot \dfrac{1}{2} = \dfrac{11}{8} = 1\dfrac{3}{8}$ cup;

    Oats: $\left(\dfrac{1}{2} + 1\right) \div 2 = \dfrac{3}{2} \cdot \dfrac{1}{2} = \dfrac{3}{4}$ cup

136. a) 1   b) $0.\overline{9}$ c) $\frac{1}{3} = 0.\overline{3}$, $\frac{2}{3} = 0.\overline{6}$, $\frac{1}{3} + \frac{2}{3} = \frac{3}{3} = 1$, $0.\overline{3} + 0.\overline{6} = 1$ d) $0.\overline{9} = 1$

136. a) $\frac{1}{8}$ b) $\frac{1}{16}$ c) 5 times   d) 5 times

Exercise Set 5.4

1. A rational number can be written as a ratio of two integers, p/q, with q not equal to zero. Numbers that cannot be written as the ratio of two integers are called irrational numbers.

2. The principal square root of a number n written $\sqrt{n}$, is the positive number that when multiplied by itself gives n.

3. A perfect square number is any number that is the square of a natural number.

4. The product rule for radical numbers: $\sqrt{a \cdot b} = \sqrt{a} \cdot \sqrt{b}$ $a \geq 0, b \geq 0$

   The quotient rule for radical numbers: $\frac{\sqrt{a}}{\sqrt{b}} = \sqrt{\frac{a}{b}}$ $a \geq 0, b \geq 0$

5. a) To add or subtract two or more square roots with the same radicand, add or subtract their coefficients and then multiply by the common radical.

   b) $3\sqrt{6} + 5\sqrt{6} - 9\sqrt{6} = 8\sqrt{6} - 9\sqrt{6} = -1\sqrt{6} = -\sqrt{6}$

6. A rationalized denominator contains no radical expressions.

7. a) Multiply both the numerator and denominator by the same number that will result in the radicand in the denominator becoming a perfect square.

   b) $\frac{7}{\sqrt{3}} = \frac{7}{\sqrt{3}} \cdot \frac{\sqrt{3}}{\sqrt{3}} = \frac{7\sqrt{3}}{\sqrt{9}} = \frac{7\sqrt{3}}{3}$

8. (a) $\left[\sqrt{} \right]$ $[\#]$ $[\text{Enter}]$

   (b) $\sqrt{7} = 2.645751311 = 2.65$

9. $\sqrt{36} = 6$ rational

10. $\sqrt{18} = \sqrt{2}\sqrt{9} = 3\sqrt{2}$ irrational

11. $\frac{2}{3}$ rational

12. Irrational; non-terminating, non-repeating decimal

13. Irrational; non-terminating, non-repeating decimal

14. Irrational; π is non-terminating, non-repeating.

15. Rational; quotient of two integers

16. Rational; terminating decimal

17. Irrational; non-terminating, non-repeating decimal

18. Rational; $\frac{\sqrt{5}}{\sqrt{5}} = 1$ 1 is an integer.

19. $\sqrt{64} = 8$

20. $\sqrt{144} = 12$

21. $\sqrt{100} = 10$

22. $-\sqrt{144} = -12$

23. $-\sqrt{169} = -13$

24. $\sqrt{25} = 5$

25. $-\sqrt{225} = -15$

26. $-\sqrt{36} = -6$

27. $-\sqrt{100} = -10$

28. $\sqrt{256} = 16$

29. 1, rational, integer, natural

30. -5, rational, integer

31. $\sqrt{25} = 5$, rat'l, integer., nat'l

32. rational

33. rational

34. rational

35. rational

36. rational

37. rational

38. irrational

39. $\sqrt{18}=\sqrt{2}\sqrt{9}=3\sqrt{2}$

40. $\sqrt{20}=\sqrt{4}\sqrt{5}=2\sqrt{5}$

41. $\sqrt{48}=\sqrt{3}\sqrt{16}=4\sqrt{3}$

42. $\sqrt{60}=\sqrt{4}\sqrt{15}=2\sqrt{15}$

43. $\sqrt{63}=\sqrt{9}\sqrt{7}=3\sqrt{7}$

44. $\sqrt{75}=\sqrt{25}\sqrt{3}=5\sqrt{3}$

45. $\sqrt{80}=\sqrt{16}\sqrt{5}=4\sqrt{5}$

46. $\sqrt{90}=\sqrt{9}\sqrt{10}=3\sqrt{10}$

47. $\sqrt{162}=\sqrt{81}\sqrt{2}=9\sqrt{2}$

48. $\sqrt{300}=\sqrt{100}\sqrt{3}=10\sqrt{3}$

49. $2\sqrt{6}+5\sqrt{6}=(2+5)\sqrt{6}=7\sqrt{6}$

50. $3\sqrt{17}+\sqrt{17}=(3+1)\sqrt{17}=4\sqrt{17}$

51.
$$5\sqrt{12}-\sqrt{75}=5\sqrt{4}\sqrt{3}-\sqrt{25}\sqrt{3}$$
$$=5\cdot2\sqrt{3}-5\sqrt{3}=10\sqrt{3}-5\sqrt{3}=5\sqrt{3}$$

52.
$$2\sqrt{5}+3\sqrt{20}=2\sqrt{5}+3\square2\sqrt{5}$$
$$=2\sqrt{5}+6\sqrt{5}=8\sqrt{5}$$

53.
$$4\sqrt{12}-7\sqrt{27}=4\sqrt{4}\sqrt{3}-7\sqrt{9}\sqrt{3}$$
$$=4\cdot2\sqrt{3}-7\cdot3\sqrt{3}=8\sqrt{3}-21\sqrt{3}$$
$$=-13\sqrt{3}$$

54.
$$2\sqrt{7}+5\sqrt{28}=2\sqrt{7}+5\cdot2\sqrt{7}$$
$$=2\sqrt{7}+10\sqrt{7}=(2+10)\sqrt{7}$$
$$=12\sqrt{7}$$

55.
$$5\sqrt{3}+7\sqrt{12}-3\sqrt{75}$$
$$=5\sqrt{3}+7\cdot2\sqrt{3}-3\cdot5\sqrt{3}$$
$$=5\sqrt{3}+14\sqrt{3}-15\sqrt{3}$$
$$=(5+14-15)\sqrt{3}=4\sqrt{3}$$

56.
$$13\sqrt{2}+2\sqrt{18}-5\sqrt{32}$$
$$=13\sqrt{2}+2\cdot3\sqrt{2}-5\cdot4\sqrt{2}$$
$$=13\sqrt{2}+6\sqrt{2}-20\sqrt{2}$$
$$=(13+6-20)\sqrt{2}=-\sqrt{2}$$

57.
$$\sqrt{8}-3\sqrt{50}+9\sqrt{32}$$
$$=2\sqrt{2}-3\cdot5\sqrt{2}+9\cdot4\sqrt{2}$$
$$=2\sqrt{2}-15\sqrt{2}+36\sqrt{2}$$
$$=(2-15+36)\sqrt{2}=-19\sqrt{2}$$

58.
$$\sqrt{63}+13\sqrt{98}-5\sqrt{112}$$
$$=3\sqrt{7}+13\cdot7\sqrt{2}-5\cdot4\sqrt{7}$$
$$=3\sqrt{7}+91\sqrt{2}-20\sqrt{7}$$
$$=-17\sqrt{7}+91\sqrt{2}$$

59.
$$\sqrt{2}\cdot\sqrt{8}=\sqrt{2}\sqrt{4}\sqrt{2}$$
$$=2\sqrt{2}\sqrt{2}=2\sqrt{4}$$
$$=2\cdot2=4$$

60.
$$\sqrt{5}\cdot\sqrt{15}=\sqrt{5}\sqrt{5}\sqrt{3}$$
$$=5\sqrt{2}$$

61.
$$\sqrt{6}\cdot\sqrt{10}=\sqrt{2}\sqrt{3}\sqrt{2}\sqrt{5}$$
$$=\sqrt{4}\sqrt{15}=2\sqrt{15}$$

62.
$$\sqrt{3}\cdot\sqrt{6}=\sqrt{18}$$
$$=\sqrt{9}\cdot\sqrt{2}=3\sqrt{2}$$

63.
$$\sqrt{10}\cdot\sqrt{20}=\sqrt{200}$$
$$=\sqrt{100}\cdot\sqrt{2}=10\sqrt{2}$$

64.
$$\sqrt{11}\cdot\sqrt{33}=\sqrt{11}\cdot\sqrt{11}\cdot\sqrt{3}$$
$$=11\sqrt{3}$$

65. $\dfrac{\sqrt{8}}{\sqrt{4}}=\sqrt{2}$

66. $\dfrac{\sqrt{125}}{\sqrt{5}}=\sqrt{25}=5$

67. $\dfrac{\sqrt{72}}{\sqrt{8}}=\sqrt{9}=3$

68. $\dfrac{\sqrt{136}}{\sqrt{8}}=\sqrt{17}$

69. $\dfrac{1}{\sqrt{2}} = \dfrac{1}{\sqrt{2}}\dfrac{\sqrt{2}}{\sqrt{2}} = \dfrac{1\sqrt{2}}{\sqrt{4}} = \dfrac{\sqrt{2}}{2}$

70. $\dfrac{3}{\sqrt{3}} = \dfrac{3}{\sqrt{3}}\cdot\dfrac{\sqrt{3}}{\sqrt{3}} = \dfrac{3\sqrt{3}}{3} = \sqrt{3}$

71. $\dfrac{\sqrt{3}}{\sqrt{7}}\cdot\dfrac{\sqrt{7}}{\sqrt{7}} = \dfrac{\sqrt{21}}{7}$

72. $\dfrac{\sqrt{3}}{\sqrt{10}} = \dfrac{\sqrt{3}}{\sqrt{10}}\cdot\dfrac{\sqrt{10}}{\sqrt{10}} = \dfrac{\sqrt{30}}{\sqrt{100}} = \dfrac{\sqrt{30}}{10}$

73. $\dfrac{\sqrt{20}}{\sqrt{3}} = \dfrac{\sqrt{20}}{\sqrt{3}}\dfrac{\sqrt{3}}{\sqrt{3}} = \dfrac{\sqrt{60}}{\sqrt{9}}$
$= \dfrac{\sqrt{4}\sqrt{15}}{3} = \dfrac{2\sqrt{15}}{3}$

74. $\dfrac{\sqrt{50}}{\sqrt{14}} = \sqrt{\dfrac{50}{14}} = \sqrt{\dfrac{25}{7}}$
$= \dfrac{\sqrt{25}}{\sqrt{7}}\dfrac{\sqrt{7}}{\sqrt{7}} = \dfrac{5\sqrt{7}}{7}$

75. $\dfrac{\sqrt{9}}{\sqrt{2}}\cdot\dfrac{\sqrt{2}}{\sqrt{2}} = \dfrac{3\sqrt{2}}{2}$

76. $\dfrac{\sqrt{15}}{\sqrt{3}} = \sqrt{5}$

77. $\dfrac{\sqrt{10}}{\sqrt{6}}\cdot\dfrac{\sqrt{6}}{\sqrt{6}} = \dfrac{\sqrt{60}}{6}$
$= \dfrac{2\sqrt{15}}{6} = \dfrac{\sqrt{15}}{3}$

78. $\dfrac{8}{\sqrt{8}} = \dfrac{8}{\sqrt{8}}\cdot\dfrac{\sqrt{2}}{\sqrt{2}} = \dfrac{8\sqrt{2}}{\sqrt{16}} = \dfrac{8\sqrt{2}}{4} = 2\sqrt{2}$

79. $\sqrt{7}$ is between 2 and 3 since $\sqrt{7}$ is between $\sqrt{4} = 2$ and $\sqrt{9} = 3$. $\sqrt{7}$ is between 2.5 and 3 since 7 is closer to 9 than to 4. Using a calculator $\sqrt{7} \approx 2.6$.

80. $\sqrt{37}$ is between 6 and 7 since $\sqrt{37}$ is between $\sqrt{36} = 6$ and $\sqrt{49} = 7$. $\sqrt{37}$ is between 6 and 6.5 since 37 is closer to 36 than to 49. Using a calculator $\sqrt{37} \approx 6.1$.

81. $\sqrt{107}$ is between 10 and 11 since $\sqrt{107}$ is between $\sqrt{100} = 10$ and $\sqrt{121} = 11$. $\sqrt{107}$ is between 10 and 10.5 since 107 is closer to 100 than to 121. Using a calculator $\sqrt{107} \approx 10.3$.

82. $\sqrt{135}$ is between 11 and 12 since $\sqrt{135}$ is between $\sqrt{121} = 11$ and $\sqrt{144} = 12$. $\sqrt{135}$ is between 11.5 and 12 since 135 is closer to 144 than to 121. Using a calculator $\sqrt{135} \approx 11.6$.

83. $\sqrt{170}$ is between 13 and 14 since $\sqrt{170}$ is between $\sqrt{169} = 13$ and $\sqrt{196} = 14$. $\sqrt{170}$ is between 13 and 13.5 since 170 is closer to 169 than to 196. Using a calculator $\sqrt{170} \approx 13.04$.

84. $\sqrt{200}$ is between 14 and 15 since $\sqrt{200}$ is between $\sqrt{196} = 14$ and $\sqrt{225} = 15$. $\sqrt{200}$ is between 14 and 14.5 since 200 is closer to 196 than to 225. Using a calculator $\sqrt{200} \approx 14.1$.

85. False. $\sqrt{p}$ is an irrational number for any prime number p.

86. False. The result may be a rational number or an irrational number.

87. True   88. True   89. False. The result may be a rational number or an irrational number.

90. False. The result may be a rational number or an irrational number.

91. $\sqrt{2} + (-\sqrt{2}) = 0$

92. $\sqrt{3} + 5\sqrt{3} = 6\sqrt{3}$

93. $\sqrt{2}\cdot\sqrt{3} = \sqrt{6}$

94. $\sqrt{3} \cdot \sqrt{3} = \sqrt{9} = 3$ 95. No. $\sqrt{3} \neq 1.732$ since $\sqrt{3}$ is an irrational number and 1.732 is a rational number.

96. $\sqrt{14} = \sqrt{7}\sqrt{2}$ $\sqrt{7}$ is irrational and $\sqrt{2}$ is irrational, therefore $\sqrt{14}$ is irrational. Because 3.742 is rational, then $\sqrt{14} \neq 3.742$.

97. No. 3.14 and $\dfrac{22}{7}$ are rational numbers, π is an irrational number.

98.
$$\sqrt{9+16} \neq \sqrt{9} + \sqrt{16}$$
$$\sqrt{25} \neq 3 + 4$$
$$5 \neq 7$$

99.
$$\sqrt{4 \cdot 16} = \sqrt{4}\sqrt{16}$$
$$\sqrt{64} = 2 \cdot 4$$
$$8 = 8$$

100.
$$T = 2\pi\sqrt{\frac{35}{980}} = 2\pi\frac{\sqrt{35}}{\sqrt{980}} = 2\pi\frac{\sqrt{5}\sqrt{7}}{\sqrt{5}\sqrt{196}}$$
$$= 2\pi\frac{\sqrt{7}}{14} = \frac{2\pi\sqrt{7}}{2 \cdot 7} = \frac{\pi\sqrt{7}}{7}$$

101. a) $s = \sqrt{\dfrac{4}{0.04}} = \sqrt{100} = 10$ mph

b) $s = \sqrt{\dfrac{16}{0.04}} = \sqrt{400} = 20$ mph

c) $s = \sqrt{\dfrac{64}{0.04}} = \sqrt{1600} = 40$ mph

d) $s = \sqrt{\dfrac{256}{0.04}} = \sqrt{6400} = 80$ mph

102. a) $t = \dfrac{\sqrt{100}}{4} = \dfrac{10}{4} = 2.5$ sec

b) $t = \dfrac{\sqrt{400}}{4} = \dfrac{20}{4} = 5$ sec

c) $t = \dfrac{\sqrt{900}}{4} = \dfrac{30}{4} = 7.5$ sec

d) $t = \dfrac{\sqrt{1600}}{4} = \dfrac{40}{4} = 10$ sec

103. a) The number is rational if the result on the calculator is a terminating or repeating decimal number. Otherwise, the number is irrational.

b) Using a calculator, $\sqrt{0.04} = 0.2$ a terminating decimal and thus it is rational.

c) Using a calculator, $\sqrt{0.07} = 0.264575131...$, thus it is irrational.

104. No. The sum of two irrational numbers may not be irrational. (i.e. $-\sqrt{3} + \sqrt{3} = 0$)

105. a) $\left(44 \div \sqrt{4}\right) \div \sqrt{4} = \left(44 \div 2\right) \div 2 = 22 \div 2 = 11$

b) $\left(44 \div 4\right) + \sqrt{4} = 11 + 2 = 13$

c) $4 + 4 + 4 + \sqrt{4} = 12 + 2 = 14$

d) $\sqrt{4}\left(4 + 4\right) + \sqrt{4} = 2(8) + 2 = 16 + 2 = 18$

Exercise Set 5.5

1. The set of real numbers is the union of the rational numbers and the irrational numbers.
2. All real numbers = R
3. If the given operation is preformed on any two elements of the set and the result is an element of the set, then the set is <u>closed</u> under the given operation.
4. The order in which two numbers are multiplied does not make a difference in the result.   Ex. $2 \cdot 3 = 3 \cdot 2$
5. The order in which two numbers are added does not make a difference in the result.   Ex. a+b = b+a

6. The associative property of addition states that when adding three real numbers, parentheses may be placed around any two adjacent numbers. (a+b)+c = a+(b+c)

7. The associative property of multiplication states that when multiplying three real numbers, parentheses may be placed around any two adjacent numbers.   Ex. $(2 \cdot 3) \cdot 4 = 2 \cdot (3 \cdot 4)$.

8. The distributive property of multiplication over addition allows you to either add first and then multiply, or multiply first and then add. a(b+c) = ab + ac

9. Closed. The sum of two natural numbers is a natural number.

10. Not closed. (i.e. $3 - 5 = -2$ is not a natural number).

11. Not closed. (i.e. $3 \div 5 = \dfrac{3}{5} = 0.6$ is not a natural number).

12. Closed. The product of two natural numbers is a natural number.

13. Closed. The difference of two integers is an integer.

14. Closed. The sum of two integers is an integer.

15. Not closed. (i.e. $2 \div 5 = \dfrac{2}{5} = 0.4 = 0.4$ is not an integer).

16. Closed. The product of two integers is an integer.

17. Closed	18. Closed	19. Closed	20. Not closed
21. Not closed	22. Not closed	23. Not closed	24. Not closed
25. Closed	26. Closed	27. Not closed	28. Closed

29. Commutative property. The order is changed from $(x) + (3+4) = (3+4) + x$.

30. $4 + (5+6) = 4 + (6+5)$;   Commutative because the only thing that has changed is the order of 5 and 6.

31. $(-4) \cdot (-5) = 20 = (-5) \cdot (-4)$

32. $(-2) + (-3) = -5 = (-3) + (-2)$

33. No. $6 \div 3 = 2$, but $3 \div 6 = \dfrac{1}{2}$

34. No. $5 - 3 = 2$, but $3 - 5 = -2$

35. $[(-3) \cdot (-5]) \cdot (-7) = (15) \cdot (-7) = -105$

$(-3) \cdot [(-5]) \cdot (-7)] = (-3) \cdot (35) = -105$

36. $[(-3) + (-5]) + (-7) = (-8) + (-7) = -15$

$(-3) + [(-5]) + (-7)] = (-3) + (-12) = -15$

37. No. $(8 \div 4) \div 2 = 2 \div 2 = 1$, but $8 \div (4 \div 2) = 8 \div 2 = 4$

38. No. $(8-7) - 12 = 1 - 12 = -11$, but $8 - (-5)$

$= 8 + 5 = 13$

39. No. $(8 \div 4) \div 2 = 2 \div 2 = 1$, but $8 \div (4 \div 2) = 8 \div 2 = 4$

40. No. $2 + (3 \cdot 4) = 2 + 12 = 14$, but $(2+3) \cdot (2+4)$

$= 5 \cdot 6 = 30$

41. Commutative property of addition

46. $5(x+3) = 5x + 5 \cdot 3$ Distributive property

43. $(7 \cdot 4) \cdot 5 = 7 \cdot (4 \cdot 5)$

Associative property of multiplication

44. $v + w = w + v$

Commutative property of addition

45. $(24+7) + 3 = 24 + (7+3)$

Associative property of addition

46. $4 \cdot (11 \cdot x) = (4 \cdot 11) \cdot x$

Associative property of multiplication

47. $\sqrt{3} \cdot 7 = 7 \cdot \sqrt{3}$

    Commutative property of multiplication

48. $\frac{3}{8} + \left(\frac{1}{8} + \frac{3}{2} \right) = \left(\frac{3}{8} + \frac{1}{8} \right) + \frac{3}{2}$

    Associative property of addition

49. $8\left(7 + \sqrt{2} \right) = 8 \cdot 7 + 8 \cdot \sqrt{2}$

    Distributive property

50. $\sqrt{5} \cdot \frac{2}{3} = \frac{2}{3} \cdot \sqrt{5}$

    Commutative property of multiplication

51. Commutative property of addition

52. Commutative property of addition

53. Distributive property

54. Commutative property of multiplication

55. Commutative property of addition

56. Commutative property of multiplication

57. $2(c + 7) = 2c + 14$

58. $-3(d - 1) = -3d + 3$

59. $\frac{2}{3}(x - 6) = \frac{2}{3}x - \frac{12}{3} = \frac{2}{3}x - 4$

60. $\frac{-5}{8}(k + 8) = \frac{-5}{8}k + \frac{-40}{8} = \frac{-5}{8}k = 5$

61. $6\left(\frac{x}{2} + \frac{2}{3} \right) = \frac{6x}{2} + \frac{12}{3} = 3x + 4$

62. $24\left(\frac{x}{3} - \frac{1}{8} \right) = \frac{24x}{3} - \frac{24}{8} = 8x - 3$

63. $32\left(\frac{1}{16}x - \frac{1}{32} \right) = \frac{32x}{16} - \frac{32}{32} = 2x - 1$

64. $15\left(\frac{2}{3}x - \frac{4}{5} \right) = \frac{30x}{3} - \frac{60}{5} = 10x - 12$

65. $3\left(5 - \sqrt{5} \right) = 15 - 3\sqrt{5}$

66. $-7\left(2 + \sqrt{11} \right) = -14 - 2\sqrt{11}$

67. $\sqrt{2}\left(\sqrt{2} + \sqrt{3} \right) = \sqrt{4} + \sqrt{6} = 2 + \sqrt{6}$

68. $\sqrt{3}\left(\sqrt{15} + \sqrt{21} \right) = \sqrt{45} + \sqrt{63} = \sqrt{9}\sqrt{5} + \sqrt{9}\sqrt{7}$
    $= 3\sqrt{5} + 3\sqrt{7}$

69. a) Distributive property
    b) Associative property of addition
    c) Combine like terms

70. a) Distributive property
    b) Associative property of addition;
    c) Combine like terms

71. a) Distributive property
    b) Associative property of addition;
    c) Commutative property of addition
    d) Associative property of addition
    e) Combine like terms

72. a) Distributive property
    b) Associative property of addition;
    c) Commutative property of addition
    d) Associative property of addition
    e) Combine like terms

73. a) Distributive property
    b) Commutative property of addition;
    c) Associative property of addition
    d) Combine like terms
    e) Commutative property of addition

74. a) Distributive property
    b) Commutative property of addition;
    c) Associative property of addition
    d) Combine like terms
    e) Commutative property of addition

75. Yes. You can either lock your door first or put on your seat belt first.

76. Yes. Can be done independently; no order needed

77. No. The clothes must be washed first before being dried.

78. No. The PC must be turned on first before you can type a term paper.

79. Yes. Can be done in either order; either fill the car with gas or wash the windshield

80. No. The lamp must be turned on first before reading a book.

81. Yes. The order of events does not matter.

82. No. The book must be read first, then write a report, then make a presentation.

83. Yes. The order does not matter.

84. Yes. The order does not matter.

85. Yes. The order does not matter

86. No. The egg cannot be poured before it is cracked.

87. Yes. The final result will be the same regardless of the order of the events.

88. Yes. The meatloaf will taste the same regardless of the order the items are mixed.

89. Baking pizzelles: mixing eggs into the batter, or mixing sugar into the batter.; Yard work: mowing the lawn, or trimming the bushes

90. Washing siding/washing windows/washing the car Writing letters to spouse, parents or friends

91. No. $0 \div a = 0$ but $a \div 0$ is undefined.

92. a) No. (Man eating) tiger is a tiger that eats men, and man (eating tiger) is a man that is eating a tiger.

b) No. (Horse riding) monkey is a monkey that rides a horse, and horse (riding monkey) is a horse that rides a monkey.

c) Answers will vary.

Exercise Set 5.6

1. 2 is the base and 3 is the exponent or power.

2. b^n is b multiplied by itself n times. $b^n = \underbrace{b \cdot b \cdot b \cdots b}_{\text{n factors of b}}$

3. a) If m and n are natural numbers and a is any real number, then $a^m a^n = a^{m+n}$

b) $2^3 \cdot 2^4 = 2^{3+4} = 2^7 = 128$

4. a) If m and n are natural numbers and a is any real number except 0, then $\dfrac{a^m}{a^n} = a^{m-n}$.

b) $\dfrac{5^6}{5^4} = 5^{6-4} = 5^2 = 25$

5. a) If a is any real number except 0, then $a^0 = 1$.

b) $7^0 = 1$

6. a) If n is a natural number and a is any real number except 0, then $a^{-n} = \dfrac{1}{a^n}$.

b) $2^{-3} = \dfrac{1}{2^3} = \dfrac{1}{8}$

7. a) If m and n are natural numbers and a is any real number, then $\left(a^m\right)^n = a^{m \cdot n}$

   b) $\left(3^2\right)^4 = 3^{2 \cdot 4} = 3^8 = 6561$

8. Since 1 raised to any power equals 1, $1^{500} = 1$.

9. a) Since 1 raised to any exponent equals +1, then $-1^{500} = (-1)\left(1^{500}\right) = (-1)(1) = 1$

   b) Since -1 raised to an even exponent equals 1, then number $(-1)^{500} = \left((-1)^2\right)^{250} = (1)^{250} = 1$

   c) In -1^{501} -1 is not raised to the 501st power, but +1 is; so $-1^{501} = (-1)\left(1^{501}\right) = (-1)(1) = -1$

   d) Since -1 is raised to a negative exponent is -1, then $(-1)^{501} = -1$

10. a) Move the decimal point in the original number to the right or left until you obtain a number greater or equal to 1 and less than 10. Count the number of places the decimal was moved. If it was moved to the left the count is a positive number and if it was moved to the right the count is a negative number. Multiply the number obtained in the first step by 10 raised to the count number.

    b) $0.000426 = 4.26 \times 10^{-4}$. note: the count number is -4

11. a) If the exponent is positive, move the decimal point in the number to the right the same number of places as the exponent adding zeros where necessary. If the exponent is negative, move the decimal point in the number to the left the same number of places as the exponent adding zeros where necessary.

    b) $5.76 \times 10^{-4} = 0.000576$

12. a) The number is greater than or equal to 10.
    b) The number is greater than or equal 1 but < 10.
    c) The number is less than 1.

13. $5^2 = 5 \cdot 5 = 25$

14. $3^4 = 3 \cdot 3 \cdot 3 \cdot 3 = 81$

15. $(-2)^4 = (-2) \cdot (-2) \cdot (-2) \cdot (-2) = 16$

16. $-2^4 = -(2) \cdot (2) \cdot (2) \cdot (2) = -16$

17. $-3^2 = -(3) \cdot (3) = -9$

18. $(-3)^2 = (-3) \cdot (-3) = 9$

19. $\left(\frac{2}{3}\right)^2 = \left(\frac{2}{3}\right)\left(\frac{2}{3}\right) = \frac{4}{9}$

20. $\left(\frac{-7}{8}\right)^2 = \left(\frac{-7}{8}\right)\left(\frac{-7}{8}\right) = \frac{49}{64}$

21. $(-5)^2 = (-5) \cdot (-5) = 25$

22. $-(5)^2 = -(5) \cdot (5) = -25$

22. $-(5)^2 = -(5) \cdot (5) = -25$

23. $2^3 \cdot 3^2 = (2) \cdot (2) \cdot (2) \cdot (3) \cdot (3) = 72$

24. $\frac{15^2}{3^2} = \frac{15 \cdot 15}{3 \cdot 3} = \frac{225}{9} = 25$

25. $\dfrac{5^7}{5^5} = 5^{7-5} = 5^2 = 5 \cdot 5 = 25$

26. $3^3 \cdot 3^4 = 3^{3+4} = 3^7 = 2187$

27. $\dfrac{7}{7^3} = 7^{1-3} = 7^{-2} = \dfrac{1}{7^2} = \dfrac{1}{7 \cdot 7} = \dfrac{1}{49}$

28. $3^4 \cdot 7^0 = (3)(3)(3)(3)(1) = 81$

29. $(-13)^0 = 1$

30. $(-3)^4 = (-3)(-3)(-3)(-3) = 81$

31. $3^4 = (3)(3)(3)(3) = 81$

32. $-3^4 = -(3)(3)(3)(3) = -81$

33. $3^{-2} = \dfrac{1}{3^2} = \dfrac{1}{9}$

34. $3^{-3} = \dfrac{1}{3^3} = \dfrac{1}{27}$

35. $\left(2^3\right)^4 = 2^{3 \cdot 4} = 2^{12} = 4096$

36. $\left(1^{12}\right)^{13} = 1^{12 \cdot 13} = 1^{156} = 1$

37. $\dfrac{11^{25}}{11^{23}} = 11^{25-23} = 11^2 = 121$

38. $5^2 \cdot 5 = 5^{2+1} = 5^3 = 125$

39. $(-4)^2 = (-4) \cdot (-4) = 16$

40. $4^{-2} = \dfrac{1}{4^2} = \dfrac{1}{16}$

41. $-(4)^2 = -(4) \cdot (4) = -16$

42. $\left(4^3\right)^2 = 4^{3 \cdot 2} = 4^6 = 4096$

43. $\left(2^2\right)^{-3} = 2^{2(-3)} = 2^{-6} = \dfrac{1}{2^6} = \dfrac{1}{64}$

44. $3^{-3} \cdot 3 = 3^{-3+1} = 3^{-2} = \dfrac{1}{3^2} = \dfrac{1}{9}$

45. $231000 = 2.31 \times 10^5$

46. $297000000 = 2.97 \times 10^8$

47. $15 = 1.5 \times 10^1$

48. $0.000034 = 3.4 \times 10^{-5}$

49. $0.56 = 5.6 \times 10^{-1}$

50. $0.00467 = 4.67 \times 10^{-3}$

51. $19000 = 1.9 \times 10^4$

52. $1260000000 = 1.26 \times 10^9$

53. $0.000186 = 1.86 \times 10^{-4}$

54. $0.0003 = 3.0 \times 10^{-4}$

55. $0.00000423 = 4.23 \times 10^{-6}$

56. $54000 = 5.4 \times 10^4$

57. $711 = 7.11 \times 10^2$

58. $0.02 = 2.0 \times 10^{-2}$

59. $0.153 = 1.53 \times 10^{-1}$

60. $416000 = 4.16 \times 10^5$

61. $8.4 \times 10^4 = 84000$

62. $2.71 \times 10^{-3} = 0.00271$

63. $1.2 \times 10^{-2} = 0.012$

64. $5.19 \times 10^5 = 519000$

65. $2.13 \times 10^{-5} = 0.0000213$

66. $2.74 \times 10^{-7} = 0.000000274$

67. $3.12 \times 10^{-1} = 0.312$

68. $4.6 \times 10^1 = 46$

69. $9.0 \times 10^6 = 9000000$

70. $7.3 \times 10^4 = 73000$

71. $2.31 \times 10^2 = 231$

72. $1.04 \times 10^{-2} = 0.0104$

73. $3.5 \times 10^4 = 35000$

74. $2.17 \times 10^{-6} = 0.00000217$

75. $1.0 \times 10^4 = 10000$

76. $1.0 \times 10^{-3} = 0.001$

77. $\left(2.0 \times 10^3\right)\left(4.0 \times 10^2\right) = 8.0 \times 10^5 = 800000$

78. $\left(4.1 \times 10^{-3}\right)\left(2.0 \times 10^3\right) = 8.2 \times 10^0 = 8.2$

79. $\left(5.1 \times 10^1\right)\left(3.0 \times 10^{-4}\right) = 15.3 \times 10^{-3} = 0.0153$

80. $\left(1.6 \times 10^{-2}\right)\left(4.0 \times 10^{-3}\right) = 6.4 \times 10^{-} = 0.000064$

81. $\dfrac{6.4 \times 10^5}{2.0 \times 10^3} = 3.2 \times 10^2 = 320$

82. $\dfrac{8.0 \times 10^{-3}}{2.0 \times 10^1} = 4.0 \times 10^{-4} = 0.0004$

83. $\dfrac{8.4 \times 10^{-6}}{4.0 \times 10^{-3}} = 2.1 \times 10^{-3} = 0.0021$

84. $\dfrac{25.0 \times 10^3}{5.0 \times 10^{-2}} = 5.0 \times 10^5 = 500000$

85. $\dfrac{4.0 \times 10^5}{2.0 \times 10^4} = 2.0 \times 10^1 = 20$

86. $\dfrac{16.0 \times 10^3}{8.0 \times 10^{-3}} = 2.0 \times 10^6 = 2000000$

87. $(300000)(2000000) = \left(3.0 \times 10^5\right)\left(2.0 \times 10^6\right)$
    $= 6.0 \times 10^{11}$

88. $\left(4.1 \times 10^{-5}\right)\left(3.0 \times 10^3\right) = 12.3 \times 10^{-2}$
    $= 1.23 \times 10^{-1}$

89. $\left(3.0 \times 10^{-3}\right)\left(1.5 \times 10^{-4}\right) = 4.5 \times 10^{-7}$

90. $\left(2.3 \times 10^5\right)\left(3.0 \times 10^3\right) = 6.9 \times 10^8$

91. $\dfrac{1.4 \times 10^6}{7.0 \times 10^2} = 0.2 \times 10^4 = 2.0 \times 10^3$

92. $\dfrac{2.0 \times 10^4}{5.0 \times 10^{-4}} = 0.4 \times 10^8 = 4.0 \times 10^7$

93. $\dfrac{4.0 \times 10^{-5}}{2.0 \times 10^2} = 2.0 \times 10^{-7}$

94. $\dfrac{1.2 \times 10^{-3}}{6 \times 10^{-6}} = 0.2 \times 10^3 = 2.0 \times 10^2$

95. $\dfrac{1.5 \times 10^5}{5.0 \times 10^{-4}} = 0.3 \times 10^9 = 3.0 \times 10^8$

96. $\dfrac{2.4 \times 10^4}{8.0 \times 10^6} = 0.3 \times 10^{-2} = 3.0 \times 10^{-3}$

97. $8.3 \times 10^{-4},\ 3.2 \times 10^{-1},\ 4.6,\ 5.8 \times 10^5$

98. $8.5 \times 10^{-5},\ 1.3 \times 10^{-1},\ 8.2 \times 10^3,\ 6.2 \times 10^4$

99. $8.3 \times 10^{-5}; 0.00079; 4.1 \times 10^3; 40{,}000;$ Note: $0.00079 = 7.9 \times 10^{-4},\ 40{,}000 = 4 \times 10^4$

100. $1{,}962{,}000;\ 4.79 \times 10^6;\ 3.14 \times 10^7;\ 267{,}000{,}000$

101. $\dfrac{\$10.1432 \times 10^{12}}{285.0 \times 10^6} = 0.3559017548 \times 10^6$ a) $\$35{,}590.18$ b) 3.559018×10^4 GDP/person

102. $\dfrac{\$4.1468 \times 10^{12}}{127.0 \times 10^6} = 0.0326514685 \times 10^6$ a) $\$32{,}651.97$ b) $\$3.2652 \times 10^4$ GDP/person

103. $\dfrac{7.69 \times 10^{33}}{36.6 \times 10^{12}} = 0.2101092896 \times 10^{21}$ a) $\$210{,}109{,}000{,}000{,}000{,}000{,}000$ b) $\$2.1011 \times 10^{20}$ seconds

104. $6.251 \times 10^9 - 1.283 \times 10^9 = 4.968 \times 10^9$ people

105. $t = \dfrac{d}{r} = \dfrac{4.5 \times 10^8}{2.5 \times 10^4} = 1.8 \times 10^4$ a) 18,000 hrs.   b) 1.8×10^4 hrs

106. $t = \dfrac{d}{r} = \dfrac{239000 \text{ mi}}{20000 \text{ mph}} = 11.95$ a) 11.95 hrs   b) 1.195×10^1 hrs

107. $(500{,}000)(40{,}000{,}000{,}000) = (5 \times 10^5)(4 \times 10^{10}) = 20 \times 10^{15} = 2 \times 10^{16}$

   a) 20,000,000,000,000,000 drops   b) 2.0×10^{16} drops

108. $(50)(5{,}800{,}000) = (5 \times 10^1)(5.8 \times 10^6) = 29 \times 10^7 = 2.9 \times 10^8$

   a) 290,000,000 cells   b) 2.9×10^8 cells

109. $\dfrac{4.5 \times 10^9}{2.5 \times 10^5} = 1.8 \times 10^4$ a) 18,000 times   b) 1.8×10^4 times

110. a) (100,000 cu.ft./sec) (60 sec/min) (60 min/hr) (24 hr) = 8,640,000,000 ft^3 b) 8.64×10^9 cu ft

111. $\dfrac{\$4.65 \times 10^{12}}{257.0 \times 10^6} = 0.0180933852 \times 10^6$ a) $\$32{,}651.97 - 18{,}093.00 = \$3{,}434.78$

112. a) 18 billion = 18,000,000,000 = 1.8×10^{10} diapers

   b) $(14)(2.38)(10^5) = 33.32 \times 10^5 = 3.332 \times 10^6$ or 3,332,000 miles

113. a) $(0.60)(1{,}200{,}000{,}000) = \$720{,}000{,}000$ b) $(0.25)(1{,}200{,}000{,}000) = \$300{,}000{,}000$

   c) $(0.10)(1{,}200{,}000{,}000) = \$120{,}000{,}000$ d) $(0.05)(1{,}200{,}000{,}000) = \$60{,}000{,}000$

114. a) $(0.40)(3{,}400{,}000{,}000) = \$1{,}360{,}000{,}000$ b) $(0.40)(3{,}400{,}000{,}000) = \$1{,}360{,}000{,}000$

   c) $(0.10)(3{,}400{,}000{,}000) = \$340{,}000{,}000$ d) $(0.10)(3{,}400{,}000{,}000) = \$340{,}000{,}000$

115. 1,000 times, since 1 meter = 10^3 millimeters = 1,000 millimeters

116. Since 1 gram = 10^3 milligrams and 1 gram = 10^{-3} kilograms, 10^{-3} kilograms = 10^3 milligrams

   $\dfrac{10^{-3}\ \text{Kilograms}}{10^{-3}} = \dfrac{10^3\ \text{milligrams}}{10^{-3}}$,   Thus, 1 kilogram = 10^6 milligrams

117. $\dfrac{2 \times 10^{30}}{6 \times 10^{24}} = 0.\overline{3} \times 10^6 = 333{,}333$ times

118. a) (2) (6 billion) = 12 billion = 12,000,000,000 people

   b) $\dfrac{6{,}000{,}000{,}000}{(35)(365)} = \dfrac{6{,}000{,}000{,}000}{12775} = 469{,}667$ people per day

119. $\dfrac{897{,}000{,}000{,}000{,}000{,}000}{3{,}900{,}000{,}000{,}000} = \dfrac{8.97 \times 10^{17}}{3.9 \times 1012} = 2.3 \times 10^5 = 230{,}000$ seconds or about 2.66 days

120. a) $1,000,000 = 1.0 \times 10^6$; $1,000,000,000 = 1.0 \times 10^9$; $1,000,000,000,000 = 1.0 \times 10^{12}$

b) $\dfrac{1.0 \times 10^6}{1.0 \times 10^3} = 1.0 \times 10^3$ days or 1,000 days = 2.74 years

c) $\dfrac{1.0 \times 10^9}{1.0 \times 10^3} = 1.0 \times 10^6$ days or 1,000,000 days = 2,739.73 years

d) $\dfrac{1.0 \times 10^{12}}{1.0 \times 10^3} = 1.0 \times 10^9$ days or 1,000,000,000 days = 2,739,726.03 years

e) $\dfrac{1 \text{ billion}}{1 \text{ million}} = \dfrac{1.0 \times 10^9}{1.0 \times 10^6} = 1.0 \times 10^3 = 1,000$ times greater

121. a) $(1.86 \times 10^5 \text{ mi/sec}) (60 \text{ sec/min}) (60 \text{ min/hr}) (24 \text{ hr/day}) (365 \text{ days/yr}) (1 \text{ yr})$

$= (1.86 \times 10^5)(6 \times 10^1)(6 \times 10^1)(2.4 \times 10^1)(3.65 \times 10^2) = 586.5696 \times 10^{10} = 5.865696 \times 10^{12}$ miles

b) $t = \dfrac{d}{r} = \dfrac{9.3 \times 10^7}{1.86 \times 10^5} = 5.0 \times 10^2 = 500$ seconds or 8 min. 20 sec.

122. a) $E(0) = 2^{10} \times 2^0 = 2^{10} \times 1 = 1024$ bacteria   b) $E(1/2) = 2^{10} \times 2^{1/2} = 2^{10.5} = 1448.2$ bacteria

Exercise Set 5.7

1. A sequence is a list of numbers that are related to each other by a given rule. One example is 2, 4, 6, 8….

2. The terms of the sequence.

3. a) An arithmetic sequence is a sequence in which each term differs from the preceding term by a constant amount. One example is 1, 4, 7, 10,....

b) A geometric sequence is one in which the ratio of any term to the term that directly precedes it is a constant. One example is 1, 3, 9, 27,....

4. a) $d = +3$,   b) $r = \dfrac{2}{1}$

5. a) $a_n = n^{st}$ term of the sequence   b) $a_1 = 1^{st}$ term of a sequence   c) d = common difference in a sequence
   d) s_n = the sum of the 1st n terms of the arithmetic sequence

6. a) $a_n = n^{st}$ term of the sequence   b) $a_1 = 1^{st}$ term of a sequence   c) r = common ratio between consecutive
   terms   d) s_n = the sum of the 1st n terms of the arithmetic sequence

7. $a_1 = 3$, $d = 2$ 3, 5, 7, 9, 11

8. $a_1 = 1$, $d = 3$ 1, 4, 7, 10, 13

9. $a_1 = -5$, $d = -3$ -5, -2, 1, 4, 7

10. $a_1 = -11$, $d = -6$ -11, -6, -1, 4, 9

11. 5, 3, 1, -1, -3

12. -3, -7, -11, -15, -19

13. 1/2, 1, 3/2, 2, 5/2

14. 5/2, 1, $-1/2$, -2, $-7/2$

15. a_6, $a_1 = 2$, $d = 3$,   2, 5, 8, 11, 17 $a_6 = 17$

16. a_9, $a_1 = 3$, $d = -2$,   3, 1, -1, -3, -5, -7, -9,
     -11, -13 $a_9 = -13$

17. a_{10}, $a_1 = -5$, $d = 2$, $-5, -3, -1, 1, 3, 5, 7, 9,$
    $11, 13$ $a_{10} = 13$

18. $a_{12} = 7 + (12 - 1)(-3) = 7 + (11)(-3)$
    $= 7 - 33 = -26$

19. $a_{20} = \dfrac{4}{5} + (19)(-1) = \dfrac{4}{5} - 19 = \dfrac{4}{5} - \dfrac{95}{5} = -\dfrac{91}{5}$

20. $\dfrac{-1}{2} + (14)(-2) = \dfrac{-1}{2} - 28 = \dfrac{-1}{2} - \dfrac{56}{2} = -\dfrac{57}{2}$

21. $a_{11} = 4 + (10)\left(\dfrac{1}{2}\right) = 4 + 5 = 9$

22. $a_{15} = \dfrac{4}{3} + (14)\left(\dfrac{1}{3}\right) = \dfrac{4}{3} + \dfrac{14}{3} = \dfrac{18}{3} = 6$

23. $a_n = n$ $a_n = 1 + (n - 1)1 = 1 + n - 1 = n$

24. $a_n = 2n - 1$ $a_n = 1 + (n - 1)2 = 1 + 2n - 2 = 2n - 1$

25. $a_n = 2n$ $a_n = 2 + (n - 1)2 = 2 + 2n - 2 = 2n$

26. $3, 1, -1, -3$ $a_n = 3 + (n - 1)(-2) = 3 - 2n + 2$
    $a_n = 5 - 2n$

27. $a_n = \dfrac{-5}{3} + (n - 1)\left(\dfrac{1}{3}\right) = \dfrac{-5}{3} + \dfrac{1}{3}n - \dfrac{1}{3} = \dfrac{1}{3}n - 2$

28. $a_n = -15 + (n - 1)(5) = 5n - 20$

29. $a_n = -3 + (n - 1)\left(\dfrac{3}{2}\right) = -3 + \dfrac{3}{2}n - \dfrac{3}{2} = \dfrac{3}{2}n - \dfrac{9}{2}$

30. $a_n = -5 + (n - 1)(3) = 3n - 8$

31. $s_n = \dfrac{n(a_1 + a_n)}{2} = \dfrac{50(1 + 50)}{2} = \dfrac{50(51)}{2}$
    $= (25)(51) = 1275$

32. $s_n = \dfrac{50(2 + 100)}{2} = \dfrac{50(102)}{2} = (25)(102) = 2550$

33. $s_n = \dfrac{50(1 + 99)}{2} = \dfrac{50(100)}{2} = (25)(100) = 2500$

34. $s_9 = \dfrac{9(-4 + (-28))}{2} = \dfrac{9 \cdot (-32)}{2} = -144$

35. $s_8 = \dfrac{8(11 + (-24))}{2} = \dfrac{8 \cdot (-13)}{2} = -52$

36. $s_{18} = \dfrac{18\left(-9 + \left(\dfrac{-1}{2}\right)\right)}{2} = \dfrac{18 \cdot \left(\dfrac{-19}{2}\right)}{2} = -\dfrac{171}{2} = -85.5$

37. $s_8 = \dfrac{8\left(\dfrac{1}{2} + \dfrac{29}{2}\right)}{2} = \dfrac{8 \cdot \left(\dfrac{30}{2}\right)}{2} = \dfrac{8 \cdot 15}{2} = 60$

38. $s_{18} = \dfrac{18\left(\dfrac{3}{5} + 4\right)}{2} = \dfrac{18 \cdot \left(\dfrac{23}{5}\right)}{2} = \dfrac{207}{5} = 41.4$

39. $a_1 = 3$, $r = 2$, $a_n = a_1 r^{n-1} = 3(2)^{n-1}$
    $3, 6, 12, 24, 48$

40. $a_1 = 6$, $r = 3$, $a_5 = 6(3)^4 = 6(81) = 486$
    $6, 18, 54, 162, 486$

41. $a_1 = 2$, $r = -2$, $a_5 = 2(-2)^4 = 2(16) = 32$
    $2, -4, 8, -16, 32$

42. $8, 4, 2, 1, \dfrac{1}{2}$

43. $-3, 3, -3, 3, -3$

44. $-6, 12, -24, 48, -96$

45. $-16, 8, -4, 2, -1$

46. $5, 3, 9/5, 27/25, 81/125$ $\dfrac{9}{5}, \dfrac{27}{25}, \dfrac{81}{125}$

47. $a_6 = 3(4)^5 = (3)(1024) = 3072$

48. $a_5 = 2(2)^4 = (2)(16) = 32$

49. $a_3 = 3\left(\dfrac{1}{2}\right)^2 = 3\left(\dfrac{1}{4}\right) = \dfrac{3}{4}$

50. $a_7 = -3(-3)^6 = -3(729) = -2187$

51. $a_5 = \left(\dfrac{1}{2}\right) \cdot 2^4 = \left(\dfrac{1}{2}\right)(16) = 8$

52. a_{25} $a_1 = 2,\ r = 2,\quad a_{25} = 1(2)^{24} = 16,777,216$

53. $a_{10} = (-2)(3)^9 = -39,366$

54. a_{18} $a_1 = -5,\ r = -2,\quad a_{18} = (-5)(-2)^{17} = 655,360$

55. $1, 2, 4, 8\quad a_n = 1(2)^{n-1} = 2^{n-1}$

56. $3, 6, 12, 24\quad a_n = 3(2)^{n-1}$

57. $3, -3, 3, -3\quad a_n = 3(-1)^{n-1}$

58. $a_n = a_1 r^{n-1} = -16\left(\dfrac{1}{2}\right)^{n-1}$

59. $a_n = a_1 r^{n-1} = \left(\dfrac{1}{4}\right)(2)^{n-1}$

60. $a_n = a_1 r^{n-1} = (-3)(-2)^{n-1}$

61. $a_n = a_1 r^{n-1} = (9)\left(\dfrac{1}{3}\right)^{n-1}$

62. $a_n = a_1 r^{n-1} = (-4)\left(\dfrac{2}{3}\right)^{n-1}$

63. $s_4 = \dfrac{a_1\left(1-r^4\right)}{1-r} = \dfrac{3\left(1-2^4\right)}{1-2} = \dfrac{3(-15)}{-1} = 45$

64. $s_5 = \dfrac{a_1\left(1-r^5\right)}{1-r} = \dfrac{2\left(1-3^5\right)}{1-3} = \dfrac{2(-242)}{-2} = 242$

65. $s_7 = \dfrac{a_1\left(1-r^7\right)}{1-r} = \dfrac{5\left(1-4^7\right)}{1-4} = \dfrac{5(-16383)}{-3}$
$= 27,305$

66. $s_9 = \dfrac{a_1\left(1-r^9\right)}{1-r} = \dfrac{-3\left(1-5^9\right)}{1-5} = \dfrac{-3(-1953124)}{-4}$
$= -1,464,843$

67. $s_{11} = \dfrac{a_1\left(1-r^{11}\right)}{1-r} = \dfrac{-7\left(1-3^{11}\right)}{1-3} = \dfrac{-7(-177146)}{-2}$
$= -620,011$

68. $n = 15,\ a_1 = -1,\ r = 2$
$s_{15} = \dfrac{a_1\left(1-r^n\right)}{1-r} = \dfrac{-1\left(1-(2)^{15}\right)}{1-2} = \dfrac{(-1)(-32768)}{-1}$
$= 1-32768 = -32767$

69. $n = 15,\ a_1 = -1,\ r = -2$
$s_{13} = \dfrac{(-1)\left(1-(-2)^{15}\right)}{1-(-2)} = \dfrac{(-1)(1+32768)}{3}$
$= \dfrac{(-1)(32769)}{3} = -10923$

70. $n = 10,\ a_1 = 512,\ r = \frac{1}{2}$
$s_{10} = \dfrac{(512)\left(1-\left(\dfrac{1}{2}\right)^{10}\right)}{1-\dfrac{1}{2}} = \dfrac{(1024)\left(1-\dfrac{1}{1024}\right)}{\dfrac{1}{2}}$
$= 1024-1 = 1023$

71. $s_{100} = \dfrac{(100)(1+100)}{2} = \dfrac{(100)(101)}{2} = 50(101) = 5050$

72. $s_{100} = \dfrac{(100)(2+200)}{2} = \dfrac{(100)(202)}{2} = 50(202) = 10100$

73. $s_{100} = \dfrac{(100)(1+199)}{2} = \dfrac{(100)(200)}{2} = 50(200) = 10000$

74. $s_{50} = \dfrac{(50)(3+150)}{2} = \dfrac{(50)(153)}{2} = 25(153) = 3825$

75. a) Using the formula $a_n = a_1 + (n-1)d$, we get
$$a_8 = 20{,}200 + (8-1)(1200) = \$28{,}600$$
   b) $\dfrac{8(20200+28600)}{2} = \dfrac{8(48800)}{2} = \$195{,}200$

76. a) $a_{12} = 96 + (11)(-3) = 96 - 33 = 63$ in.
   b) $\dfrac{[12(96+63)]}{2} = \dfrac{(12)(159)}{2} = (6)(159) = 954$ in.

77. $a_{11} = 72 + (10)(-6) = 72 - 60 = 12$ in.

78. $s_{12} = \dfrac{12(1+12)}{2} = \dfrac{12(13)}{2} = \dfrac{156}{2} = 78$ times

79. 1, 2, 3,.... n=31
$$s_{31} = \dfrac{31(1+31)}{2} = \dfrac{31(32)}{2} = 31(16) = 496 \text{ PCs}$$

80. $a_n = a_1 r^{n-1}$ $a_{10} = (8000)(1.08)^9 = 15992$ students

81. $a_6 = 200(0.8)^6 = 200(0.262144)^1 = 52.4288$ g

82. $a_{15} = a_1 r^{15} = 1(2)^{15} = 32{,}768$ layers

83. $a_{15} = 20{,}000(1.06)^{14} = \$45{,}218$

84. $a_5 = 30(0.8)^4 = 12.288$ ft.

85. This is a geometric sequence where $a_1 = 2000$ and $r = 3$. In ten years the stock will triple its value 5 times.
$$a_6 = a_1 r^{6-1} = 2000(3)^5 = \$486{,}000$$

86. The sequence of bets during a losing streak is geometric.

   a) $a_6 = a_1 r^{n-1} = 1(2)^{6-1} = 1(32) = \$32$ $s_5 = \dfrac{a_1(1-r^n)}{1-r} = \dfrac{1(1-2^5)}{1-2} = \dfrac{-31}{-1} = \$31$

   b) $a_6 = a_1 r^{n-1} = 10(2)^{6-1} = 10(32) = \$320$ $s_5 = \dfrac{a_1(1-r^n)}{1-r} = \dfrac{10(1-2^5)}{1-2} = \dfrac{10(-31)}{-1} = \$310$

   c) $a_{11} = a_1 r^{n-1} = 1(2)^{11-1} = 1(1024) = \$1{,}024$ $s_{10} = \dfrac{a_1(1-r^n)}{1-r} = \dfrac{1(1-2^{10})}{1-2} = \dfrac{1(-1023)}{-1} = \$1{,}023$

   d) $a_{11} = a_1 r^{n-1} = 10(2)^{11-1} = 10(1024) = \$10{,}240$ $s_{10} = \dfrac{a_1(1-r^n)}{1-r} = \dfrac{10(1-2^{10})}{1-2} = \dfrac{10(-1023)}{-1} = \$10{,}230$

   e) If you lose too many times in a row, then you will run out of money.

87. $\dfrac{82[1-(1/2)^6]}{1-(1/2)} = \dfrac{82[1-(1/64)]}{1/2} = \dfrac{82}{1} \cdot \dfrac{63}{64} \cdot \dfrac{2}{1} = 161.4375$

88. The arithmetic sequence $180^0, 360^0, 540^0, 720^0, \ldots$ has a common difference of 180.
   Thus, $a_n = 180(n-2) = 180n - 360$, $n \geq 3$

89. 12, 18, 24, ... ,1608 is an arithmetic sequence with $a_1 = 12$ and $d = 6$. Using the expression for the nth term of an arithmetic sequence $a_n = a_1 + (n-1)d$ or $1608 = 12 + (n-1)6$ and dividing both sides by 6 gives $268 = 2 + n - 1$ or $n = 267$

90. Since $a_5 = a_1 r^4$ and $a_2 = a_1 r$, $a_5/a_2 = r^3$. Thus $r^3 = 648/24 = 27$ or $r = 3$.
   Then $24 = a_2 = a_1 r = a_1(3)$ or $a_1 = 24/3 = 8$.

91. The total distance is 30 plus twice the sum of the terms of the geometric sequence having $a_1 = (30)(0.8) = 24$

and r = 0.8.   Thus $s_5 = \dfrac{24[1-(0.8)^5]}{(1-0.8)} = \dfrac{24[1-0.32768]}{0.2} = \dfrac{24(0.67232)}{0.2} = 80.6784$.

So the total distance is 30 + 2(80.6784) = 191.3568 ft.

92. The sequence of bets during a losing streak is geometric.

a) $a_6 = a_1 r^{n-1} = 1(2)^{6-1} = 1(32) = \$32$ $s_5 = \dfrac{a_1(1-r^n)}{1-r} = \dfrac{1(1-2^5)}{1-2} = \dfrac{-31}{-1} = \$31$

b) $a_6 = a_1 r^{n-1} = 10(2)^{6-1} = 10(32) = \$320$ $s_5 = \dfrac{a_1(1-r^n)}{1-r} = \dfrac{10(1-2^5)}{1-2} = \dfrac{10(-31)}{-1} = \$310$

c) $a_{11} = a_1 r^{n-1} = 1(2)^{11-1} = 1(1024) = \$1,024$ $s_{10} = \dfrac{a_1(1-r^n)}{1-r} = \dfrac{1(1-2^{10})}{1-2} = \dfrac{1(-1023)}{-1} = \$1,023$

d) $a_{11} = a_1 r^{n-1} = 10(2)^{11-1} = 10(1024) = \$10,240$ $s_{10} = \dfrac{a_1(1-r^n)}{1-r} = \dfrac{10(1-2^{10})}{1-2} = \dfrac{10(-1023)}{-1} = \$10,230$

e) If you lose too many times in a row, then you will run out of money.

Exercise Set 5.8

1. Begin with the numbers 1, 1, then add 1 and 1 to get 2 and continue to add the previous two numbers in the sequence to get the next number in the sequence.

2. a) 1,2,3,5,8,13,21,34,55,89   b) $\dfrac{55}{34} = 1.61764 \rightarrow 1.619$ c) $\dfrac{89}{55} = 1.61818 \rightarrow 1.618$

d) $\dfrac{8}{5} = 1.6 \rightarrow 1.600$ e) $\dfrac{5}{3} = 1.\overline{6} \rightarrow 1.667$ f) $\dfrac{21}{13} = 1.61538 \rightarrow 1.615$

3. a) Golden number = $\dfrac{\sqrt{5}+1}{2}$

b) 1.618 = golden ratio   When a line segment AB is divided at a point C, such that the ratio of the whole, AB, to the larger part, AC, is equal to the ratio of the larger part, AC, to the smaller part, CB, then each

of the ratios $\dfrac{AB}{AC}$ and $\dfrac{AC}{CB}$ is known as the golden ratio.

c) The golden proportion is: $\dfrac{AB}{AC} = \dfrac{AC}{CB}$

d) The golden rectangle: $\dfrac{L}{W} = \dfrac{a+b}{a} = \dfrac{a}{b} = \dfrac{\sqrt{5}+1}{2} = $ golden number

4. All are essentially the same number when rounded.   5. a) Flowering head of a sunflower b) Great Pyramid
6. a) Petals on daisies   b) Parthenon in Athens

7. a) $\dfrac{\sqrt{5}+1}{2} = 1.618033989$ b) $\dfrac{\sqrt{5}-1}{2} = .6180339887$

c) Differ by 1

8. 89, $\dfrac{1}{89} = .01\underline{12359}551$, part of Fibonacci sequence

9. 1/1 = 1, 2/1 = 2, 3/2 = 1.5, 5/3 = 1.6, 8/5 = 1.6, 13/8 = 1.625, 21/13 = 1.6154, 34/21 = 1.619, 55/34 = 1.6176
    89/55 = 1.61818. The consecutive ratios alternate increasing then decreasing about the golden ratio.

10. The ratio of the second to the first and the fourth to the third estimates the golden ratio.

11.

Fib. No.	prime factors	Fib. No.	prime factors
1	-------	34	$2 \bullet 17$
1	-------	55	$5 \bullet 11$
2	prime	89	prime
3	prime	144	$2^4 \bullet 3^2$
5	prime	233	prime
8	2^3	377	$13 \bullet 29$
13	prime	610	$2 \bullet 5 \bullet 61$

12. If the first ten are selected; $\dfrac{1+1+2+3+5+8+13+21+34+55}{11} = \dfrac{143}{11} = 13$

13. If 5 is selected the result is $2(5) - 8 = 10 - 8 = 2$ which is the second number preceding 5.

14. If 2, 3, 5, and 8 are selected the result is $5^2 - 3^2 = 2 \bullet 8 \quad \rightarrow \quad 25 - 9 = 16 \quad \rightarrow \quad 16 = 16$

15. Answers will vary.   16. 6/4 = 1.5 which is a little < 1.6.   17. Answers will vary.
18. Answers will vary.   19. Answers will vary.   20. Answers will vary.
21. Answers will vary.   22. Answers will vary.

23. Fibonacci type; 11 + 18 = 29   18 + 29 = 47
24. Not Fibonacci. Each term is not the sum of the two preceding terms.
25. Not Fibonacci. Each term is not the sum of the two preceding terms.
26. Fibonacci type: 1 + 2 = 3, 2 + 3 = 5 Each term is the sum of the two preceding terms.

27. Fibonacci type; 40 + 65 = 105;   65 + 105 = 170
    28. Fibonacci type; $1\frac{1}{4} + 2 = 3\frac{1}{4}$; $2 + 3\frac{1}{4} = 5\frac{1}{4}$

29. Fibonacci type; -1 + 0 = -1;   0 + (-1) = -1   30. Fibonacci type; 7 + 13 = 20;   13 + 20 = 33

31. a) If 6 and 10 are selected the sequence is 6, 10, 16, 26, 42, 68, 110, …
    b) 10/6 = 1.666, 16/10 = 1.600, 26/16 = 1.625, 42/26 = 1.615, 68/42 = 1.619, 110/68 = 1.618, …

32. a) If 5 and 7 are selected the sequence is 5, 7, 12, 19, 31, 50, 81, …
    b) 7/5 = 1.4, 12/7 = 1.714, 19/12 = 1.583, 31/19 = 1.623, 50/31 = 1.613, 81/50 = 1.62, …

33. a) If 5, 8, and 13 are selected the result is $8^2 - (5)(13) = 64 - 65 = -1$.

    b) If 21, 34, and 55 are selected the result is $34^2 - (21)(55) = 1156 - 1155 = 1$.

    c) The square of the middle term of three consecutive terms in a Fibonacci sequence differs from the
       product of the 1st and 2nd term by 1.

34. The sum of the numbers along the diagonals parallel to the one shown is a Fibonacci number.

35. a) Lucas sequence: 1, 3, 4, 7, 11, 18, 29, 47, ...   b) $8 + 21 = 29$; $13 + 34 = 47$
    c) The first column is a Fibonacci-type sequence.

36. $-10, x, -10 + x, -10 + 2x, -20 + 3x, -30 + 5x, -50 + 8x, -80 + 13x, -130 + 21x, -210 + 34x$
    a) $-10, 4, -6, -2, -8, -10, -18, -28, -46, -74$
    b) $-10, 5, -5, 0, -5, -5, -10, -15, -25, -40$
    c) $-10, 6, -4, 2, -2, 0, -2, -2, -4, -6$
    d) $-10, 7, -3, 4, 1, 5, 6, 11, 17, 28$
    e) $-10, 8, -2, 6, 4, 10, 14, 24, 38, 62$
    f) Yes, because each multiple causes the x term to be greater than the number term.

37. $\dfrac{(a+b)}{a} = \dfrac{a}{b}$ Let $x = \dfrac{a}{b}$ $\dfrac{b}{a} = \dfrac{1}{x}$ $1 + \dfrac{b}{a} = \dfrac{a}{b}$ $1 + \dfrac{1}{x} = x$ multiply by x $x\left(1 + \dfrac{1}{x}\right) = x(x)$

$x + 1 = x^2$ $x^2 - x - 1 = 0$ $a = 1, b = -1, c = -1$

Solve for x using the quadratic formula, $x = \dfrac{-b \pm \sqrt{b^2 - 4ac}}{2a} = \dfrac{1 \pm \sqrt{1 - 4(1)(-1)}}{2(1)} = \dfrac{1 \pm \sqrt{5}}{2}$

38. $\dfrac{5 - x}{5} = \dfrac{5}{x}$ $x(5 - x) = 25$ $5x - x^2 = 25$ $x^2 - 5x + 25 = 0$ $a = 1, b = -5, c = 25$

Solve for x using the quadratic formula, $x = \dfrac{-b \pm \sqrt{b^2 - 4ac}}{2a} = \dfrac{5 \pm \sqrt{25 - 4(1)(25)}}{2(1)} = \dfrac{5 \pm \sqrt{-75}}{2} = \dfrac{5 \pm 5\sqrt{3}i}{2}$

39. Answers will vary. {5, 12, 13]   {16, 30, 34]   {105, 208, 233}   {272, 546, 610}

40. a) 3 reflections, 5 paths   b) 4 reflections, 8 paths   c) 5 reflections, 13 paths

Review Exercises

1. Use the divisibility rules in section 5.1.
   670,920 is divisible by 2, 3, 4, 5, 6, and 9.

2. Use the divisibility rules in section 5.1.
   400,644 is divisible by 2, 3, 4, 6, and 9

3.
```
2 | 252
2 | 126
3 | 63
3 | 21
    7
```
$252 = 2^2 \bullet 3^2 \bullet 7$

4.
```
5 | 385
7 | 77
    11
```
$385 = 5 \bullet 7 \bullet 11$

5.
```
2 | 840
2 | 420
2 | 210
5 | 105
3 | 21
    7
```
$840 = 2^3 \bullet 3 \bullet 5 \bullet 7$

6.
$$
\begin{array}{r|r}
2 & 882 \\
3 & 441 \\
3 & 147 \\
7 & 49 \\
\hline & 7
\end{array}
$$
$882 = 2 \cdot 3^2 \cdot 7^2$

7.
$$
\begin{array}{r|r}
2 & 1452 \\
2 & 726 \\
3 & 363 \\
11 & 121 \\
\hline & 11
\end{array}
$$
$1452 = 2^2 \cdot 3 \cdot 11^2$

8. $15 = 3 \cdot 5$, $60 = 2^2 \cdot 3 \cdot 5$
    gcd = 15   lcm = 60

9. $63 = 3 \cdot 3 \cdot 5$, $108 = 3 \cdot 4 \cdot 9$
    gcd = 9;   lcm = 756

10. $45 = 3^2 \cdot 5$, $250 = 2 \cdot 5^3$; gcd = 5; lcm = $2 \cdot 3^2 \cdot 5^3 = 2250$

11. $840 = 2^3 \cdot 3 \cdot 5 \cdot 7$, $320 = 2^6 \cdot 5$; gcd = $2^3 \cdot 5 = 40$; lcm = $2^6 \cdot 3 \cdot 5 \cdot 7 = 6720$

12. $60 = 2^2 \cdot 3 \cdot 5$, $40 = 2^3 \cdot 5$, $96 = 2^5 \cdot 3$; gcd = $2^2 = 4$; lcm = $2^5 \cdot 3 \cdot 5 = 480$

13. $36 = 2^2 \cdot 3^2$, $108 = 2^2 \cdot 3^3$, $144 = 2^4 \cdot 3^2$; gcd = $2^2 \cdot 3^2 = 36$; lcm = $2^4 \cdot 3^3 = 432$

14. $15 = 3 \cdot 5$, $9 = 3^2$; lcm = $3^2 \cdot 5 = 45$. In 45 days the train will stop in both cities.

15. $-2 + 5 = 3$

16. $4 + (-7) = -3$

17. $4 - 8 = 4 + (-8) = -4$

18. $(-2) + (-4) = -6$

19. $-5 - 4 = -5 + (-4) = -9$

20. $-3 - (-6) = -3 + 6 = 3$

21. $(-3 + 7) - 4 = 4 + (-4) = 0$

22. $-1 + (9 - 4) = -1 + 5 = 4$

23. $(-3)(-11) = 33$

24. $(-4)(9) = -36$

25. $14(-3) = -56$

26. $-35/-7 = 5$

27. $12/-6 = -2$

28. $[8 \div (-4)](-3) = (-2)(-3) = 6$

29. $[(-4)(-3)] \div 2 = 12 \div 2 = 6$

30. $[-30 \div (10)] \div -1) = -3 \div (-1) = 3$

31. $3/10 = 0.3$

32. $3/5 = 0.6$

33. $15/40 = 3/8 = 0.375$

34. $13/4 = 3.25$

35. $3/7 = 0.\overline{428571}$

36. $7/12 = 0.58\overline{3}$

37. $3/8 = 0.375$

38. $7/8 = 0.875$

39. $5/7 = 0.\overline{714285}$

40. $0.225 = \dfrac{225}{1000} = \dfrac{45}{200} = \dfrac{9}{40}$

41. $4.5 = 4\dfrac{5}{10} = \dfrac{45}{10} = \dfrac{9}{2}$

42. 0.6666 $10n = 6.6666....$
$$
\begin{array}{l}
10n = 6.\overline{6} \\
-\,n = 0.\overline{6} \\
\hline
9n = 6.0
\end{array}
\qquad
\begin{array}{l}
\dfrac{9n}{9} = \dfrac{6}{9} \\
n = \dfrac{2}{3}
\end{array}
$$

43. 2.373737 $100n = 237.373737....$
$$
\begin{array}{l}
100n = 237.\overline{37} \\
-\,n = \ \ \ 2.\overline{37} \\
\hline
99n = 235.00
\end{array}
\qquad
\dfrac{99n}{99} = \dfrac{235}{99} = n
$$

44. $0.083 = \dfrac{83}{1000}$

45. $0.0042 = \dfrac{42}{10000} = \dfrac{21}{5000}$

46. 2.344444 $100n = 234.444444....$
$$
\begin{array}{l}
100n = 234.\overline{4} \\
-10n = \ \ 23.\overline{4} \\
\hline
90n = 211.00
\end{array}
\qquad
\dfrac{90n}{90} = \dfrac{211}{90} = n
$$

47. $2\dfrac{5}{7} = \dfrac{19}{7}$

48. $4\dfrac{1}{6} = \dfrac{25}{6}$

49. $-3\frac{1}{4} = \frac{((-3)(4))-1}{4} = \frac{-13}{4}$

50. $-35\frac{3}{8} = \frac{((-35)(8))-3}{8} = \frac{-283}{8}$

51. $\frac{11}{5} = \frac{2 \cdot 5 + 1}{5} = 2\frac{1}{5}$

52. $\frac{27}{15} = \frac{1 \cdot 15 + 12}{15} = 1\frac{12}{15} = 1\frac{4}{5}$

53. $\frac{-12}{7} = \frac{(-1)(7)-5}{7} = -1\frac{5}{7}$

54. $\frac{-136}{5} = \frac{(-27)(5)-1}{5} = -27\frac{1}{5}$

55. $\frac{1}{2} + \frac{4}{5} = \frac{1}{2} \cdot \frac{5}{5} + \frac{4}{5} \cdot \frac{2}{2} = \frac{5}{10} + \frac{8}{10} = \frac{13}{10}$

56. $\frac{7}{8} - \frac{3}{4} = \frac{7}{8} - \frac{3}{4} \cdot \frac{2}{2} = \frac{7}{8} - \frac{6}{8} = \frac{1}{8}$

57. $\frac{1}{6} + \frac{5}{4} = \frac{1}{6} \cdot \frac{2}{2} + \frac{5}{4} \cdot \frac{3}{3} = \frac{2}{12} + \frac{15}{12} = \frac{17}{12}$

58. $\frac{4}{5} \cdot \frac{15}{16} = \frac{60}{80} = \frac{6}{8} = \frac{3}{4}$

59. $\frac{5}{9} \div \frac{6}{7} = \frac{5}{9} \div \frac{7}{6} = \frac{35}{54}$

60. $\left(\frac{4}{5} + \frac{5}{7}\right) \div \frac{4}{5} = \frac{28+25}{35} \cdot \frac{5}{4} = \frac{53}{35} \cdot \frac{5}{4} = \frac{53}{28}$

61. $\left(\frac{2}{3} \square \frac{1}{7}\right) \div \frac{4}{7} = \frac{2}{21} \sqcup \frac{7}{4} = \frac{1}{6}$

62. $\left(\frac{1}{5} + \frac{2}{3}\right) \cdot \frac{3}{8} = \frac{3+10}{15} \cdot \frac{3}{8} = \frac{13}{15} \cdot \frac{3}{8} = \frac{13}{40}$

63. $\left(\frac{1}{5}\right)\left(\frac{2}{3}\right) + \left(\frac{1}{5} \div \frac{1}{2}\right) = \frac{2}{15} + \left(\frac{1}{5}\right)\left(\frac{2}{1}\right) = \frac{2}{15} + \frac{2}{5}$

$= \frac{2}{15} + \frac{6}{15} = \frac{8}{15}$

64. $\left(\frac{1}{8}\right)\left(17\frac{3}{4}\right) = \left(\frac{1}{8}\right)\left(\frac{71}{4}\right) = \frac{71}{32} = 2\frac{7}{32}$ teaspoons

65. $\sqrt{50} = \sqrt{25 \cdot 2} = \sqrt{25} \cdot \sqrt{2} = 5\sqrt{2}$

66. $\sqrt{200} = \sqrt{100 \cdot 2} = \sqrt{100} \cdot \sqrt{2} = 10\sqrt{2}$

67. $\sqrt{5} + 7\sqrt{5} = 8\sqrt{5}$

68. $\sqrt{3} - 4\sqrt{3} = -3\sqrt{3}$

69. $\sqrt{8} + 6\sqrt{2} = 2\sqrt{2} + 6\sqrt{2} = 8\sqrt{2}$

70. $\sqrt{3} - 7\sqrt{27} = \sqrt{3} - 21\sqrt{3} = -20\sqrt{3}$

71. $\sqrt{75} + \sqrt{27} = 5\sqrt{3} + 3\sqrt{3} = 8\sqrt{3}$

72. $\sqrt{3} \cdot \sqrt{6} = \sqrt{18} = \sqrt{9 \cdot 2} = \sqrt{9} \cdot \sqrt{2} = 3\sqrt{2}$

73. $\sqrt{8} \cdot \sqrt{6} = \sqrt{48} = \sqrt{16 \cdot 3} = \sqrt{16} \cdot \sqrt{3} = 4\sqrt{3}$

74. $\frac{\sqrt{18}}{\sqrt{2}} = \sqrt{\frac{18}{2}} = \sqrt{9} = 3$

75. $\frac{\sqrt{56}}{\sqrt{2}} = \sqrt{\frac{56}{2}} = \sqrt{28} = 2\sqrt{7}$

76. $\frac{4}{\sqrt{3}} \cdot \frac{\sqrt{3}}{\sqrt{3}} = \frac{4\sqrt{3}}{3}$

77. $\frac{\sqrt{3}}{\sqrt{5}} \cdot \frac{\sqrt{5}}{\sqrt{5}} = \frac{\sqrt{15}}{5}$

78. $3(2 + \sqrt{7}) = 6 + 3\sqrt{7}$

79. $\sqrt{3}(4 + \sqrt{6}) = 4\sqrt{3} + \sqrt{18} = 4\sqrt{3} + 3\sqrt{2}$

80. $\sqrt{3}(\sqrt{6} + \sqrt{15}) = \sqrt{18} + \sqrt{45} = 3\sqrt{2} + 3\sqrt{5}$

81. $x + 2 = 2 + x$ Commutative property of addition

82. $5 - m = m - 5$ Commutative property of multiplication

83. Associative property of addition

84. Distributive property

85. Commutative property of addition

86. Commutative property of addition

87. Associative property of multiplication

88. Commutative property of multiplication

89. Distributive property

90. Commutative property of multiplication

91. Natural numbers – closed for addition $3 + 4 = 7$

92. Whole numbers – not closed for subtraction
$$3 - 2 = 2 - 3 \qquad 1 \neq -1$$

93. Not closed; $1 \div 2$ is not an integer

94. Closed

95. Not closed; $\sqrt{2} \cdot \sqrt{2} = 2$ is not irrational

96. Not closed; $1 \div 0$ is undefined

97. $3^2 = 3 \bullet 3 = 9$

98. $3\,3^{-2} = \dfrac{1}{3^2} = \dfrac{1}{3 \bullet 3} = \dfrac{1}{9}$

99. $\dfrac{9^5}{9^3} = 9^{5-3} = 9^2 = 81$

100. $5^2 \bullet 5^1 = 5^3 = 125$

101. $7^0 = 1$

102. $4^{-3} = \dfrac{1}{4^3} = \dfrac{1}{64}$

103. $(2^3)^2 = 2^{3 \bullet 2} = 2^6 = 64$

104. $(3^2)^2 = 3^{2 \bullet 2} = 3^4 = 81$

105. $230,000 = 2.3 \times 10^5$

106. $0.0000158 = 1.58 \times 10^{-5}$

107. $0.00275 = 2.75 \times 10^{-3}$

108. $4,950,000 = 4.95 \times 10^6$

109. $4.3 \times 10^7 = 43,000,000$

110. $1.39 \times 10^{-4} = 0.000139$

111. $1.75 \times 10^{-4} = 0.000175$

112. $1 \times 10^5 = 100,000$

113. a) $(7 \times 10^3)(2 \times 10^{-5})$
$(14) \times 10^{-2} = 1.4 \times 10^{-1}$

114. a) $(4 \times 10^2)(2.5 \times 10^2)$
$(4)(2.5) \times (10^2 \sqcup 10^2)$
$10.0 \times 10^4 = 1.0 \times 10^5$

115. $\dfrac{8.4 \times 10^3}{4 \times 10^2} = \dfrac{8.4}{4} \times \dfrac{10^3}{10^2} = 2.1 \times 10^1$

116. $\dfrac{1.5 \times 10^{-3}}{5 \times 10^{-4}} = \dfrac{1.5}{5} \times \dfrac{10^{-3}}{10^{-4}} = 0.3 \times 10^1 = 3.0 \times 10^0$

117. a) $(4,000,000)(2,000) = (4.0 \times 10^6)(2.0 \times 10^3)$
$= (4)(2) \times 10^6 \bullet 10^3 = 8.0 \times 10^9$
   b) $8.0 \, \text{E} \, 09$

118. a) $(35,000)(0.00002) = (3.5 \times 10^4)(2.0 \times 10^{-5})$
$= (3.5)(2) \times 10^4 \bullet 10^{-5} = 7.0 \times 10^{-1} = 0.7$
   b) $7.0 \, \text{E} \, {-01}$

119. $\dfrac{9600000}{3000} = \dfrac{9.6 \times 10^6}{3 \times 10^3} = 3.2 \times 10^3 = 3,200$

120. $\dfrac{0.000002}{0.0000004} = \dfrac{2 \times 10^{-6}}{4 \times 10^{-7}} = 0.5 \times 10^1 = 5.0$

121. $\dfrac{1.49 \times 10^{11}}{3.84 \times 10^8} = .3880208333 \times 10^3 = 388.02$
388 times

122. $\dfrac{20,000,000}{3,600} = \dfrac{2.0 \times 10^7}{3.6 \times 10^3}$
$\approx 0.555556 \times 10^4 = \$5,555.56$

123. Arithmetic   14, 17

124. Geometric   8, 16

125. Arithmetic $-15, -18$

126. Geometric   1/32, 1/64

127. Arithmetic   16, 19

128. Geometric $-2, 2$

129. 3, 7, 11, 15 $a_4 = 15$

130. $-4, -10, -14, -18, -22, -26, -30, -34$ $a_8 = -34$

131. $-20, -15, -10, -5, 0, 5, 10, 15, 20$ $a_{10} = 25$

132. 3, 6, 12, 24, 48 $a_4 = 48$

133. $a_5 = 4(1/2)^{5-1} = 4(1/2)^4 = 4(1/16) = 1/4$

134. $a_4 = -6(2)^{4-1} = -6(2)^3 = -6(8) = -48$

135. $s_{30} = \dfrac{30(2+89)}{2} = (15)(91) = 1365$

136. $s_8 = \dfrac{8\left(-4+\left(-2\frac{1}{4}\right)\right)}{2} = \dfrac{(8)\left(-6\frac{1}{4}\right)}{2} = -25$

137. $s_8 = \dfrac{8(100+58)}{2} = \dfrac{(8)(158)}{2} = 632$

138. $s_{20} = \dfrac{20(0.5+5.25)}{2} = \dfrac{(20)(5.75)}{2} = 57.5$

139. $s_3 = \dfrac{5\left(1-3^4\right)}{1-3} = \dfrac{(5)(1-81)}{-2} = \dfrac{(5)(-80)}{-2} = 200$

140. $s_4 = \dfrac{2\left(1-3^4\right)}{1-3} = \dfrac{(2)(1-81)}{-2} = \dfrac{(2)(-80)}{-2} = 80$

141. $s_5 = \dfrac{3\left(1-(-2)^5\right)}{1-(-2)} = \dfrac{(3)(1+32)}{3} = \dfrac{(3)(33)}{3} = 33$

142. $s_6 = \dfrac{1\left(1-(-2)^6\right)}{1-(-2)} = \dfrac{(1)(1-64)}{3} = \dfrac{(1)(-63)}{3} = -21$

143. Arithmetic: $a_n = -3n + 10$

144. Arithmetic: $a_n = 3 + (n-1)3 = 3 + 3n - 3 = 3n$

145. Arithmetic: $a_n = -(3/2)n + (11/2)$

146. Geometric: $a_n = 3(2)^{n-1}$

147. Geometric: $a_n = 2(-1)^{n-1}$

148. Geometric: $a_n = 5(1/3)^{n-1}$

149. Yes;   13, 21

150. Yes;   17, 28

151. No;   1,4,3,-1,-4,-5

152. No

Chapter Test

1. 38,610 is divisible by: 2, 3, 5, 6, 9, 10

2.

```
    2 | 840
    2 | 420
    2 | 210
    5 | 105
    3 | 35
          7
```
$840 = 2^3 \bullet 3 \bullet 5 \bullet 7$

3. $[(-6) + (-9)] + 8 = -15 + 8 = -7$

4. $-7 - 13 = -20$

5. $[(-70)(-5)] \div (8-10) = 350 \div [8 + (-10)]$
   $= 350 \div (-2) = -175$

6. $4\frac{5}{8} = \dfrac{(8)(4)+5}{8} = \dfrac{32+5}{8} = \dfrac{37}{8}$

7. $\dfrac{176}{9} = \dfrac{(19)(9)+5}{9} = 19\frac{5}{9}$

8. $\dfrac{5}{8} = 0.625$

9. $6.45 = \dfrac{645}{100} = \dfrac{129}{20}$

10. $\left(\dfrac{5}{16} \div 3\right) + \left(\dfrac{4}{5} \bullet \dfrac{1}{2}\right) = \left(\dfrac{5}{16} \bullet \dfrac{1}{3}\right) + \dfrac{4}{10}$

$= \dfrac{5}{48} + \dfrac{4}{10} = \dfrac{50}{480} + \dfrac{192}{480} = \dfrac{242}{480} = \dfrac{141}{240}$

11. $\frac{11}{12} - \frac{3}{8} = \left(\frac{11}{12}\right)\left(\frac{2}{2}\right) - \left(\frac{3}{8}\right)\left(\frac{3}{3}\right) = \frac{22}{24} - \frac{9}{24} = \frac{13}{24}$

12. $\sqrt{75} + \sqrt{48} = \sqrt{25}\sqrt{3} + \sqrt{16}\sqrt{3} = 5\sqrt{3} + 4\sqrt{3} = 9\sqrt{3}$

13. $\frac{\sqrt{2}}{\sqrt{7}} = \frac{\sqrt{2}}{\sqrt{7}} \cdot \frac{\sqrt{7}}{\sqrt{7}} = \frac{\sqrt{14}}{\sqrt{49}} = \frac{\sqrt{14}}{7}$

14. The integers are closed under multiplication since the product of two integers is always an integer.

15. Associative property of addition

16. Distributive property

17. $\frac{4^5}{4^2} = 4^{5-2} = 4^3 = 64$

18. $4^3 \bullet 4^2 = 4^5 = 4 \cdot 4 \cdot 4 \cdot 4 \cdot 4 = 1024$

19. $3^{-4} = \frac{1}{3^4} = \frac{1}{81}$

20. $\frac{7.2 \times 10^6}{9.0 \times 10^{-6}} = 0.8 \times 10^{12} = 8.0 \times 10^{11}$

21. $a_n = -4n + 2$

22. $\frac{11\left[-2 + (-32)\right]}{2} = \frac{11(-34)}{2} = -187$

23. $a_5 = 3(3)^4 = 3^5 = 243$

24. $\frac{3\left(1 - 4^5\right)}{1 - 4} = \frac{3(1 - 1024)}{-3} = 1023$

25. $a_n = 3 \bullet (2)^{n-1}$

26. 1, 1, 2, 3, 5, 8, 13, 21, 34, 55

Group Projects

1. In this exercise, you may obtain different answers depending upon how you work the problem.
    1) a) 2 servings   Rice: 2/3 cup, Salt: 1/4 tsp., Butter: 1 tsp.
       b) 1 serving   Rice: 1/3 cup, Salt: 1/8 tsp., Butter: 1/2 tsp.
       c) 29 servings Rice: 5 cup, Salt: 1 7/8 tsp., Butter: 7 1/2 tsp.

2. a) Area of triangle 1 = $A_1 = \frac{1}{2}bh = \frac{1}{2}(5)\left(2\sqrt{5}\right) = 5\sqrt{5}$

    Area of triangle 2 = $A_1 = \frac{1}{2}bh = \frac{1}{2}(5)\left(2\sqrt{5}\right) = 5\sqrt{5}$

    Area of rectangle = $A_R = bh = (10)\left(2\sqrt{5}\right) = 20\sqrt{5}$

   b) Area of trapezoid = $A_T = \frac{1}{2}h(b_1 + b_2) = \frac{1}{2}\left(2\sqrt{5}\right)(10 + 21) = 31\sqrt{5}$

   c) Yes, same

3. Co-pay for prescriptions = 50%   Co=pay for office visits = $10   Co-pay for medical tests = 20%

01/10: $10 + .50 ($44) = $32.00

02/27: $10 + .20 (348) = $47.60

04/19: $10 + .20 (348) + .50 (76) = $117.60

a) Total = $197.20

b) .50 (44) + .80 (188) + .80 (348) + ,50 (76) = $488.80

c) $500.00 − 197.20 = $302.80

4. a) 1 branch   b) 8 branches   c) 512 branches   d) Yes

CHAPTER SIX

ALGEBRA, GRAPHS, AND FUNCTIONS

Exercise Set 6.1

1. **Variables** are letters of the alphabet used to represent numbers.

2. A symbol that represents a specific quantity is called a **constant**.

3. The **solution** to an equation is the number or numbers that replace the variable to make the equation a true statement.

4. An **algebraic expression** is a collection of variables, numbers, parentheses, and operation symbols. An example is $5x^2y - 11$.

5. a) Base: 4, exponent: 5
   b) Multiply 4 by itself 5 times.

6. First: Perform all operations within parentheses or other grouping symbols. Next: Perform all exponential operations. Next: Perform all multiplication and division from left to right. Finally: Perform all addition and subtraction from left to right.

7. $8 + 16 \div 4 = 8 + 4 = 12$

8. $9 + 6 \cdot 3 = 9 + 18 = 27$

9. $x = 7, \ x^2 = (7)^2 = 49$

10. $x = -8, \ x^2 = (-8)^2 = 64$

11. $x = -3, -x^2 = -(-3)^2 = -9$

12. $x = -5, -x^2 = -(-5)^2 = -25$

13. $x = -7, -2x^3 = -2(-7)^3 = -2(-343) = 686$

14. $x = -4, -x^3 = -(-4)^3 = -(-64) = 64$

15. $x = 4, x - 7 = 4 - 7 = -3$

16. $x = \dfrac{5}{2}, 8x - 3 = 8\left(\dfrac{5}{2}\right) - 3 = 20 - 3 = 17$

17. $x = -2, \ -7x + 4 = -7(-2) + 4 = 14 + 4 = 18$

18. $x = 5, \ x^2 - 3x + 8 = (5)^2 - 3(5) + 8 = 25 - 15 + 8 = 18$

19. $x = -2, \ -x^2 + 5x - 13 = -(-2)^2 + 5(-2) - 13$
    $$= -4 - 10 - 13 = -27$$

20. $x = -1, 5x^2 + 7x - 11 = 5(-1)^2 + 7(-1) - 11$
    $$= 5 - 7 - 11 = -13$$

21. $x = \dfrac{2}{3}, \dfrac{1}{2}x^2 - 5x + 2 = \dfrac{1}{2}\left(\dfrac{2}{3}\right)^2 - 5\left(\dfrac{2}{3}\right) + 2$

$$= \dfrac{1}{2}\left(\dfrac{4}{9}\right) - \dfrac{10}{3} + 2$$

$$= \dfrac{4}{18} - \dfrac{10}{3} + 2$$

$$= \dfrac{4}{18} - \dfrac{60}{18} + \dfrac{36}{18} = -\dfrac{20}{18} = -\dfrac{10}{9}$$

22. $x = \dfrac{1}{2}, \dfrac{2}{3}x^2 + x - 1 = \dfrac{2}{3}\left(\dfrac{1}{2}\right)^2 + \dfrac{1}{2} - 1$

$$= \dfrac{2}{3}\left(\dfrac{1}{4}\right) + \dfrac{1}{2} - 1$$

$$= \dfrac{2}{12} + \dfrac{1}{2} - 1$$

$$= \dfrac{2}{12} + \dfrac{6}{12} - \dfrac{12}{12} = -\dfrac{4}{12} = -\dfrac{1}{3}$$

23. $x = \frac{1}{2}, 8x^3 - 4x^2 + 7 = 8\left(\frac{1}{2}\right)^3 - 4\left(\frac{1}{2}\right)^2 + 7$

$$= 8\left(\frac{1}{8}\right) - 4\left(\frac{1}{4}\right) + 7$$
$$= 1 - 1 + 7 = 7$$

24. $x = 2, y = 3, -x^2 + 4xy = -(2)^2 + 4(2)(3)$

$$= -4 + 24 = 20$$

25. $x = -2, \ y = 1, \ 2x^2 + xy + 3y^2$

$$= 2(-2)^2 + (-2)(1) + 3(1)^2 = 8 - 2 + 3 = 9$$

26. $x = 2, \ y = 5, \ 3x^2 + \frac{2}{5}xy - \frac{1}{5}y^2$

$$= 3(2)^2 + \frac{2}{5}(2)(5) - \frac{1}{5}(5)^2$$
$$= 12 + 4 - 5 = 11$$

27. $x = 3, \ y = 2, \ 4x^2 - 12xy + 9y^2$

$$= 4(3)^2 - 12(3)(2) + 9(2)^2 = 36 - 72 + 36 = 0$$

28. $x = 4, y = -3, (x + 3y)^2 = [4 + 3(-3)]^2$

$$= (-5)^2 = 25$$

29. $7x + 3 = 23, \ x = 3$

$$7(3) + 3 = 21 + 3 = 24$$
$24 \neq 23, x = 3$ is not a solution.

30. $5x - 7 = -27, \ x = -4$

$$5(-4) - 7 = -20 - 7 = -27$$
$-27 = -27, x = -4$ is a solution.

31. $x - 3y = 0, \ x = 6, \ y = 3$

$$6 - 3(3) = 6 - 9 = -3$$
$-3 \neq 0, x = 6, y = 3$ is not a solution.

32. $4x + 2y = -2, \ x = -2, \ y = 3$

$$4(-2) + 2(3) = -8 + 6 = -2$$
$-2 = -2, x = -2, y = 3$ is a solution.

33. $x^2 + 3x - 4 = 5, x = 2$

$$(2)^2 + 3(2) - 4 = 4 + 6 - 4 = 6$$
$6 \neq 5, x = 2$ is not a solution.

34. $2x^2 - x - 5 = 0, x = 3$

$$2(3)^2 - 3 - 5 = 2(9) - 3 - 5 = 10$$
$10 \neq 0, x = 3$ is not a solution.

35. $2x^2 + x = 28, x = -4$

$$2(-4)^2 + (-4) = 2(16) - 4 = 32 - 4 = 28$$
$28 = 28, x = -4$ is a solution.

36. $y = x^2 + 3x - 5, x = 1, y = -1$

$$(1)^2 + 3(1) - 5 = 1 + 3 - 5 = -1$$
$-1 = -1, x = 1, y = -1$ is a solution.

37. $y = -x^2 + 3x - 1, \ x = 3, \ y = -1$

$$-(3)^2 + 3(3) - 1 = -9 + 9 - 1 = -1$$
$-1 = -1, x = 3, y = -1$ is a solution.

38. $y = x^3 - 3x^2 + 1, x = 2, y = -3$

$$(2)^3 - 3(2)^2 + 1 = 8 - 12 + 1 = -3$$
$-3 = -3, x = 2, y = -3$ is a solution.

39. $d = \$175, 0.07d = 0.07(\$175) = \$12.25$

40. $t = 3, 0.5t = 0.5(3) = 1.5$ ft

41. $x = 75, 220 + 2.75x = 220 + 2.75(75)$

$$= 220 + 206.25 = \$426.25$$

42. $x = 60, 25x - 0.2x^2 = 25(60) - 0.2(60)^2$

$$= 1500 - 0.2(3600)$$
$$= 1500 - 720$$
$$= 780 \text{ baskets of oranges}$$

43. $n = 8,000,000,000,000$

$0.000002n = 0.000002(8,000,000,000,000)$

$\qquad = 16,000,000 \text{ sec}$

44. $h = 0.60, 2h^2 + 80h + 40 = 2(0.60)^2 + 80(0.60) + 40$

$\qquad = 2(0.36) + 48 + 40$

$\qquad = 0.72 + 48 + 40$

$\qquad = 88.72 \text{ min}$

45. $R = 2, T = 70, 0.2R^2 + 0.003RT + 0.0001T^2 = 0.2(2)^2 + 0.003(2)(70) + 0.0001(70)^2 = 0.8 + 0.42 + 0.49 = 1.71 \text{ in.}$

46. $(-1)^n = 1$ for any even number, n, since there will be an even number of factors of (-1), and when these are multiplied, the product will always be 1.

47.

x	y	$(x+y)^2$	$x^2 + y^2$
2	3	$5^2 = 25$	$4 + 9 = 13$
−2	−3	$(-5)^2 = 25$	$4 + 9 = 13$
−2	3	$1^2 = 1$	$4 + 9 = 13$
2	−3	$(-1)^2 = 1$	$4 + 9 = 13$

The two expressions are not equal.

48. $1^n = 1$ for all natural numbers since 1 multiplied by itself any number of times will always be 1.

Exercise Set 6.2

1. The parts that are added or subtracted in an algebraic expression are called **terms**.

   In $3x - 2y$, the $3x$ and $-2y$ are terms.

2. **Like terms** are terms that have the same variables with the same exponents on the variables.

   $3x^2$ and $4x^2$ are like terms.

3. The numerical part of a term is called its **numerical coefficient.**

   For the term $3x$, 3 is the numerical coefficient.

4. A **linear equation** is one in which the exponent on the variable is 1. Example: $4x + 6 = 10$

5. To **simplify** an expression means to combine like terms by using the commutative, associative, and distributive properties. Example: $12 + x + 7 - 3x = x - 3x + 12 + 7 = -2x + 19$

6. If $a = b$, then $a + c = b + c$ for all real numbers a, b, and c. Example: If $x - 5 = 2$, then $x - 5 + 5 = 2 + 5$.

7. If $a = b$, then $a - c = b - c$ for all real numbers a, b, and c. Example: If $2x + 3 = 5$, then $2x + 3 - 3 = 5 - 3$.

8. If $a = b$, then $a \cdot c = b \cdot c$ for all real numbers a, b, and c, where $c \neq 0$. Example: If $\dfrac{x}{3} = 2$, then $3\left(\dfrac{x}{3}\right) = 3(2)$.

9. If $a = b$, then $\dfrac{a}{c} = \dfrac{b}{c}$ for all real numbers a, b, and c, where $c \neq 0$. Example: If $4x = 8$ then $\dfrac{4x}{4} = \dfrac{8}{4}$.

10. An **algorithm** is a general procedure for accomplishing a task.

11. A **ratio** is a quotient of two quantities. Example: $\dfrac{7}{9}$

12. A **proportion** is a statement of equality between two ratios. Example: $\dfrac{3}{7} = \dfrac{x}{10}$

13. Yes. They have the same variable and the same exponent on the variable.

14. No. They do not have the same variable.

15. $2x + 9x = 11x$

16. $-4x - 7x = -11x$

17. $5x - 3x + 12 = 2x + 12$

18. $-6x + 3x + 21 = -3x + 21$

19. $7x+3y-4x+8y=3x+11y$

20. $x-4x+3=-3x+3$

21. $-3x+2-5x=-8x+2$

22. $-3x+4x-2+5=x+3$

23. $2-3x-2x+1=-5x+3$

24. $-0.2x+1.7x-4=1.5x-4$

25. $6.2x-8.3+7.1x=13.3x-8.3$

26. $\dfrac{2}{3}x+\dfrac{1}{6}x-5=\dfrac{4}{6}x+\dfrac{1}{6}x-5=\dfrac{5}{6}x-5$

27. $\dfrac{1}{5}x-\dfrac{1}{3}x-4=\dfrac{3}{15}x-\dfrac{5}{15}x-4=-\dfrac{2}{15}x-4$

28. $7t+5s+9-3t-2s-12=4t+3s-3$

29. $5x-4y-3y+8x+3=13x-7y+3$

30. $3(p+2)-4(p+3)=3p+6-4p-12=-p-6$

31. $2(s+3)+6(s-4)+1=2s+6+6s-24+1=8s-17$

32. $6(r-3)-2(r+5)+10=6r-18-2r-10+10$
$=4r-18$

33. $0.3(x+2)+1.2(x-4)=0.3x+0.6+1.2x-4.8$
$=1.5x-4.2$

34. $\dfrac{1}{5}(x+2)-\dfrac{1}{10}x=\dfrac{1}{5}x+\dfrac{2}{5}-\dfrac{1}{10}x$

$=\dfrac{2}{10}x-\dfrac{1}{10}x+\dfrac{2}{5}=\dfrac{1}{10}x+\dfrac{2}{5}$

35. $\dfrac{2}{3}x+\dfrac{3}{7}-\dfrac{1}{4}x=\dfrac{8}{12}x-\dfrac{3}{12}x+\dfrac{3}{7}=\dfrac{5}{12}x+\dfrac{3}{7}$

36. $n-\dfrac{3}{4}+\dfrac{5}{9}n-\dfrac{1}{6}=\dfrac{9}{9}n+\dfrac{5}{9}n-\dfrac{9}{12}-\dfrac{2}{12}=\dfrac{14}{9}n-\dfrac{11}{12}$

37. $0.5(2.6x-4)+2.3(1.4x-5)=1.3x-2+3.22-11.5$
$=4.52x-13.5$

38. $\dfrac{2}{3}(3x+9)-\dfrac{1}{4}(2x+5)=2x+6-\dfrac{1}{2}x-\dfrac{5}{4}$

$=\dfrac{4}{2}x-\dfrac{1}{2}x+\dfrac{24}{4}-\dfrac{5}{4}=\dfrac{3}{2}x+\dfrac{19}{4}$

39.

$$y+8=13$$
$$y+8-8=13-8 \qquad \text{Subtract 8 from both sides of the equation.}$$
$$y=5$$

40.

$$2y-7=17$$
$$2y-7+7=17+7 \qquad \text{Add 7 to both sides of the equation.}$$
$$2y=24$$
$$\dfrac{2y}{2}=\dfrac{24}{2} \qquad \text{Divide both sides of the equation by 2.}$$
$$y=12$$

41.

$$9=12-3x$$
$$9-12=12-12-3x \qquad \text{Subtract 12 from both sides of the equation.}$$
$$-3=-3x$$
$$\dfrac{-3}{-3}=\dfrac{-3x}{-3} \qquad \text{Divide both sides of the equation by -3.}$$
$$1=x$$

42.

$$14=3x+5$$
$$14-5=3x+5-5 \qquad \text{Subtract 5 from both sides of the equation.}$$
$$9=3x$$
$$\dfrac{9}{3}=\dfrac{3x}{3} \qquad \text{Divide both sides of the equation by 3.}$$
$$3=x$$

43.

$$\frac{3}{x} = \frac{7}{8}$$

$3(8) = 7x$ Cross multiplication

$24 = 7x$

$$\frac{24}{7} = \frac{7x}{7}$$ Divide both sides of the equation by 7.

$$\frac{24}{7} = x$$

44.

$$\frac{x-1}{5} = \frac{x+5}{15}$$

$15(x-1) = 5(x+5)$ Cross multiplication

$15x - 15 = 5x + 25$ Distributive Property

$15x - 5x - 15 = 5x - 5x + 25$ Subtract $5x$ from both sides of the equation.

$10x - 15 = 25$

$10x - 15 + 15 = 25 + 15$ Add 15 to both sides of the equation.

$10x = 40$

$$\frac{10x}{10} = \frac{40}{10}$$ Divide both sides of the equation by 10.

$x = 4$

45.

$$\frac{1}{2}x + \frac{1}{3} = \frac{2}{3}$$

$$6\left(\frac{1}{2}x + \frac{1}{3}\right) = 6\left(\frac{2}{3}\right)$$ Multiply both sides of the equation by the LCD.

$3x + 2 = 4$ Distributive Property

$3x + 2 - 2 = 4 - 2$ Subtract 2 from both sides of the equation.

$3x = 2$

$$\frac{3x}{3} = \frac{2}{3}$$ Divide both sides of the equation by 3.

$$x = \frac{2}{3}$$

46.

$$\frac{1}{2}y + \frac{1}{3} = \frac{1}{4}$$

$$12\left(\frac{1}{2}y + \frac{1}{3}\right) = 12\left(\frac{1}{4}\right)$$ Multiply both sides of the equation by the LCD.

$6y + 4 = 3$ Distributive Property

$6y + 4 - 4 = 3 - 4$ Subtract 4 from both sides of the equation.

$6y = -1$

$$\frac{6y}{6} = \frac{-1}{6}$$ Divide both sides of the equation by 6.

$$y = -\frac{1}{6}$$

47.
$$0.7x - 0.3 = 1.8$$
$$0.7x - 0.3 + 0.3 = 1.8 + 0.3 \qquad \text{Add 0.3 to both sides of the equation.}$$
$$0.7x = 2.1$$
$$\frac{0.7x}{0.7} = \frac{2.1}{0.7} \qquad \text{Divide both sides of the equation by 0.7.}$$
$$x = 3$$

48.
$$5x + 0.050 = -0.732$$
$$5x + 0.050 - 0.050 = -0.732 - 0.050 \qquad \text{Subtract 0.050 from both sides of the equation.}$$
$$5x = -0.782$$
$$\frac{5x}{5} = \frac{-0.782}{5} \qquad \text{Divide both sides of the equation by 5.}$$
$$x = -0.1564$$

49.
$$6t - 8 = 4t - 2$$
$$6t - 4t - 8 = 4t - 4t - 2 \qquad \text{Subtract } 4t \text{ from both sides of the equation.}$$
$$2t - 8 = -2$$
$$2t - 8 + 8 = -2 + 8 \qquad \text{Add 8 to both sides of the equation.}$$
$$2t = 6$$
$$\frac{2t}{2} = \frac{6}{2} \qquad \text{Divide both sides of the equation by 2.}$$
$$t = 3$$

50.
$$\frac{x}{4} + 2x = \frac{1}{3}$$
$$12\left(\frac{x}{4} + 2x\right) = 12\left(\frac{1}{3}\right) \qquad \text{Mulitply both sides of the equation by the LCD.}$$
$$3x + 24x = 4 \qquad \text{Distributive Property}$$
$$27x = 4$$
$$\frac{27x}{27} = \frac{4}{27} \qquad \text{Divide both sides of the equation by 27.}$$
$$x = \frac{4}{27}$$

51.
$$\frac{x-3}{2} = \frac{x+4}{3}$$
$$3(x-3) = 2(x+4) \qquad \text{Cross multiplication}$$
$$3x - 9 = 2x + 8 \qquad \text{Distributive Property}$$
$$3x - 2x - 9 = 2x - 2x + 8 \qquad \text{Subtract } 2x \text{ from both sides of the equation.}$$
$$x - 9 = 8$$
$$x - 9 + 9 = 8 + 9 \qquad \text{Add 9 to both sides of the equation.}$$
$$x = 17$$

52.
$$\frac{x-5}{4}=\frac{x-9}{3}$$
$$3(x-5)=4(x-9) \quad \text{Cross multiplication}$$
$$3x-15=4x-36 \quad \text{Distributive Property}$$
$$3x-3x-15=4x-3x-36 \quad \text{Subtract } 3x \text{ from both sides of the equation.}$$
$$-15=x-36$$
$$-15+36=x-36+36 \quad \text{Add 36 to both sides of the equation.}$$
$$21=x$$

53.
$$6t-7=8t+9$$
$$6t-6t-7=8t-6t+9 \quad \text{Subtract } 6t \text{ from both sides of the equation.}$$
$$-7=2t+9$$
$$-7-9=2t+9-9 \quad \text{Subtract 9 from both sides of the equation.}$$
$$-16=2t$$
$$\frac{-16}{2}=\frac{2t}{2} \quad \text{Divide both sides of the equation by 2.}$$
$$-8=t$$

54.
$$12x-1.2=3x+1.5$$
$$12x-3x-1.2=3x-3x+1.5 \quad \text{Subtract } 3x \text{ from both sides of the equation.}$$
$$9x-1.2=1.5$$
$$9x-1.2+1.2=1.5+1.2 \quad \text{Add 1.2 to both sides of the equation.}$$
$$9x=2.7$$
$$\frac{9x}{9}=\frac{2.7}{9} \quad \text{Divide both sides of the equation by 9.}$$
$$x=0.3$$

55.
$$2(x+3)-4=2(x-4)$$
$$2x+6-4=2x-8 \quad \text{Distributive Property}$$
$$2x+2=2x-8$$
$$2x-2x+2=2x-2x-8 \quad \text{Subtract } 2x \text{ from both sides of the equation.}$$
$$2=-8 \quad \text{False}$$
$$\text{No solution}$$

56.
$$3(x+2)+2(x-1)=5x-7$$
$$3x+6+2x-2=5x-7 \quad \text{Distributive Property}$$
$$5x+4=5x-7$$
$$5x-5x+4=5x-5x-7 \quad \text{Subtract } 5x \text{ from both sides of the equation.}$$
$$4=-7 \quad \text{False}$$
$$\text{No solution}$$

57.

$$4(x-4)+12=4(x-1)$$
$$4x-16+12=4x-4 \qquad \text{Distributive Property}$$
$$4x-4=4x-4$$

This equation is an identity. Therefore, the solution is all real numbers.

58.

$$\frac{x}{3}+4=\frac{2x}{5}-6$$
$$15\left(\frac{x}{3}+4\right)=15\left(\frac{2x}{5}-6\right) \qquad \text{Multiply both sides of the equation by the LCD.}$$
$$5x+60=6x-90 \qquad \text{Distributive Property}$$
$$5x-5x+60=6x-5x-90 \qquad \text{Subtract } 5x \text{ from both sides of the equation.}$$
$$60=x-90$$
$$60+90=x-90+90 \qquad \text{Add 90 to both sides of the equation.}$$
$$150=x$$

59.

$$\frac{1}{4}(x+4)=\frac{2}{5}(x+2)$$
$$20\left(\frac{1}{4}\right)(x+4)=20\left(\frac{2}{5}\right)(x+2) \qquad \text{Multiply both sides of the equation by the LCD.}$$
$$5(x+4)=8(x+2)$$
$$5x+20=8x+16 \qquad \text{Distributive Property}$$
$$5x-8x+20=8x-8x+16 \qquad \text{Subtract } 8x \text{ from both sides of the equation.}$$
$$-3x+20=16$$
$$-3x+20-20=16-20 \qquad \text{Subtract 20 from both sides of the equation.}$$
$$-3x=-4$$
$$\frac{-3x}{-3}=\frac{-4}{-3} \qquad \text{Divide both sides of the equation by -3.}$$
$$x=\frac{4}{3}$$

60.

$$\frac{2}{3}(x+5)=\frac{1}{4}(x+2)$$
$$12\left(\frac{2}{3}\right)(x+5)=12\left(\frac{1}{4}\right)(x+2) \qquad \text{Multiply both sides of the equation by the LCD.}$$
$$8(x+5)=3(x+2)$$
$$8x+40=3x+6 \qquad \text{Distributive Property}$$
$$8x-3x+40=3x-3x+6 \qquad \text{Subtract } 3x \text{ from both sides of the equation.}$$
$$5x+40=6$$
$$5x+40-40=6-40 \qquad \text{Subtract 40 from both sides of the equation.}$$
$$5x=-34$$
$$\frac{5x}{5}=\frac{-34}{5} \qquad \text{Divide both sides of the equation by 5.}$$
$$x=-\frac{34}{5}$$

61.

$$3x+2-6x=-x-15+8-5x$$
$$-3x+2=-6x-7$$
$$-3x+6x+2=-6x+6x-7 \qquad \text{Add } 6x \text{ to both sides of the equation.}$$
$$3x+2=-7$$
$$3x+2-2=-7-2 \qquad \text{Subtract 2 from both sides of the equation.}$$
$$3x=-9$$
$$\frac{3x}{3}=\frac{-9}{3} \qquad \text{Divide both sides of the equation by 3.}$$
$$x=-3$$

62.

$$6x+8-22x=28+14x-10+12x$$
$$-16x+8=26x+18$$
$$-16x-26x+8=26x-26x+18 \qquad \text{Subtract } 26x \text{ from both sides of the equation.}$$
$$-42x+8=18$$
$$-42x+8-8=18-8 \qquad \text{Subtract 8 from both sides of the equation.}$$
$$-42x=10$$
$$\frac{-42x}{-42}=\frac{10}{-42} \qquad \text{Divide both sides of the equation by -42.}$$
$$x=-\frac{10}{42}=-\frac{5}{21}$$

63.

$$2(x-3)+2=2(2x-6)$$
$$2x-6+2=4x-12 \qquad \text{Distributive Property}$$
$$2x-4=4x-12$$
$$2x-4x-4=4x-4x-12 \qquad \text{Subtract } 4x \text{ from both sides of the equation.}$$
$$-2x-4=-12$$
$$-2x-4+4=-12+4 \qquad \text{Add 4 to both sides of the equation.}$$
$$-2x=-8$$
$$\frac{-2x}{-2}=\frac{-8}{-2} \qquad \text{Divide both sides of the equation by -2.}$$
$$x=4$$

64.

$$5.7x-3.1(x+5)=7.3$$
$$5.7x-3.1x-15.5=7.3 \qquad \text{Distributive Property}$$
$$2.6x-15.5=7.3$$
$$2.6x-15.5+15.5=7.3+15.5 \qquad \text{Add 15.5 to both sides of the equation.}$$
$$2.6x=22.8$$
$$\frac{2.6x}{2.6}=\frac{22.8}{2.6} \qquad \text{Divide both sides of the equation by 2.6.}$$
$$x=\frac{22.8}{2.6}=\frac{228}{26}=\frac{114}{13} \text{ or } x\approx 8.7692$$

65.
$$\frac{2.05}{1000} = \frac{x}{35,300}$$
$$2.05(35,300) = 1000x$$
$$72,365 = 1000x$$
$$\frac{72,365}{1000} = \frac{1000x}{1000}$$
$$x = 72.365 \approx \$72.37$$

66.
$$\frac{2.05}{1000} = \frac{40.68}{x}$$
$$2.05x = 40.68(1000)$$
$$2.05x = 40,680$$
$$\frac{2.05x}{2.05} = \frac{40,680}{2.05}$$
$$x = 19,843.90244 \approx 19,844 \text{ gal}$$

67.
$$\frac{x}{354} = \frac{1}{6}$$
$$6x = 354$$
$$\frac{6x}{6} = \frac{354}{6}$$
$$x = 59 \text{ times}$$

68. a)
$$\frac{6}{9} = \frac{16}{x}$$
$$6x = 9(16)$$
$$6x = 144$$
$$\frac{6x}{6} = \frac{144}{6}$$
$$x = 24 \text{ oz}$$

b)
$$\frac{x}{32} = \frac{6}{16}$$
$$16x = 32(6)$$
$$16x = 192$$
$$\frac{16x}{16} = \frac{192}{16}$$
$$x = 12 \text{ servings}$$

69.
$$\frac{1}{1,022,000} = \frac{20.3}{x}$$
$$x = 1,022,000(20.3)$$
$$x = 20,746,600 \text{ households}$$

70. a)
$$\frac{20}{10,000} = \frac{x}{140,000}$$
$$20(140,000) = 10,000x$$
$$2,800,000 = 10,000x$$
$$\frac{2,800,000}{10,000} = \frac{10,000x}{10,000}$$
$$x = 280 \text{ lb}$$

b)
$$\frac{280}{20} = 14 \text{ bags}$$

71. a)
$$\frac{50}{80} = \frac{1}{x}$$
$$50x = 80$$
$$\frac{50x}{50} = \frac{80}{50}$$
$$x = 1.6 \text{ kph}$$

b)
$$\frac{50}{80} = \frac{x}{90}$$
$$80x = 50(90)$$
$$80x = 4500$$
$$\frac{80x}{80} = \frac{4500}{80}$$
$$x = 56.25 \text{ mph}$$

72.
$$\frac{40}{0.6} = \frac{250}{x}$$
$$40x = 0.6(250)$$
$$40x = 150$$
$$\frac{40x}{40} = \frac{150}{40}$$
$$x = 3.75 \text{ mm}$$

73. $\dfrac{40}{1}=\dfrac{12}{x}$

$40x = 12$

$\dfrac{40x}{40}=\dfrac{12}{40}$

$x = 0.3$ cc

74. $\dfrac{40}{1}=\dfrac{35}{x}$

$40x = 35$

$\dfrac{40x}{40}=\dfrac{35}{40}$

$x = 0.875$ cc

75. a) Answers will vary.

b) $2(x+3)=4x+3-5x$

$2x+6=-x+3$ Distributive Property

$2x+x+6=-x+x+3$ Add x to both sides of the equation.

$3x+6=3$

$3x+6-6=3-6$ Subtract 6 from both sides of the equation.

$3x=-3$

$\dfrac{3x}{3}=\dfrac{-3}{3}$ Divide both sides of the equation by 3.

$x=-1$

76. a) An **identity** is an equation that has an infinite number of solutions.

b) When solving an equation, if you have the same expressions on both sides of the equal sign, the equation is an identity.

77. a) An **inconsistent equation** is an equation that has no solution.

b) When solving an equation, if you obtain a false statement, then the equation is inconsistent.

78. a) $P = 14.70 + 0.43x$

$148 = 14.70 + 0.43x$ Given $P = 148$, find x.

$148 - 14.70 = 14.70 - 14.70 + 0.43x$ Subtract 14.70 from both sides of the equation.

$133.3 = 0.43x$

$\dfrac{133.3}{0.43}=\dfrac{0.43x}{0.43}$ Divide both sides of the equation by 0.43.

$x = 310$ ft

b) $P = 14.70 + 0.43x$

$128.65 = 14.70 + 0.43x$ Given $P = 128.65$, find x.

$128.65 - 14.70 = 14.70 - 14.70 + 0.43x$ Subtract 14.70 from both sides of the equation.

$113.95 = 0.43x$

$\dfrac{113.95}{0.43}=\dfrac{0.43x}{0.43}$ Divide both sides of the equation by 0.43.

$x = 265$ ft down

79. a) 2:5; There are 2 males and a total of $2 + 3 = 5$ students.

b) $m : m + n$

Exercise Set 6.3

1. A **formula** is an equation that typically has a real-life application.

2. To **evaluate a formula**, substitute the given values for their respective variables, then evaluate.

3. **Subscripts** are numbers (or letters) placed below and to the right of variables. They are used to help clarify a formula.

4. $i = prt$

5. An **exponential equation** is of the form $y = a^x, a > 0, a \neq 1$.

6. a) $a > 0, a \neq 1$

   b) P_0 represents the original amount present.

7. $P = 4s = 4(5) = 20$

8. $P = a + b + c = 25 + 53 + 32 = 110$

9. $P = 2l + 2w$
   $P = 2(12) + 2(16) = 24 + 32 = 56$

10. $F = MA$
    $40 = M(5)$
    $\dfrac{40}{5} = \dfrac{5M}{5}$
    $8 = M$

11. $E = mc^2$
    $400 = m(4)^2$
    $400 = 16m$
    $\dfrac{400}{16} = \dfrac{16m}{16}$
    $25 = m$

12. $p = i^2 r$
    $62{,}500 = (5)^2 r$
    $62{,}500 = 25r$
    $\dfrac{62{,}500}{25} = \dfrac{25r}{25}$
    $2500 = r$

13. $A = \pi(R^2 - r^2)$
    $A = 3.14\left((6)^2 - (4)^2\right)$
    $A = 3.14(36 - 16)$
    $A = 3.14(20)$
    $A = 62.8$

14. $B = \dfrac{703w}{h^2}$
    $B = \dfrac{703(130)}{(67)^2}$
    $B = \dfrac{91{,}390}{4489} = 20.35865449 \approx 20.36$

15. $z = \dfrac{x - \mu}{\sigma}$
    $\dfrac{2.5}{1} = \dfrac{42.1 - \mu}{2}$
    $2.5(2) = 42.1 - \mu$
    $5 = 42.1 - \mu$
    $5 - 42.1 = 42.1 - 42.1 - \mu$
    $-37.1 = -\mu$
    $\dfrac{-37.1}{-1} = \dfrac{-\mu}{-1}$
    $37.1 = \mu$

16. $S = B + \dfrac{1}{2}Ps$
    $300 = 100 + \dfrac{1}{2}P(10)$
    $300 = 100 + 5P$
    $300 - 100 = 100 - 100 + 5P$
    $200 = 5P$
    $\dfrac{200}{5} = \dfrac{5P}{5}$
    $40 = P$

17. $T = \dfrac{PV}{k}$
    $\dfrac{80}{1} = \dfrac{P(20)}{0.5}$
    $80(0.5) = 20P$
    $40 = 20P$
    $\dfrac{40}{20} = \dfrac{20P}{20}$
    $2 = P$

18. $m = \dfrac{a + b + c}{3}$
    $70 = \dfrac{a + 60 + 90}{3}$
    $\dfrac{70}{1} = \dfrac{a + 150}{3}$
    $70(3) = a + 150$
    $210 = a + 150$
    $210 - 150 = a + 150 - 150$
    $60 = a$

19.
$$A = P(1+rt)$$
$$3600 = P(1+0.04(5))$$
$$3600 = P(1+0.2)$$
$$3600 = 1.2P$$
$$\frac{3600}{1.2} = \frac{1.2P}{1.2}$$
$$3000 = P$$

20.
$$m = \frac{a+b}{2}$$
$$70 = \frac{a+77}{2}$$
$$70(2) = a+77$$
$$140 = a+77$$
$$140-77 = a+77-77$$
$$63 = a$$

21.
$$V = \frac{1}{2}at^2$$
$$576 = \frac{1}{2}a(12)^2$$
$$\frac{576}{1} = \frac{144a}{2}$$
$$576(2) = 144a$$
$$1152 = 144a$$
$$\frac{1152}{144} = \frac{144a}{144}$$
$$8 = a$$

22.
$$F = \frac{9}{5}C+32$$
$$F = \frac{9}{5}(7)+32$$
$$F = \frac{63}{5}+32 = 12.6+32 = 44.6$$

23.
$$C = \frac{5}{9}(F-32)$$
$$C = \frac{5}{9}(77-32)$$
$$C = \frac{5}{9}(45) = 25$$

24.
$$K = \frac{F-32}{1.8}+273.1$$
$$K = \frac{100-32}{1.8}+273.1$$
$$K = \frac{68}{1.8}+273.1$$
$$K = 37.\overline{7}+273.1 = 310.8\overline{7} \approx 310.88$$

25.
$$m = \frac{y_2 - y_1}{x_2 - x_1}$$
$$m = \frac{8-(-4)}{-3-(-5)}$$
$$m = \frac{8+4}{-3+5} = \frac{12}{2} = 6$$

26.
$$z = \frac{\overline{x}-\mu}{\frac{\sigma}{\sqrt{n}}}$$
$$z = \frac{66-60}{\frac{15}{\sqrt{25}}}$$
$$z = \frac{6}{\frac{15}{5}} = \frac{6}{3} = 2$$

27.
$$S = R-rR$$
$$186 = 1R-0.07R$$
$$186 = 0.93R$$
$$\frac{186}{0.93} = \frac{0.93R}{0.93}$$
$$200 = R$$

28.
$$S = C+rC$$
$$115 = 1C+0.15C$$
$$115 = 1.15C$$
$$\frac{115}{1.15} = \frac{1.15C}{1.15}$$
$$100 = C$$

29.
$$E = a_1 p_1 + a_2 p_2 + a_3 p_3$$
$$E = 5(0.2) + 7(0.6) + 10(0.2)$$
$$E = 1 + 4.2 + 2 = 7.2$$

30.
$$x = \frac{-b + \sqrt{b^2 - 4ac}}{2a}$$
$$x = \frac{-(-5) + \sqrt{(-5)^2 - 4(2)(-12)}}{2(2)}$$
$$x = \frac{5 + \sqrt{25 + 96}}{4}$$
$$x = \frac{5 + \sqrt{121}}{4} = \frac{5 + 11}{4} = \frac{16}{4} = 4$$

31.
$$s = -16t^2 + v_0 t + s_0$$
$$s = -16(4)^2 + 30(4) + 150$$
$$s = -16(16) + 120 + 150$$
$$s = -256 + 120 + 150 = 14$$

32.
$$R = O + (V - D)r$$
$$670 = O + (100 - 10)(4)$$
$$670 = O + 360$$
$$670 - 360 = O + 360 - 360$$
$$310 = O$$

33.
$$P = \frac{f}{1 + i}$$
$$3000 = \frac{f}{1 + 0.08}$$
$$\frac{3000}{1} = \frac{f}{1.08}$$
$$3000(1.08) = f$$
$$3240 = f$$

34.
$$c = \sqrt{a^2 + b^2}$$
$$c = \sqrt{(5)^2 + (12)^2}$$
$$c = \sqrt{25 + 144}$$
$$c = \sqrt{169} = 13$$

35.
$$F = \frac{Gm_1 m_2}{r^2}$$
$$625 = \frac{G(100)(200)}{(4)^2}$$
$$625 = 1250G$$
$$\frac{625}{1250} = \frac{1250G}{1250}$$
$$0.5 = G$$

36.
$$P = \frac{nRT}{V}$$
$$12 = \frac{(10)(60)(8)}{V}$$
$$\frac{12}{1} = \frac{4800}{V}$$
$$12V = 4800$$
$$\frac{12V}{12} = \frac{4800}{12}$$
$$V = 400$$

37.
$$S_n = \frac{a_1(1 - r^n)}{1 - r}$$
$$S_n = \frac{8\left(1 - \left(\frac{1}{2}\right)^3\right)}{1 - \frac{1}{2}}$$
$$S_n = \frac{8\left(1 - \frac{1}{8}\right)}{1 - \frac{1}{2}}$$
$$S_n = \frac{8\left(\frac{7}{8}\right)}{\frac{1}{2}} = \frac{7}{\frac{1}{2}} = 7(2) = 14$$

38.
$$A = P\left(1 + \frac{r}{n}\right)^{nt}$$
$$A = 100\left(1 + \frac{0.06}{1}\right)^{1(3)}$$
$$A = 100(1 + 0.06)^3$$
$$A = 100(1.06)^3$$
$$A = 100(1.191016)$$
$$A = 119.1016 \approx 119.10$$

39.
$$10x - 4y = 13$$
$$10x - 10x - 4y = -10x + 13 \qquad \text{Subtract } 10x \text{ from both sides of the equation.}$$
$$-4y = -10x + 13$$
$$\frac{-4y}{-4} = \frac{-10x + 13}{-4} \qquad \text{Divide both sides of the equation by -4.}$$
$$y = \frac{-10x + 13}{-4} = \frac{-(-10x + 13)}{4}$$
$$= \frac{10x - 13}{4} = \frac{10x}{4} - \frac{13}{4} = \frac{5}{2}x - \frac{13}{4}$$

40.
$$8x - 6y = 21$$
$$8x - 8x - 6y = -8x + 21 \qquad \text{Subtract } 8x \text{ from both sides of the equation.}$$
$$-6y = -8x + 21$$
$$\frac{-6y}{-6} = \frac{-8x + 21}{-6} \qquad \text{Divide both sides of the equation by -6.}$$
$$y = \frac{-8x + 21}{-6} = \frac{-(-8x + 21)}{6} = \frac{8x - 21}{6} = \frac{8x}{6} - \frac{21}{6} = \frac{4}{3}x - \frac{7}{2}$$

41.
$$4x + 7y = 14$$
$$-4x + 4x + 7y = -4x + 14 \qquad \text{Subtract } 4x \text{ from both sides of the equation.}$$
$$7y = -4x + 14$$
$$\frac{7y}{7} = \frac{-4x + 14}{7} \qquad \text{Divide both sides of the equation by 7.}$$
$$y = \frac{-4x + 14}{7} = \frac{-4x}{7} + \frac{14}{7} = -\frac{4}{7}x + 2$$

42.
$$-2x + 4y = 9$$
$$-2x + 2x + 4y = 2x + 9 \qquad \text{Add } 2x \text{ to both sides of the equation.}$$
$$4y = 2x + 9$$
$$\frac{4y}{4} = \frac{2x + 9}{4} \qquad \text{Divide both sides of the equation by 4.}$$
$$y = \frac{2x + 9}{4} = \frac{2x}{4} + \frac{9}{4} = \frac{1}{2}x + \frac{9}{4}$$

43.
$$2x - 3y + 6 = 0$$
$$2x - 3y + 6 - 6 = 0 - 6 \qquad \text{Subtract 6 from both sides of the equation.}$$
$$2x - 3y = -6$$
$$-2x + 2x - 3y = -2x - 6 \qquad \text{Subtract } 2x \text{ from both sides of the equation.}$$
$$-3y = -2x - 6$$
$$\frac{-3y}{-3} = \frac{-2x - 6}{-3} \qquad \text{Divide both sides of the equation by -3.}$$
$$y = \frac{-2x - 6}{-3} = \frac{-(-2x - 6)}{3} = \frac{2x + 6}{3} = \frac{2x}{3} + \frac{6}{3} = \frac{2}{3}x + 2$$

44.

$$3x + 4y = 0$$

$$-3x + 3x + 4y = -3x + 0$$ Subtract $3x$ from boths sides of the equation.

$$4y = -3x$$

$$\frac{4y}{4} = \frac{-3x}{4}$$ Divide both sides of the equation by 4.

$$y = -\frac{3}{4}x$$

45.

$$-2x + 3y + z = 15$$

$$-2x + 2x + 3y + z = 2x + 15$$ Add $2x$ to both sides of the equation.

$$3y + z = 2x + 15$$

$$3y + z - z = 2x - z + 15$$ Subtract z from both sides of the equation.

$$3y = 2x - z + 15$$

$$\frac{3y}{3} = \frac{2x - z + 15}{3}$$ Divide both sides of the equation by 3.

$$y = \frac{2x - z + 15}{3} = \frac{2}{3}x - \frac{1}{3}z + 5$$

46.

$$5x + 3y - 2z = 22$$

$$5x - 5x + 3y - 2z = -5x + 22$$ Subtract $5x$ from both sides of the equation.

$$3y - 2z = -5x + 22$$

$$3y - 2z + 2z = -5x + 2z + 22$$ Add $2z$ to both sides of the equation.

$$3y = -5x + 2z + 22$$

$$\frac{3y}{3} = \frac{-5x + 2z + 22}{3}$$ Divide both sides of the equation by 3.

$$y = \frac{-5x + 2z + 22}{3} = -\frac{5}{3}x + \frac{2}{3}z + \frac{22}{3}$$

47.

$$9x + 4z = 7 + 8y$$

$$9x + 4z - 7 = 7 - 7 + 8y$$ Subtract 7 from both sides of the equation.

$$9x + 4z - 7 = 8y$$

$$\frac{9x + 4z - 7}{8} = \frac{8y}{8}$$ Divide both sides of the equation by 8.

$$y = \frac{9x + 4z - 7}{8} = \frac{9}{8}x + \frac{1}{2}z - \frac{7}{8}$$

48.

$$2x - 3y + 5z = 0$$

$$2x - 3y + 3y + 5z = 0 + 3y$$ Add $3y$ to both sides of the equation.

$$2x + 5z = 3y$$

$$\frac{2x + 5z}{3} = \frac{3y}{3}$$ Divide both sides of the equation by 3.

$$y = \frac{2x + 5z}{3} = \frac{2}{3}x + \frac{5}{3}z$$

49. $E = IR$

$$\frac{E}{I} = \frac{IR}{I}$$　　　　　　　　　　　Divide both sides of the equation by I.

$$R = \frac{E}{I}$$

50. $p = irt$

$$\frac{p}{i} = \frac{irt}{i}$$　　　　　　　　　　　Divide both sides of the equation by i.

$$\frac{p}{i} = rt$$

$$\frac{p}{ir} = \frac{rt}{r}$$　　　　　　　　　　　Divide both sides of the equation by r.

$$t = \frac{p}{ir}$$

51. $p = a + b + c$

$p - b = a + b - b + c$　　　　　　Subtract b from both sides of the equation.

$p - b = a + c$

$p - b - c = a + c - c$　　　　　　Subtract c from both sides of the equation.

$a = p - b - c$

52. $p = a + b + s_1 + s_2$

$p - a = a - a + b + s_1 + s_2$　　　　Subtract a from both sides of the equation.

$p - a = b + s_1 + s_2$

$p - a - b = b - b + s_1 + s_2$　　　　Subtract b from both sides of the equation.

$p - a - b = s_1 + s_2$

$p - a - b - s_2 = s_1 + s_2 - s_2$　　　　Subtract s_2 from both sides of the equation.

$s_1 = p - a - b - s_2$

53. $V = \dfrac{1}{3}Bh$

$3V = 3\left(\dfrac{1}{3}Bh\right)$　　　　　　　Multiply both sides of the equation by 3.

$3V = Bh$

$$\frac{3V}{h} = \frac{Bh}{h}$$　　　　　　　　　Divide both sides of the equation by h.

$$B = \frac{3V}{h}$$

54.
$$V = \pi r^2 h$$

$$\frac{V}{\pi} = \frac{\pi r^2 h}{\pi}$$ Divide both sides of the equation by π.

$$\frac{V}{\pi} = r^2 h$$

$$\frac{V}{\pi r^2} = \frac{r^2 h}{r^2}$$ Divide both sides of the equation by r^2.

$$h = \frac{V}{\pi r^2}$$

55.
$$C = 2\pi r$$

$$\frac{C}{2} = \frac{2\pi r}{2}$$ Divide both sides of the equation by 2.

$$\frac{C}{2} = \pi r$$

$$\frac{C}{2\pi} = \frac{\pi r}{\pi}$$ Divide both sides of the equation by π.

$$r = \frac{C}{2\pi}$$

56.
$$r = \frac{2gm}{c^2}$$

$$rc^2 = \left(\frac{2gm}{c^2}\right) c^2$$ Multiply both sides of the equation by c^2.

$$rc^2 = 2gm$$

$$\frac{rc^2}{2} = \frac{2gm}{2}$$ Divide both sides of the equation by 2.

$$\frac{rc^2}{2} = gm$$

$$\frac{rc^2}{2g} = \frac{gm}{g}$$ Divide both sides of the equation by g.

$$m = \frac{rc^2}{2g}$$

57.
$$y = mx + b$$

$$y - mx = mx - mx + b$$ Subtract mx from both sides of the equation.

$$b = y - mx$$

58.
$$y = mx + b$$

$$y - b = mx + b - b$$ Subtract b from both sides of the equation.

$$y - b = mx$$

$$\frac{y - b}{x} = \frac{mx}{x}$$ Divide both sides of the equation by x.

$$m = \frac{y - b}{x}$$

59.
$$P = 2l + 2w$$
$$P - 2l = 2l - 2l + 2w$$ Subtract $2l$ from both sides of the equation.
$$P - 2l = 2w$$
$$\frac{P - 2l}{2} = \frac{2w}{2}$$ Divide both sides of the equation by 2.
$$w = \frac{P - 2l}{2}$$

60.
$$A = \frac{d_1 d_2}{2}$$
$$2A = 2\left(\frac{d_1 d_2}{2}\right)$$ Multiply both sides of the equation by 2.
$$2A = d_1 d_2$$
$$\frac{2A}{d_1} = \frac{d_1 d_2}{d_1}$$ Divide both sides of the equation by d_1.
$$d_2 = \frac{2A}{d_1}$$

61.
$$A = \frac{a + b + c}{3}$$
$$3A = 3\left(\frac{a + b + c}{3}\right)$$ Multiply both sides of the equation by 3.
$$3A = a + b + c$$
$$3A - a = a - a + b + c$$ Subtract a from both sides of the equation.
$$3A - a = b + c$$
$$3A - a - b = b - b + c$$ Subtract b from both sides of the equation.
$$c = 3A - a - b$$

62.
$$A = \frac{1}{2}bh$$
$$2A = 2\left(\frac{1}{2}bh\right)$$ Multiply both sides of the equation by 2.
$$2A = bh$$
$$\frac{2A}{h} = \frac{bh}{h}$$ Divide both sides of the equation by h.
$$b = \frac{2A}{h}$$

63.
$$P = \frac{KT}{V}$$

$$PV = \left(\frac{KT}{V}\right)V \qquad \text{Multiply both sides of the equation by } V.$$

$$PV = KT$$

$$\frac{PV}{K} = \frac{KT}{K} \qquad \text{Divide both sides of the equation by } K.$$

$$T = \frac{PV}{K}$$

64.
$$\frac{P_1V_1}{T_1} = \frac{P_2V_2}{T_2}$$

$$P_1V_1T_2 = P_2V_2T_1 \qquad \text{Cross multiplication}$$

$$\frac{P_1V_1T_2}{T_1} = \frac{P_2V_2T_1}{T_1} \qquad \text{Divide both sides of the equation by } T_1.$$

$$\frac{P_1V_1T_2}{T_1} = P_2V_2$$

$$\frac{P_1V_1T_2}{T_1P_2} = \frac{P_2V_2}{P_2} \qquad \text{Divide both sides of the equation by } P_2.$$

$$V_2 = \frac{P_1V_1T_2}{T_1P_2}$$

65.
$$F = \frac{9}{5}C + 32$$

$$F - 32 = \frac{9}{5}C + 32 - 32 \qquad \text{Subtract 32 from both sides of the equation.}$$

$$F - 32 = \frac{9}{5}C$$

$$\frac{5}{9}(F - 32) = \frac{5}{9}\left(\frac{9}{5}C\right) \qquad \text{Multiply both sides of the equation by } \frac{5}{9}.$$

$$C = \frac{5}{9}(F - 32)$$

66.
$$C = \frac{5}{9}(F - 32)$$

$$\frac{9}{5}C = \frac{9}{5}\left(\frac{5}{9}\right)(F - 32) \qquad \text{Multiply both sides of the equation by } \frac{9}{5}.$$

$$\frac{9}{5}C = F - 32$$

$$\frac{9}{5}C + 32 = F - 32 + 32 \qquad \text{Add 32 to both sides of the equation.}$$

$$F = \frac{9}{5}C + 32$$

67.
$$S = \pi r^2 + \pi rs$$

$$S - \pi r^2 = \pi r^2 - \pi r^2 + \pi rs$$ Subtract πr^2 from both sides of the equation.

$$S - \pi r^2 = \pi rs$$

$$\frac{S - \pi r^2}{\pi} = \frac{\pi rs}{\pi}$$ Divide both sides of the equation by π.

$$\frac{S - \pi r^2}{\pi} = rs$$

$$\frac{S - \pi r^2}{\pi r} = \frac{rs}{r}$$ Divide both sides of the equation by r.

$$s = \frac{S - \pi r^2}{\pi r}$$

68.
$$A = \frac{1}{2} h \left(b_1 + b_2 \right)$$

$$2A = 2 \left(\frac{1}{2} h \left(b_1 + b_2 \right) \right)$$ Multiply both sides of the equation by 2.

$$2A = h \left(b_1 + b_2 \right)$$

$$\frac{2A}{h} = \frac{h \left(b_1 + b_2 \right)}{h}$$ Divide both sides of the equation by h.

$$\frac{2A}{h} = b_1 + b_2$$

$$\frac{2A}{h} - b_1 = b_1 - b_1 + b_2$$ Subtract b_1 from both sides of the equation.

$$b_2 = \frac{2A}{h} - b_1$$

69. a) $i = prt$

$i = 600(0.02)(1) = \$12$

b) $\$600 + \$12 = \$612$

70.
$$i = prt$$

$$128 = 800(r)(2)$$

$$128 = 1600r$$

$$\frac{128}{1600} = \frac{1600r}{1600}$$

$$r = 0.08 = 8\%$$

71. Radius $= \dfrac{2.5}{2} = 1.25$ in.

$V = \pi r^2 h$

$V = \pi (1.25)^2 (3.75)$

$V = \pi (1.5625)(3.75)$

$V = 18.40776945$ in.$^3 \approx 18.4$ in.^{3}

72. a) 6 ft $= 6(12) = 72$ in.

$B = \dfrac{703w}{h^2}$

$B = \dfrac{703(200)}{(72)^2}$

$B = \dfrac{140,600}{5184} = 27.12191358 \approx 27.12$

b) $B = \dfrac{703w}{h^2}$

$26 = \dfrac{703w}{(72)^2}$

$26 = \dfrac{703w}{5184}$

$134,784 = 703w$

$\dfrac{134,784}{703} = \dfrac{703w}{703}$

$w = 191.7268848$ lb

He would have to lose $200 - 191.7268848$

$= 8.2731152 \approx 8.27$ lb

73. $y = 2000(3)^x$

$y = 2000(3)^5$

$y = 2000(243)$

$y = 486,000$ bacteria

74. $P_n = P(1+r)^n$

$P_n = 8(1+0.03)^{10}$

$P_n = 8(1.03)^{10}$

$P_n = 8(1.343916379)$

$P_n = \$10.75133103 \approx \$10.75$

75. $V = 24e^{0.08t}$

$V = 24e^{0.08(377)}$

$V = 24e^{30.16}$

$V = \$300,976,658,300,000$

76. $S = S_0 e^{-0.028t}$

$S = 1000e^{-0.028(30)}$

$S = 1000e^{-0.84}$

$S = 1000(0.4317105234)$

$S = 431.7105234$ g ≈ 431.71 g

77. $V = lwh - \pi r^2 h$

$V = 12(8)(12) - \pi (2)^2 (8)$

$V = 1152 - 100.5309649$

$V = 1051.469035$ in.$^3 \approx 1051.47$ in.^{3}

78.

Exercise Set 6.4

1. A **mathematical expression** is a collection of variables, numbers, parentheses, and operation symbols. An **equation** is two algebraic expressions joined by an equal sign.

2. Expression: $2x + 3y$; equation: $2x + 3y = 16$

3. $4 + 3x$

4. $6x - 2$

5. $6r + 5$

6. $10s - 13$

7. $15 - 2r$

8. $x + 6$

9. $2m + 9$

10. $8 + 5x$

11. $\dfrac{18 - s}{4}$

12. $\dfrac{8 + t}{2}$

13. $(5y - 6) + 3$

14. $\dfrac{8}{y} - 3x$

15.
$$\text{Let } x = \text{the number}$$
$$x - 6 = \text{the number decreased by 6}$$
$$x - 6 = 5$$
$$x - 6 + 6 = 5 + 6$$
$$x = 11$$

16.
$$\text{Let } x = \text{the number}$$
$$x + 7 = \text{the sum of the number and 7}$$
$$x + 7 = 15$$
$$x + 7 - 7 = 15 - 7$$
$$x = 8$$

17.
$$\text{Let } x = \text{the number}$$
$$x - 4 = \text{the difference between the number and 4}$$
$$x - 4 = 20$$
$$x - 4 + 4 = 20 + 4$$
$$x = 24$$

18.
$$\text{Let } x = \text{the number}$$
$$7x = \text{the number multiplied by 7}$$
$$7x = 42$$
$$\frac{7x}{7} = \frac{42}{7}$$
$$x = 6$$

19.
$$\text{Let } x = \text{the number}$$
$$12 + 5x = 12 \text{ increased by 5 times the number}$$
$$12 + 5x = 47$$
$$12 - 12 + 5x = 47 - 12$$
$$5x = 35$$
$$\frac{5x}{5} = \frac{35}{5}$$
$$x = 7$$

20.
$$\text{Let } x = \text{the number}$$
$$4x - 10 = 4 \text{ times the number decreased by 10}$$
$$4x - 10 = 42$$
$$4x - 10 + 10 = 42 + 10$$
$$4x = 52$$
$$\frac{4x}{4} = \frac{52}{4}$$
$$x = 13$$

21.
$$\text{Let } x = \text{the number}$$
$$8x + 16 = 16 \text{ more than 8 times the number}$$
$$8x + 16 = 88$$
$$8x + 16 - 16 = 88 - 16$$
$$8x = 72$$
$$\frac{8x}{8} = \frac{72}{8}$$
$$x = 9$$

22.
$$\text{Let } x = \text{the number}$$
$$5x + 6 = 6 \text{ more than 5 times the number}$$
$$7x - 18 = 7 \text{ times the number decreased by 18}$$
$$5x + 6 = 7x - 18$$
$$5x - 7x + 6 = 7x - 7x - 18$$
$$-2x + 6 = -18$$
$$-2x + 6 - 6 = -18 - 6$$
$$-2x = -24$$
$$\frac{-2x}{-2} = \frac{-24}{-2}$$
$$x = 12$$

23.
$$\text{Let } x = \text{the number}$$
$$x + 11 = \text{the number increased by 11}$$
$$3x + 1 = 1 \text{ more than 3 times the number}$$
$$x + 11 = 3x + 1$$
$$x - x + 11 = 3x - x + 1$$
$$11 = 2x + 1$$
$$11 - 1 = 2x + 1 - 1$$
$$10 = 2x$$
$$\frac{10}{2} = \frac{2x}{2}$$
$$5 = x$$

24.
$$\text{Let } x = \text{the number}$$
$$\frac{x}{3} = \text{the number divided by 3}$$
$$x - 4 = 4 \text{ less than the number}$$
$$\frac{x}{3} = x - 4$$
$$3\left(\frac{x}{3}\right) = 3(x - 4)$$
$$x = 3x - 12$$
$$x - 3x = 3x - 3x - 12$$
$$-2x = -12$$
$$\frac{-2x}{-2} = \frac{-12}{-2}$$
$$x = 6$$

25.
$$\text{Let } x = \text{the number}$$
$$x + 10 = \text{the number increased by 10}$$
$$2(x + 3) = 2 \text{ times the sum of the number and 3}$$
$$x + 10 = 2(x + 3)$$
$$x + 10 = 2x + 6$$
$$x - x + 10 = 2x - x + 6$$
$$10 = x + 6$$
$$10 - 6 = x + 6 - 6$$
$$4 = x$$

26.
$$\text{Let } x = \text{the number}$$
$$2x = \text{the product of 2 and the number}$$
$$2x - 3 = \text{the product of 2 and the number,}$$
$$\text{decreased by 3}$$
$$x + 4 = 4 \text{ more than the number}$$
$$2x - 3 = x + 4$$
$$2x - x - 3 = x - x + 4$$
$$x - 3 = 4$$
$$x - 3 + 3 = 4 + 3$$
$$x = 7$$

27.
$$\text{Let } x = \text{the number of tickets sold}$$
$$\text{to nonstudents}$$
$$3x = \text{the number of tickets sold}$$
$$\text{to students}$$
$$x + 3x = 600$$
$$4x = 600$$
$$\frac{4x}{4} = \frac{600}{4}$$
$$x = 150 \text{ tickets to nonstudents}$$
$$3x = 3(150) = 450 \text{ tickets to students}$$

28.
$$\text{Let } x = \text{cost of cheaper pair}$$
$$x + 10 = \text{cost of more expensive pair}$$
$$x + (x + 10) = 60$$
$$2x + 10 = 60$$
$$2x + 10 - 10 = 60 - 10$$
$$2x = 50$$
$$\frac{2x}{2} = \frac{50}{2}$$
$$x = \$25 \text{ for the cheaper pair}$$
$$x + 10 = 25 + 10$$
$$= \$35 \text{ for the more expensive pair}$$

29.   Let x = the number filing electronically

           in 1999

$$0.116x = \text{the amount of the increase}$$
$$x + 0.116x = 34.20$$
$$1.116x = 34.20$$
$$\frac{1.116x}{1.116} = \frac{34.20}{1.116}$$
$$x = 30.64516129$$
$$\approx 30.65 \text{ million taxpayers}$$

30.   Let x = Vinny's dollar sales

$$0.06x = \text{the amount Vinny made}$$
           on commission
$$400 + 0.06x = 790$$
$$400 - 400 + 0.06x = 790 - 400$$
$$0.06x = 390$$
$$\frac{0.06x}{0.06} = \frac{390}{0.06}$$
$$x = \$6500$$

31.   Let x = the original price before tax

$$0.10x = \text{the amount saved on}$$
           spending x dollars
$$x - 0.10x = 15.72$$
$$0.9x = 15.72$$
$$\frac{0.9x}{0.9} = \frac{15.72}{0.9}$$
$$x = 17.4\overline{6} \approx \$17.47$$

32.   Let x = the number of copies Ronnie must make

$$0.08x = \text{the amount spent on } x \text{ copies}$$
$$0.08x = 250$$
$$\frac{0.08x}{0.08} = \frac{250}{0.08}$$
$$x = 3125 \text{ copies}$$

33.   Let x = the number of compact discs

           for Samantha

$$3x = \text{the number of compact discs}$$
           for Josie
$$x + 3x = 12$$
$$4x = 12$$
$$\frac{4x}{4} = \frac{12}{4}$$
$$x = 3 \text{ compact discs for Samantha}$$
$$3x = 3(3) = 9 \text{ compact discs for Josie}$$

34.   Let x = the amount donated for Business

$$3x = \text{the amount donated for}$$
           Liberal Arts
$$x + 3x = 1000$$
$$4x = 1000$$
$$\frac{4x}{4} = \frac{1000}{4}$$
$$x = \$250 \text{ for Business}$$
$$3x = 3(250) = \$750 \text{ for Liberal Arts}$$

35.   Let x = the amount charged

           to each homeowner

$$50x = \text{the total amount charged}$$
           to homeowners
$$2000 + 50x = \text{the total cost for}$$
           the repairs
$$2000 + 50x = 13,350$$
$$2000 - 2000 + 50x = 13,350 - 2000$$
$$50x = 11,350$$
$$\frac{50x}{50} = \frac{11,350}{50}$$
$$x = \$227$$

36.   Let w = the width

$$w + 3 = \text{the length}$$
$$2w + 2(w + 3) = P$$
$$2w + 2(w + 3) = 54$$
$$2w + 2w + 6 = 54$$
$$4w + 6 = 54$$
$$4w + 6 - 6 = 54 - 6$$
$$4w = 48$$
$$\frac{4w}{4} = \frac{48}{4}$$
$$\text{width} = 12 \text{ ft}$$
$$\text{length} = w + 3 = 12 + 3 = 15 \text{ ft}$$

37. a)   Let x = area of smaller ones
            $3x$ = area of largest one
$$x + x + 3x = 45,000$$
$$5x = 45,000$$
$$\frac{5x}{5} = \frac{45,000}{5}$$
$$x = 9000 \text{ ft}^2 \text{ for the two smaller barns}$$
$$3x = 3(9000)$$
$$= 27,000 \text{ ft}^2 \text{ for the largest barn}$$
    b) Yes

38.   Let x = average per capita income
            in Mississippi
$2x - 1346$ = average per capita income
            in Connecticut
$$x + 2x - 1346 = 61,663$$
$$3x - 1346 = 61,663$$
$$3x - 1346 + 1346 = 61,663 + 1346$$
$$3x = 63,009$$
$$\frac{3x}{3} = \frac{63,009}{3}$$
$$x = \$21,003 \text{ in Mississippi}$$
$$2x - 1346 = 2(21,003) - 1346 = 42,006 - 1346$$
$$= \$40,660 \text{ in Connecticut}$$

39.   Let x = the number of vacation days
            in the U.S.
$3x + 3$ = the number of vacation days
            in Italy
$$x + 3x + 3 = 55$$
$$4x + 3 = 55$$
$$4x + 3 - 3 = 55 - 3$$
$$4x = 52$$
$$\frac{4x}{4} = \frac{52}{4}$$
$$x = 13 \text{ in the U.S.}$$
$$3x + 3 = 3(13) + 3 = 39 + 3 = 42 \text{ in Italy}$$

40.   Let x = the cost of the car before tax
$0.05x$ = 5% of the cost of the car (tax)
$$x + 0.05x = 14,512$$
$$1.05x = 14,512$$
$$\frac{1.05x}{1.05} = \frac{14,512}{1.05}$$
$$x = \$13,820.95238 \approx \$13,820.95$$

41.   Let w = width
            $2w$ = length of entire enclosed region
$$3w + 2(2w) = \text{total amount of fencing}$$
$$3w + 2(2w) = 140$$
$$3w + 4w = 140$$
$$7w = 140$$
$$\frac{7w}{7} = \frac{140}{7}$$
$$\text{width} = 20 \text{ ft}$$
$$\text{length} = 2w = 2(20) = 40 \text{ ft}$$

42.   Let l = length of a shelf
            $l + 2$ = height of the bookcase
$$4l + 2(l + 2) = \text{total amount of wood}$$
$$4l + 2(l + 2) = 32$$
$$4l + 2l + 4 = 32$$
$$6l + 4 = 32$$
$$6l + 4 - 4 = 32 - 4$$
$$6l = 28$$
$$\frac{6l}{6} = \frac{28}{6}$$
$$\text{length} = 4\frac{4}{6} = 4\frac{2}{3} \text{ ft} = 4 \text{ ft } 8 \text{ in.}$$
$$\text{height} = l + 2 = 4 \text{ ft } 8 \text{ in.} + 2 \text{ ft} = 6 \text{ ft } 8 \text{ in.}$$

43. Let x = the number of months

$70x$ = cost of laundry for x months

$70x = 760$

$$\frac{70x}{70} = \frac{760}{70}$$

$x = 10.85714286$ months ≈ 11 months

44. Let x = the number of visits per month

56 = the cost of Plan A for 1 month

$3x + 20$ = the cost of Plan B for 1 month

$3x + 20 = 56$

$3x + 20 - 20 = 56 - 20$

$3x = 36$

$$\frac{3x}{3} = \frac{36}{3}$$

$x = 12$ visits per month

45. Let r = regular fare

$\dfrac{r}{2}$ = half off regular fare

$0.07r$ = tax on regular fare

$$\frac{r}{2} + 0.07r = 257$$

$$2\left(\frac{r}{2} + 0.07r\right) = 2(257)$$

$r + 0.14r = 514$

$1.14r = 514$

$$\frac{1.14r}{1.14} = \frac{514}{1.14}$$

$r = \$450.877193$

$\approx \$450.88$

46. Let x = the number of miles in one day

$35 + 0.20x$ = U-Haul charge per day

$25 + 0.32x$ = Ryder charge per day

$35 + 0.20x = 25 + 0.32x$

$35 + 0.20x - 25 = 25 - 25 + 0.32x$

$10 + 0.20x = 0.32x$

$10 + 0.20x - 0.20x = 0.32x - 0.20x$

$10 = 0.12x$

$$\frac{10}{0.12} = \frac{0.12x}{0.12}$$

$x = 83.\overline{3}$ mi $= 83\dfrac{1}{3}$ mi

47. Let x = amount of tax reduction to be deducted from Mr. McAdam's income

$3640 - x$ = amount of tax reduction to be deducted from Mrs. McAdam's income

$24,200 - x = 26,400 - (3640 - x)$

$24,200 - x = 26,400 - 3640 + x$

$24,200 - x = 22,760 + x$

$24,200 - x + x = 22,760 + x + x$

$24,200 = 22,760 + 2x$

$24,200 - 22,760 = 22,760 - 22,760 + 2x$

$1440 = 2x$

$$\frac{1440}{2} = \frac{2x}{2}$$

$x = \$720$ deducted from Mr. McAdam's income

$3640 - x = 3640 - 720 = \$2920$ deducted from Mrs. McAdam's income

48. a) A number increased by 3 is 13.

b) 3 times a number increased by 5 is 8.

c) 3 times a number decreased by 8 is 7.

49.   Let x = the first integer
        $x + 1$ = the second integer
        $x + 2$ = the third integer (the largest)

$$x + (x+1) + (x+2) = 3(x+2) - 3$$
$$3x + 3 = 3x + 6 - 3$$
$$3x + 3 = 3x + 3$$

50. a)   Let x = the number of years for the amount
            saved to equal the price of the course

$$0.10(600) = \$60 \text{ saved per year}$$
$$60x = 45$$
$$\frac{60x}{60} = \frac{45}{60}$$
$$x = \frac{3}{4} \text{ year} = \frac{3}{4}(12) = 9 \text{ months}$$

b) $25 - 18 = 7$ years
     $7(60) = \$420$ saved before paying for course
     $\$420 - \$45 = \$375$ total savings

51. $F = \dfrac{9}{5}C + 32$

The thermometers will read the same when $F = C$.

Substitute C for F in the above equation.

$$C = \frac{9}{5}C + 32$$
$$5C = 5\left(\frac{9}{5}C + 32\right)$$
$$5C = 9C + 160$$
$$5C - 9C = 9C - 9C + 160$$
$$-4C = 160$$
$$\frac{-4C}{-4} = \frac{160}{-4}$$
$$C = -40°$$

Exercise Set 6.5

1. **Inverse variation** - As one variable increases, the other decreases and vice versa.
2. **Direct variation** - As one variable increases, so does the other, and as one variable decreases, so does the other.
3. **Joint variation** - One quantity varies directly as the product of two or more other quantities.
4. **Combined variation** uses at least two forms of variation.

5. Direct	6. Inverse	7. Inverse	8. Direct
9. Direct	10. Direct	11. Inverse	12. Direct
13. Inverse	14. Direct	15. Inverse	16. Direct
17. Direct	18. Inverse	19. Direct	20. Inverse

21. Answers will vary.

22. Answers will vary.

23. a) $y = kx$

    b) $y = 3(5) = 15$

24. a) $x = \dfrac{k}{y}$

    b) $x = \dfrac{15}{12} = 1.25$

25. a) $m = \dfrac{k}{n^2}$

    b) $m = \dfrac{16}{(8)^2} = \dfrac{16}{64} = 0.25$

26. a) $r = ks^2$

    b) $r = 13(2)^2 = 13(4) = 52$

27. a) $R = \dfrac{k}{W}$

    b) $R = \dfrac{8}{160} = 0.05$

28. a) $D = \dfrac{kJ}{C}$

    b) $D = \dfrac{5(10)}{25} = \dfrac{50}{25} = 2$

29. a) $F = kDE$
    b) $F = 7(3)(10) = 210$

30. a) $A = \dfrac{kR_1 R_2}{L^2}$

    b) $A = \dfrac{\frac{3}{2}(120)(8)}{(5)^2} = \dfrac{(1.5)(120)(8)}{25} = \dfrac{1440}{25} = 57.6$

31. a) $t = \dfrac{kd^2}{f}$

    b) $192 = \dfrac{k(8)^2}{4}$

    $192 = \dfrac{64k}{4}$

    $768 = 64k$

    $\dfrac{768}{64} = \dfrac{64k}{64}$

    $k = 12$

    $t = \dfrac{12d^2}{f}$

    $t = \dfrac{12(10)^2}{6} = \dfrac{12(100)}{6} = \dfrac{1200}{6} = 200$

32. a) $y = \dfrac{k\sqrt{t}}{s}$

    b) $12 = \dfrac{k\sqrt{36}}{2}$

    $12 = \dfrac{6k}{2}$

    $24 = 6k$

    $\dfrac{24}{6} = \dfrac{6k}{6}$

    $k = 4$

    $y = \dfrac{4\sqrt{t}}{s}$

    $y = \dfrac{4\sqrt{81}}{4} = \dfrac{4(9)}{4} = \dfrac{36}{4} = 9$

33. a) $Z = kWY$
    b) $12 = k(9)(4)$

    $12 = 36k$

    $\dfrac{12}{36} = \dfrac{36k}{36}$

    $k = \dfrac{1}{3}$

    $Z = \dfrac{1}{3}WY$

    $Z = \dfrac{1}{3}(50)(6) = \dfrac{300}{3} = 100$

34. a) $y = kR^2$
    b) $4 = k(4)^2$

    $4 = 16k$

    $\dfrac{4}{16} = \dfrac{16k}{16}$

    $k = 0.25$

    $y = 0.25R^2$

    $y = 0.25(8)^2 = 0.25(64) = 16$

35. a) $H = kL$

b) $15 = k(50)$

$$\frac{15}{50} = \frac{50k}{50}$$

$$k = 0.3$$

$$H = 0.3L$$

$$H = 0.3(10) = 3$$

36. a) $C = \dfrac{k}{J}$

b) $7 = \dfrac{k}{0.7}$

$$k = 7(0.7) = 4.9$$

$$C = \frac{4.9}{J}$$

$$C = \frac{4.9}{12} = 0.408\overline{3} \approx 0.41$$

37. a) $A = kB^2$

b) $245 = k(7)^2$

$$245 = 49k$$

$$\frac{245}{49} = \frac{49k}{49}$$

$$k = 5$$

$$A = 5B^2$$

$$A = 5(12)^2 = 5(144) = 720$$

38. a) $F = \dfrac{kM_1 M_2}{d^2}$

b) $20 = \dfrac{k(5)(10)}{(0.2)^2}$

$$20 = \frac{50k}{0.04}$$

$$50k = 0.8$$

$$k = \frac{0.8}{50} = 0.016$$

$$F = \frac{0.016 M_1 M_2}{d^2}$$

$$F = \frac{0.016(10)(20)}{(0.4)^2} = \frac{3.2}{0.16} = 20$$

39. a) $F = \dfrac{kq_1 q_2}{d^2}$

b) $8 = \dfrac{k(2)(8)}{(4)^2}$

$$8 = \frac{16k}{16}$$

$$k = 8$$

$$F = \frac{8q_1 q_2}{d^2}$$

$$F = \frac{8(28)(12)}{(2)^2} = \frac{2688}{4} = 672$$

40. a) $S = k\,I\,T^2$

b) $8 = k(20)(4)^2$

$$8 = 320k$$

$$k = \frac{8}{320} = 0.025$$

$$S = 0.025IT^2$$

$$S = 0.025(2)(2)^2 = 0.025(2)(4) = 0.2$$

41. a) $R = kL$

b) $0.24 = k(30)$

$$\frac{0.24}{30} = \frac{30k}{30}$$

$$k = 0.008$$

$$R = 0.008L$$

$$R = 0.008(40) = 0.32 \text{ ohm}$$

42. a) $I = k\,r$

b) $40 = k(0.04)$

$$k = \frac{40}{0.04} = 1000$$

$$I = 1000r$$

$$I = 1000(0.06) = \$60$$

43. a) $l = \dfrac{k}{d^2}$

b) $20 = \dfrac{k}{(6)^2}$

$k = 20(36) = 720$

$l = \dfrac{720}{d^2}$

$l = \dfrac{720}{(3)^2} = \dfrac{720}{9} = 80 \text{ dB}$

44. a) $t = \dfrac{k}{n}$

b) $16 = \dfrac{k}{2}$

$k = 16(2) = 32$

$t = \dfrac{32}{n}$

$t = \dfrac{32}{4} = 8 \text{ hours}$

45. a) $R = \dfrac{kA}{P}$

b) $4800 = \dfrac{k(600)}{3}$

$600k = 14,400$

$k = \dfrac{14,400}{600} = 24$

$R = \dfrac{24A}{P}$

$R = \dfrac{24(700)}{3.50} = \dfrac{16,800}{3.50} = 4800 \text{ tapes}$

46. a) $a = kd^2$

b) $100 = k(25)^2$

$100 = 625k$

$\dfrac{100}{625} = \dfrac{625k}{625}$

$k = \dfrac{100}{625} = 0.16$

$a = 0.16d^2$

$a = 0.16(40)^2 = 0.16(1600)$

$= 256 \text{ square feet}$

47. a) $s = kwd^2$

b) $2250 = k(2)(10)^2$

$2250 = 200k$

$\dfrac{2250}{200} = \dfrac{200k}{200}$

$k = \dfrac{2250}{200} = 11.25$

$s = 11.25wd^2$

$s = 11.25(4)(12)^2 = 11.25(4)(144)$

$= 6480 \text{ pounds per square inch}$

48. a) $R = \dfrac{kL}{A}$

b) $0.2 = \dfrac{k(200)}{0.05}$

$200k = 0.01$

$k = \dfrac{0.01}{200} = 0.00005$

$R = \dfrac{0.00005L}{A}$

$R = \dfrac{0.00005(5000)}{0.01} = \dfrac{0.25}{0.01} = 25 \text{ ohms}$

49. a) $N = \dfrac{kp_1 p_2}{d}$

 b) $\quad 100,000 = \dfrac{k(60,000)(200,000)}{300}$

 $12,000,000,000k = 30,000,000$

 $k = \dfrac{30,000,000}{12,000,000,000} = 0.0025$

 $N = \dfrac{0.0025\, p_1 p_2}{d}$

 $N = \dfrac{0.0025(125,000)(175,000)}{450}$

 $N = \dfrac{54,687,500}{450}$

 $= 121,527.7778 \approx 121,528 \text{ calls}$

50. a) $y = kx$

 $y = 2x$

 $\dfrac{y}{2} = \dfrac{2x}{2}$

 $x = \dfrac{y}{2} = 0.5y$

 Directly

 b) $k = 0.5$

51. a) $\quad y = \dfrac{k}{x}$

 $y = \dfrac{0.3}{x}$

 $xy = 0.3$

 $\dfrac{xy}{y} = \dfrac{0.3}{y}$

 $x = \dfrac{0.3}{y}$

 Inversely

 b) k stays 0.3

52. $\quad I = \dfrac{k}{d^2}$

 $\dfrac{1}{16} = \dfrac{k}{(4)^2}$

 $\dfrac{1}{16} = \dfrac{k}{16}$

 $k = 1$

 $I = \dfrac{1}{d^2}$

 $I = \dfrac{1}{(3)^2} = \dfrac{1}{9}$

53. $\quad W = \dfrac{kTA\sqrt{F}}{R}$

 $72 = \dfrac{k(78)(1000)\sqrt{4}}{5.6}$

 $156,000k = 403.2$

 $k = \dfrac{403.2}{156,000} = 0.0025846154$

 $W = \dfrac{0.0025846154\,TA\sqrt{F}}{R}$

 $W = \dfrac{0.0025846154(78)(1500)\sqrt{6}}{5.6}$

 $W = \dfrac{740.7256982}{5.6} = 132.2724461 \approx \$132.27$

Exercise Set 6.6

1. $a < b$ means that a is less than b, $a \le b$ means that a is less than or equal to b, $a > b$ means that a is greater than b, $a \ge b$ means that a is greater than or equal to b.

2. a) An **inequality** consists of two (or more) expressions joined by an inequality sign.
   b) $2 < 7$, $3 > -1$, $5x + 2 \ge 9$

3. When both sides of an inequality are multiplied or divided by a negative number, the direction of the inequality symbol must be reversed.

4. Yes, the inequality symbol points to the x in both cases.

5. Yes, the inequality symbol points to the -3 in both cases.

6. You should use an open circle if the solution does not include the number. You should use a closed circle if the solution includes the number.

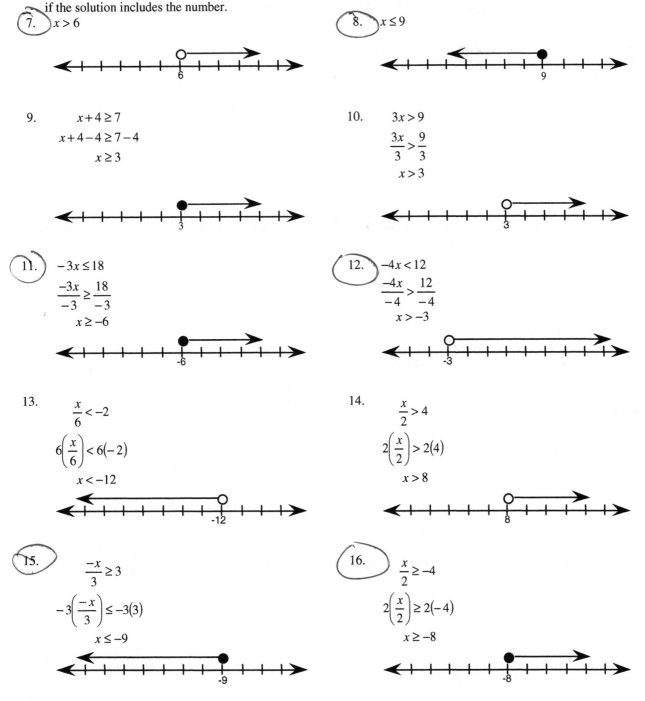

7. $x > 6$

8. $x \le 9$

9. $x + 4 \ge 7$
$x + 4 - 4 \ge 7 - 4$
$x \ge 3$

10. $3x > 9$
$\dfrac{3x}{3} > \dfrac{9}{3}$
$x > 3$

11. $-3x \le 18$
$\dfrac{-3x}{-3} \ge \dfrac{18}{-3}$
$x \ge -6$

12. $-4x < 12$
$\dfrac{-4x}{-4} > \dfrac{12}{-4}$
$x > -3$

13. $\dfrac{x}{6} < -2$
$6\left(\dfrac{x}{6}\right) < 6(-2)$
$x < -12$

14. $\dfrac{x}{2} > 4$
$2\left(\dfrac{x}{2}\right) > 2(4)$
$x > 8$

15. $\dfrac{-x}{3} \ge 3$
$-3\left(\dfrac{-x}{3}\right) \le -3(3)$
$x \le -9$

16. $\dfrac{x}{2} \ge -4$
$2\left(\dfrac{x}{2}\right) \ge 2(-4)$
$x \ge -8$

17.
$$2x + 6 \geq 14$$
$$2x + 6 - 6 \geq 14 - 6$$
$$2x \geq 8$$
$$\frac{2x}{2} \geq \frac{8}{2}$$
$$x \geq 4$$

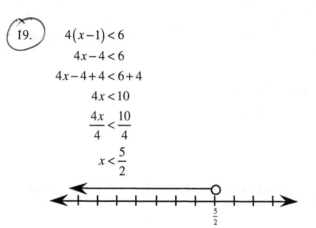

18.
$$3x + 12 < 5x + 14$$
$$3x - 5x + 12 < 5x - 5x + 14$$
$$-2x + 12 < 14$$
$$-2x + 12 - 12 < 14 - 12$$
$$-2x < 2$$
$$\frac{-2x}{-2} > \frac{2}{-2}$$
$$x > -1$$

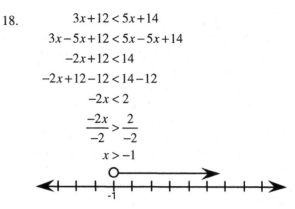

19.
$$4(x - 1) < 6$$
$$4x - 4 < 6$$
$$4x - 4 + 4 < 6 + 4$$
$$4x < 10$$
$$\frac{4x}{4} < \frac{10}{4}$$
$$x < \frac{5}{2}$$

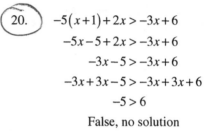

20.
$$-5(x + 1) + 2x > -3x + 6$$
$$-5x - 5 + 2x > -3x + 6$$
$$-3x - 5 > -3x + 6$$
$$-3x + 3x - 5 > -3x + 3x + 6$$
$$-5 > 6$$
False, no solution

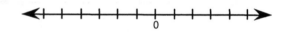

21.
$$3(x + 4) - 2 < 3x + 10$$
$$3x + 12 - 2 < 3x + 10$$
$$3x + 10 < 3x + 10$$
False, no solution

22. $-2 \leq x \leq 1$

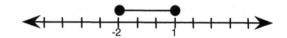

23.
$$3 < x - 7 \leq 6$$
$$3 + 7 < x - 7 + 7 \leq 6 + 7$$
$$10 < x \leq 13$$

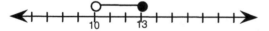

24.
$$\frac{1}{2} < \frac{x + 4}{2} \leq 4$$
$$2\left(\frac{1}{2}\right) < 2\left(\frac{x + 4}{2}\right) \leq 2(4)$$
$$1 < x + 4 \leq 8$$
$$1 - 4 < x + 4 - 4 \leq 8 - 4$$
$$-3 < x \leq 4$$

25. $x \geq 2$

26. $-3 < x$
$$x > -3$$

27. $-3x \le 27$

$\dfrac{-3x}{-3} \ge \dfrac{27}{-3}$

$x \ge -9$

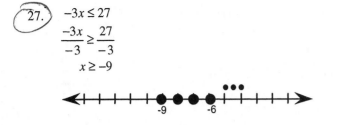

28. $3x \ge 27$

$\dfrac{3x}{3} \ge \dfrac{27}{3}$

$x \ge 9$

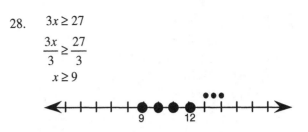

29. $x - 2 < 4$

$x - 2 + 2 < 4 + 2$

$x < 6$

30. $-5x \le 15$

$\dfrac{-5x}{-5} \ge \dfrac{15}{-5}$

$x \ge -3$

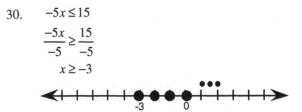

31. $\dfrac{x}{3} \le -2$

$3\left(\dfrac{x}{3}\right) \le 3(-2)$

$x \le -6$

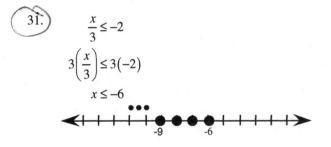

32. $\dfrac{x}{4} \ge -3$

$4\left(\dfrac{x}{4}\right) \ge 4(-3)$

$x \ge -12$

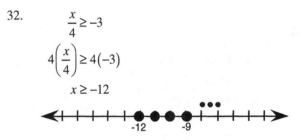

33. $\dfrac{-x}{6} \ge 3$

$-6\left(\dfrac{-x}{6}\right) \le -6(3)$

$x \le -18$

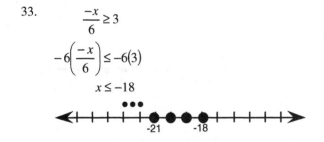

34. $\dfrac{2x}{3} \le 4$

$\dfrac{3}{2}\left(\dfrac{2x}{3}\right) \le \dfrac{3}{2}(4)$

$x \le 6$

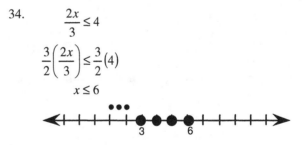

35. $-11 < -5x + 4$

$-11 - 4 < -5x + 4 - 4$

$-15 < -5x$

$\dfrac{-15}{-5} > \dfrac{-5x}{-5}$

$3 > x$

$x < 3$

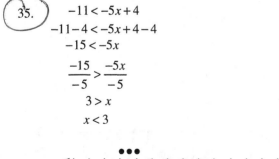

36. $2x + 5 < -3 + 6x$

$2x - 6x + 5 < -3 + 6x - 6x$

$-4x + 5 < -3$

$-4x + 5 - 5 < -3 - 5$

$-4x < -8$

$\dfrac{-4x}{-4} > \dfrac{-8}{-4}$

$x > 2$

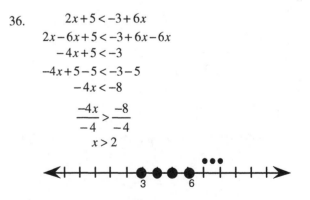

37.
$$3(x+4) \geq 4x+13$$
$$3x+12 \geq 4x+13$$
$$3x-4x+12 \geq 4x-4x+13$$
$$-x+12 \geq 13$$
$$-x+12-12 \geq 13-12$$
$$-x \geq 1$$
$$\frac{-x}{-1} \leq \frac{1}{-1}$$
$$x \leq -1$$

38.
$$-2(x-1) < 3(x-4)+5$$
$$-2x+2 < 3x-12+5$$
$$-2x+2 < 3x-7$$
$$-2x-3x+2 < 3x-3x-7$$
$$-5x+2 < -7$$
$$-5x+2-2 < -7-2$$
$$-5x < -9$$
$$\frac{-5x}{-5} > \frac{-9}{-5}$$
$$x > \frac{9}{5}$$

39.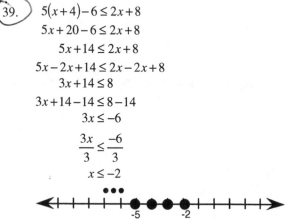
$$5(x+4)-6 \leq 2x+8$$
$$5x+20-6 \leq 2x+8$$
$$5x+14 \leq 2x+8$$
$$5x-2x+14 \leq 2x-2x+8$$
$$3x+14 \leq 8$$
$$3x+14-14 \leq 8-14$$
$$3x \leq -6$$
$$\frac{3x}{3} \leq \frac{-6}{3}$$
$$x \leq -2$$

40. $-3 \leq x < 5$

41.
$$1 > -x > -5$$
$$\frac{1}{-1} < \frac{-x}{-1} < \frac{-5}{-1}$$
$$-1 < x < 5$$

42.
$$-2 < 2x+3 < 6$$
$$-2-3 < 2x+3-3 < 6-3$$
$$-5 < 2x < 3$$
$$\frac{-5}{2} < \frac{2x}{2} < \frac{3}{2}$$
$$-\frac{5}{2} < x < \frac{3}{2}$$

43.
$$0.2 \le \frac{x-4}{10} \le 0.4$$
$$10(0.2) \le 10\left(\frac{x-4}{10}\right) \le 10(0.4)$$
$$2 \le x-4 \le 4$$
$$2+4 \le x-4+4 \le 4+4$$
$$6 \le x \le 8$$

44.
$$-\frac{1}{3} < \frac{x-2}{12} \le \frac{1}{4}$$
$$12\left(-\frac{1}{3}\right) < 12\left(\frac{x-2}{12}\right) \le 12\left(\frac{1}{4}\right)$$
$$-4 < x-2 \le 3$$
$$-4+2 < x-2+2 \le 3+2$$
$$-2 < x \le 5$$

45. a) 2000, 2001
b) 1997, 1998
c) 1997, 1998, 1999, 2000
d) 1998, 1999, 2000, 2001

46. a) 1990, 2000
b) 1890, 1910, 1950, 1970
c) 2000
d) 2000

47. Let x = the number of videos

No Fee Plan cost: $2.99x$

Annual Fee Plan: $30+1.49x$
$$2.99x < 30+1.49x$$
$$2.99x-1.49x < 30+1.49x-1.49x$$
$$1.50x < 30$$
$$\frac{1.50x}{1.50} < \frac{30}{1.50}$$
$$x < 20$$

The maximum number of videos that can be rented for the No Fee Plan to cost less than the Annual Fee Plan is 19.

48. Let x = the dollar amount of weekly sales

Plan A: $500+0.06x$

Plan B: $400+0.08x$
$$400+0.08x > 500+0.06x$$
$$400+0.08x-0.06x > 500+0.06x-0.06x$$
$$400+0.02x > 500$$
$$400-400+0.02x > 500-400$$
$$0.02x > 100$$
$$\frac{0.02x}{0.02} > \frac{100}{0.02}$$
$$x > 5000$$

The dollar amount of weekly sales that would result in Bobby earning more by Plan B than by Plan A is more than $5000.

49. Let x = the number of miles

$110+0.25x$ = cost of renting from Fred's
$$110+0.25x < 200$$
$$110-110+0.25x < 200-110$$
$$0.25x < 90$$
$$\frac{0.25x}{0.25} < \frac{90}{0.25}$$
$$x < 360 \text{ mi}$$

50. a) Let x = the number of boxes of books

$60x$ = the weight of x boxes of books
$$180+60x \le 1200$$
b) $$180-180+60x \le 1200-180$$
$$60x \le 1020$$
$$\frac{60x}{60} \le \frac{1020}{60}$$
$$x \le 17$$

The maximum number of boxes is 17.

51. Let x = the cost of the meal

$0.07x$ = the tax on the meal

$0.15x$ = the tip on the meal
$$x+0.07x+0.15x \le 19$$
$$1.22x \le 19$$
$$\frac{1.22x}{1.22} \le \frac{19}{1.22}$$
$$x \le 15.57377049$$

Mrs. Franklin can select a meal for $x \le \$15.57$.

52.
$$12x > 2x+2000$$
$$12x-2x > 2x-2x+2000$$
$$10x > 2000$$
$$\frac{10x}{10} > \frac{2000}{10}$$
$$x > 200$$

More than 200 books must be sold weekly to make a profit.

53. $$36 < 84 - 32t < 68$$
$$36 - 84 < 84 - 84 - 32t < 68 - 84$$
$$-48 < -32t < -16$$
$$\frac{-48}{-32} > \frac{-32t}{-32} > \frac{-16}{-32}$$
$$1.5 > t > 0.5$$
$$0.5 < t < 1.5$$

The velocity will be between $36\frac{\text{ft}}{\text{sec}}$ and $68\frac{\text{ft}}{\text{sec}}$ when t is between $0.5\,\text{sec}$ and $1.5\,\text{sec}$.

54. Let x = the number of miles
distance = rate × time
$$40(4) \le x \le 55(4)$$
$$160 \le x \le 220$$

55. Let x = Devon's grade on the fifth test
$$80 \le \frac{78 + 64 + 88 + 76 + x}{5} < 90$$
$$80 \le \frac{306 + x}{5} < 90$$
$$5(80) \le 5\left(\frac{306 + x}{5}\right) < 5(90)$$
$$400 \le 306 + x < 450$$
$$400 - 306 \le 306 - 306 + x < 450 - 306$$
$$94 \le x < 144$$

Devon must have a score of $94 \le x \le 100$, assuming 100 is the highest grade possible.

56. Let x = the number of tents rented
$$950 \le 325 + 125x \le 1200$$
$$950 - 325 \le 325 - 325 + 125x \le 1200 - 325$$
$$625 \le 125x \le 875$$
$$\frac{625}{125} \le \frac{125x}{125} \le \frac{875}{125}$$
$$5 \le x \le 7$$

Minimum: 5   Maximum: 7

57. Let x = the number of gallons
$$250x = 2750 \text{ and } 400x = 2750$$
$$x = \frac{2750}{250}, \; x = \frac{2750}{400}$$
$$x = 11, \; x = 6.875$$
$$6.875 \le x \le 11$$

58. Let $x =$ the final exam grade

The semester average $= \dfrac{86+74+68+96+72}{5} = \dfrac{396}{5} = 79.2$

The final grade is found by taking $\dfrac{2}{3}$ of the semester average and adding this to $\dfrac{1}{3}$ of the final exam. The final

grade is $\dfrac{2}{3}(79.2)+\dfrac{1}{3}x = 52.8+\dfrac{1}{3}x$. In order for Teresa to receive a final grade of B in the course, she must have an

average greater than or equal to 80 and less than 90.

$$80 \le 52.8+\dfrac{1}{3}x < 90$$

$$80-52.8 \le 52.8-52.8+\dfrac{1}{3}x < 90-52.8$$

$$27.2 \le \dfrac{1}{3}x < 37.2$$

$$3(27.2) \le 3\left(\dfrac{1}{3}x\right) < 3(37.2)$$

$$81.6 \le x < 111.6$$

Thus, Teresa must receive $81.6 \le x \le 100$, assuming that 100 is the highest grade possible.

59. Student's answer: $-\dfrac{1}{3}x \le 4$

$$-3\left(-\dfrac{1}{3}x\right) \le -3(4)$$

$$x \le -12$$

Correct answer: $-\dfrac{1}{3}x \le 4$

$$-3\left(-\dfrac{1}{3}x\right) \ge -3(4)$$

$$x \ge -12$$

Yes, -12 is in both solution sets.

Exercise Set 6.7

1. A **graph** is an illustration of all the points whose coordinates satisfy an equation.

2. To find the **x-intercept**, set $y = 0$ and solve the equation for x.

3. To find the **y-intercept**, set $x = 0$ and solve the equation for y.

4. The **slope of a line** is a ratio of the vertical change to the horizontal change for any two points on the line.

5. a) Divide the difference between the $y-$coordinates by the difference between the $x-$coordinates.

   b) $m = \dfrac{5-2}{-3-6} = \dfrac{3}{-9} = -\dfrac{1}{3}$

6. Plotting points, using intercepts, and using the slope and $y-$intercept

7. a) First
    b) Second

8. Two

9. - 16.

17. - 24.

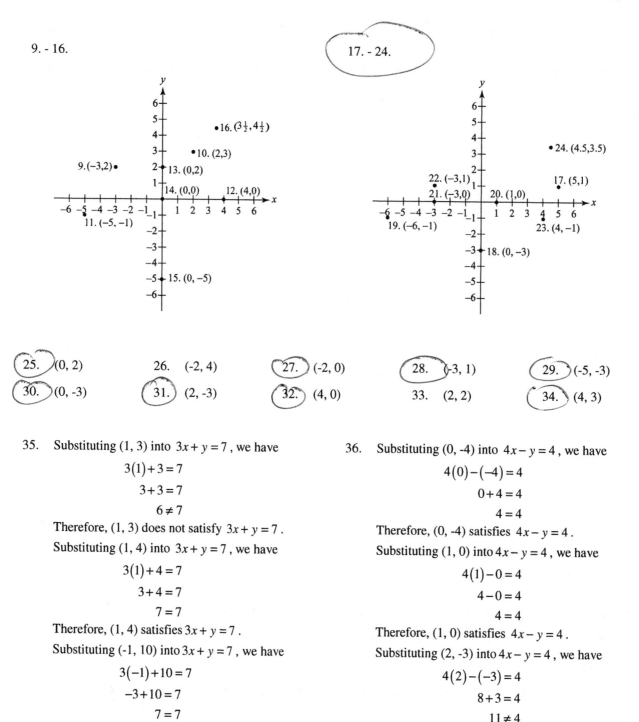

25. (0, 2) 26. (-2, 4) 27. (-2, 0) 28. (-3, 1) 29. (-5, -3)

30. (0, -3) 31. (2, -3) 32. (4, 0) 33. (2, 2) 34. (4, 3)

35. Substituting (1, 3) into $3x + y = 7$, we have

$$3(1) + 3 = 7$$
$$3 + 3 = 7$$
$$6 \neq 7$$

Therefore, (1, 3) does not satisfy $3x + y = 7$.

Substituting (1, 4) into $3x + y = 7$, we have

$$3(1) + 4 = 7$$
$$3 + 4 = 7$$
$$7 = 7$$

Therefore, (1, 4) satisfies $3x + y = 7$.

Substituting (-1, 10) into $3x + y = 7$, we have

$$3(-1) + 10 = 7$$
$$-3 + 10 = 7$$
$$7 = 7$$

Therefore, (-1, 10) satisfies $3x + y = 7$.

36. Substituting (0, -4) into $4x - y = 4$, we have

$$4(0) - (-4) = 4$$
$$0 + 4 = 4$$
$$4 = 4$$

Therefore, (0, -4) satisfies $4x - y = 4$.

Substituting (1, 0) into $4x - y = 4$, we have

$$4(1) - 0 = 4$$
$$4 - 0 = 4$$
$$4 = 4$$

Therefore, (1, 0) satisfies $4x - y = 4$.

Substituting (2, -3) into $4x - y = 4$, we have

$$4(2) - (-3) = 4$$
$$8 + 3 = 4$$
$$11 \neq 4$$

Therefore, (2, -3) does not satisfy $4x - y = 4$.

37. Substituting $(5, 0)$ into $2x - 3y = 10$, we have

$$2(5) - 3(0) = 10$$
$$10 - 0 = 10$$
$$10 = 10$$

Therefore, $(5, 0)$ satisfies $2x - 3y = 10$.

Substituting $(0, 3)$ into $2x - 3y = 10$, we have

$$2(0) - 3(3) = 10$$
$$0 - 9 = 10$$
$$-9 \neq 10$$

Therefore, $(0, 3)$ does not satisfy $2x - 3y = 10$.

Substituting $\left(0, -\dfrac{10}{3}\right)$ into $2x - 3y = 10$, we have

$$2(0) - 3\left(-\dfrac{10}{3}\right) = 10$$
$$0 + 10 = 10$$
$$10 = 10$$

Therefore, $\left(0, -\dfrac{10}{3}\right)$ satisfies $2x - 3y = 10$.

38. Substituting $(2, 1)$ into $3y = 4x + 2$, we have

$$3(1) = 4(2) + 2$$
$$3 = 8 + 2$$
$$3 \neq 10$$

Therefore, $(2, 1)$ does not satisfy $3y = 4x + 2$.

Substituting $(1, 2)$ into $3y = 4x + 2$, we have

$$3(2) = 4(1) + 2$$
$$6 = 4 + 2$$
$$6 = 6$$

Therefore, $(1, 2)$ satisfies $3y = 4x + 2$.

Substituting $\left(0, \dfrac{2}{3}\right)$ into $3y = 4x + 2$, we have

$$3\left(\dfrac{2}{3}\right) = 4(0) + 2$$
$$2 = 0 + 2$$
$$2 = 2$$

Therefore, $\left(0, \dfrac{2}{3}\right)$ satisfies $3y = 4x + 2$.

39. Substituting $(1, -1)$ into $7y = 3x - 5$, we have

$$7(-1) = 3(1) - 5$$
$$-7 = 3 - 5$$
$$-7 \neq -2$$

Therefore, $(1, -1)$ does not satisfy $7y = 3x - 5$.

Substituting $(-3, -2)$ into $7y = 3x - 5$, we have

$$7(-2) = 3(-3) - 5$$
$$-14 = -9 - 5$$
$$-14 = -14$$

Therefore, $(-3, -2)$ satisfies $7y = 3x - 5$.

Substituting $(2, 5)$ into $7y = 3x - 5$, we have

$$7(5) = 3(2) - 5$$
$$35 = 6 - 5$$
$$35 \neq 1$$

Therefore, $(2, 5)$ does not satisfy $7y = 3x - 5$.

40. Substituting $\left(0, \dfrac{4}{3}\right)$ into $\dfrac{x}{2} + 3y = 4$, we have

$$\dfrac{0}{2} + 3\left(\dfrac{4}{3}\right) = 4$$
$$0 + 4 = 4$$
$$4 = 4$$

Therefore, $\left(0, \dfrac{4}{3}\right)$ satisfies $\dfrac{x}{2} + 3y = 4$.

Substituting $(8, 0)$ into $\dfrac{x}{2} + 3y = 4$, we have

$$\dfrac{8}{2} + 3(0) = 4$$
$$4 + 0 = 4$$
$$4 = 4$$

Therefore, $(8, 0)$ satisfies $\dfrac{x}{2} + 3y = 4$.

Substituting $(10, -2)$ into $\dfrac{x}{2} + 3y = 4$, we have

$$\dfrac{10}{2} + 3(-2) = 4$$
$$5 - 6 = 4$$
$$-1 \neq 4$$

Therefore, $(10, -2)$ does not satisfy $\dfrac{x}{2} + 3y = 4$.

41. Substituting $\left(0, \dfrac{8}{3}\right)$ into $\dfrac{x}{2} + \dfrac{3y}{4} = 2$, we have

$$\dfrac{0}{2} + \dfrac{8}{4} = 2$$
$$0 + 2 = 2$$
$$2 = 2$$

Therefore, $\left(0, \dfrac{8}{3}\right)$ satisfies $\dfrac{x}{2} + \dfrac{3y}{4} = 2$.

Substituting $\left(1, \dfrac{11}{4}\right)$ into $\dfrac{x}{2} + \dfrac{3y}{4} = 2$, we have

$$\dfrac{1}{2} + \dfrac{33}{16} = 2$$
$$\dfrac{8}{16} + \dfrac{33}{16} = 2$$
$$\dfrac{41}{16} \neq 2$$

Therefore, $\left(1, \dfrac{11}{4}\right)$ does not satisfy $\dfrac{x}{2} + \dfrac{3y}{4} = 2$.

Substituting $(4, 0)$ into $\dfrac{x}{2} + \dfrac{3y}{4} = 2$, we have

$$\dfrac{4}{2} + \dfrac{0}{4} = 2$$
$$2 + 0 = 2$$
$$2 = 2$$

Therefore, $(4, 0)$ satisfies $\dfrac{x}{2} + \dfrac{3y}{4} = 2$.

42. Substituting $(2, 1)$ into $2x - 5y = -7$, we have

$$2(2) - 5(1) = -7$$
$$4 - 5 = -7$$
$$-1 \neq -7$$

Therefore, $(2, 1)$ does not satisfy $2x - 5y = -7$.

Substituting $(-1, 1)$ into $2x - 5y = -7$, we have

$$2(-1) - 5(1) = -7$$
$$-2 - 5 = -7$$
$$-7 = -7$$

Therefore, $(-1, 1)$ satisfies $2x - 5y = -7$.

Substituting $(4, 3)$ into $2x - 5y = -7$, we have

$$2(4) - 5(3) = -7$$
$$8 - 15 = -7$$
$$-7 = -7$$

Therefore, $(4, 3)$ satisfies $2x - 5y = -7$.

43. Since the line is vertical, its slope is undefined.

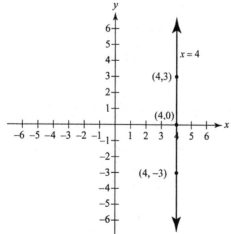

44. Since the line is vertical, its slope is undefined.

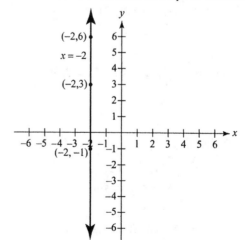

45. Since the line is horizontal, its slope is 0.

46. Since the line is horizontal, its slope is 0.

47.

48.

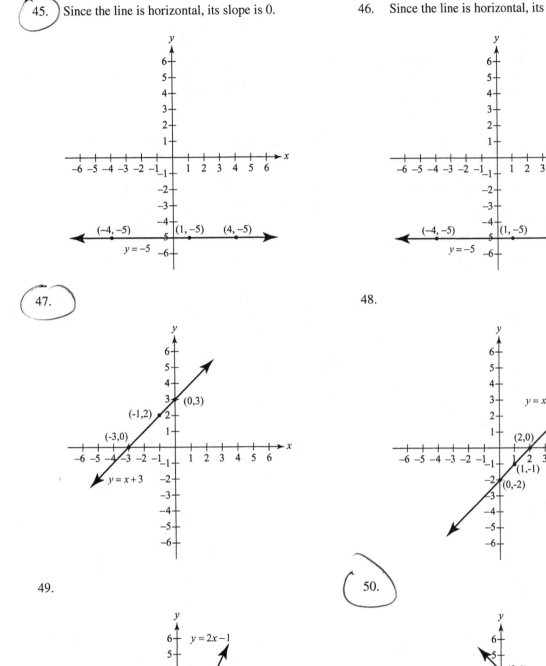

49.

50.

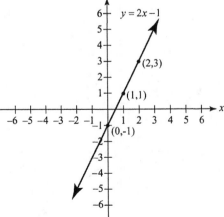

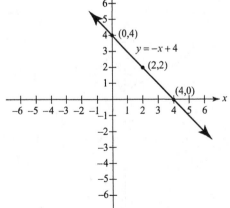

51.

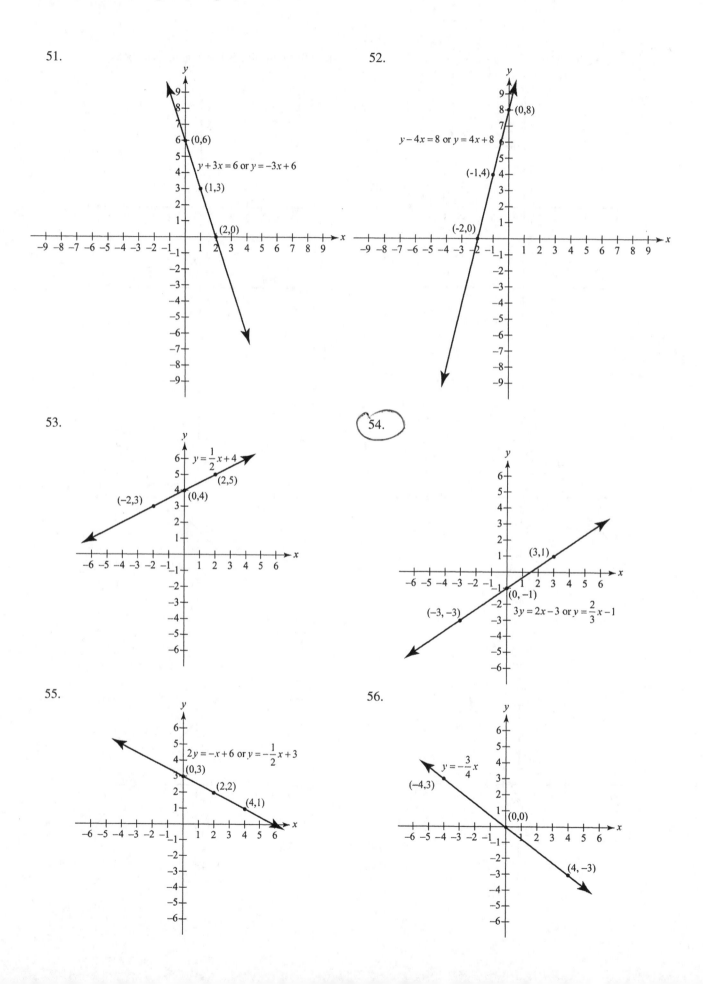

$y + 3x = 6$ or $y = -3x + 6$

(0,6)

(1,3)

(2,0)

52.

(0,8)

$y - 4x = 8$ or $y = 4x + 8$

(-1,4)

(-2,0)

53.

$y = \frac{1}{2}x + 4$

(2,5)

(-2,3)   (0,4)

54.

(3,1)

(0,-1)

(-3,-3) $3y = 2x - 3$ or $y = \frac{2}{3}x - 1$

55.

$2y = -x + 6$ or $y = -\frac{1}{2}x + 3$

(0,3)

(2,2)

(4,1)

56.

$y = -\frac{3}{4}x$

(-4,3)

(0,0)

(4,-3)

57.

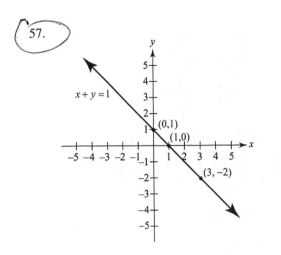

58.

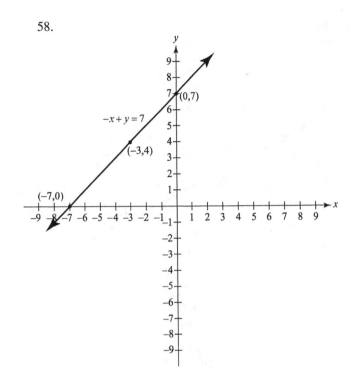

59.

60.

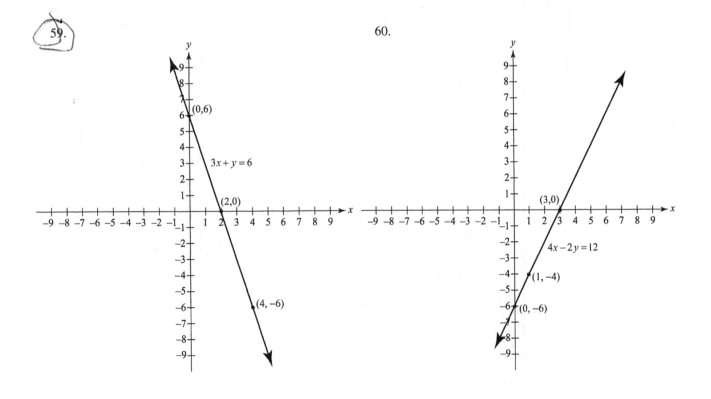

61.

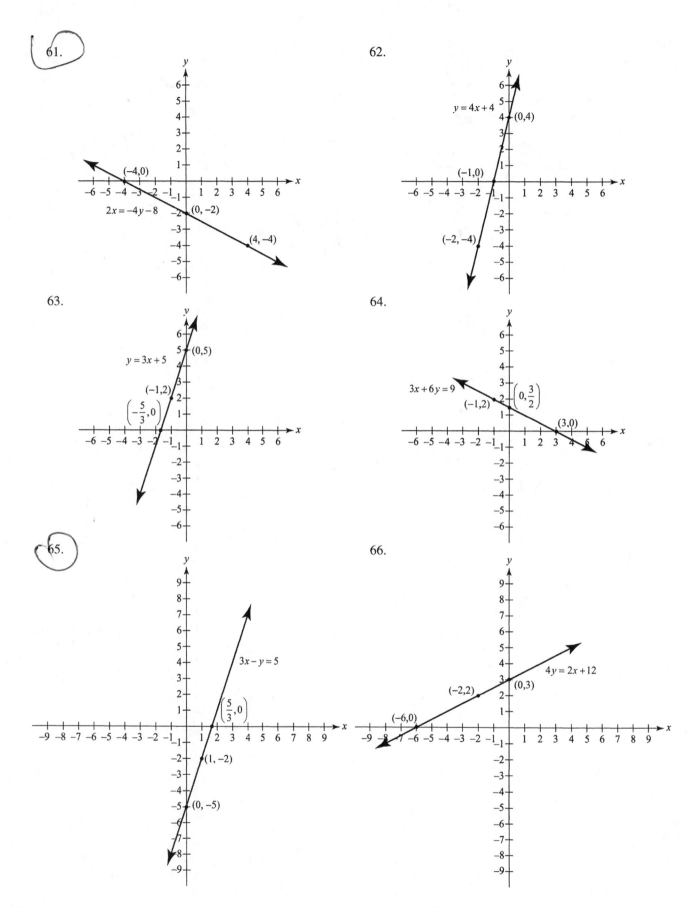

67. $(3,7),(10,21)$ $m = \dfrac{21-7}{10-3} = \dfrac{14}{7} = 2$

68. $(4,1),(1,4)$ $m = \dfrac{4-1}{1-4} = \dfrac{3}{-3} = -1$

69. $(2,6),(-5,-9)$ $m = \dfrac{-9-6}{-5-2} = \dfrac{-15}{-7} = \dfrac{15}{7}$

70. $(-5,6),(7,-9)$ $m = \dfrac{-9-6}{7-(-5)} = \dfrac{-15}{12} = -\dfrac{5}{4}$

71. $(5,2),(-3,2)$ $m = \dfrac{2-2}{-3-5} = \dfrac{0}{-8} = 0$

72. $(-3,-5),(-1,-2)$ $m = \dfrac{-2-(-5)}{-1-(-3)} = \dfrac{3}{2}$

73. $(8,-3),(8,3)$ $m = \dfrac{3-(-3)}{8-8} = \dfrac{6}{0}$ Undefined

74. $(2,6),(2,-3)$ $m = \dfrac{-3-6}{2-2} = \dfrac{-9}{0}$ Undefined

75. $(-2,3),(1,-1)$ $m = \dfrac{-1-3}{1-(-2)} = \dfrac{-4}{3} = -\dfrac{4}{3}$

76. $(-7,-5),(5,-6)$ $m = \dfrac{-6-(-5)}{5-(-7)} = \dfrac{-1}{12} = -\dfrac{1}{12}$

77.

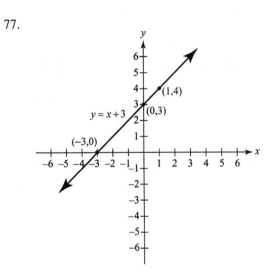

78.

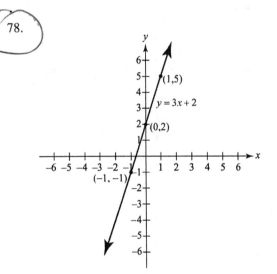

79.

80.

81.

82.

83.

84.

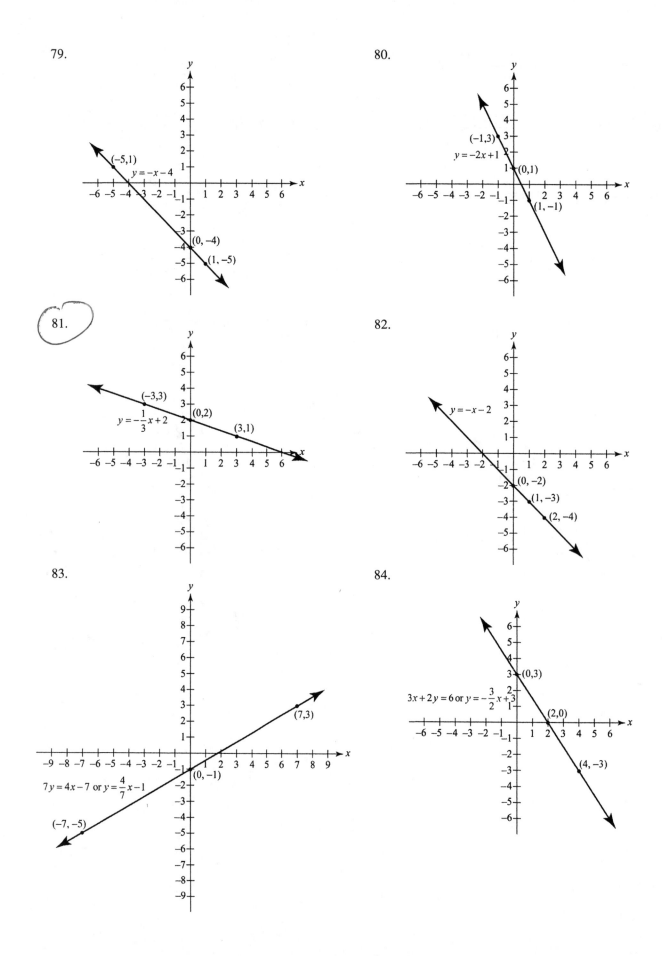

85.

86.

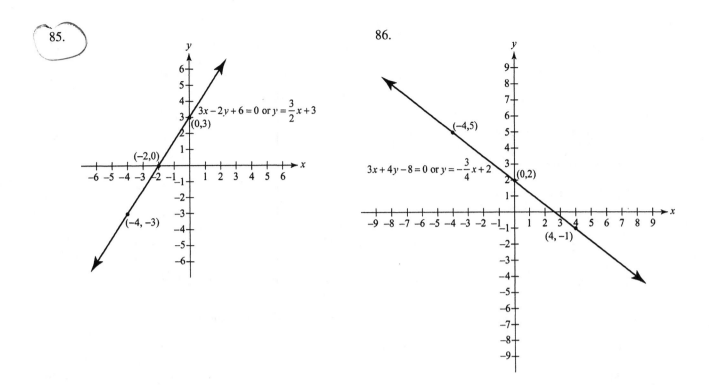

87. The y-intercept is 3; thus $b = 3$. The slope is negative since the graph falls from left to right. The change in y is 3, while the change in x is 4. Thus m, the slope, is $-\dfrac{3}{4}$. The equation is $y = -\dfrac{3}{4}x + 3$.

88. The y-intercept is 3; thus $b = 3$. The slope is positive since the graph rises from left to right. The change in y is 3, while the change in x is 2. Thus m, the slope, is $\dfrac{3}{2}$. The equation is $y = \dfrac{3}{2}x + 3$.

89. The y-intercept is 2; thus $b = 2$. The slope is positive since the graph rises from left to right. The change in y is 3, while the change in x is 1. Thus m, the slope, is $\dfrac{3}{1} = 3$. The equation is $y = 3x + 2$.

90. The y-intercept is 1; thus $b = 1$. The slope is negative since the graph falls from left to right. The change in y is 2, while the change in x is 1. Thus m, the slope, is -2. The equation is $y = -2x + 1$.

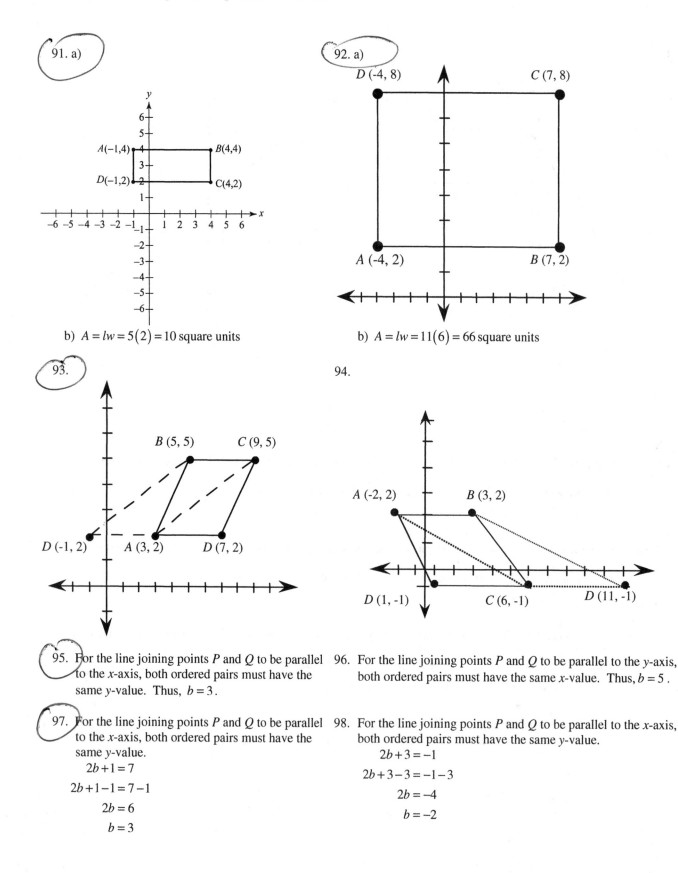

91. a)

b) $A = lw = 5(2) = 10$ square units

92. a)

b) $A = lw = 11(6) = 66$ square units

93.

94.

95. For the line joining points P and Q to be parallel to the x-axis, both ordered pairs must have the same y-value. Thus, $b = 3$.

96. For the line joining points P and Q to be parallel to the y-axis, both ordered pairs must have the same x-value. Thus, $b = 5$.

97. For the line joining points P and Q to be parallel to the x-axis, both ordered pairs must have the same y-value.

$$2b + 1 = 7$$
$$2b + 1 - 1 = 7 - 1$$
$$2b = 6$$
$$b = 3$$

98. For the line joining points P and Q to be parallel to the x-axis, both ordered pairs must have the same y-value.

$$2b + 3 = -1$$
$$2b + 3 - 3 = -1 - 3$$
$$2b = -4$$
$$b = -2$$

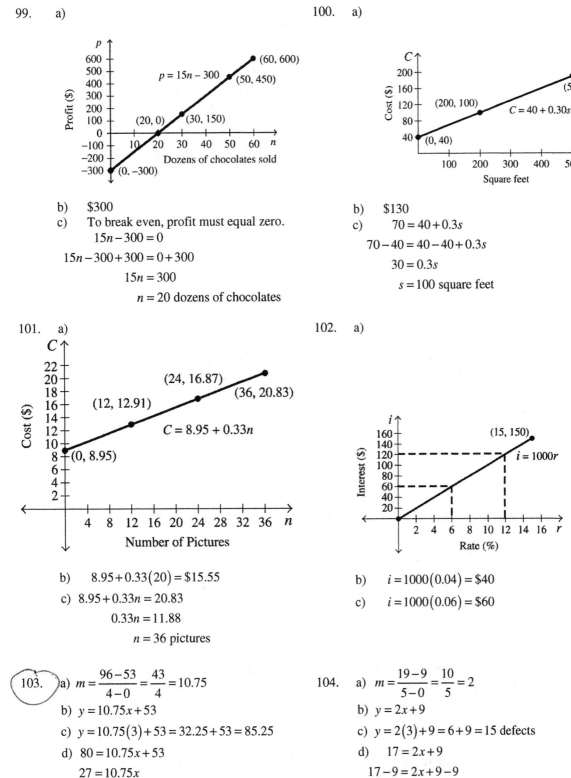

99.   a)

b) $300
c)   To break even, profit must equal zero.
$$15n - 300 = 0$$
$$15n - 300 + 300 = 0 + 300$$
$$15n = 300$$
$$n = 20 \text{ dozens of chocolates}$$

100.   a)

b) $130
c) $$70 = 40 + 0.3s$$
$$70 - 40 = 40 - 40 + 0.3s$$
$$30 = 0.3s$$
$$s = 100 \text{ square feet}$$

101.   a)

b) $$8.95 + 0.33(20) = \$15.55$$
c) $$8.95 + 0.33n = 20.83$$
$$0.33n = 11.88$$
$$n = 36 \text{ pictures}$$

102.   a)

b) $$i = 1000(0.04) = \$40$$
c) $$i = 1000(0.06) = \$60$$

103.   a) $$m = \frac{96 - 53}{4 - 0} = \frac{43}{4} = 10.75$$

b) $$y = 10.75x + 53$$

c) $$y = 10.75(3) + 53 = 32.25 + 53 = 85.25$$

d) $$80 = 10.75x + 53$$
$$27 = 10.75x$$
$$x = 2.511627907 \approx 2.5 \text{ hours}$$

104.   a) $$m = \frac{19 - 9}{5 - 0} = \frac{10}{5} = 2$$

b) $$y = 2x + 9$$

c) $$y = 2(3) + 9 = 6 + 9 = 15 \text{ defects}$$

d) $$17 = 2x + 9$$
$$17 - 9 = 2x + 9 - 9$$
$$8 = 2x$$
$$x = 4 \text{ workers}$$

105. a) $m = \dfrac{24-40}{30-0} = \dfrac{-16}{30} = -\dfrac{8}{15}$

   b) $y = -\dfrac{8}{15}x + 40$

   c) $y = -\dfrac{8}{15}(15) + 40 = -8 + 40$

   $= 32\%$

   d) $30 = -\dfrac{8}{15}x + 40$

   $30 - 40 = -\dfrac{8}{15}x + 40 - 40$

   $-10 = -\dfrac{8}{15}x$

   $-10\left(-\dfrac{15}{8}\right) = -\dfrac{8}{15}x\left(-\dfrac{15}{8}\right)$

   $x = \dfrac{150}{8}$

   $= 18.75$ years after 1970, or in 1988

106. a) $m = \dfrac{25,000 - 17,000}{9-0} = \dfrac{8000}{9} = 888.\overline{8} \approx 888.89$ million

   b) $y = 888.89x + 17,000$ (numbers in millions)

   c) $y = 888.89(4) + 17,000 = 20,555.56 \approx \$20,556$ million

   d) $20,000 = 888.89x + 17,000$

   $3000 = 888.89x$

   $x = 3.374995781$

   ≈ 3.37 years after 1994, or in 1997

107. a) Solve the equations for y to put them in slope-intercept form. Then compare the slopes and y-intercepts. If the slopes are equal but the y-intercepts are different, then the lines are parallel.

   b) $2x - 3y = 6$

   $2x - 2x - 3y = -2x + 6$

   $-3y = -2x + 6$

   $\dfrac{-3y}{-3} = \dfrac{-2x}{-3} + \dfrac{6}{-3}$

   $y = \dfrac{2}{3}x - 2$

   $4x = 6y + 6$

   $4x - 6 = 6y + 6 - 6$

   $4x - 6 = 6y$

   $\dfrac{4x}{6} - \dfrac{6}{6} = \dfrac{6y}{6}$

   $\dfrac{2}{3}x - 1 = y$

   Since the two equations have the same slope, $m = \dfrac{2}{3}$, the graphs of the equations are parallel lines.

108. Quadrants 1, 2, and 4. The graph of the line $x + y = 1$ is in quadrants 1, 2, and 4; therefore, the set of points that satisfy the equation is in these quadrants.

Exercise Set 6.8

1. (1) Mentally substitute the equal sign for the inequality sign and plot points as if you were graphing the equation. (2) If the inequality is < or >, draw a dashed line through the points. If the inequality is ≤ or ≥, draw a solid line through the points. (3) Select a test point not on the line and substitute the x- and y- coordinates into the inequality. If the substitution results in a true statement, shade in the area on the same side of the line as the test point. If the substitution results in a false statement, shade in the area on the opposite side of the line as the test point.

2. To indicate that the line is part of the solution set, we draw a solid line. To indicate that the line is not part of the solution set, we draw a dashed line.

3. Graph $x = 1$. Since the original statement is less than or equal to, a solid line is drawn. Since the point (0, 0) satisfies the inequality $x \leq 1$, all points on the line and in the half-plane to the left of the line $x = 1$ are in the solution set.

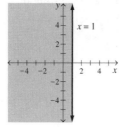

4. Graph $y = -2$. Since the original statement is greater than or equal to, a solid line is drawn. Since the point (0, 0) satisfies the inequality $y \geq -2$, all points on the line and in the half-plane above the line $y = -2$ are in the solution set.

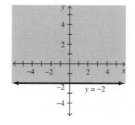

5. Graph $y = x + 3$. Since the original statement is strictly greater than, a dashed line is drawn. Since the point (0, 0) does not satisfy the inequality $y > x + 3$, all points in the half-plane above the line $y = x + 3$ are in the solution set.

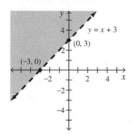

6. Graph $y = x - 5$. Since the original statement is strictly less than, a dashed line is drawn. Since the point (0, 0) does not satisfy the inequality $y < x - 5$, all points in the half-plane below the line $y = x - 5$ are in the solution set.

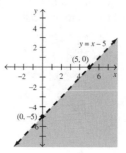

7. Graph $y = 2x - 6$. Since the original statement is greater than or equal to, a solid line is drawn. Since the point (0, 0) satisfies the inequality $y \geq 2x - 6$, all points on the line and in the half-plane above the line $y = 2x - 6$ are in the solution set.

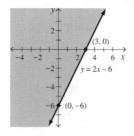

8. Graph $y = -2x + 2$. Since the original statement is strictly less than, a dashed line is drawn. Since the point (0, 0) satisfies the inequality $y < -2x + 2$, all points in the half-plane below the line $y = -2x + 2$ are in the solution set.

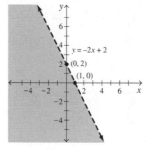

9. Graph $3x - 4y = 12$. Since the original statement is strictly greater than, a dashed line is drawn. Since the point (0, 0) does not satisfy the inequality $3x - 4y > 12$, all points in the half-plane below the line $3x - 4y = 12$ are in the solution set.

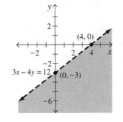

10. Graph $x + 2y = 4$. Since the original statement is strictly greater than, a dashed line is drawn. Since the point (0, 0) does not satisfy the inequality $x + 2y > 4$, all points in the half-plane above the line $x + 2y = 4$ are in the solution set.

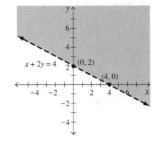

11. Graph $3x - 4y = 9$. Since the original statement is less than or equal to, a solid line is drawn. Since the point (0, 0) satisfies the inequality $3x - 4y \leq 9$, all points on the line and in the half-plane above the line $3x - 4y = 9$ are in the solution set.

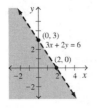

12. Graph $4y - 3x = 9$. Since the original statement is greater than or equal to, a solid line is drawn. Since the point (0, 0) does not satisfy the inequality $4y - 3x \geq 9$, all points on the line and in the half-plane above the line $4y - 3x = 9$ are in the solution set.

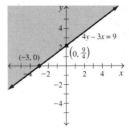

13. Graph $3x + 2y = 6$. Since the original statement is strictly less than, a dashed line is drawn. Since the point (0, 0) satisfies the inequality $3x + 2y < 6$, all points in the half-plane to the left of the line $3x + 2y = 6$ are in the solution set.

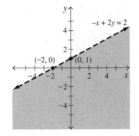

14. Graph $-x + 2y = 2$. Since the original statement is strictly less than, a dashed line is drawn. Since the point (0, 0) satisfies the inequality $-x + 2y < 2$, all points in the half-plane below the line $-x + 2y = 2$ are in the solution set.

15. Graph $x + y = 0$. Since the original statement is strictly greater than, a dashed line is drawn. Since the point (1, 1) satisfies the inequality $x + y > 0$, all points in the half-plane above the line $x + y = 0$ are in the solution set.

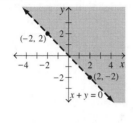

16. Graph $x + 2y = 0$. Since the original statement is less than or equal to, a solid line is drawn. Since the point (1, 1) does not satisfy the inequality $x + 2y \le 0$, all points on the line $x + 2y = 0$ and in the half-plane below the line are in the solution set.

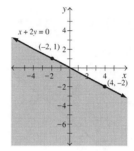

17. Graph $5x - 2y = 10$. Since the original statement is less than or equal to, a solid line is drawn. Since the point (0, 0) satisfies the inequality $5x - 2y \le 10$, all points on the line and in the half-plane above the line $5x - 2y = 10$ are in the solution set.

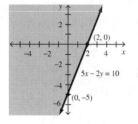

18. Graph $y = -2x + 1$. Since the original statement is greater than or equal to, a solid line is drawn. Since the point (0, 0) does not satisfy the inequality $y \ge -2x + 1$, all points on the line and in the half-plane above the line $y = -2x + 1$ are in the solution set.

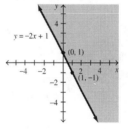

19. Graph $3x + 2y = 12$. Since the original statement is strictly greater than, a dashed line is drawn. Since the point (0, 0) does not satisfy the inequality $3x + 2y > 12$, all points in the half-plane above the line $3x + 2y = 12$ are in the solution set.

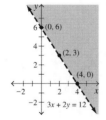

20. Graph $y = 3x - 4$. Since the original statement is less than or equal to, a solid line is drawn. Since the point (0, 0) does not satisfy the inequality $y \le 3x - 4$, all points on the line and in the half-plane below the line $y = 3x - 4$ are in the solution set.

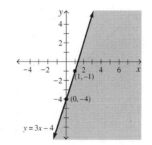

21. Graph $\frac{2}{5}x - \frac{1}{2}y = 1$. Since the original statement is less than or equal to, a solid line is drawn. Since the point (0, 0) satisfies the inequality $\frac{2}{5}x - \frac{1}{2}y \le 1$, all points on the line and in the half-plane above the line $\frac{2}{5}x - \frac{1}{2}y = 1$ are in the solution set.

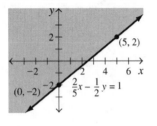

22. Graph $0.1x + 0.3y = 0.4$. Since the original statement is less than or equal to, a solid line is drawn. Since the point (0, 0) satisfies the inequality $0.1x + 0.3y \le 0.4$, all points on the line and in the half-plane below the line $0.1x + 0.3y = 0.4$ are in the solution set.

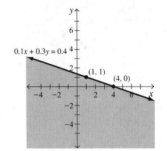

23. Graph $0.2x + 0.5y = 0.3$. Since the original statement is less than or equal to, a solid line is drawn. Since the point (0, 0) satisfies the inequality $0.2x + 0.5y \le 0.3$, all points on the line and in the half-plane below the line $0.2x + 0.5y = 0.3$ are in the solution set.

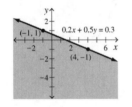

24. Graph $\frac{1}{3}x + \frac{3}{4}y = 1$. Since the original statement is greater than or equal to, a solid line is drawn. Since the point (0, 0) does not satisfy the inequality $\frac{1}{3}x + \frac{3}{4}y \ge 1$, all points on the line and in the half-plane above the line $\frac{1}{3}x + \frac{3}{4}y = 1$ are in the solution set.

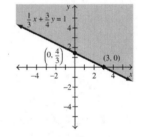

25. a) $x + y \le 300$
    b)

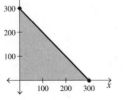

26. a) $2l + 2w \le 40, 0 \le l \le 20, 0 \le w \le 20$
    b)

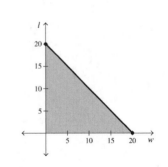

27. a) x = the number of acres of land, y = the number of square feet in the house

    b)

    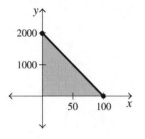

    c) $1500x + 75(1950) = 150,000$

    $1500x + 146,250 = 150,000$

    $1500x = 3750$

    $x = 2.5$ acres or less

    d) $1500(5) + 75y = 150,000$

    $7500 + 75y = 150,000$

    $75y = 142,500$

    $y = 1900$ ft^2 or less

28. a) No, you cannot have a negative number of shirts.

    b)

    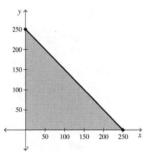

    c) Answers will vary.

29. a)
$$3x - y < 6$$
$$3x - 3x - y < -3x + 6$$
$$-y < -3x + 6$$
$$\frac{-y}{-1} > \frac{-3x}{-1} + \frac{6}{-1}$$
$$y > 3x - 6$$

b)
$$-3x + y > -6$$
$$-3x + 3x + y > 3x - 6$$
$$y > 3x - 6$$

c)
$$3x - 2y < 12$$
$$3x - 3x - 2y < -3x + 12$$
$$-2y < -3x + 12$$
$$\frac{-2y}{-2} > \frac{-3x}{-2} + \frac{12}{-2}$$
$$y > \frac{3}{2}x - 6$$

d) $y > 3x - 6$

a, b, and d

Exercise Set 6.9

1. A **binomial** is an expression that contains two terms in which each exponent that appears on the variable is a whole number. Examples: $2x+3$, $x-7$, x^2-9

2. A **trinomial** is an expression containing three terms in which each exponent that appears on the variable is a whole number. Examples: $x+y+2$, $2x+y+1$, x^2-y^4-1

3. The **foil method** is a method that obtains the products of the First, Outer, Inner, and Last terms of the binomials.

4. If the product of two factors is 0, then one or both of the factors must have a value of 0.

5. $ax^2+bx+c=0$, $a\neq 0$

6. $x=\dfrac{-b\pm\sqrt{b^2-4ac}}{2a}$

7. $x^2+9x+18=(x+6)(x+3)$

8. $x^2+5x+4=(x+4)(x+1)$

9. $x^2-x-6=(x-3)(x+2)$

10. $x^2+x-6=(x+3)(x-2)$

11. $x^2+2x-24=(x+6)(x-4)$

12. $x^2-6x+8=(x-4)(x-2)$

13. $x^2-2x-3=(x+1)(x-3)$

14. $x^2-5x-6=(x-6)(x+1)$

15. $x^2-10x+21=(x-7)(x-3)$

16. $x^2-81=(x-9)(x+9)$

17. $x^2-25=(x-5)(x+5)$

18. $x^2-x-20=(x-5)(x+4)$

19. $x^2+3x-28=(x+7)(x-4)$

20. $x^2+4x-32=(x-4)(x+8)$

21. $x^2+2x-63=(x+9)(x-7)$

22. $x^2-2x-48=(x+6)(x-8)$

23. $2x^2-x-10=(2x-5)(x+2)$

24. $3x^2-2x-5=(3x-5)(x+1)$

25. $4x^2+13x+3=(4x+1)(x+3)$

26. $2x^2-11x-21=(2x+3)(x-7)$

27. $5x^2+12x+4=(5x+2)(x+2)$

28. $2x^2-9x+10=(2x-5)(x-2)$

29. $4x^2+11x+6=(4x+3)(x+2)$

30. $4x^2+20x+21=(2x+7)(2x+3)$

31. $4x^2-11x+6=(4x-3)(x-2)$

32. $6x^2-11x+4=(3x-4)(2x-1)$

33. $3x^2-14x-24=(3x+4)(x-6)$

34. $6x^2+5x+1=(2x+1)(3x+1)$

35. $(x-1)(x+2)=0$

   $x-1=0$ or $x+2=0$

   $x=1$ $x=-2$

36. $(2x+5)(x-1)=0$

   $2x+5=0$ or $x-1=0$

   $2x=-5$ $x=1$

   $x=-\dfrac{5}{2}$

37. $(3x+4)(2x-1)=0$

   $3x+4=0$ or $2x-1=0$

   $3x=-4$ $2x=1$

   $x=-\dfrac{4}{3}$ $x=\dfrac{1}{2}$

38. $(x-6)(5x-4)=0$

   $x-6=0$ or $5x-4=0$

   $x=6$ $5x=4$

   $x=\dfrac{4}{5}$

39. $x^2 + 10x + 21 = 0$

     $(x+7)(x+3) = 0$

     $x+7 = 0$ or $x+3 = 0$

     $x = -7$ $x = -3$

40. $x^2 + 4x - 5 = 0$

     $(x+5)(x-1) = 0$

     $x+5 = 0$ or $x-1 = 0$

     $x = -5$ $x = 1$

41. $x^2 - 4x + 3 = 0$

     $(x-3)(x-1) = 0$

     $x-3 = 0$ or $x-1 = 0$

     $x = 3$ $x = 1$

42. $x^2 - 5x - 24 = 0$

     $(x-8)(x+3) = 0$

     $x-8 = 0$ or $x+3 = 0$

     $x = 8$ $x = -3$

43. $x^2 - 15 = 2x$

     $x^2 - 2x - 15 = 0$

     $(x-5)(x+3) = 0$

     $x-5 = 0$ or $x+3 = 0$

     $x = 5$ $x = -3$

44. $x^2 - 7x = -6$

     $x^2 - 7x + 6 = 0$

     $(x-6)(x-1) = 0$

     $x-6 = 0$ or $x-1 = 0$

     $x = 6$ $x = 1$

45. $x^2 = 4x - 3$

     $x^2 - 4x + 3 = 0$

     $(x-3)(x-1) = 0$

     $x-3 = 0$ or $x-1 = 0$

     $x = 3$ $x = 1$

46. $x^2 - 13x + 40 = 0$

     $(x-5)(x-8) = 0$

     $x-5 = 0$ or $x-8 = 0$

     $x = 5$ $x = 8$

47. $x^2 - 81 = 0$

     $(x-9)(x+9) = 0$

     $x-9 = 0$ or $x+9 = 0$

     $x = 9$ $x = -9$

48. $x^2 - 64 = 0$

     $(x-8)(x+8) = 0$

     $x-8 = 0$ or $x+8 = 0$

     $x = 8$ $x = -8$

49. $x^2 + 5x - 36 = 0$

     $(x+9)(x-4) = 0$

     $x+9 = 0$ or $x-4 = 0$

     $x = -9$ $x = 4$

50. $x^2 + 12x + 20 = 0$

     $(x+10)(x+2) = 0$

     $x+10 = 0$ or $x+2 = 0$

     $x = -10$ $x = -2$

51. $3x^2 + 10x = 8$

     $3x^2 + 10x - 8 = 0$

     $(3x-2)(x+4) = 0$

     $3x-2 = 0$ or $x+4 = 0$

     $3x = 2$ $x = -4$

     $x = \dfrac{2}{3}$

52. $3x^2 - 5x = 2$

     $3x^2 - 5x - 2 = 0$

     $(3x+1)(x-2) = 0$

     $3x+1 = 0$ or $x-2 = 0$

     $3x = -1$ $x = 2$

     $x = -\dfrac{1}{3}$

53. $5x^2 + 11x = -2$

    $5x^2 + 11x + 2 = 0$

    $(5x + 1)(x + 2) = 0$

    $5x + 1 = 0$ or $x + 2 = 0$

    $5x = -1 \qquad x = -2$

    $x = -\dfrac{1}{5}$

54. $2x^2 = -5x + 3$

    $2x^2 + 5x - 3 = 0$

    $(2x - 1)(x + 3) = 0$

    $2x - 1 = 0$ or $x + 3 = 0$

    $2x = 1 \qquad x = -3$

    $x = \dfrac{1}{2}$

55. $3x^2 - 4x = -1$

    $3x^2 - 4x + 1 = 0$

    $(3x - 1)(x - 1) = 0$

    $3x - 1 = 0$ or $x - 1 = 0$

    $3x = 1 \qquad x = 1$

    $x = \dfrac{1}{3}$

56. $5x^2 + 16x + 12 = 0$

    $(5x + 6)(x + 2) = 0$

    $5x + 6 = 0$ or $x + 2 = 0$

    $5x = -6 \qquad x = -2$

    $x = -\dfrac{6}{5}$

57. $4x^2 - 9x + 2 = 0$

    $(4x - 1)(x - 2) = 0$

    $4x - 1 = 0$ or $x - 2 = 0$

    $4x = 1 \qquad x = 2$

    $x = \dfrac{1}{4}$

58. $6x^2 + x - 2 = 0$

    $(2x - 1)(3x + 2) = 0$

    $2x - 1 = 0$ or $3x + 2 = 0$

    $2x = 1 \qquad 3x = -2$

    $x = \dfrac{1}{2} \qquad x = -\dfrac{2}{3}$

59. $x^2 + 2x - 15 = 0$

    $a = 1, \quad b = 2, \quad c = -15$

    $x = \dfrac{-2 \pm \sqrt{(2)^2 - 4(1)(-15)}}{2(1)}$

    $x = \dfrac{-2 \pm \sqrt{4 + 60}}{2} = \dfrac{-2 \pm \sqrt{64}}{2} = \dfrac{-2 \pm 8}{2}$

    $x = \dfrac{6}{2} = 3$ or $x = \dfrac{-10}{2} = -5$

60. $x^2 + 12x + 27 = 0$

    $a = 1, \quad b = 12, \quad c = 27$

    $x = \dfrac{-12 \pm \sqrt{(12)^2 - 4(1)(27)}}{2(1)}$

    $x = \dfrac{-12 \pm \sqrt{144 - 108}}{2} = \dfrac{-12 \pm \sqrt{36}}{2} = \dfrac{-12 \pm 6}{2}$

    $x = \dfrac{-6}{2} = -3$ or $x = \dfrac{-18}{2} = -9$

61. $x^2 - 3x - 18 = 0$

    $a = 1, \quad b = -3, \quad c = -18$

    $x = \dfrac{-(-3) \pm \sqrt{(-3)^2 - 4(1)(-18)}}{2(1)}$

    $x = \dfrac{3 \pm \sqrt{9 + 72}}{2} = \dfrac{3 \pm \sqrt{81}}{2} = \dfrac{3 \pm 9}{2}$

    $x = \dfrac{12}{2} = 6$ or $x = \dfrac{-6}{2} = -3$

62. $x^2 - 6x - 16 = 0$

    $a = 1, \quad b = -6, \quad c = -16$

    $x = \dfrac{-(-6) \pm \sqrt{(-6)^2 - 4(1)(-16)}}{2(1)}$

    $x = \dfrac{6 \pm \sqrt{36 + 64}}{2} = \dfrac{6 \pm \sqrt{100}}{2} = \dfrac{6 \pm 10}{2}$

    $x = \dfrac{16}{2} = 8$ or $x = \dfrac{-4}{2} = -2$

63. $x^2 - 8x = 9$

    $x^2 - 8x - 9 = 0$

    $a = 1,\ b = -8,\ c = -9$

    $$x = \frac{-(-8) \pm \sqrt{(-8)^2 - 4(1)(-9)}}{2(1)}$$

    $$x = \frac{8 \pm \sqrt{64 + 36}}{2} = \frac{8 \pm \sqrt{100}}{2} = \frac{8 \pm 10}{2}$$

    $$x = \frac{18}{2} = 9 \ \text{ or } \ x = \frac{-2}{2} = -1$$

64. $x^2 = -8x + 15$

    $x^2 + 8x - 15 = 0$

    $a = 1,\ b = 8,\ c = -15$

    $$x = \frac{-8 \pm \sqrt{(8)^2 - 4(1)(-15)}}{2(1)}$$

    $$x = \frac{-8 \pm \sqrt{64 + 60}}{2} = \frac{-8 \pm \sqrt{124}}{2} = \frac{-8 \pm 2\sqrt{31}}{2}$$

    $$x = -4 \pm \sqrt{31}$$

65. $x^2 - 2x + 3 = 0$

    $a = 1,\ b = -2,\ c = 3$

    $$x = \frac{-(-2) \pm \sqrt{(-2)^2 - 4(1)(3)}}{2(1)}$$

    $$x = \frac{2 \pm \sqrt{4 - 12}}{2} = \frac{2 \pm \sqrt{-8}}{2}$$

    No real solution

66. $2x^2 - x - 3 = 0$

    $a = 2,\ b = -1,\ c = -3$

    $$x = \frac{-(-1) \pm \sqrt{(-1)^2 - 4(2)(-3)}}{2(2)}$$

    $$x = \frac{1 \pm \sqrt{1 + 24}}{4} = \frac{1 \pm \sqrt{25}}{4} = \frac{1 \pm 5}{4}$$

    $$x = \frac{6}{4} = \frac{3}{2} \ \text{ or } \ x = \frac{-4}{4} = -1$$

67. $x^2 - 4x + 2 = 0$

    $a = 1,\ b = -4,\ c = 2$

    $$x = \frac{-(-4) \pm \sqrt{(-4)^2 - 4(1)(2)}}{2(1)}$$

    $$x = \frac{4 \pm \sqrt{16 - 8}}{2} = \frac{4 \pm \sqrt{8}}{2} = \frac{4 \pm 2\sqrt{2}}{2}$$

    $$x = 2 \pm \sqrt{2}$$

68. $2x^2 - 5x - 2 = 0$

    $a = 2,\ b = -5,\ c = -2$

    $$x = \frac{-(-5) \pm \sqrt{(-5)^2 - 4(2)(-2)}}{2(2)}$$

    $$x = \frac{5 \pm \sqrt{25 + 16}}{4} = \frac{5 \pm \sqrt{41}}{4}$$

69. $3x^2 - 8x + 1 = 0$

    $a = 3,\ b = -8,\ c = 1$

    $$x = \frac{-(-8) \pm \sqrt{(-8)^2 - 4(3)(1)}}{2(3)}$$

    $$x = \frac{8 \pm \sqrt{64 - 12}}{6} = \frac{8 \pm \sqrt{52}}{6} = \frac{8 \pm 2\sqrt{13}}{6}$$

    $$x = \frac{4 \pm \sqrt{13}}{3}$$

70. $2x^2 + 4x + 1 = 0$

    $a = 2,\ b = 4,\ c = 1$

    $$x = \frac{-4 \pm \sqrt{(4)^2 - 4(2)(1)}}{2(2)}$$

    $$x = \frac{-4 \pm \sqrt{16 - 8}}{4} = \frac{-4 \pm \sqrt{8}}{4} = \frac{-4 \pm 2\sqrt{2}}{4}$$

    $$x = \frac{-2 \pm \sqrt{2}}{2}$$

71. $4x^2 - x - 1 = 0$

$a = 4,\ b = -1,\ c = -1$

$$x = \frac{-(-1) \pm \sqrt{(-1)^2 - 4(4)(-1)}}{2(4)}$$

$$x = \frac{1 \pm \sqrt{1 + 16}}{8} = \frac{1 \pm \sqrt{17}}{8}$$

72. $4x^2 - 5x - 3 = 0$

$a = 4,\ b = -5,\ c = -3$

$$x = \frac{-(-5) \pm \sqrt{(-5)^2 - 4(4)(-3)}}{2(4)}$$

$$x = \frac{5 \pm \sqrt{25 + 48}}{8} = \frac{5 \pm \sqrt{73}}{8}$$

73. $2x^2 + 7x + 5 = 0$

$a = 2,\ b = 7,\ c = 5$

$$x = \frac{-7 \pm \sqrt{(7)^2 - 4(2)(5)}}{2(2)}$$

$$x = \frac{-7 \pm \sqrt{49 - 40}}{4} = \frac{-7 \pm \sqrt{9}}{4} = \frac{-7 \pm 3}{4}$$

$$x = \frac{-4}{4} = -1 \text{ or } x = \frac{-10}{4} = -\frac{5}{2}$$

74. $3x^2 = 9x - 5$

$3x^2 - 9x + 5 = 0$

$a = 3,\ b = -9,\ c = 5$

$$x = \frac{-(-9) \pm \sqrt{(-9)^2 - 4(3)(5)}}{2(3)}$$

$$x = \frac{9 \pm \sqrt{81 - 60}}{6} = \frac{9 \pm \sqrt{21}}{6}$$

75. $3x^2 - 10x + 7 = 0$

$a = 3,\ b = -10,\ c = 7$

$$x = \frac{-(-10) \pm \sqrt{(-10)^2 - 4(3)(7)}}{2(3)}$$

$$x = \frac{10 \pm \sqrt{100 - 84}}{6} = \frac{10 \pm \sqrt{16}}{6} = \frac{10 \pm 4}{6}$$

$$x = \frac{14}{6} = \frac{7}{3} \text{ or } x = \frac{6}{6} = 1$$

76. $4x^2 + 7x - 1 = 0$

$a = 4,\ b = 7,\ c = -1$

$$x = \frac{-7 \pm \sqrt{(7)^2 - 4(4)(-1)}}{2(4)}$$

$$x = \frac{-7 \pm \sqrt{49 + 16}}{8} = \frac{-7 \pm \sqrt{65}}{8}$$

77. $4x^2 - 11x + 13 = 0$

$a = 4,\ b = -11,\ c = 13$

$$x = \frac{-(-11) \pm \sqrt{(-11)^2 - 4(4)(13)}}{2(4)}$$

$$x = \frac{11 \pm \sqrt{121 - 208}}{8} = \frac{11 \pm \sqrt{-87}}{8}$$

No real solution

78. $5x^2 + 9x - 2 = 0$

$a = 5,\ b = 9,\ c = -2$

$$x = \frac{-9 \pm \sqrt{(9)^2 - 4(5)(-2)}}{2(5)}$$

$$x = \frac{-9 \pm \sqrt{81 + 40}}{10} = \frac{-9 \pm \sqrt{121}}{10} = \frac{-9 \pm 11}{10}$$

$$x = \frac{2}{10} = \frac{1}{5} \text{ or } x = \frac{-20}{10} = -2$$

79. Area of backyard $= lw = 30(20) = 600 \text{ m}^2$

Let x = width of grass around all sides of the flower garden

Width of flower garden $= 20 - 2x$

Length of flower garden $= 30 - 2x$

Area of flower garden $= lw = (30 - 2x)(20 - 2x)$

Area of grass $= 600 - (30 - 2x)(20 - 2x)$

$600 - (30 - 2x)(20 - 2x) = 336$

$600 - (600 - 100x + 4x^2) = 336$

$600 - 600 + 100x - 4x^2 = 336$

$-4x^2 + 100x = 336$

$4x^2 - 100x + 336 = 0$

$x^2 - 25x + 84 = 0$

$(x - 21)(x - 4) = 0$

$x - 21 = 0 \quad \text{or} \quad x - 4 = 0$

$x = 21 \qquad\qquad x = 4$

$x \neq 21$ since the width of the backyard is 20 m.

Width of grass = 4 m

Width of flower garden

$= 20 - 2x = 20 - 2(4) = 20 - 8 = 12$ m

Length of flower garden

$= 30 - 2x = 30 - 2(4) = 30 - 8 = 22$ m

80. $45,000 = x^2 + 15x - 100$

$x^2 + 15x - 45,100 = 0$

$(x + 220)(x - 205) = 0$

$x + 220 = 0 \quad \text{or} \quad x - 205 = 0$

$x = -220 \qquad\qquad x = 205$

Cannot produce a negative number of air conditioners. Thus, $x = 205$ air conditioners.

81. a) Since the equation is equal to 6 and not 0, the zero-factor property cannot be used.

b) $(x - 4)(x - 7) = 6$

$x^2 - 11x + 28 = 6$

$x^2 - 11x + 22 = 0$

$a = 1, \ b = -11, \ c = 22$

$x = \dfrac{-(-11) \pm \sqrt{(-11)^2 - 4(1)(22)}}{2(1)}$

$x = \dfrac{11 \pm \sqrt{121 - 88}}{2} = \dfrac{11 \pm \sqrt{33}}{2}$

$x \approx 8.37 \ \text{or} \ x \approx 2.63$

82. $x = \dfrac{-b \pm \sqrt{b^2 - 4ac}}{2a}$

The $b^2 - 4ac$ is the radicand in the quadratic formula, the part under the square root sign.

a) If $b^2 - 4ac > 0$, then you are taking the square root of a positive number and there are two solutions.

These solutions are $x = \dfrac{-b + \sqrt{b^2 - 4ac}}{2a}$ and $x = \dfrac{-b - \sqrt{b^2 - 4ac}}{2a}$.

b) If $b^2 - 4ac = 0$, then you are taking the square root of zero and there is one solution. This solution

is $x = \dfrac{-b \pm \sqrt{0}}{2a} = \dfrac{-b}{2a}$.

c) If $b^2 - 4ac < 0$, then you are taking the square root of a negative number and there is no real solution.

83. $(x+1)(x-3) = 0$

$x^2 - 2x - 3 = 0$

84.

```
R  I  A  A  S  N  E  Y
A  V  T  R  R  I  U  P
U  I  I  O  N  R  A  B
R  R  L  U  A  L  G  D
U  S  O  T  N  A  E  C
A  L  U  I  O  R  B  U
X  S  M  S  E  I  I  N
F  O  R  M  M  O  N  S
C  G  E  M  I  A  L  P
```

Exercise Set 6.10

1. A **function** is a special type of relation where each value of the independent variable corresponds to a unique value of the dependent variable.

2. A **relation** is any set of ordered pairs.

3. The **domain** of a function is the set of values that can be used for the independent variable.

4. The **range** of a function is the set of values obtained for the dependent variable.

5. The vertical line test can be used to determine if a graph represents a function. If a vertical line can be drawn so that it intersects the graph at more than one point, then each value of x does not have a unique value of y and the graph does not represent a function. If a vertical line cannot be made to intersect the graph in at least two different places, then the graph represents a function.

6. The area of a square is a function of the length of a side, the average stopping distance of a car is a function of its speed, the cost of apples is a function of the number of apples

7. Not a function since $x = 2$ is not paired with a unique value of y.

8. Function since each value of x is paired with a unique value of y.

    D: $x = -2, -1, 1, 2, 3$ R: $y = -1, 1, 2, 3$

9. Function since each vertical line intersects the graph at only one point.
    D: all real numbers   R: all real numbers

10. Function since each vertical line intersects the graph at only one point.
    D: all real numbers   R: all real numbers

11. Function since each vertical line intersects the graph at only one point.
    D: all real numbers   R: $y = 2$

12. Not a function since $x = -1$ is not paired with a unique value of y.

13. Function since each vertical line intersects the graph at only one point.
    D: all real numbers   R: $y \geq -4$

14. Function since each vertical line intersects the graph at only one point.
    D: all real numbers   R: $y \leq 10$

15. Not a function since it is possible to draw a vertical line that intersects the graph at more than one point.

16. Function since each vertical line intersects the graph at only one point.
    D: $0 \leq x \leq 8$ R: $-1 \leq y \leq 1$

17. Function since each vertical line intersects the graph at only one point.
    D: $0 \leq x < 12$ R: $y = 1, 2, 3$

18. Function since each vertical line intersects the graph at only one point.
    D: all real numbers   R: all real numbers

19. Not a function since it is possible to draw a vertical line that intersects the graph at more than one point.

20. Function since each vertical line intersects the graph at only one point.
    D: all real numbers   R: all real numbers

21. Function since each vertical line intersects the graph at only one point.
    D: all real numbers   R: $y > 0$

22. Function since each vertical line intersects the graph at only one point.
    D: $0 \leq x \leq 10$ R: $-1 \leq y \leq 3$

23. Not a function since it is possible to draw a vertical line that intersects the graph at more than one point.

24. Function since each vertical line intersects the graph at only one point.
    D: all real numbers   R: $y \geq 0$

25. Function since each value of x is paired with a unique value of y.

26. Function since each value of x is paired with a unique value of y.

27. Not a function since $x = 4$ is paired with two different values of y.

28. Not a function since $x = 3$ is paired with three different values of y.

29. Function since each value of x is paired with a unique value of y.

30. Not a function since $x = 1$ is paired with three different values of y.

31. $f(x) = x + 3, \quad x = 2$
    $f(2) = 2 + 3 = 5$

32. $f(x) = 2x + 5, \quad x = 4$
    $f(4) = 2(4) + 5 = 8 + 5 = 13$

33. $f(x) = -2x - 7, \quad x = -4$
    $f(-4) = -2(-4) - 7 = 8 - 7 = 1$

34. $f(x) = -5x + 3, \quad x = -1$
    $f(-1) = -5(-1) + 3 = 5 + 3 = 8$

35. $f(x) = 10x - 6, \quad x = 0$
    $f(0) = 10(0) - 6 = 0 - 6 = -6$

36. $f(x) = 7x - 6, \quad x = 4$
    $f(4) = 7(4) - 6 = 28 - 6 = 22$

37. $f(x) = x^2 - 3x + 1, \quad x = 4$

$f(4) = (4)^2 - 3(4) + 1 = 16 - 12 + 1 = 5$

38. $f(x) = x^2 - 5, \quad x = 7$

$f(7) = (7)^2 - 5 = 49 - 5 = 44$

39. $f(x) = 2x^2 - 2x - 8, \quad x = -2$

$f(-2) = 2(-2)^2 - 2(-2) - 8 = 8 + 4 - 8 = 4$

40. $f(x) = -x^2 + 3x + 7, \quad x = 2$

$f(2) = -(2)^2 + 3(2) + 7 = -4 + 6 + 7 = 9$

41. $f(x) = -3x^2 + 5x + 4, \quad x = -3$

$f(-3) = -3(-3)^2 + 5(-3) + 4 = -27 - 15 + 4 = -38$

42. $f(x) = 5x^2 + 2x + 5, \quad x = 4$

$f(4) = 5(4)^2 + 2(4) + 5 = 80 + 8 + 5 = 93$

43. $f(x) = -5x^2 + 3x - 9, \quad x = -1$

$f(-1) = -5(-1)^2 + 3(-1) - 9 = -5 - 3 - 9 = -17$

44. $f(x) = -3x^2 - 6x + 10, \quad x = -2$

$f(-2) = -3(-2)^2 - 6(-2) + 10 = -12 + 12 + 10 = 10$

45.

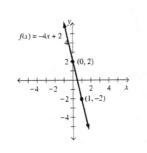

46.

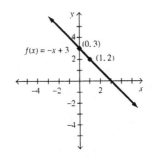

47.

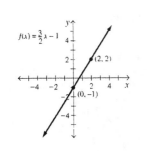

48.

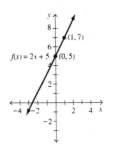

49.

50.

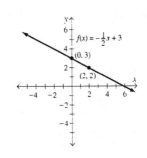

51. $y = x^2 - 16$

a) $a = 1 > 0$, opens upward

b) $x = 0$ c) $(0, -16)$ d) $(0, -16)$

e) $(4, 0), (-4, 0)$

f)

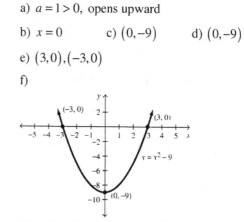

g) D: all real numbers   R: $y \geq -16$

52. $y = x^2 - 9$

a) $a = 1 > 0$, opens upward

b) $x = 0$ c) $(0, -9)$ d) $(0, -9)$

e) $(3, 0), (-3, 0)$

f)

g) D: all real numbers   R: $y \geq -9$

53. $y = -x^2 + 4$

a) $a = -1 < 0$, opens downward

b) $x = 0$ c) $(0, 4)$ d) $(0, 4)$

e) $(-2, 0), (2, 0)$

f)

g) D: all real numbers   R: $y \leq 4$

54. $y = -x^2 + 16$

a) $a = -1 < 0$, opens downward

b) $x = 0$ c) $(0, 16)$ d) $(0, 16)$

e) $(-4, 0), (4, 0)$

f)

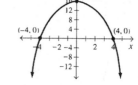

g) D: all real numbers   R: $y \leq 16$

55. $f(x) = -x^2 - 4$

a) $a = -1 < 0$, opens downward

b) $x = 0$ c) $(0, -4)$ d) $(0, -4)$

e) no x-intercepts

f)

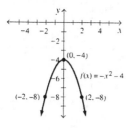

g) D: all real numbers   R: $y \leq -4$

56. $y = -2x^2 - 8$

a) $a = -2 < 0$, opens downward

b) $x = 0$ c) $(0, -8)$ d) $(0, -8)$

e) no x-intercepts

f)

g) D: all real numbers   R: $y \leq -8$

57. $y = 2x^2 - 3$

a) $a = 2 > 0,$ opens upward

b) $x = 0$ c) $(0, -3)$ d) $(0, -3)$

e) $(-1.22, 0), (1.22, 0)$

f)

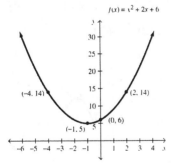

g) D: all real numbers   R: $y \geq -3$

58. $f(x) = -3x^2 - 6$

a) $a = -3 < 0,$ opens downward

b) $x = 0$ c) $(0, -6)$ d) $(0, -6)$

e) no x-intercepts

f)

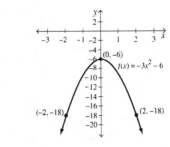

g) D: all real numbers   R: $y \leq -6$

59. $f(x) = x^2 + 2x + 6$

a) $a = 1 > 0,$ opens upward

b) $x = -1$ c) $(-1, 5)$ d) $(0, 6)$

e) no x-intercepts

f)

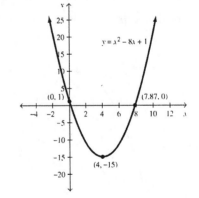

g) D: all real numbers   R: $y \geq 5$

60. $y = x^2 - 8x + 1$

a) $a = 1 > 0,$ opens upward

b) $x = 4$ c) $(4, -15)$ d) $(0, 1)$

e) $(7.87, 0), (0.13, 0)$

f)

g) D: all real numbers   R: $y \geq -15$

61. $y = x^2 + 5x + 6$

   a) $a = 1 > 0$, opens upward

   b) $x = -\dfrac{5}{2}$ c) $(-2.5, -0.25)$ d) $(0, 6)$

   e) $(-3, 0), (-2, 0)$

   f)

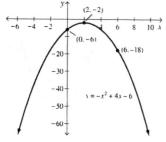

   g) D: all real numbers   R: $y \geq -0.25$

62. $y = x^2 - 7x - 8$

   a) $a = 1 > 0$, opens upward

   b) $x = \dfrac{7}{2}$ c) $\left(\dfrac{7}{2}, -\dfrac{81}{4}\right)$ d) $(0, -8)$

   e) $(-1, 0), (8, 0)$

   f)

   g) D: all real numbers   R: $y \geq -\dfrac{81}{4}$

63. $y = -x^2 + 4x - 6$

   a) $a = -1 < 0$, opens downward

   b) $x = 2$ c) $(2, -2)$ d) $(0, -6)$

   e) no x-intercepts

   f)

   g) D: all real numbers   R: $y \leq -2$

64. $y = -x^2 + 8x - 8$

   a) $a = -1 < 0$, opens downward

   b) $x = 4$ c) $(4, 8)$ d) $(0, -8)$

   e) $(1.17, 0), (6.83, 0)$

   f)

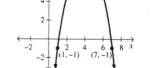

   g) D: all real numbers   R: $y \leq 8$

65. $y = -3x^2 + 14x - 8$

a) $a = -3 < 0$, opens downward

b) $x = \dfrac{7}{3}$ c) $\left(\dfrac{7}{3}, \dfrac{25}{3}\right)$ d) $(0, -8)$

e) $\left(\dfrac{2}{3}, 0\right), (4, 0)$

f)

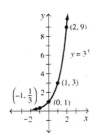

g) D: all real numbers   R: $y \le \dfrac{25}{3}$

66. $y = 2x^2 - x - 6$

a) $a = 2 > 0$, opens upward

b) $x = \dfrac{1}{4}$ c) $\left(\dfrac{1}{4}, -6\frac{1}{8}\right)$ d) $(0, -6)$

e) $(2, 0), \left(-\dfrac{3}{2}, 0\right)$

f)

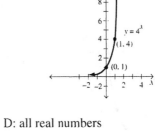

g) D: all real numbers   R: $y \ge -6\frac{1}{8}$

67.

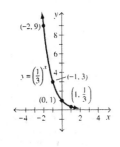

D: all real numbers   R: $y > 0$

68.

D: all real numbers   R: $y > 0$

69.

D: all real numbers   R: $y > 0$

70.

D: all real numbers   R: $y > 0$

71.

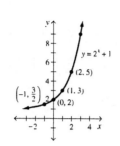

D: all real numbers   R: $y > 1$

72.

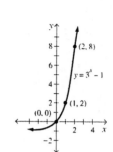

D: all real numbers   R: $y > -1$

73.

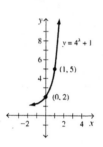

D: all real numbers   R: $y > 1$

74.

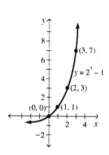

D: all real numbers   R: $y > -1$

75.

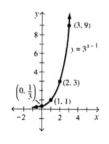

D: all real numbers   R: $y > 0$

76.

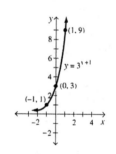

D: all real numbers   R: $y > 0$

77.

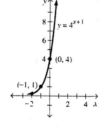

D: all real numbers   R: $y > 0$

78.

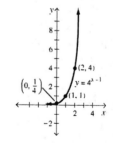

D: all real numbers   R: $y > 0$

79. $m(s) = 300 + 0.10s$

$$m(20,000) = 300 + 0.10(20,000)$$
$$= 300 + 2000 = \$2300$$

80. $d(t) = 60t$

a) $t = 3$, $d(3) = 60(3) = 180$ mi

b) $t = 7$, $d(7) = 60(7) = 420$ mi

81. a) $2000 \rightarrow x = 8$

$$f(8) = 0.56(8)^2 - 5.43(8) + 59.83$$
$$= 35.84 - 43.44 + 59.83 = 52.23\%$$

b) 1997

c) $x = \dfrac{-b}{2a} = \dfrac{-(-5.43)}{2(0.56)} = \dfrac{5.43}{1.12} = 4.848214286$

≈ 4.85

$$f(4.85) = 0.56(4.85)^2 - 5.43(4.85) + 59.83$$
$$= 13.1726 - 26.3355 + 59.83$$
$$= 46.6671 \approx 46.67\%$$

82. a) $1999 \rightarrow x = 5$

$$l(5) = -4.25(5)^2 + 30.32(5) + 150.14$$
$$= -106.25 + 151.6 + 150.14 = 195.49$$
$$\approx 195,490 \text{ free lunches}$$

b) 1998

c) $x = \dfrac{-b}{2a} = \dfrac{-30.32}{2(-4.25)} = \dfrac{-30.32}{-8.5} = 3.567058824$

≈ 3.57

$$l(3.57) = -4.25(3.57)^2 + 30.32(3.57) + 150.14$$
$$= -4.25(12.7449) + 108.2424 + 150.14$$
$$= -54.165825 + 108.2424 + 150.14$$
$$= 204.216575 = 204,216.575$$
$$\approx 204,217 \text{ free lunches}$$

83. $P(x) = 4000(1.3)^{0.1x}$

a) $x = 10$, $P(10) = 4000(1.3)^{0.1(10)}$

$$= 4000(1.3) = 5200 \text{ people}$$

b) $x = 50$, $P(50) = 4000(1.3)^{0.1(50)}$

$$= 4000(3.71293)$$
$$= 14,851.72 \approx 14,852 \text{ people}$$

84. $P = P_0 e^{-0.00003t}$

$$P = 2000e^{-0.00003(50)}$$
$$= 2000e^{-0.0015}$$
$$= 2000(0.9985011244)$$
$$= 1997.002249 \approx 1997 \text{ g}$$

85. a) Yes

b) ≈ 6500 scooter injuries

86. a) No, the average cost is increasing and then decreasing.

b) $\approx \$1200$

87. $d = (21.9)(2)^{(20-x)/12}$

  a) $x = 19$, $d = (21.9)(2)^{(20-19)/12}$

  $\quad = (21.9)(1.059463094) \approx 23.2 \text{ cm}$

  b) $x = 4$, $d = (21.9)(2)^{(20-4)/12}$

  $\quad = (21.9)(2.519842099) \approx 55.2 \text{ cm}$

  c) $x = 0$, $d = (21.9)(2)^{(20-0)/12}$

  $\quad = (21.9)(3.174802105) \approx 69.5 \text{ cm}$

88. $f(n) = 85,000(1.04)^n$

  a) $f(8) = 85,000(1.04)^8$

  $\quad = 85,000(1.36856905) \approx 116,328$;

  The value of the house after 8 years is about \$116,328.

  b) 15 years since $f(14) \approx \$147,192.50$ and $f(15) \approx \$153,080.20$

89. $f(x) = -0.85x + 187$

  a) $f(20) = -0.85(20) + 187 = 170$ beats per minute

  b) $f(30) = -0.85(30) + 187 = 161.5 \approx 162$ beats per minute

  c) $f(50) = -0.85(50) + 187 = 144.5 \approx 145$ beats per minute

  d) $f(60) = -0.85(60) + 187 = 136$ beats per minute

  e) $-0.85x + 187 = 85$

  $-0.85x = -102$

  $x = 120$ years of age

90. $d(t) = 186,000t$

  a) $t = 1.3$, $d(1.3) = 186,000(1.3) = 241,800 \text{ mi}$

  b) $186,000 \dfrac{\text{m}}{\text{sec}} \times 60 \dfrac{\text{sec}}{\text{min}} = 11,160,000 \dfrac{\text{m}}{\text{min}}$

  $d(t) = 11,160,000t$

  c) $d(8.3) = 11,160,000(8.3) = 92,628,000 \text{ mi}$

Review Exercises

1. $x = 3, \quad x^2 + 12 = (3)^2 + 12 = 9 + 12 = 21$

2. $x = -1, \quad -x^2 - 9 = -(-1)^2 - 9 = -1 - 9 = -10$

3. $x = 2, 4x^2 - 2x + 5 = 4(2)^2 - 2(2) + 5$
$= 16 - 4 + 5 = 17$

4. $x = \dfrac{1}{2}, \quad -x^2 + 7x - 3 = -\left(\dfrac{1}{2}\right)^2 + 7\left(\dfrac{1}{2}\right) - 3$
$= -\dfrac{1}{4} + \dfrac{14}{4} - \dfrac{12}{4} = \dfrac{1}{4}$

5. $x = -2, 4x^3 - 7x^2 + 3x + 1$
$= 4(-2)^3 - 7(-2)^2 + 3(-2) + 1$
$= -32 - 28 - 6 + 1 = -65$

6. $x = 1, y = -2, \quad 3x^2 - xy + 2y^2$
$= 3(1)^2 - 1(-2) + 2(-2)^2$
$= 3 + 2 + 8 = 13$

7. $3x - 4 + x + 5 = 4x + 1$

8. $3x + 4(x-2) + 6x = 3x + 4x - 8 + 6x = 13x - 8$

9. $4(x-1) + \dfrac{1}{3}(9x+3) = 4x - 4 + 3x + 1 = 7x - 3$

10.
$$4s + 10 = -30$$
$$4s + 10 - 10 = -30 - 10$$
$$4s = -40$$
$$\frac{4s}{4} = \frac{-40}{4}$$
$$s = -10$$

11.
$$3t + 8 = 6t - 13$$
$$3t - 3t + 8 = 6t - 3t - 13$$
$$8 = 3t - 13$$
$$8 + 13 = 3t - 13 + 13$$
$$21 = 3t$$
$$\frac{21}{3} = \frac{3t}{3}$$
$$7 = t$$

12.
$$\frac{x+5}{6} = \frac{x-3}{3}$$
$$3(x+5) = 6(x-3)$$
$$3x + 15 = 6x - 18$$
$$3x - 3x + 15 = 6x - 3x - 18$$
$$15 = 3x - 18$$
$$15 + 18 = 3x - 18 + 18$$
$$33 = 3x$$
$$\frac{33}{3} = \frac{3x}{3}$$
$$11 = x$$

13.
$$4(x-2) = 3 + 5(x+4)$$
$$4x - 8 = 3 + 5x + 20$$
$$4x - 8 = 5x + 23$$
$$4x - 4x - 8 = 5x - 4x + 23$$
$$-8 = x + 23$$
$$-8 - 23 = x + 23 - 23$$
$$-31 = x$$

14.
$$\frac{x}{4} + \frac{3}{5} = 7$$
$$20\left(\frac{x}{4} + \frac{3}{5}\right) = 20(7)$$
$$5x + 12 = 140$$
$$5x + 12 - 12 = 140 - 12$$
$$5x = 128$$
$$\frac{5x}{5} = \frac{128}{5}$$
$$x = \frac{128}{5}$$

15. $\dfrac{2}{\dfrac{1}{3}} = \dfrac{3}{x}$

$2x = 3\left(\dfrac{1}{3}\right)$

$2x = 1$

$\dfrac{2x}{2} = \dfrac{1}{2}$

$x = \dfrac{1}{2}$ cup

16. 1 hr 40 min = 60 min + 40 min = 100 min

$\dfrac{120}{100} = \dfrac{300}{x}$

$120x = 100(300)$

$120x = 30{,}000$

$\dfrac{120x}{120} = \dfrac{30{,}000}{120}$

$x = 250$ min, or 4 hr 10 min

17. $A = bh$

$A = 12(4) = 48$

18. $V = 2\pi R^2 r^2$

$V = 2(3.14)(3)^2 (1.75)^2$

$V = 2(3.14)(9)(3.0625)$

$V = 173.0925 \approx 173.1$

19. $Z = \dfrac{\bar{x} - \mu}{\dfrac{\sigma}{\sqrt{n}}}$

$2 = \dfrac{\bar{x} - 100}{\dfrac{3}{\sqrt{16}}}$

$\dfrac{2}{1} = \dfrac{\bar{x} - 100}{\dfrac{3}{4}}$

$2\left(\dfrac{3}{4}\right) = 1(\bar{x} - 100)$

$\dfrac{3}{2} = \bar{x} - 100$

$\dfrac{3}{2} + 100 = \bar{x} - 100 + 100$

$\dfrac{3}{2} + \dfrac{200}{2} = \bar{x}$

$\dfrac{203}{2} = \bar{x}$

$101.5 = \bar{x}$

20. $K = \dfrac{1}{2}mv^2$

$4500 = \dfrac{1}{2}m(30)^2$

$4500 = 450m$

$\dfrac{4500}{450} = \dfrac{450m}{450}$

$10 = m$

21. $3x - 9y = 18$

$3x - 3x - 9y = -3x + 18$

$-9y = -3x + 18$

$\dfrac{-9y}{-9} = \dfrac{-3x + 18}{-9}$

$y = \dfrac{-3x + 18}{-9} = \dfrac{-(-3x + 18)}{9}$

$= \dfrac{3x - 18}{9} = \dfrac{3x}{9} - \dfrac{18}{9} = \dfrac{1}{3}x - 2$

22. $2x + 5y = 12$

$2x - 2x + 5y = -2x + 12$

$5y = -2x + 12$

$\dfrac{5y}{5} = \dfrac{-2x + 12}{5} = -\dfrac{2}{5}x + \dfrac{12}{5}$

23. $2x - 3y + 52 = 30$

$2x - 2x - 3y + 52 = -2x + 30$

$-3y + 52 = -2x + 30$

$-3y + 52 - 52 = -2x + 30 - 52$

$-3y = -2x - 22$

$\dfrac{-3y}{-3} = \dfrac{-2x - 22}{-3}$

$y = \dfrac{-2x - 22}{-3} = \dfrac{2x + 22}{3} = \dfrac{2}{3}x + \dfrac{22}{3}$

24. $-3x - 4y + 5z = 4$

$-3x + 3x - 4y + 5z = 3x + 4$

$-4y + 5z = 3x + 4$

$-4y + 5z - 5z = 3x - 5z + 4$

$-4y = 3x - 5z + 4$

$\dfrac{-4y}{-4} = \dfrac{3x - 5z + 4}{-4}$

$y = \dfrac{3x - 5z + 4}{-4}$

$= \dfrac{-(3x - 5z + 4)}{4}$

$= \dfrac{-3x + 5z - 4}{4}$

$= -\dfrac{3}{4}x + \dfrac{5}{4}z - 1$

25. $A = lw$

$\dfrac{A}{l} = \dfrac{lw}{l}$

$\dfrac{A}{l} = w$

26. $P = 2l + 2w$

$P - 2l = 2l - 2l + 2w$

$P - 2l = 2w$

$\dfrac{P - 2l}{2} = \dfrac{2w}{2}$

$\dfrac{P - 2l}{2} = w$

27. $L = 2(wh + lh)$

$L = 2wh + 2lh$

$L - 2wh = 2wh - 2wh + 2lh$

$L - 2wh = 2lh$

$\dfrac{L - 2wh}{2h} = \dfrac{2lh}{2h}$

$\dfrac{L - 2wh}{2h} = l$ or $l = \dfrac{L}{2h} - \dfrac{2wh}{2h} = \dfrac{L}{2h} - w$

28. $a_n = a_1 + (n - 1)d$

$a_n - a_1 = a_1 - a_1 + (n - 1)d$

$a_n - a_1 = (n - 1)d$

$\dfrac{a_n - a_1}{n - 1} = \dfrac{(n - 1)d}{n - 1}$

$\dfrac{a_n - a_1}{n - 1} = d$

29. $8 + 2x$

30. $3y - 7$

31. $10 + 3r$

32. $\dfrac{8}{q} - 11$

33.
$$\text{Let } x = \text{the number}$$
$$3x = 3 \text{ times the number}$$
$$4 + 3x = 4 \text{ increased by 3 times the number}$$
$$4 + 3x = 22$$
$$4 - 4 + 3x = 22 - 4$$
$$3x = 18$$
$$\frac{3x}{3} = \frac{18}{3}$$
$$x = 6$$

34.
$$\text{Let } x = \text{the number}$$
$$3x = \text{the product of 3 and a number}$$
$$3x + 8 = \text{the product of 3 and a number}$$
$$\text{increased by 8}$$
$$x - 6 = 6 \text{ less than the number}$$
$$3x + 8 = x - 6$$
$$3x - x + 8 = x - x - 6$$
$$2x + 8 = -6$$
$$2x + 8 - 8 = -6 - 8$$
$$2x = -14$$
$$\frac{2x}{2} = \frac{-14}{2}$$
$$x = -7$$

35.
$$\text{Let } x = \text{the number}$$
$$x - 4 = \text{the difference of a number and 4}$$
$$5(x - 4) = 5 \text{ times the difference of a number}$$
$$\text{and 4}$$
$$5(x - 4) = 45$$
$$5x - 20 = 45$$
$$5x - 20 + 20 = 45 + 20$$
$$5x = 65$$
$$\frac{5x}{5} = \frac{65}{5}$$
$$x = 13$$

36.
$$\text{Let } x = \text{the number}$$
$$10x = 10 \text{ times a number}$$
$$10x + 14 = 14 \text{ more than 10 times a number}$$
$$x + 12 = \text{the sum of a number and 12}$$
$$8(x + 12) = 8 \text{ times the sum of a number and 12}$$
$$10x + 14 = 8(x + 12)$$
$$10x + 14 = 8x + 96$$
$$10x - 8x + 14 = 8x - 8x + 96$$
$$2x + 14 = 96$$
$$2x + 14 - 14 = 96 - 14$$
$$2x = 82$$
$$\frac{2x}{2} = \frac{82}{2}$$
$$x = 41$$

37.
$$\text{Let } x = \text{the amount invested in bonds}$$
$$2x = \text{the amount invested in mutual funds}$$
$$x + 2x = 15,000$$
$$3x = 15,000$$
$$\frac{3x}{3} = \frac{15,000}{3}$$
$$x = \$5000 \text{ in bonds}$$
$$2x = 2(5000) = \$10,000 \text{ in mutual funds}$$

38.
$$\text{Let } x = \text{number of lawn chairs}$$
$$9.50x = \text{variable cost per lawn chair}$$
$$9.50x + 15,000 = 95,000$$
$$9.50x + 15,000 - 15,000 = 95,000 - 15,000$$
$$9.50x = 80,000$$
$$\frac{9.50x}{9.50} = \frac{80,000}{9.50}$$
$$x = 8421.052632 \approx 8421 \text{ lawn chairs}$$

39. $\qquad$ Let x = the number of species at the
$\qquad$ Philadelphia Zoo

$\qquad$ $2x+140$ = the number of species at the
$\qquad$ San Diego Zoo

$$x+2x+140=1130$$
$$3x+140=1130$$
$$3x+140-140=1130-140$$
$$3x=990$$
$$\frac{3x}{3}=\frac{990}{3}$$
$$x=330 \text{ species at the}$$
$$\text{Philadelphia Zoo}$$
$$2x+140=2(330)+140=660+140$$
$$=800 \text{ species at the San Diego Zoo}$$

40. Let x = profit at restaurant B

$\qquad$ $x+12,000$ = profit at restaurant A

$$x+(x+12,000)=68,000$$
$$2x+12,000=68,000$$
$$2x+12,000-12,000=68,000-12,000$$
$$2x=56,000$$
$$\frac{2x}{2}=\frac{56,000}{2}$$
$$x=\$28,000 \text{ for restaurant B}$$

$x+12,000=28,000+12,000=\$40,000$ for restaurant A

41. $$s=\frac{k}{t}$$
$$10=\frac{k}{3}$$
$$k=10(3)=30$$
$$s=\frac{30}{t}$$
$$s=\frac{30}{5}=6$$

42. $$J=kA^2$$
$$32=k(4)^2$$
$$32=16k$$
$$\frac{32}{16}=\frac{16k}{16}$$
$$k=2$$
$$J=2A^2$$
$$J=2(7)^2=2(49)=98$$

43. $$W=\frac{kL}{A}$$
$$80=\frac{k(100)}{20}$$
$$100k=1600$$
$$\frac{100k}{100}=\frac{1600}{100}$$
$$k=16$$
$$W=\frac{16L}{A}$$
$$W=\frac{16(50)}{40}=\frac{800}{40}=20$$

44. $$z=\frac{kxy}{r^2}$$
$$12=\frac{k(20)(8)}{(8)^2}$$
$$160k=768$$
$$\frac{160k}{160}=\frac{768}{160}$$
$$k=4.8$$
$$z=\frac{4.8xy}{r^2}$$
$$z=\frac{4.8(10)(80)}{(3)^2}=\frac{3840}{9}=426.\overline{6}\approx426.7$$

45. a) $$\frac{30 \text{ lb}}{2500 \text{ ft}^2} = \frac{x \text{ lb}}{12,500 \text{ ft}^2}$$

$$30(12,500) = 2500x$$

$$375,000 = 2500x$$

$$\frac{375,000}{2500} = \frac{2500x}{2500}$$

$$x = 150 \text{ lb}$$

b) $\dfrac{150}{30} = 5$ bags

46. $$\frac{1 \text{ in.}}{30 \text{ mi}} = \frac{x \text{ in.}}{120 \text{ mi}}$$

$$30x = 120$$

$$\frac{30x}{30} = \frac{120}{30}$$

$$x = 4 \text{ in.}$$

47. $$\frac{1 \text{ } kWh}{\$0.162} = \frac{740 \text{ } kWh}{x}$$

$$x = \$119.88$$

48. $d = kt^2$

$$16 = k(1)^2$$

$$k = 16$$

$$d = 16t^2$$

$$d = 16(5)^2 = 16(25) = 400 \text{ ft}$$

49. $$5 + 9x \le 7x - 7$$
$$5 - 5 + 9x \le 7x - 7 - 5$$
$$9x \le 7x - 12$$
$$9x - 7x \le 7x - 7x - 12$$
$$2x \le -12$$
$$\frac{2x}{2} \le \frac{-12}{2}$$
$$x \le -6$$

50. $$2x + 8 \ge 4x + 10$$
$$2x - 4x + 8 \ge 4x - 4x + 10$$
$$-2x + 8 \ge 10$$
$$-2x + 8 - 8 \ge 10 - 8$$
$$-2x \ge 2$$
$$\frac{-2x}{-2} \le \frac{2}{-2}$$
$$x \le -1$$

51. $$3(x + 9) \le 4x + 11$$
$$3x + 27 \le 4x + 11$$
$$3x - 4x + 27 \le 4x - 4x + 11$$
$$-x + 27 \le 11$$
$$-x + 27 - 27 \le 11 - 27$$
$$-x \le -16$$
$$\frac{-x}{-1} \ge \frac{-16}{-1}$$
$$x \ge 16$$

52. $$-3 \le x + 1 < 7$$
$$-3 - 1 \le x + 1 - 1 < 7 - 1$$
$$-4 \le x < 6$$

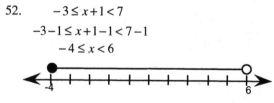

53. $2 + 5x > -8$

$2 - 2 + 5x > -8 - 2$

$5x > -10$

$\dfrac{5x}{5} > \dfrac{-10}{5}$

$x > -2$

54. $5x + 13 \geq -22$

$5x + 13 - 13 \geq -22 - 13$

$5x \geq -35$

$\dfrac{5x}{5} \geq \dfrac{-35}{5}$

$x \geq -7$

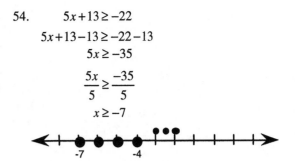

55. $-1 < x \leq 9$

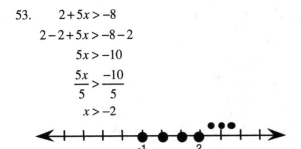

56. $-8 \leq x + 2 \leq 7$

$-8 - 2 \leq x + 2 - 2 \leq 7 - 2$

$-10 \leq x \leq 5$

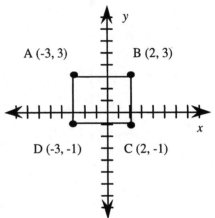

57. - 60.

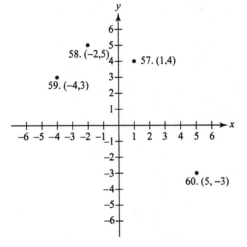

61.

Area $= lw = 5(4) = 20$ square units

62.

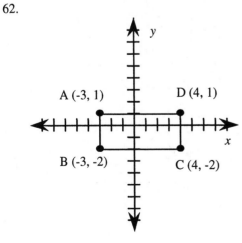

Area $= lw = 7(3) = 21$ square units

63.

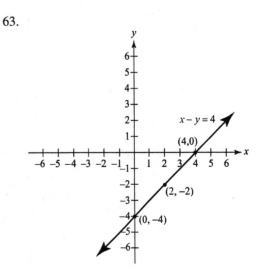

64.

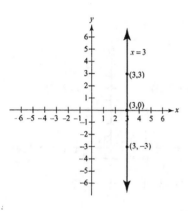

65.

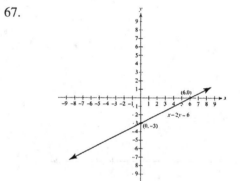

66.

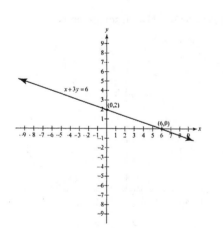

67.

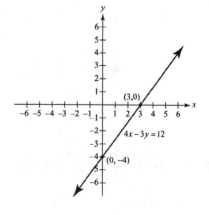

68.

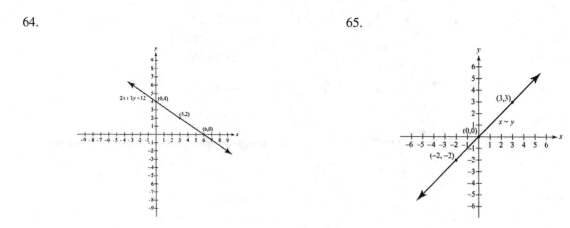

69.

70.

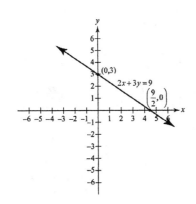

71. $m = \dfrac{5-3}{6-1} = \dfrac{2}{5}$

72. $m = \dfrac{-4-(-1)}{5-3} = \dfrac{-4+1}{5-3} = -\dfrac{3}{2}$

73. $m = \dfrac{3-(-4)}{2-(-1)} = \dfrac{3+4}{2+1} = \dfrac{7}{3}$

74. $m = \dfrac{-2-2}{6-6} = \dfrac{-4}{0}$ Undefined

75.

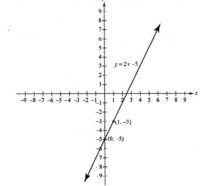

76.

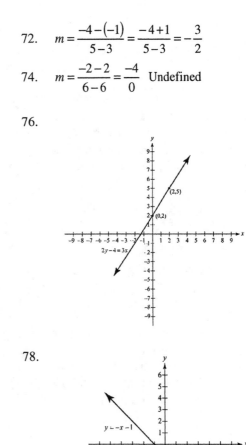

77.

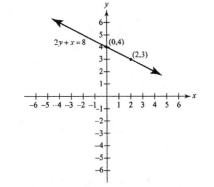

78.

79. The y-intercept is 4, thus $b = 4$. Since the graph rises from left to right, the slope is positive. The change in y is 4 units while the change in x is 2. Thus, m, the slope is $\dfrac{4}{2}$ or 2. The equation is $y = 2x + 4$.

80. The y-intercept is 1, thus $b = 1$. Since the graph falls from left to right, the slope is negative. The change in y is 3 units while the change in x is 3. Thus, m, the slope is $\dfrac{-3}{3}$ or -1. The equation is $y = -x + 1$.

81. a)

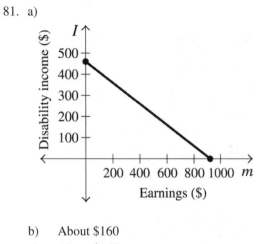

    b)   About $160

    c)   About $160

82. a)

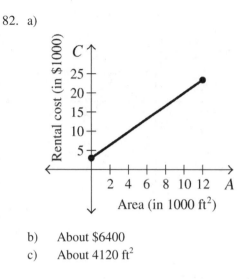

    b)   About $6400

    c)   About 4120 ft^2

83. Graph $4x + 3y = 12$. Since the original inequality is less than or equal to, a solid line is drawn. Since the point (0, 0) satisfies the inequality $4x + 3y \leq 12$, all points on the line and in the half-plane below the line $4x + 3y = 12$ are in the solution set.

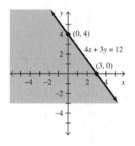

84. Graph $3x + 2y = 12$. Since the original inequality is greater than or equal to, a solid line is drawn. Since the point (0, 0) does not satisfy the inequality $3x + 2y \geq 12$, all points in the half plane above the line $3x + 2y = 12$ are in the solution set.

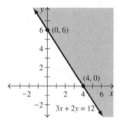

85. Graph $2x - 3y = 12$. Since the original inequality is strictly greater than, a dashed line is drawn. Since the point (0, 0) does not satisfy the inequality $2x - 3y > 12$, all points in the half-plane below the line $2x - 3y = 12$ are in the solution set.

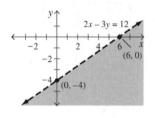

86. Graph $-7x - 2y = 14$. Since the original inequality is strictly less than, a dashed line is drawn. Since the point (0, 0) satisfies the inequality $-7x - 2y < 14$, all points in the half-plane to the right of the line $-7x - 2y = 14$ are in the solution set.

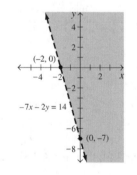

87. $x^2 + 9x + 18 = (x + 3)(x + 6)$

88. $x^2 + x - 20 = (x + 5)(x - 4)$

89. $x^2 - 10x + 24 = (x - 6)(x - 4)$

90. $x^2 - 9x + 20 = (x - 5)(x - 4)$

91. $6x^2 + 7x - 3 = (3x - 1)(2x + 3)$

92. $2x^2 + 13x - 7 = (2x - 1)(x + 7)$

93. $x^2 + 3x + 2 = 0$

$(x+1)(x+2) = 0$

$x+1 = 0$ or $x+2 = 0$

$x = -1 \qquad x = -2$

94. $x^2 - 5x = -4$

$x^2 - 5x + 4 = 0$

$(x-1)(x-4) = 0$

$x-1 = 0$ or $x-4 = 0$

$x = 1 \qquad x = 4$

95. $3x^2 - 17x + 10 = 0$

$(3x-2)(x-5) = 0$

$3x-2 = 0$ or $x-5 = 0$

$3x = 2 \qquad x = 5$

$x = \dfrac{2}{3}$

96. $3x^2 = -7x - 2$

$3x^2 + 7x + 2 = 0$

$(x+2)(3x+1) = 0$

$x+2 = 0$ or $3x+1 = 0$

$x = -2 \qquad 3x = -1$

$x = -\dfrac{1}{3}$

97. $x^2 - 4x - 1 = 0$

$a = 1, \quad b = -4, \quad c = -1$

$x = \dfrac{-(-4) \pm \sqrt{(-4)^2 - 4(1)(-1)}}{2(1)}$

$x = \dfrac{4 \pm \sqrt{16+4}}{2} = \dfrac{4 \pm \sqrt{20}}{2} = \dfrac{4 \pm 2\sqrt{5}}{2} = 2 \pm \sqrt{5}$

98. $x^2 - 3x + 2 = 0$

$a = 1, \, b = -3, \, c = 2$

$x = \dfrac{-(-3) \pm \sqrt{(-3)^2 - 4(1)(2)}}{2(1)}$

$x = \dfrac{3 \pm \sqrt{9-8}}{2} = \dfrac{3 \pm \sqrt{1}}{2} = \dfrac{3 \pm 1}{2}$

$x = \dfrac{4}{2} = 2$ or $x = \dfrac{2}{2} = 1$

99. $2x^2 - 3x + 4 = 0$

$a = 2, \, b = -3, \, c = 4$

$x = \dfrac{-(-3) \pm \sqrt{(-3)^2 - 4(2)(4)}}{2(2)}$

$x = \dfrac{3 \pm \sqrt{9-32}}{4} = \dfrac{3 \pm \sqrt{-23}}{4}$

No real solution

100. $2x^2 - x - 3 = 0$

$a = 2, \, b = -1, \, c = -3$

$x = \dfrac{-(-1) \pm \sqrt{(-1)^2 - 4(2)(-3)}}{2(2)}$

$x = \dfrac{1 \pm \sqrt{1+24}}{4} = \dfrac{1 \pm \sqrt{25}}{4} = \dfrac{1 \pm 5}{4}$

$x = \dfrac{6}{4} = \dfrac{3}{2}$ or $x = \dfrac{-4}{4} = -1$

101. Function since each value of x is paired with a unique value of y.

D: $x = $ -2, -1, 2, 3 \qquad R: $y = $ -1, 0, 2

102. Not a function since it is possible to draw a vertical line that intersects the graph at more than one point.

103. Not a function since it is possible to draw a vertical line that intersects the graph at more than one point.

104. Function since each vertical line intersects the graph at only one point.

D: all real numbers \qquad R: all real numbers

105. $f(x) = 5x - 2, \quad x = 4$

$f(4) = 5(4) - 2 = 20 - 2 = 18$

106. $f(x) = -2x + 7, \quad x = -3$

$f(-3) = -2(-3) + 7 = 6 + 7 = 13$

107. $f(x) = 2x^2 - 3x + 4,\ x = 5$

  $f(5) = 2(5)^2 - 3(5) + 4 = 50 - 15 + 4 = 39$

108. $f(x) = -4x^2 + 7x + 9,\ x = 4$

  $f(4) = -4(4)^2 + 7(4) + 9 = -64 + 28 + 9 = -27$

109. $y = -x^2 - 4x + 21$

  a) $a = -1 < 0$, opens downward

  b) $x = -2$ c) $(-2, 25)$ d) $(0, 21)$

  e) $(-7, 0), (3, 0)$

  f)

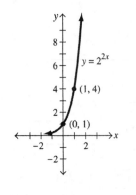

  g) D: all real numbers   R: $y \le 25$

110. $f(x) = 2x^2 - 8x + 10$

  a) $a = 2 > 0$, opens upward

  b) $x = 2$ c) $(2, 2)$ d) $(0, 10)$

  e) no x-intercepts

  f)

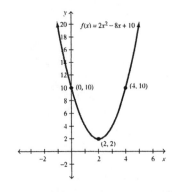

  g) D: all real numbers   R: $y \ge 2$

111.

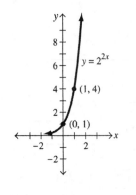

  D: all real numbers   R: $y > 0$

112.

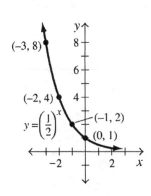

  D: all real numbers   R: $y > 0$

113. $m = 30 - 0.002n^2,\ n = 60$

  $m = 30 - 0.002(60)^2 = 30 - 0.002(3600)$

  $= 30 - 7.2 = 22.8$ mpg

114. $n = 2a^2 - 80a + 5000$

  a) $a = 18$

  $n = 2(18)^2 - 80(18) + 5000$

  $= 648 - 1440 + 5000 = 4208$

  b) $a = 25$

  $n = 2(25)^2 - 80(25) + 5000$

  $= 1250 - 2000 + 5000 = 4250$

115. $P = 100(0.92)^x,\ x = 4.5$

  $P = 100(0.92)^{4.5}$

  $= 100(0.6871399881) = 68.71399881 \approx 68.7\%$

Chapter Test

1. $3x^2 + 4x - 1, \quad x = -2$

$3(-2)^2 + 4(-2) - 1 = 12 - 8 - 1 = 3$

2.
$$3x + 5 = 2(4x - 7)$$
$$3x + 5 = 8x - 14$$
$$3x - 8x + 5 = 8x - 8x - 14$$
$$-5x + 5 = -14$$
$$-5x + 5 - 5 = -14 - 5$$
$$-5x = -19$$
$$\frac{-5x}{-5} = \frac{-19}{-5}$$
$$x = \frac{19}{5}$$

3. $-2(x - 3) + 6x = 2x + 3(x - 4)$
$$-2x + 6 + 6x = 2x + 3x - 12$$
$$4x + 6 = 5x - 12$$
$$4x - 5x + 6 = 5x - 5x - 12$$
$$-x + 6 = -12$$
$$-x + 6 - 6 = -12 - 6$$
$$-x = -18$$
$$\frac{-x}{-1} = \frac{-18}{-1}$$
$$x = 18$$

4.
Let x = the number

$2x$ = the product of the number and 2

$2x + 7$ = the product of the number and 2, increased by 7

$$2x + 7 = 25$$
$$2x + 7 - 7 = 25 - 7$$
$$2x = 18$$
$$\frac{2x}{2} = \frac{18}{2}$$
$$x = 9$$

5.
Let x = the cost of the car before tax

$0.07x$ = the amount of the sales tax

$$x + 0.07x = 26,750$$
$$1.07x = 26,750$$
$$\frac{1.07x}{1.07} = \frac{26,750}{1.07}$$
$$x = \$25,000$$

6. $L = ah + bh + ch; \ a = 3, \ b = 4, \ c = 5, \ h = 7$
$$L = 3(7) + 4(7) + 5(7)$$
$$= 21 + 28 + 35 = 84$$

7.
$$3x + 5y = 11$$
$$3x - 3x + 5y = -3x + 11$$
$$5y = -3x + 11$$
$$\frac{5y}{5} = \frac{-3x + 11}{5}$$
$$y = \frac{-3x + 11}{5} = -\frac{3}{5}x + \frac{11}{5}$$

8. $L = \dfrac{kMN}{P}$
$$12 = \frac{k(8)(3)}{2}$$
$$24k = 24$$
$$k = \frac{24}{24} = 1$$
$$L = \frac{(1)MN}{P}$$
$$L = \frac{(1)(10)(5)}{15} = \frac{50}{15} = 3.\overline{3} = 3\frac{1}{3}$$

9. $l = \dfrac{k}{w}$

$15 = \dfrac{k}{9}$

$k = 15(9) = 135$

$l = \dfrac{135}{w}$

$l = \dfrac{135}{20} = 6.75$ ft

10. $\qquad -3x + 11 \le 5x + 35$

$-3x - 5x + 11 \le 5x - 5x + 35$

$-8x + 11 \le 35$

$-8x + 11 - 11 \le 35 - 11$

$-8x \le 24$

$\dfrac{-8x}{-8} \ge \dfrac{24}{-8}$

$x \ge -3$

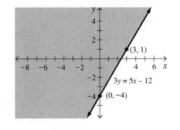

11. $m = \dfrac{12 - 5}{7 - (-3)} = \dfrac{12 - 5}{7 + 3} = \dfrac{7}{10}$

12.

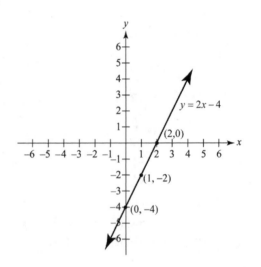

13.

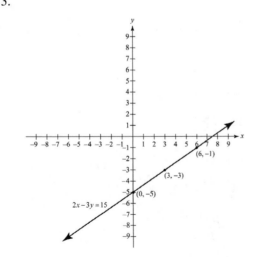

14. Graph $3y = 5x - 12$. Since the original statement is greater than or equal to, a solid line is drawn. Since the point (0, 0) satisfies the inequality $3y \ge 5x - 12$, all points on the line and in the half-plane above the line $3y = 5x - 12$ are in the solution set.

15. $x^2 - 3x = 28$

$x^2 - 3x - 28 = 0$

$(x - 7)(x + 4) = 0$

$x - 7 = 0$ or $x + 4 = 0$

$x = 7$ $x = -4$

16. $3x^2 + 2x = 8$

$3x^2 + 2x - 8 = 0$

$a = 3, \ b = 2, \ c = -8$

$x = \dfrac{-2 \pm \sqrt{(2)^2 - 4(3)(-8)}}{2(3)}$

$x = \dfrac{-2 \pm \sqrt{4 + 96}}{6} = \dfrac{-2 \pm \sqrt{100}}{6} = \dfrac{-2 \pm 10}{6}$

$x = \dfrac{8}{6} = \dfrac{4}{3}$ or $x = \dfrac{-12}{6} = -2$

17. Function since each vertical line intersects the graph at only one point.

18. $f(x) = -4x^2 - 11x + 5, \ x = -2$

$f(-2) = -4(-2)^2 - 11(-2) + 5$

$\quad = -16 + 22 + 5 = 11$

19. $y = x^2 - 2x + 4$

a) $a = 1 > 0$, opens upward

b) $x = 1$ c) $(1, 3)$ d) $(0, 4)$

e) no x-intercepts

f)

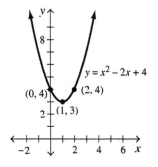

g) D: all real numbers   R: $y \geq 3$

Group Projects

1. a) - b) Answers will vary.

c) $h = 3.14H + 64.98 = 3.14(29.42) + 64.98 = 157.3588$ cm. ≈ 157.36 cm.

Yes

d) $h = 2.53T + 72.57$

$167.64 = 2.53T + 72.57$

$95.07 = 2.53T$

$T = 37.5770751$ cm. ≈ 37.58 cm.

e) i) $h = 3.14H + 64.98$

$168 = 3.14H + 64.98$

$103.02 = 3.14H$

$H = 32.8089172 \approx 32.81$ cm.

ii) $H = 32.81 - 0.06(30) = 32.81 - 1.8 = 31.01$ cm.

f) Answers will vary.

2. a) - e) Answers will vary.

CHAPTER SEVEN

SYSTEMS OF LINEAR EQUATIONS AND INEQUALITIES

Exercise Set 7.1

1. Two or more linear equations form a system of linear equations.

2. A solution to a system of linear equations is the ordered pair or pairs that satisfy all equations in the system.

3. A consistent system of equations is a system that has a solution.

4. A dependent system of equations is a system that has an infinite number of solutions.

5. An inconsistent system of equations is a system that has no solution.

6. Graph each equation on the same axes. The point(s) of intersection of the graphs is (are) the solution(s) to the system.

7. The graphs of the system of equations are parallel and do not intersect.

8. The graphs of the system of equations intersect at one point.

9. The graphs of the system of equations are in fact the same line.

10. No. If no solution, the graphs are parallel; if they intersect, there is one; or they are the same line.

11. $(3, 0)$
$$y = 2x - 6 \qquad y = -x + 3$$
$$(0) = 2(3) - 6 \qquad (0) = -(3) + 3$$
$$0 = 6 - 6 \qquad\qquad 0 = 0$$
$$0 = 0$$
Therefore, $(3, 0)$ is a solution.

12. $(-2, 4)$
$$x + 2y = 6 \qquad x - y = -6$$
$$(-2) + 2(4) = 6 \qquad (-2) - (4) = -6$$
$$-2 + 8 = 6 \qquad\qquad -6 = -6$$
$$6 = 6$$
Therefore, $(-2, 4)$ is a solution.

13.

14.

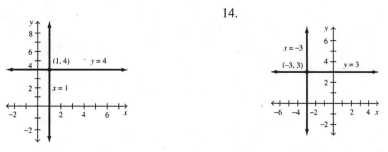

15.

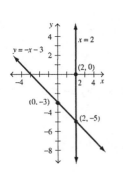

16.

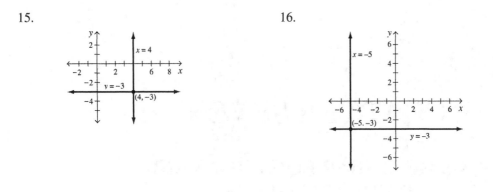

17.

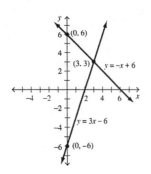

18.

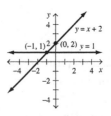

19.

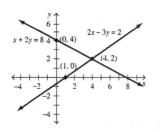

20.

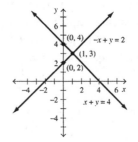

21.

22.

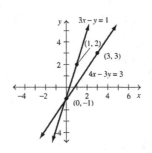

23.

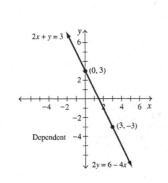

24.

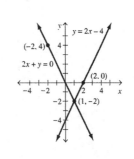

25.

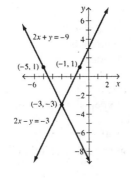

26.

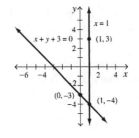

27.

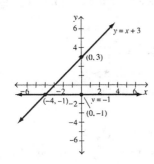

28.

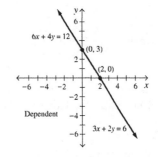

29.

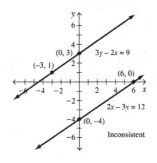

30.

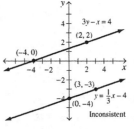

31.

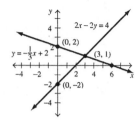

32.

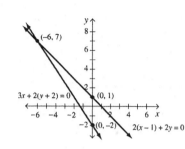

33. a) Two lines with different slopes are not
      parallel, and therefore have exactly one point
      of intersection giving one solution.
   b) Two lines with the same slope and different
      y- intercepts are distinct parallel lines and
      have no solution.
   c) Two lines with the same slopes and
      y-intercepts have infinitely many solutions,
      each point on the line.

34. a) Consistent; the system has one solution.
   b) Inconsistent; the system has no solution.
   c) Dependent; the system has infinitely many
      solutions.

35. $2x - y = 6$ $y = 2x - 6$
    same slope, same y-intercept;
    infinite number of solutions

36. $3x + 4y = 8$ $8y = -6x + 4$
    same slopes, diff. y-intercepts; no solution

37. $3x - 4y = 5$ $y = -3x + 8$
    different slopes, different y-intercepts; 1
    solution

38. $x + 3y = 6$ $3x + y = 4$
    diff. slope, different y-intercepts; 1 solution

39. $3x + y = 7$ $y = -3x + 9$
    same slope, diff. y-intercepts; no solution

40. $x + 4y = 12$ $x = 4y + 3$
    diff. slopes, diff. y-intercepts; 1 solution

41. $2x - 3y = 6$ $x - (3/2)y = 3$
    same slopes, same y-intercepts;
    infinite number of solutions

42. $x - 2y = 6$ $x + 2y = 4$
    diff. slopes, diff. y-intercepts; 1 solution

43. $3x = 6y + 5$ $y = (1/2)x - 3$
    same slope, diff. y-intercepts; no solution

44. $3y = 6x + 4$ $-2x + y = 4/3$
    same slopes, same y-intercepts;
    infinite number of solutions

45. $12x - 5y = 4$ $3x + 4y = 6$
    diff. slopes, diff. y-intercepts; 1 solution

46. $4x + 7y = 2$ $4x = 6 + 7y$
    diff. slopes, diff. y-intercepts; 1 solution

47. $5y - 2x = 15$ $2y - 5x = 2$
    slopes are not negative reciprocals,
    not perpendicular ($\nwarrow$)

48. $4y - x = 6$ $y = x + 8$
    slopes are not negative reciprocals, not $\nwarrow$

49. $2x + y = 3$ $2y - x = 5$
    slopes are negative reciprocals, $\nwarrow$

50. $6x + 5y = 3$ $-10x = 2 + 12y$
    slopes are not negative reciprocals, not $\nwarrow$

51. a) Let x = rate per hour
    y = cost
    Cost for Tom's $y_T = 60x + 200$
    Cost for Lawn Perfect $y_{LP} = 25x + 305$

    c) $60x + 200 = 25x + 305$

    $\underline{-25x \quad -200 \quad -25x \quad -200}$

    $35x \qquad = \qquad 105$

    $\underline{35x} = \underline{105}$ → x = 3 hours
    $35 \qquad 35$

b)

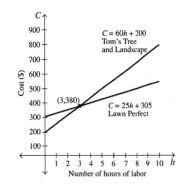

52. a) Let c = cost , x = number of months
    Cost for ABC: $c_{ABC} = 18x + 3380$
    Cost for Safe Homes: $c_S = 29x + 2302$

    b)

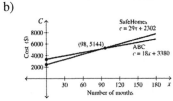

52. c) $18x + 3380 = 29x + 2302$

    $\underline{-18x \quad -2302 \quad -18x \quad -2302}$

    $1078 = 11x$

    $\underline{11x} = \underline{1078}$ → x = 98 months
    $11 \qquad 11$

    d) ABC would be less expensive for 10 years.

53. a) Let C = cost , R = revenue
    $C(x) = 15x + 400$
    $R(x) = 25x$

    b)

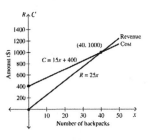

53. c) $25x = 15x + 400$

    $\underline{-15x \quad -15x}$

    $10x = 400$

    $\underline{10x} = \underline{400}$ → x = 40 backpacks
    $10 \qquad 10$

    d) $P = R(x) - C(x) = 25x - (15x + 400)$
       $P = 10x - 400$
    e) $P = 10(30) - 400 = 300 - 400 = -\$100$ (loss)
    f) $1000 = 10x - 400$ → $10x = 1400$
       $x = 140$ BPs

54. a) MDA: $M = .08s + 40$
       AHA: $A = .18s + 15$
    c) $.08s + 40 = .18s + 15$

    $\underline{-.08x \quad -15 \quad -.08x \quad -15}$

    $25 \qquad = .10x$

    $\underline{.10x} = \underline{25}$ → x = 250 shares
    $.10 \quad .10$

    d) For 300 shares, MDA would be less
       expensive. $M = .08(300) + 40 = 64$
                    $A = .18(300) + 15 = 69$

54. b)

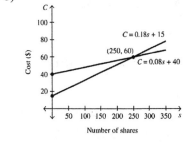

55. a) Let $C(x) = $ cost , $R(x) = $ revenue

$C(x) = 155x + 8400$

$R(x) = 225x$

b)

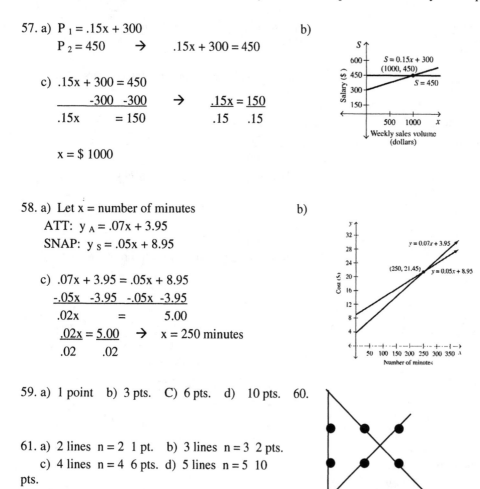

55. c) $225x = 155x + 8400$

$\underline{-155x \quad -155x}$

$70x = 8400$

$\dfrac{70x}{70} = \dfrac{8400}{70}$ → $x = 120$ units

d) $P = R(x) - C(x) = 225x - (155x + 8400)$

$P = 70x - 8400$

e) $P = 70(100) - 8400 = 7000 - 8400$

$= - \$1400$ (loss)

f) $1260 = 70x - 8400$ → $70x = 9660$

$x = 138$ units

56. Two systems are: consistent if they have different slopes; dependent if they have the same slopes and same y-intercepts; and inconsistent if they have same slopes and different y-intercepts.

57. a) $P_1 = .15x + 300$

$P_2 = 450$ → $.15x + 300 = 450$

c) $.15x + 300 = 450$

$\underline{\quad \quad -300 \quad -300}$ → $\underline{.15x} = \underline{150}$

$.15x \quad = 150 \qquad \qquad .15 \quad .15$

$x = \$1000$

b)

58. a) Let $x = $ number of minutes

ATT: $y_A = .07x + 3.95$

SNAP: $y_S = .05x + 8.95$

c) $.07x + 3.95 = .05x + 8.95$

$\underline{-.05x \quad -3.95 \quad -.05x \quad -3.95}$

$.02x \quad = \quad 5.00$

$\dfrac{.02x}{.02} = \dfrac{5.00}{.02}$ → $x = 250$ minutes

b)

59. a) 1 point   b) 3 pts.   C) 6 pts.   d) 10 pts.   60.

61. a) 2 lines $n = 2$ 1 pt.   b) 3 lines $n = 3$ 2 pts.

c) 4 lines $n = 4$ 6 pts.   d) 5 lines $n = 5$ 10 pts.

Exercise Set 7.2

1. Write the equations with the variables on one side and the constants on the other side. If necessary multiply one or both equations by a constant(s) so that when the equations are added one of the variables will be eliminated. Solve for the remaining variable and then substitute that value into one of the original equations to solve for the other variable.

2. Solve one of the equations for one of the variables in terms of the other variable. Then substitute that expression into the other equation and solve for the variable. Substitute the value found into one of the original equations and solve for the other variable.

3. The system is dependent if the result is of the form a = a.

4. The system is inconsistent if the result is a false statement.

5. Solve one equation for the variable that is most readily manipulated, then substitute into the other equation.

$$x + 3y = 3$$
$$\underline{-3y-3y} \quad \rightarrow \quad 3(3-3y) + 4y = 9$$
$$x = 3 - 3y$$

6. Manipulate the coefficient of one variable to equate it with the negative coefficient of the same variable in the other equation, then add.

7. $y = x - 6$
   $y = -x + 4$
   Substitute (x - 6) in place of y in the second equation.
   $x - 6 = -x + 4$ (solve for x)
   $\underline{+x+x}$
   $2x - 6 = 4$
   $\underline{+6+6}$
   $2x = 10$
   $\dfrac{2x}{2} = \dfrac{10}{2} \quad x = 5$
   Now substitute 5 for x in an equation
   $y = x - 6$
   $y = (5) - 6 = -1$
   The solution is (5, -1). Consistent

8. $y = 3x + 7$
   $y = -2x - 3$
   Equate both equations.

   $3x + 7 = -2x - 3$ (solve for x)
   $\underline{+2x -7+2x -7}$
   $5x = -10$
   $\dfrac{5x}{5} = \dfrac{-10}{5} \quad x = -2$
   Now substitute -2 for x in an equation
   $y = 3(-2) + 7$
   $y = -6 + 7 = 1$
   The solution is (-2, 1). Consistent

9. $2x + 4y = 8 \quad \rightarrow \quad x = -2y + 4$
   $2x - y = -2$
   Substitute (-2y + 4) in place of x in the second equation.
   $2(-2y + 4) - y = -2$ (solve for x)
   $-4y + 8 - y = -2$
   $-5y = -10 \quad y = 2$

   Now substitute 2 for y in the 1ˢᵗ equation
   $2x + 4(2) = 8$
   $2x = 0 \quad x = 0$
   The solution is (0, 2). Consistent

10. $y + 3x = 7 \quad \rightarrow \quad y = -3x + 7$
    $2x + 3y = 14$
    Substitute (-3x + 7) in place of y in the second equation.
    $2x + 3(-3x + 7) = 14$ (solve for x)
    $-7x + 21 = 14$
    $-7x = -7 \quad x = 1$

    Now substitute 1 for x in the 1ˢᵗ equation
    $y + 3(1) = 7$
    $y = 4$
    The solution is (1, 4). Consistent

11. $y - x = 4$

    $x - y = 3$

    Solve the first equation for y.

    $y - x + x = x + 4$

    $y = x + 4$

    Substitute $(x + 4)$ for y in the second equation.

    $x - (x + 4) = 3$ (combine like terms)

    -4 3   False

    Since -4 does not equal 3, there is no solution to this system. The equations are inconsistent.

12. $x + y = 3$

    $y + x = 5$

    Solve the second equation for y.

    $y + x - x = -x + 5$

    $y = -x + 5$

    Substitute $(-x + 5)$ for y in the first equation.

    $x + (-x + 5) = 3$

    5   3   False

    Since 5 does not equal 3, there is no solution to this system. The equations are inconsistent.

13. $3y + 2x = 4$

    $y = 6 - x$

    Solve the second equation for x.

    $3y = 6 - x$

    $3y - 6 = 6 - 6 - x$

    $3y - 6 = -x$

    $-3y + 6$ $= x$

    Now substitute $(-3y + 6)$ for x in the 1st eq'n.

    $3y + 2(6 - 3y) = 4$ (solve for y)

    $3y + 12 - 6y = 4$

    $-3y = -8$ (div. by -3) $y = 8/3$

    Substitute 8/3 for y in the 2nd eq'n.

    $3(8/3) n = 6 - x$

    $8 = 6 - x$ $x = -2$

    The solution is $(-2, 8/3)$.   Consistent

14. $x = 5y - 12$

    $x - y = 0$

    Substitute $(5y - 12)$ for x in the second equation.

    $5y - 12 - y = 0$ (solve for y)

    $4y - 12$ $= 0$

    $4y = 12$ (div. by 4) $y = 3$

    Now substitute 3 for y in the second equation.

    $x - 3 = 0$

    $x = 3$

    The solution is $(3,3)$.   Consistent

15. $y - 2x = 3$

    $2y = 4x + 6$

    Solve the first equation for y.

    $y - 2x + 2x = 2x + 3$

    $y = 2x + 3$

    Now substitute $(2x + 3)$ for y in the 2nd eq'n.

    $2(2x + 3) = 4x + 6$

    $4x + 6 = 4x + 6$

    $4x - 4x + 6 = 4x - 4x + 6$

    $6 = 6$

    This statement is true for all values of x.
    The system is dependent.

16. $y = 2$

    $y + x + 3 = 0$

    Substitute 2 in place of y in the second equation.

    $2 + x + 3 = 0$

    $x + 5 = 0$

    $x + 5 - 5 = 0 - 5$ $x = -5$

    The solution is $(-5, 2)$.   Consistent

17. x = y + 3
    x = − 3
    Substitute − 3 in place of x in the first
     equation.
    − 3 = y + 3
    − 3 − 3 = y + 3 − 3
    − 6 = y

    The solution is (− 3,− 6).   Consistent

18. x + 2y = 6
    y = 2x + 3
    Substitute (2x + 3) for y in the first equation.
    x + 2(2x + 3) = 6
    x + 4x + 6 = 6
    5x + 6 − 6 = 6 − 6
    5x = 0
    $\dfrac{5x}{5} = \dfrac{0}{5}$ x = 0

    Now substitute 0 for x in the second equation.
    y = 2(0) + 3 = 0 + 3 = 3

    The solution is (0,3).   Consistent

19. y + 3x − 4 = 0
    2x − y = 7
    Solve the first equation for y.
    y + 3x − 4 = 0
    y = 4 − 3x
    Substitute 4 − 3x for y in the second eq.
    2x − (4 − 3x) = 7 (solve for x)
    2x − 4 + 3x = 7
    5x = 11   x = 11/5
    Substitute 11/5 for x in the second eq'n.
    2(11/5) − y = 7 (solve for y)
    22/5 − y = 7
    − y = 13/5   y = − 13/5
    The solution is (11/5, − 13/5).   Consistent

20. x + 4y = 7
    2x + 3y = 5
    Solve the first equation for x.
    x = 7 − 4y
    Substitute (7 − 4y) for x in the second equation.
    2(7 − 4y) + 3y = 5 (solve for y)
    14 − 8y + 3y = 5
    − 5y = − 9   y = 9/5
    Now substitute (9/5) for y in the eq'n. x + 4y = 7.
    x +4(9/5) = 7
    x + 36/5 = 35/5   x = − 1/5
    The solution is (− 1/5,9/5).   Consistent

21. x = 2y + 3
    y = 3x − 1
    Substitute (3x − 1) for y in the first equation.
    x = 2(3x − 1) + 3
    x = 6x − 2 + 3
    x = 6x + 1
    x − 6x = 6x − 6x + 1
    − 5x = 1
    $\dfrac{-5x}{-5} = \dfrac{1}{-5}$ x= − 1/5
    Substitute − 1/5 for x in the second equation.
    y = 3(− 1/5) − 1 = − 3/5 − 5/5 = − 8/5
    The solution is (− 1/5, − 8/5).   Consistent

22. x + 4y = 9
    2x − y − 6 = 0
    Solve the first equation for x.
    x + 4y − 4y = 9 − 4y
    x = 9 − 4y
    Substitute (9 − 4y) for x in the second equation.
    2(9 − 4y) − y − 6 = 0
    18 − 8y − y − 6 = 0
    12 − 9y = 0
    12 − 9y + 9y = 0 + 9y
    12 = 9y   12/9 = y
    Substitute (12/9) = (4/3) for y in the equation.
    x = 9 − 4y
    x = 9 − 4(4/3) = 27/3 − 16/3 = 11/3
    The solution is (11/3,4/3).   Consistent

23. $y = -2x + 3$
$4x + 2y = 12$
Substitute $-2x + 3$ for y in the 2nd equation.
$4x + 2(-2x + 3) = 12$
$4x - 4x + 6 = 12$

6   12   False

Since 6 does not equal 12, there is no solution.
The equations are inconsistent.

24. $2x + y = 12$
$x = (-1/2)y + 6$
Substitute $(-1/2)x + 6$ for x in the 1st equation.
$2(-1/2y + 6) + y = 12$
$-y + 12 + y = 12$
$12 = 12$
This statement is true for all values of x.
The system is dependent.

25. $3x + y = 10$
$4x - y = 4$
Add the equations to eliminate y.
$7x = 14$ $x = 2$
Substitute 2 for x in either eq'n.
$3(2) + y = 10$ (solve for y)
$6 + y = 10$ $y = 4$

The solution is (2, 4)   Consistent

26. $x + 2y = 9$
$x - 2y = -3$
Add the equations to eliminate y.
$2x = 6$ $x = 3$
Substitute 3 for x in either eq'n.
$(3) + 2y = 9$
$2y = 6$ $y = 3$

The solution is (3, 3)   Consistent

27. $x + y = 10$
$x - 2y = -2$
Multiply the 1st eq'n. by 2, then add the eq'ns.
To eliminate y.
$2x + 2y = 20$
$x - 2y = -2$
$3x = 18$ $x = 6$
Substitute 6 for x in either eq'n.
$(6) + y = 10$ (solve for y) $y = 4$

The solution is (6, 4)   Consistent

28. $3x + y = 10$
$-3x + 2y = -16$
Add the equations to eliminate x.
$3y = -6$ $y = -2$
Substitute -2 for y in either eq'n.
$3x + (-2) = 10$ (solve for x)
$3x = 12$ $x = 4$

The solution is (4, -2)   Consistent

29. $2x - y = -4$
$-3x - y = 6$
Multiply the second equation by -1,
$2x - y = -4$
$3x + y = -6$ add the equations to eliminate y
$5x = -10$ $x = -2$
Substitute -2 in place of x in the first
   equation.
$2(-2) - y = -4$
$-4 - y = -4$
$-y = 0$ $y = 0$

The solution is (-2, 0).   Consistent

30. $x + y = 6$
$-2x + y = -3$
Multiply the second equation by -1,
$x + y = 6$
$2x - y = 3$ add the equations to eliminate y
$3x = 9$ $x = 3$
Substitute 3 for x in the first equation.
$3 + y = 6$ $y = 3$

The solution is (3, 3).   Consistent

31. $4x + 3y = -1$
$2x - y = -13$
Multiply the second equation by 3,
$4x + 3y = -1$
$6x - 3y = -39$ add the equations to eliminate y
$10x = -40$ $x = -4$
Substitute -4 for x in the 2nd equation.
$2(-4) - y = -13$
$-8 - y = -13$ $y = 5$
The solution is $(-4, 5)$.   Consistent

32. $2x + y = 6$
$3x + y = 5$
Multiply the first equation by -1,
$-2x - y = -6$
$3x + y = 5$__add the equations to eliminate y
$x = -1$
Substitute -1 in place of x in the first equation.
$2(-1) + y = 6$
$-2 + y = 6$ $y = 8$
The solution is $(-1, 8)$.   Consistent

33. $2x + y = 11$
$x + 3y = 18$
Multiply the second equation by -2,
$2x + y = 11$
$-2x - 6y = -36$ add the equations to elim. x
$-5y = -25$ $y = 5$
Substitute 5 for y in the 2nd equation.
$x + 3(5) = 18$
$x + 15 = 18$ $x = 3$
The solution is $(3, 5)$.

34. $5x - 2y = 11$
$-3x + 2y = 1$ add the equations to eliminate y
$2x = 12$ $x = 6$
Substitute 6 for x in the second equation.
$-3(6) + 2y = 1$
$-18 + 2y = 1$
$2y = 19$ $y = 19/2$
The solution is $(6, 19/2)$.

35. $3x - 4y = 11$
$3x + 5y = -7$
Multiply the first equation by (-1),
$-3x + 4y = -11$
$3x + 5y = -7$ add the equations to elim. x
$9y = -18$ $y = -2$
Substitute -2 for y in the first equation.
$3x - 4(-2) = 11$
$3x = 3$ $x = 1$
The solution is $(1, -2)$.   Consistent

36. $4x - 2y = 6$
$4y = 8x - 12$ or $8x - 4y = 12$
Multiply the first equation by (-2),
$-8x + 4y = -12$
$8x - 4y = -12$ add the equations to elim. y
$0 = 0$ True
This statement is true for all values of x.
This system is dependent.

37. $4x + y = 6$
$-8x - 2y = 13$
Multiply the first equation by 2,
$8x + 2y = 12$
$-8x - 2y = 13$ add the equations to elim. y
0 25 False
Since this statement is not true for any values of x and y, the equations are inconsistent.

38. $2x + 3y = 6$
$5x - 4y = -8$
Multiply the first equation by 5, and the second equation by (-2),
$10x + 15y = 30$
$-10x + 8y = 16$ add the equations to elim. x
$23y = 46$ $y = 2$
Substitute 2 for y in the first equation.
$2x + 3(2) = 6$
$2x = 0$ $x = 0$
The solution is $(0, 2)$.   Consistent

39. $3x - 4y = 10$
    $5x + 3y = 7$
    Multiply the first equation by 3, and the second equation by 4,
    $9x - 12y = 30$
    $20x + 12y = 28$ add the equations to elim. y
    $29x = 58$ $x = 2$
    Substitute 2 for x in the second equation.
    $5(2) + 3y = 7$
    $10 + 3y = 7$
    $3y = -3$ $y = -1$
    The solution is $(2, -1)$.   Consistent

40. $6x + 3y = 7$
    $5x + 2y = 9$
    Multiply the first equation by 2, and the second equation by -3,
    $12x + 6y = 14$
    $-15x - 6y = -27$ add the equations to elim. y
    $-3x = -13$ $x = 13/3$
    Substitute 13/3 for x in the 1st equation.
    $6(13/3) + 3y = 7$
    $26 + 3y = 7$
    $3y = -19$ $y = -19/3$
    The solution is $(13/3, -19/3)$.   Consistent

41. $S_1 = .15p + 12000$
    $S_2 = .05p + 27000$

    $.15p + 12000 = .05p + 27000$
    $-.05p -12000   -.05p -12000$
    $.10p = 15000$
    $p = \$ 150,000.00$

42. Let $C_1 = 780n + 1600$ $C_2 = 980n$
    a) $980n = 780n + 1600$
       $\underline{-780n \quad -780n}$
       $200n = 1600$ $n = 8$ months

    b) $C_1 = 780(60) + 1600 = 48400$
       $C_2 = 980(60) = 58800$
       The new refinanced mortgage plan would cost less.

43. Let x = # of medium pizzas
        $50 - x$ = # of large pizzas

    $10.95x + 14.95(50-x) = 663.50$
    $10.95x + 747.50 - 14.95x = 663.50$
    $-4.00x = -84.00$ $x = 21$
    Substitute 21 for x in 2nd let statement
    $50 - x = 50 - (21) = 29$

    21 medium pizzas and 29 large pizzas

44. Let x = no. of 2-pointers y = no. of 3-pointers
    $x + y = 45$ $y = -x + 45$
    $2x + 3y = 101$
    Substitute $-x + 45$ for y in 2nd eq'n.
    $2x + 3(-x + 45) = 101$
    $2x - 3x + 135 = 101$
    $-x = -34$ $x = 34$
    Substitute 34 for x in 1st eq'n.
    $34 + y = 45$ $y = -34 + 45 = 11$
    34 two pointers and 11 three pointers

45. Let x = # of liters at 25%
        $10 - x$ = # of liters at 50%

    $.25x + .50(10 - x) = .40(10)$
    $.25x + 5 - .50x = 4$
    $-.25x = -1$ $x = 4$
    Substitute 4 for x in 2nd let statement
    $10 - x = 10 - (4) = 6$

    4 liters of 25% solution and
    6 liters of 50% solution

46. Let b = gallons of milk with butter fat
        s = gallons of skim milk

    $b + s = 100$
    $0.05b + 0.0s = 100(0.035)$
    $0.05b = 3.5$
    $b = 3.5/0.05 = 70$
    $s = 100 - b = 100 - 70 = 30$
    Thus, Gina should mix 70 gallons of milk with 5% butter fat with 30 gallons of skim milk.

47. Let c = monthly cost
      x = number of copies

Eco. Sales: $c = 18 + 0.02x$
Office Sup.: $c = 24 + 0.015x$ set eq'ns.
equal
$18 + 0.02x = 24 + 0.015x$
$0.005x = 6$ $x = 1200$

1200 copies per month

48. CHP = $.45x + 30$
    VACP = $.20x + 35$

a) $.45x + 30 = .20x + 35$
   $\underline{-.20x\ -30\ \ -.20x\ -30}$
   $.25x\ \ \ \ \ \ \ =\ \ \ \ 5$
   $\underline{.25x} = \underline{5.00}$ x = 20 minutes/month
   $.25\ \ \ \ \ \ .25$

b) CHP: $.45(50) + 30 = 22.50 + 30 = 52.50$
   VACP: $.20(50) + 35 = 10 + 35 = 45.00$

Verizon America offers the cheaper plan.

49. Let x = no. of pounds of nuts
      y = no. of pounds of pretzels

$x + y = 20$ $y = -x + 20$
$3x + 1y = 30$
Substitute $(20 - x)$ for y in the 2nd equation.
$3x + (20 - x) = 30$
$3x + 20 - x = 30$
$2x = 10$ $x = 5$ Solve for y
$y = 20 - 5 = 15$

Mix 5 lbs. of nuts with 15 lbs. of pretzels

50. Let a = number of grams of Mix A
      b = number of grams of Mix B

Protein: $0.10a + 0.20b = 20$
Carbohydrates: $0.06a + 0.02b = 6$
Multiply the 2nd equation by (-10),
$-0.60a - 0.20b = -60$
$0.10a + 0.20b = 20$ add to eliminate b
$-0.50a = -40$ $a = 80$
Substitute 80 for a in the first equation.
$0.10(80) + 0.20b = 20$
$8 + 0.20b = 20$
$0.20b = 12$ $b = 60$

a = 80 grams of Mix A
b = 60 grams of Mix B

51. Let x = no. of students
      y = no. of adults

$x + y = 250$ $x = -y + 250$
$2x + 5y = 950$
Substitute $(-y + 250)$ for x in the 2nd equation.
$2(250 - y) + 5y = 950$
$500 - 2y + 5y = 950$
$3y = 450$ $y = 150$
Substitute 150 for y in the 1st eq'n.
$x + (150) = 250$ $x = 100$

100 students and 150 adults

52. Let c = total cost
      r = no. of rounds of golf.

Oakwood: $O = 3000 + 18r$
Pinecrest: $P = 2500 + 20r$
a) $3000 + 18r = 2500 + 20r$
   $500 = 2r$ $250 = r$
A golfer must play 250 rounds for the cost to be
the same at both clubs.

b) Oakwood: $O = 3000 + 18(30) = \$3540$
   Pinecrest: $P = 2500 + 20(30) = \$3100$

Ms. Sestini can play 30 rounds cheaper at
Pinecrest.

53. $y_1 = -.58x + 31$
    $y_5 = .32x + 7$

    $-.58x + 31 = .32x + 7$
    $\underline{.58x \quad -7 \quad .58x \ -7}$
    $ 24 = .90x$

    $\underline{.90x = 24} \qquad x = 26.666\ldots \qquad 27 \text{ years}$
    $.90 \quad .90$
    $1981 + 27 = 2008 \qquad \text{During } 2007$

54. $y_C = -1.13x + 27$
    $y_W = 0.38x + 9$

    $-1.13x + 27 = 0.38x + 9$
    $\underline{-.38x \quad -27 \quad -.38x \ -27}$
    $-1.51x = -18$

    $\underline{-1.51x = -18} \qquad x = 11.92 \qquad 12 \text{ years}$
    $-1.51 \quad -1.51$
    $1992 + 12 = 2004 \qquad \text{During } 2003$

55. $(1/u) + (2/v) = 8$
    $(3/u) - (1/v) = 3$
    Substitute x for $\frac{1}{u}$ and y for $\frac{1}{v}$.
    (1) $x + 2y = 8$
    (2) $3x - y = 3$
    Multiply eq'n. (2) by 2,
    $x + 2y = 8$
    $6x - 2y = 6$ add to eliminate y
    $7x = 14 \qquad x = 2$, thus $u = \frac{1}{2}$
    Substitute 2 for x in eq. (1).
    $(2) + 2y = 8$
    $2y = 6 \qquad y = 3$, thus $v = 1/3$

    Answer: $(1/2, 1/3)$

56. Determine the equations of two lines that pass through (6,5) and another point.
    Example: $y = 5$
    $ y = (5/6)x$

57. a) $(2) + (1) + (4) = 7 \qquad (2) - (1) + 2(4) = 9$
    $ 7 = 7 9 = 9$
    $ -(2) + 2(1) + (4) = 4$
    $ 4 = 4 (2,1,4)$ is a solution.

   b) Add eq'ns. 1 and 2 to yield eq'n. 4
       Multiply eq'n. 2 by 2, then add eq'ns. 2 and 3 to yield eq'n. 5
       Combine eq'ns. 4 and 5 to find one variable.
       Substitute back into various equations to find the other 2 variables.

58. $y = 2x - 7$
    $y = 2x + 5$
    The system of equations has no solution because their slopes are equal ($m = 2$), which means that they never intersect.

59. $y = 3x + 3$
    $(1/3)y = x + 1$
    If we multiply the 2nd eq'n. by 3, we get the eq'n. $y = 3x + 1$, the same as eq'n. # 1.
    2 lines that line on top of on another have an infinite number of solutions.

60. a) $(0, 0)$ b) $(1, 0)$ c) $(0, 1)$ d) $(1, 1)$

Exercise Set 7.3

1. A matrix is a rectangular array of elements.
2. The dimensions of a matrix are determined by the number of rows and columns.
3. A square matrix contains the same number of rows as columns.
4. A 4 x 3 matrix has 4 rows.
5. A 3 x 2 matrix has 2 columns.
6. They must have the same dimensions (the number of rows must be the same and the number of columns must be the same).

7. a) Add numbers in the same positions to produce an entry in that position.

 b) $\begin{bmatrix} 1 & 4 & -1 \\ 3 & 2 & 5 \end{bmatrix} + \begin{bmatrix} 3 & 5 & -6 \\ -1 & 2 & 4 \end{bmatrix} = \begin{bmatrix} 1+3 & 4+5 & -1+(-6) \\ 3+(-1) & 2+2 & 5+4 \end{bmatrix} = \begin{bmatrix} 4 & 9 & -7 \\ 2 & 4 & 9 \end{bmatrix}$

8. a) Subtract the entry in each position in the 2nd matrix from the # in the same position in the 1st matrix.

 b) $\begin{bmatrix} 3 & -5 & 6 \\ -2 & 3 & 4 \end{bmatrix} - \begin{bmatrix} 8 & 4 & 2 \\ 0 & -2 & 4 \end{bmatrix} = \begin{bmatrix} 3-8 & -5-4 & 6-2 \\ -2-0 & 3-(-2) & 4-4 \end{bmatrix} = \begin{bmatrix} -5 & -9 & 4 \\ -2 & 5 & 0 \end{bmatrix}$

9. a) The number of rows of the first matrix must be the same as the number of columns of the second matrix.

 b) The dimensions of the resulting matrix will have the same number of rows as the first matrix and the same number of columns as the second matrix. The product of a 2 x 2 with a 2 x 3 matrix will yield a 2 x 3 matrix.

10. a) The numbers in the first row of the first matrix are multiplied by the numbers in the first column of the second matrix and the results are added together to produce the first entry of the result. Continue this procedure with each row of the first matrix and each column of the second matrix to obtain all the entries in the result matrix.

 b) $\begin{bmatrix} 6 & -1 \\ 5 & 0 \end{bmatrix}\begin{bmatrix} 2 & -3 \\ 1 & -4 \end{bmatrix} = \begin{bmatrix} 6(2)+(-1)(1) & 6(-3)+(-1)(-4) \\ 5(2)+0(1) & 5(-3)+0(-4) \end{bmatrix} = \begin{bmatrix} 11 & -14 \\ 10 & -15 \end{bmatrix}$

11. a) Identity matrix for 2x2 $\begin{bmatrix} 1 & 0 \\ 0 & 1 \end{bmatrix}$ b) Identity matrix for 3x3 $\begin{bmatrix} 1 & 0 & 0 \\ 0 & 1 & 0 \\ 0 & 0 & 1 \end{bmatrix}$

12. $\begin{bmatrix} E & 110 & 232 & 103 & 190 & 212 \\ W & 107 & 250 & 135 & 203 & 189 \\ C & 115 & 218 & 122 & 192 & 210 \end{bmatrix}$

13. $A = \begin{bmatrix} 1 & 3 \\ 5 & 7 \end{bmatrix}$ $B = \begin{bmatrix} -5 & -1 \\ 7 & 2 \end{bmatrix}$ $A+B = \begin{bmatrix} 1+(-5) & 3+(-1) \\ 5+7 & 7+2 \end{bmatrix} = \begin{bmatrix} -4 & 2 \\ 12 & 9 \end{bmatrix}$

14. $A+B = \begin{bmatrix} 2 & 3 & -7 \\ 4 & 0 & -1 \end{bmatrix} + \begin{bmatrix} -4 & -3 & 8 \\ 6 & 5 & 0 \end{bmatrix} = \begin{bmatrix} 2+(-4) & 3+(-3) & -7+8 \\ 4+6 & 0+5 & -1+0 \end{bmatrix} = \begin{bmatrix} -2 & 0 & 1 \\ 10 & 5 & -1 \end{bmatrix}$

15. $A+B = \begin{bmatrix} 3 & 1 \\ 0 & 4 \\ 6 & 0 \end{bmatrix} + \begin{bmatrix} -3 & 3 \\ 4 & 0 \\ -1 & -1 \end{bmatrix} = \begin{bmatrix} 3+(-3) & 1+3 \\ 0+4 & 4+0 \\ 6+(-1) & 0+(-1) \end{bmatrix} = \begin{bmatrix} 0 & 4 \\ 4 & 4 \\ 5 & -1 \end{bmatrix}$

16. $A+B = \begin{bmatrix} 2 & 6 & 3 \\ -1 & -6 & 4 \\ 3 & 0 & 5 \end{bmatrix} + \begin{bmatrix} -1 & 3 & 1 \\ 7 & -2 & 1 \\ 2 & 3 & 8 \end{bmatrix} = \begin{bmatrix} 2+(-1) & 6+3 & 3+1 \\ -1+7 & -6+(-2) & 4+1 \\ 3+2 & 0+3 & 5+8 \end{bmatrix} = \begin{bmatrix} 1 & 9 & 4 \\ 6 & -8 & 5 \\ 5 & 3 & 13 \end{bmatrix}$

17. $A - B = \begin{bmatrix} 4 & -2 \\ -3 & 5 \end{bmatrix} - \begin{bmatrix} -2 & 5 \\ 9 & 1 \end{bmatrix} = \begin{bmatrix} 4-(-2) & -2-(-5) \\ -3-(9) & 5-1 \end{bmatrix} = \begin{bmatrix} 6 & -7 \\ -12 & 4 \end{bmatrix}$

18. $A - B = \begin{bmatrix} 8 & 1 \\ 0 & 2 \\ -3 & -9 \end{bmatrix} - \begin{bmatrix} 3 & 3 \\ -4 & 5 \\ -2 & 6 \end{bmatrix} = \begin{bmatrix} 8-3 & 1-3 \\ 0-(-4) & 2-5 \\ -3-(-2) & -9-(6) \end{bmatrix} = \begin{bmatrix} 5 & -2 \\ 4 & -3 \\ -1 & -15 \end{bmatrix}$

19. $A - B = \begin{bmatrix} -4 & 3 \\ 6 & 2 \\ 1 & -5 \end{bmatrix} - \begin{bmatrix} -6 & -8 \\ -10 & -11 \\ 3 & -7 \end{bmatrix} = \begin{bmatrix} -4+6 & 3+8 \\ 6+10 & 2+11 \\ 1-3 & -5+7 \end{bmatrix} = \begin{bmatrix} 2 & 11 \\ 16 & 13 \\ -2 & 2 \end{bmatrix}$

20. $A - B = \begin{bmatrix} 5 & 3 & -1 \\ 7 & 4 & 2 \\ 6 & -1 & -5 \end{bmatrix} - \begin{bmatrix} 4 & 3 & 6 \\ -2 & -4 & 9 \\ 0 & -2 & 4 \end{bmatrix} = \begin{bmatrix} 5-4 & 3-3 & -1-6 \\ 7+2 & 4+4 & 2-9 \\ 6-0 & -1+2 & -5-4 \end{bmatrix} = \begin{bmatrix} 1 & 0 & -7 \\ 9 & 8 & -7 \\ 6 & 1 & -9 \end{bmatrix}$

21. $2B = 2\begin{bmatrix} 3 & 2 \\ 5 & 0 \end{bmatrix} = \begin{bmatrix} 2(3) & 2(2) \\ 2(5) & 2(0) \end{bmatrix} = \begin{bmatrix} 6 & 4 \\ 10 & 0 \end{bmatrix}$ 22. $3B = -3\begin{bmatrix} 3 & 2 \\ 5 & 0 \end{bmatrix} = \begin{bmatrix} -3(3) & -3(2) \\ -3(5) & -3(0) \end{bmatrix} = \begin{bmatrix} -9 & -6 \\ -15 & 0 \end{bmatrix}$

23. $2B + 3C = 2\begin{bmatrix} 3 & 2 \\ 5 & 0 \end{bmatrix} + 3\begin{bmatrix} -2 & 3 \\ 4 & 0 \end{bmatrix} = \begin{bmatrix} 6 & 4 \\ 10 & 0 \end{bmatrix} + \begin{bmatrix} -6 & 9 \\ 12 & 0 \end{bmatrix} = \begin{bmatrix} 6-6 & 4+9 \\ 10+12 & 0+0 \end{bmatrix} = \begin{bmatrix} 0 & 13 \\ 22 & 0 \end{bmatrix}$

24. $2B + 3A = 2\begin{bmatrix} 3 & 2 \\ 5 & 0 \end{bmatrix} + 3\begin{bmatrix} 1 & 2 \\ 0 & 5 \end{bmatrix} = \begin{bmatrix} 6 & 4 \\ 10 & 0 \end{bmatrix} + \begin{bmatrix} 3 & 6 \\ 0 & 15 \end{bmatrix} = \begin{bmatrix} 6+3 & 4+6 \\ 10+0 & 0+15 \end{bmatrix} = \begin{bmatrix} 9 & 10 \\ 10 & 15 \end{bmatrix}$

25. $3B - 2C = 3\begin{bmatrix} 3 & 2 \\ 5 & 0 \end{bmatrix} - 2\begin{bmatrix} -2 & 3 \\ 4 & 0 \end{bmatrix} = \begin{bmatrix} 9 & 6 \\ 15 & 0 \end{bmatrix} - \begin{bmatrix} -4 & 6 \\ 8 & 0 \end{bmatrix} = \begin{bmatrix} 9+4 & 6-6 \\ 15-8 & 0-0 \end{bmatrix} = \begin{bmatrix} 13 & 0 \\ 7 & 0 \end{bmatrix}$

26. $4C - 2A = 4\begin{bmatrix} -2 & 3 \\ 4 & 0 \end{bmatrix} - 2\begin{bmatrix} 1 & 2 \\ 0 & 5 \end{bmatrix} = \begin{bmatrix} -8 & 12 \\ 16 & 0 \end{bmatrix} - \begin{bmatrix} 2 & 4 \\ 0 & 10 \end{bmatrix} = \begin{bmatrix} -8-(2) & 12-(4) \\ 16-(0) & 0-(10) \end{bmatrix} = \begin{bmatrix} -10 & 8 \\ 16 & -10 \end{bmatrix}$

27. $A \times B = \begin{bmatrix} 2 & 0 \\ 3 & 1 \end{bmatrix}\begin{bmatrix} 2 & 6 \\ 8 & 4 \end{bmatrix} = \begin{bmatrix} 2(2)+0(8) & 2(6)+0(4) \\ 3(2)+1(8) & 3(6)+1(4) \end{bmatrix} = \begin{bmatrix} 4 & 12 \\ 14 & 22 \end{bmatrix}$

28. $A \times B = \begin{bmatrix} 1 & -1 \\ 2 & 6 \end{bmatrix}\begin{bmatrix} 4 & -2 \\ -3 & -2 \end{bmatrix} = \begin{bmatrix} 1(4)+(-1)(-3) & 1(-2)+(-1)(-2) \\ 2(4)+6(-3) & 2(-2)+6(-2) \end{bmatrix} = \begin{bmatrix} 7 & 0 \\ -10 & -16 \end{bmatrix}$

29. $A \times B = \begin{bmatrix} 2 & 3 & -1 \\ 0 & 4 & 6 \end{bmatrix}\begin{bmatrix} 2 \\ 4 \\ 1 \end{bmatrix} = \begin{bmatrix} 2(2)+3(4)-1(1) \\ 0(2)+4(4)+6(1) \end{bmatrix} = \begin{bmatrix} 15 \\ 22 \end{bmatrix}$

30. $A \times B = \begin{bmatrix} 1 & 1 \\ 1 & 1 \end{bmatrix}\begin{bmatrix} 1 & -1 \\ -1 & 2 \end{bmatrix} = \begin{bmatrix} 1(1)+1(-1) & 1(-1)+1(2) \\ 1(1)+1(-1) & 1(-1)+1(2) \end{bmatrix} = \begin{bmatrix} 0 & 1 \\ 0 & 1 \end{bmatrix}$

31. $A \times B = \begin{bmatrix} 4 & 7 & 6 \\ -2 & 3 & 1 \\ 5 & 1 & 2 \end{bmatrix}\begin{bmatrix} 1 & 0 & 0 \\ 0 & 1 & 0 \\ 0 & 0 & 1 \end{bmatrix} = \begin{bmatrix} 4+0+0 & 0+7+0 & 0+0+6 \\ -2+0+0 & 0+3+0 & 0+0+1 \\ 5+0+0 & 0+1+0 & 0+0+2 \end{bmatrix} = \begin{bmatrix} 4 & 7 & 6 \\ -2 & 3 & 1 \\ 5 & 1 & 2 \end{bmatrix}$

32. $A \times B = \begin{bmatrix} -3 & 1 \\ 2 & 7 \end{bmatrix}\begin{bmatrix} 4 & 0 \\ 1 & 6 \end{bmatrix} = \begin{bmatrix} -3(4)+1(1) & -3(0)+1(6) \\ 2(4)+7(1) & 2(0)+7(6) \end{bmatrix} = \begin{bmatrix} -11 & 6 \\ 15 & 42 \end{bmatrix}$

33. $A + B = \begin{bmatrix} 1 & 3 & -2 \\ 4 & 0 & 3 \end{bmatrix} + \begin{bmatrix} 5 & -1 & 3 \\ 2 & -2 & 1 \end{bmatrix} = \begin{bmatrix} 1+5 & 3+(-1) & -2+3 \\ 4+2 & 0+(-2) & 3+1 \end{bmatrix} = \begin{bmatrix} 6 & 2 & 1 \\ 6 & -2 & 4 \end{bmatrix}$

$A \times B = \begin{bmatrix} 1 & 3 & -2 \\ 4 & 0 & 3 \end{bmatrix}\begin{bmatrix} 5 & -1 & 3 \\ 2 & -2 & 1 \end{bmatrix} = $ Operation cannot be performed because # of columns ≠ # of rows

34. $A = \begin{bmatrix} 6 & 4 & -1 \\ 2 & 3 & 4 \end{bmatrix}$ A+B cannot be performed because the # of columns   # of columns

$B = \begin{bmatrix} 1 & 0 \\ 4 & -1 \end{bmatrix}$ AxB cannot be performed because the # of columns   # of rows

35. Matrices A and B cannot be added because they do not have the same dimensions.

$A \times B = \begin{bmatrix} 4 & 5 & 3 \\ 6 & 2 & 1 \end{bmatrix} \times \begin{bmatrix} 3 & 2 \\ 4 & 6 \\ -2 & 0 \end{bmatrix} = \begin{bmatrix} 4(3)+5(4)+3(-2) & 4(2)+5(6)+3(0) \\ 6(3)+2(4)+1(-2) & 6(2)+2(6)+1(0) \end{bmatrix} = \begin{bmatrix} 26 & 38 \\ 24 & 24 \end{bmatrix}$

36. $A + B = \begin{bmatrix} 1 & 2 \\ 3 & 4 \\ 5 & 6 \end{bmatrix} + \begin{bmatrix} 1 & 2 \\ 3 & 4 \\ 5 & 6 \end{bmatrix} = \begin{bmatrix} 1+1 & 2+2 \\ 3+3 & 4+4 \\ 5+5 & 6+6 \end{bmatrix} = \begin{bmatrix} 2 & 4 \\ 6 & 8 \\ 10 & 12 \end{bmatrix}$

A and B cannot be multiplied because the # of columns in A is not equal to the number of rows in B.

37. A and B cannot be added because they do not have the same dimensions.

$A \times B = \begin{bmatrix} 1 & 2 \\ 3 & 4 \end{bmatrix}\begin{bmatrix} -3 \\ 2 \end{bmatrix} = \begin{bmatrix} 1(-3)+2(2) \\ 3(-3)+4(2) \end{bmatrix} = \begin{bmatrix} 1 \\ -1 \end{bmatrix}$

38. $A + B = \begin{bmatrix} 5 & -1 \\ 6 & -2 \end{bmatrix} + \begin{bmatrix} 1 & 2 \\ 3 & 4 \end{bmatrix} = \begin{bmatrix} 5+1 & -1+2 \\ 6+3 & -2+4 \end{bmatrix} = \begin{bmatrix} 6 & 1 \\ 9 & 2 \end{bmatrix}$

$A \times B = \begin{bmatrix} 5 & -1 \\ 6 & -2 \end{bmatrix} \times \begin{bmatrix} 1 & 2 \\ 3 & 4 \end{bmatrix} = \begin{bmatrix} 5(1)+(-1)(3) & 5(2)+(-1)(4) \\ 6(1)+(-2)(3) & 6(2)+(-2)(4) \end{bmatrix} = \begin{bmatrix} 2 & 6 \\ 0 & 4 \end{bmatrix}$

39. $A + B = \begin{bmatrix} 1 & 2 \\ 2 & -3 \end{bmatrix} + \begin{bmatrix} 4 & 5 \\ 6 & 7 \end{bmatrix} = \begin{bmatrix} 1+4 & 2+5 \\ 2+6 & -3+7 \end{bmatrix} = \begin{bmatrix} 5 & 7 \\ 8 & 4 \end{bmatrix}$

$B + A = \begin{bmatrix} 4 & 5 \\ 6 & 7 \end{bmatrix} + \begin{bmatrix} 1 & 2 \\ 2 & -3 \end{bmatrix} = \begin{bmatrix} 4+1 & 5+2 \\ 6+2 & 7+(-3) \end{bmatrix} = \begin{bmatrix} 5 & 7 \\ 8 & 4 \end{bmatrix}$ Thus $A + B = B + A$.

40. $A + B = \begin{bmatrix} 9 & 4 \\ 1 & 7 \end{bmatrix} + \begin{bmatrix} 0 & 6 \\ -1 & 5 \end{bmatrix} = \begin{bmatrix} 9+0 & 4+6 \\ 1+(-1) & 7+5 \end{bmatrix} = \begin{bmatrix} 9 & 10 \\ 0 & 12 \end{bmatrix}$

$B + A = \begin{bmatrix} 0 & 6 \\ -1 & 5 \end{bmatrix} + \begin{bmatrix} 9 & 4 \\ 1 & 7 \end{bmatrix} = \begin{bmatrix} 0+9 & 6+4 \\ -1+1 & 5+7 \end{bmatrix} = \begin{bmatrix} 9 & 10 \\ 0 & 12 \end{bmatrix}$ Thus $A + B = B + A$.

41. $A + B = \begin{bmatrix} 0 & -1 \\ 3 & -4 \end{bmatrix} + \begin{bmatrix} 8 & 1 \\ 3 & -4 \end{bmatrix} = \begin{bmatrix} 0+8 & -1+1 \\ 3+3 & -4+(-4) \end{bmatrix} = \begin{bmatrix} 8 & 0 \\ 6 & -8 \end{bmatrix}$

$B + A = \begin{bmatrix} 8 & 1 \\ 3 & -4 \end{bmatrix} + \begin{bmatrix} 0 & -1 \\ 3 & -4 \end{bmatrix} = \begin{bmatrix} 8+0 & 1+(-1) \\ 3+3 & -4+(-4) \end{bmatrix} = \begin{bmatrix} 8 & 0 \\ 6 & -8 \end{bmatrix}$ Thus $A + B = B + A$.

42. $A + B = \begin{bmatrix} 1 & 2 \\ 3 & 2 \end{bmatrix} + \begin{bmatrix} 5 & 6 \\ 6 & 5 \end{bmatrix} = \begin{bmatrix} 1+5 & 2+6 \\ 3+6 & 2+5 \end{bmatrix} = \begin{bmatrix} 6 & 8 \\ 9 & 7 \end{bmatrix}$

$B + A = \begin{bmatrix} 5 & 6 \\ 6 & 5 \end{bmatrix} + \begin{bmatrix} 1 & 2 \\ 3 & 2 \end{bmatrix} = \begin{bmatrix} 5+1 & 6+2 \\ 6+3 & 5+2 \end{bmatrix} = \begin{bmatrix} 6 & 8 \\ 9 & 7 \end{bmatrix}$ Thus $A + B = B + A$.

43. $(A + B) + C = \left(\begin{bmatrix} 5 & 2 \\ 3 & 6 \end{bmatrix} + \begin{bmatrix} 3 & 4 \\ -2 & 7 \end{bmatrix} \right) + \begin{bmatrix} -1 & 4 \\ 5 & 0 \end{bmatrix} = \begin{bmatrix} 8 & 6 \\ 1 & 13 \end{bmatrix} + \begin{bmatrix} -1 & 4 \\ 5 & 0 \end{bmatrix} = \begin{bmatrix} 7 & 10 \\ 6 & 13 \end{bmatrix}$

$A + (B + C) = \begin{bmatrix} 5 & 2 \\ 3 & 6 \end{bmatrix} + \left(\begin{bmatrix} 3 & 4 \\ -2 & 7 \end{bmatrix} + \begin{bmatrix} -1 & 4 \\ 5 & 0 \end{bmatrix} \right) = \begin{bmatrix} 5 & 2 \\ 3 & 6 \end{bmatrix} + \begin{bmatrix} 2 & 8 \\ 3 & 7 \end{bmatrix} = \begin{bmatrix} 7 & 10 \\ 6 & 13 \end{bmatrix}$

Thus, $(A + B) + C = A + (B + C)$.

44. $(A + B) + C = \left(\begin{bmatrix} 4 & 1 \\ 6 & 7 \end{bmatrix} + \begin{bmatrix} -9 & 1 \\ -7 & 2 \end{bmatrix} \right) + \begin{bmatrix} -6 & -3 \\ 3 & 6 \end{bmatrix} = \begin{bmatrix} -5 & 2 \\ -1 & 9 \end{bmatrix} + \begin{bmatrix} -6 & -3 \\ 3 & 6 \end{bmatrix} = \begin{bmatrix} -11 & -1 \\ 2 & 15 \end{bmatrix}$

$A + (B + C) = \begin{bmatrix} 4 & 1 \\ 6 & 7 \end{bmatrix} + \left(\begin{bmatrix} -9 & 1 \\ -7 & 2 \end{bmatrix} + \begin{bmatrix} -6 & -3 \\ 3 & 6 \end{bmatrix} \right) = \begin{bmatrix} 4 & 1 \\ 6 & 7 \end{bmatrix} + \begin{bmatrix} -15 & -2 \\ -4 & 8 \end{bmatrix} = \begin{bmatrix} -11 & -1 \\ 2 & 15 \end{bmatrix}$

Thus, $(A + B) + C = A + (B + C)$.

45. $(A + B) + C = \left(\begin{bmatrix} 7 & 4 \\ 9 & -36 \end{bmatrix} + \begin{bmatrix} 5 & 6 \\ -1 & -4 \end{bmatrix} \right) + \begin{bmatrix} -7 & -5 \\ -1 & 3 \end{bmatrix} = \begin{bmatrix} 12 & 10 \\ 8 & -40 \end{bmatrix} + \begin{bmatrix} -7 & -5 \\ -1 & 3 \end{bmatrix} = \begin{bmatrix} 5 & 5 \\ 7 & -37 \end{bmatrix}$

$A + (B + C) = \begin{bmatrix} 7 & 4 \\ 9 & -36 \end{bmatrix} + \left(\begin{bmatrix} 5 & 6 \\ -1 & -4 \end{bmatrix} + \begin{bmatrix} -7 & -5 \\ -1 & 3 \end{bmatrix} \right) = \begin{bmatrix} 7 & 4 \\ 9 & -36 \end{bmatrix} + \begin{bmatrix} -2 & 1 \\ -2 & -1 \end{bmatrix} = \begin{bmatrix} 5 & 5 \\ 7 & -37 \end{bmatrix}$

Thus, $(A + B) + C = A + (B + C)$.

46. $A = \begin{bmatrix} 1 \\ 1 \end{bmatrix}$ ..$B = \begin{bmatrix} 2 \\ 0 \end{bmatrix}$ ..$C = \begin{bmatrix} 3 \\ 3 \end{bmatrix}$ (Your choices may be different)

$(A + B) + C = \left(\begin{bmatrix} 1 \\ 1 \end{bmatrix} + \begin{bmatrix} 2 \\ 0 \end{bmatrix} \right) + \begin{bmatrix} 3 \\ 3 \end{bmatrix} = \begin{bmatrix} 3 \\ 1 \end{bmatrix} + \begin{bmatrix} 3 \\ 3 \end{bmatrix} = \begin{bmatrix} 6 \\ 4 \end{bmatrix}$

$A + (B + C) = \begin{bmatrix} 1 \\ 1 \end{bmatrix} + \left(\begin{bmatrix} 2 \\ 0 \end{bmatrix} + \begin{bmatrix} 3 \\ 3 \end{bmatrix} \right) = \begin{bmatrix} 1 \\ 1 \end{bmatrix} + \begin{bmatrix} 5 \\ 3 \end{bmatrix} = \begin{bmatrix} 6 \\ 4 \end{bmatrix}$ Thus, $(A+B) + C = A + (B+C)$.

47. $A \times B = \begin{bmatrix} 1 & 2 \\ 4 & -3 \end{bmatrix} \begin{bmatrix} -1 & -3 \\ 2 & 4 \end{bmatrix} = \begin{bmatrix} 1(-1) + 2(2) & 1(-3) + 2(4) \\ 4(-1) + (-3)(2) & 4(-3) + (-3)(4) \end{bmatrix} = \begin{bmatrix} 3 & 5 \\ -10 & -24 \end{bmatrix}$

$B \times A = \begin{bmatrix} -1 & -3 \\ 2 & 4 \end{bmatrix} \begin{bmatrix} 1 & 2 \\ 4 & -3 \end{bmatrix} = \begin{bmatrix} -1(1) + (-3)4 & -1(2) + (-3)(-3) \\ 2(1) + 4(4) & 2(2) + 4(-3) \end{bmatrix} = \begin{bmatrix} -13 & 7 \\ 180 & -8 \end{bmatrix}$ Thus, $A \times B \neq B \times A$.

48. $A \times B = \begin{bmatrix} 3 & 1 \\ 6 & 6 \end{bmatrix} \begin{bmatrix} 1 & 0 \\ 0 & 1 \end{bmatrix} = \begin{bmatrix} 3(1) + 1(0) & 3(0) + 1(1) \\ 6(1) + 6(0) & 6(0) + 6(1) \end{bmatrix} = \begin{bmatrix} 3 & 1 \\ 6 & 6 \end{bmatrix}$

$B \times A = \begin{bmatrix} 1 & 0 \\ 0 & 1 \end{bmatrix} \begin{bmatrix} 3 & 1 \\ 6 & 6 \end{bmatrix} = \begin{bmatrix} 1(3) + 0(6) & 0(3) + 1(1) \\ 1(6) + 0(6) & 0(6) + 1(6) \end{bmatrix} = \begin{bmatrix} 3 & 1 \\ 6 & 6 \end{bmatrix}$ Thus, $A \times B \neq B \times A$.

49. $A \times B = \begin{bmatrix} 4 & 2 \\ 1 & -3 \end{bmatrix} \begin{bmatrix} 2 & 4 \\ -3 & 1 \end{bmatrix} = \begin{bmatrix} 4(2) + 2(-3) & 4(4) + 2(1) \\ 1(2) + (-3)(-3) & 1(4) + (-3)(1) \end{bmatrix} = \begin{bmatrix} 2 & 18 \\ 11 & 1 \end{bmatrix}$

$B \times A = \begin{bmatrix} 2 & 4 \\ -3 & 1 \end{bmatrix} \times \begin{bmatrix} 4 & 2 \\ 1 & -3 \end{bmatrix} = \begin{bmatrix} 2(4) + 4(1) & 2(2) + 4(-3) \\ -3(4) + 1(1) & -3(2) + 1(-3) \end{bmatrix} = \begin{bmatrix} 12 & -8 \\ -11 & -9 \end{bmatrix}$ Thus, $A \times B \neq B \times A$.

50. $A \times B = \begin{bmatrix} -3 & 2 \\ 6 & -5 \end{bmatrix} \times \begin{bmatrix} -5/3 & -2/3 \\ -2 & -1 \end{bmatrix} = \begin{bmatrix} -3(-5/3) + 2(-2) & -3(-2/3) + 2(-1) \\ 6(-5/3) - 5(-2) & 6(-2/3) - 5(-1) \end{bmatrix} = \begin{bmatrix} 1 & 0 \\ 0 & 1 \end{bmatrix}$

$B \times A = \begin{bmatrix} -5/3 & -2/3 \\ -2 & -1 \end{bmatrix} \times \begin{bmatrix} -3 & 2 \\ 6 & -5 \end{bmatrix} = \begin{bmatrix} (-5/3)(-3) - (2/3)(6) & -(5/3)(2) - (2/3)(5) \\ -2(-3) - 1(6) & -2(2) - 1(-5) \end{bmatrix} = \begin{bmatrix} 1 & 0 \\ 0 & 1 \end{bmatrix}$

Thus, $A \times B = B \times A$

51. Since $B = I$, (the identity matrix), and $A \times I = I \times A = A$, we can conclude that $A \times B = B \times A$.

52. $A = \begin{bmatrix} 1 & 1 \\ 0 & 2 \end{bmatrix}$ $B = \begin{bmatrix} 2 & 3 \\ 2 & 3 \end{bmatrix}$ (Your choices may be different)

$A \times B = \begin{bmatrix} 1 & 1 \\ 0 & 2 \end{bmatrix} \begin{bmatrix} 2 & 3 \\ 2 & 3 \end{bmatrix} = \begin{bmatrix} 1(2) + 1(2) & 1(3) + 1(3) \\ 0(2) + 2(2) & 0(3) + 2(3) \end{bmatrix} = \begin{bmatrix} 4 & 6 \\ 4 & 6 \end{bmatrix}$

$B \times A = \begin{bmatrix} 2 & 3 \\ 2 & 3 \end{bmatrix} \times \begin{bmatrix} 1 & 1 \\ 0 & 2 \end{bmatrix} = \begin{bmatrix} 2(1) + 3(0) & 2(1) + 3(2) \\ 2(1) + 3(0) & 2(1) + 3(2) \end{bmatrix} = \begin{bmatrix} 2 & 5 \\ 2 & 5 \end{bmatrix}$ Thus, $A \times B \neq B \times A$.

53. $(A \times B) \times C = \left(\begin{bmatrix} 1 & 3 \\ 4 & 0 \end{bmatrix} \begin{bmatrix} 4 & 2 \\ 3 & 1 \end{bmatrix} \right) \begin{bmatrix} 2 & 1 \\ 3 & 0 \end{bmatrix} = \begin{bmatrix} 13 & 5 \\ 16 & 8 \end{bmatrix} \begin{bmatrix} 2 & 1 \\ 3 & 0 \end{bmatrix} = \begin{bmatrix} 41 & 13 \\ 56 & 16 \end{bmatrix}$

$A \times (B \times C) = \begin{bmatrix} 1 & 3 \\ 4 & 0 \end{bmatrix} \left(\begin{bmatrix} 4 & 2 \\ 3 & 1 \end{bmatrix} \begin{bmatrix} 2 & 1 \\ 3 & 0 \end{bmatrix} \right) = \begin{bmatrix} 1 & 3 \\ 4 & 0 \end{bmatrix} \begin{bmatrix} 14 & 4 \\ 9 & 3 \end{bmatrix} = \begin{bmatrix} 41 & 13 \\ 56 & 16 \end{bmatrix}$ Thus, $(A \times B) \times C = A \times (B \times C)$.

54. $(A \times B) \times C = \left(\begin{bmatrix} -2 & 3 \\ 0 & 4 \end{bmatrix} \begin{bmatrix} 4 & 0 \\ 3 & 5 \end{bmatrix} \right) \begin{bmatrix} 3 & 4 \\ -2 & 5 \end{bmatrix} = \begin{bmatrix} 1 & 15 \\ 12 & 20 \end{bmatrix} \begin{bmatrix} 3 & 4 \\ -2 & 5 \end{bmatrix} = \begin{bmatrix} -27 & 79 \\ -4 & 148 \end{bmatrix}$

$A \times (B \times C) = \begin{bmatrix} -2 & 3 \\ 0 & 4 \end{bmatrix} \left(\begin{bmatrix} 4 & 0 \\ 3 & 5 \end{bmatrix} \begin{bmatrix} 3 & 4 \\ -2 & 5 \end{bmatrix} \right) = \begin{bmatrix} -2 & 3 \\ 0 & 4 \end{bmatrix} \begin{bmatrix} 12 & 16 \\ -1 & 37 \end{bmatrix} = \begin{bmatrix} -27 & 79 \\ -4 & 148 \end{bmatrix}$

Thus, $(A \times B) \times C = A \times (B \times C)$.

55. $(A \times B) \times C = \left(\begin{bmatrix} 4 & 3 \\ -6 & 2 \end{bmatrix} \begin{bmatrix} 1 & 2 \\ 0 & 1 \end{bmatrix} \right) \begin{bmatrix} 4 & 3 \\ 0 & -2 \end{bmatrix} = \begin{bmatrix} 4 & 11 \\ -6 & -10 \end{bmatrix} \begin{bmatrix} 4 & 3 \\ 0 & -2 \end{bmatrix} = \begin{bmatrix} 16 & -10 \\ -24 & 2 \end{bmatrix}$

$A \times (B \times C) = \begin{bmatrix} 4 & 3 \\ -6 & 2 \end{bmatrix} \left(\begin{bmatrix} 1 & 2 \\ 0 & 1 \end{bmatrix} \begin{bmatrix} 4 & 3 \\ 0 & -2 \end{bmatrix} \right) = \begin{bmatrix} 4 & 3 \\ -6 & 2 \end{bmatrix} \begin{bmatrix} 4 & -1 \\ 0 & -2 \end{bmatrix} = \begin{bmatrix} 16 & -10 \\ -24 & 2 \end{bmatrix}$

Thus, $(A \times B) \times C = A \times (B \times C)$.

56. $(A \times B) \times C = (A \times I) \times C = A \times C$, and $A \times (B \times C) = A \times (I \times C) = A \times C$, thus $(A \times B) \times C = A \times (B \times C)$.

57. $(A \times B) \times C = \left(\begin{bmatrix} 3 & 4 \\ -1 & -2 \end{bmatrix} \begin{bmatrix} 0 & 1 \\ 1 & 0 \end{bmatrix} \right) \begin{bmatrix} 2 & 0 \\ 3 & 0 \end{bmatrix} = \begin{bmatrix} 4 & 3 \\ -2 & -1 \end{bmatrix} \begin{bmatrix} 2 & 0 \\ 3 & 0 \end{bmatrix} = \begin{bmatrix} 17 & 0 \\ -7 & 0 \end{bmatrix}$

$A \times (B \times C) = \begin{bmatrix} 3 & 4 \\ -1 & -2 \end{bmatrix} \left(\begin{bmatrix} 0 & 1 \\ 1 & 0 \end{bmatrix} \begin{bmatrix} 2 & 0 \\ 3 & 0 \end{bmatrix} \right) = \begin{bmatrix} 3 & 4 \\ -1 & -2 \end{bmatrix} \begin{bmatrix} 3 & 0 \\ 2 & 0 \end{bmatrix} = \begin{bmatrix} 17 & 0 \\ -7 & 0 \end{bmatrix}$

Thus, $(A \times B) \times C = A \times (B \times C)$.

58. $(A \times B) \times C = A \times (B \times C)$ for any choices of A, B, and C that can be multiplied.

59. $A \times B = \begin{bmatrix} 2 & 2 & .5 & 1 \\ 3 & 2 & 1 & 2 \\ 0 & 1 & 0 & 3 \\ .5 & 1 & 0 & 0 \end{bmatrix} \begin{bmatrix} 10 & 12 \\ 5 & 8 \\ 8 & 8 \\ 4 & 6 \end{bmatrix} = \begin{bmatrix} 2 \cdot 10 + 2 \cdot 5 + .5 \cdot 8 + 1 \cdot 4 & 2 \cdot 12 + 2 \cdot 8 + .5 \cdot 8 + 1 \cdot 6 \\ 3 \cdot 10 + 2 \cdot 5 + 1 \cdot 8 + 2 \cdot 4 & 3 \cdot 12 + 2 \cdot 8 + 1 \cdot 8 + 2 \cdot 6 \\ 0 \cdot 10 + 1 \cdot 5 + 0 \cdot 8 + 3 \cdot 4 & 0 \cdot 12 + 1 \cdot 8 + 0 \cdot 8 + 3 \cdot 6 \\ .5 \cdot 10 + 1 \cdot 5 + 0 \cdot 8 + 0 \cdot 4 & .5 \cdot 12 + 1 \cdot 8 + 0 \cdot 8 + 0 \cdot 6 \end{bmatrix} = \begin{bmatrix} 38 & 50 \\ 56 & 72 \\ 17 & 26 \\ 10 & 14 \end{bmatrix}$

60. a) Let $C = [40 \ 30 \ 12 \ 20]$.

b) $C \times A = [40 \ 30 \ 12 \ 20] \begin{bmatrix} 2 & 2 & .5 & 1 \\ 3 & 2 & 1 & 2 \\ 0 & 1 & 0 & 3 \\ .5 & 1 & 0 & 0 \end{bmatrix} = [\underset{sug.}{180} \ \underset{flr.}{172} \ \underset{mlk.}{50} \ \underset{eggs}{136}]$

61. $C(A \times B) = [40 \quad 30 \quad 12 \quad 20] \begin{bmatrix} 38 & 50 \\ 56 & 72 \\ 17 & 26 \\ 10 & 14 \end{bmatrix} = [36.04 \quad 47.52]$ cents   small \$36.04, large \$47.52

62. $A \times B = \begin{bmatrix} 52 & 50 & 75 \\ 48 & 43 & 60 \\ 62 & 57 & 81 \end{bmatrix} \begin{bmatrix} .30 & .75 \\ .25 & .50 \\ .15 & .45 \end{bmatrix} = \begin{bmatrix} 39.35 & 97.75 \\ 34.15 & 84.50 \\ 45.00 & 111.45 \end{bmatrix}$

63. $A + B = \begin{bmatrix} 6 & 3 \\ 4 & -2 \end{bmatrix} + \begin{bmatrix} -6 & -3 \\ -2 & -4 \end{bmatrix} = \begin{bmatrix} 6+(-6) & 3+(-3) \\ 4+(-2) & -2+(-4) \end{bmatrix} = \begin{bmatrix} 0 & 0 \\ 2 & -6 \end{bmatrix}$

Since $A + B \neq I$, where I is the additive identity matrix, A and B are not additive inverses.

64. $A + B = \begin{bmatrix} 4 & 6 & 3 \\ 2 & 3 & -1 \\ -1 & 0 & 6 \end{bmatrix} \begin{bmatrix} -4 & -6 & -3 \\ -2 & -3 & 1 \\ 1 & 0 & -6 \end{bmatrix} = \begin{bmatrix} 4+(-4) & 6+(-6) & 3+(-3) \\ 2+(-2) & 3+(-3) & -1+(1) \\ -1+(1) & 0+0 & 6+(-6) \end{bmatrix} = \begin{bmatrix} 0 & 0 & 0 \\ 0 & 0 & 0 \\ 0 & 0 & 0 \end{bmatrix}$

$B + A = \begin{bmatrix} -4 & -6 & -3 \\ -2 & -3 & 1 \\ 1 & 0 & -6 \end{bmatrix} \begin{bmatrix} 4 & 6 & 3 \\ 2 & 3 & -1 \\ -1 & 0 & 6 \end{bmatrix} = \begin{bmatrix} -4+(4) & -6+(6) & -3+(3) \\ -2+(2) & -3+(3) & 1+(-1) \\ 1+(-1) & 0+(0) & -6+(6) \end{bmatrix} = \begin{bmatrix} 0 & 0 & 0 \\ 0 & 0 & 0 \\ 0 & 0 & 0 \end{bmatrix}$

Thus, A and B are additive inverses.

65. $A \times B = \begin{bmatrix} 5 & -2 \\ -2 & 1 \end{bmatrix} \begin{bmatrix} 1 & 2 \\ 2 & 5 \end{bmatrix} = \begin{bmatrix} 5(1)-2(2) & 5(2)-2(5) \\ -2(1)+1(2) & -2(2)+1(5) \end{bmatrix} = \begin{bmatrix} 1 & 0 \\ 0 & 1 \end{bmatrix}$

$B \times A = \begin{bmatrix} 1 & 2 \\ 2 & 5 \end{bmatrix} \begin{bmatrix} 5 & -2 \\ -2 & 1 \end{bmatrix} = \begin{bmatrix} 1(5)+2(-2) & 1(-2)+2(1) \\ 2(5)+5(-2) & 2(-2)+5(1) \end{bmatrix} = \begin{bmatrix} 1 & 0 \\ 0 & 1 \end{bmatrix}$

Thus, A and B are multiplicative inverses.

66. $A \times B = \begin{bmatrix} 7 & 3 \\ 2 & 1 \end{bmatrix} \begin{bmatrix} 1 & -3 \\ -2 & 7 \end{bmatrix} = \begin{bmatrix} 7(1)+3(-2) & 7(-3)+3(7) \\ 2(1)+1(-2) & 2(-3)+1(7) \end{bmatrix} = \begin{bmatrix} 1 & 0 \\ 0 & 1 \end{bmatrix}$

$B \times A = \begin{bmatrix} 1 & -3 \\ -2 & 7 \end{bmatrix} \begin{bmatrix} 7 & 3 \\ 2 & 1 \end{bmatrix} = \begin{bmatrix} 1(7)-3(2) & 1(3)-3(1) \\ -2(7)+7(2) & -2(3)+7(1) \end{bmatrix} = \begin{bmatrix} 1 & 0 \\ 0 & 1 \end{bmatrix}$

Thus, A and B are multiplicative inverses.

67. False. Let $A = [1 \quad 3]$ and $B = [2 \quad 1]$. Then $A - B = [-1, 2]$ and $B - A = [1, -2]$ $A - B \neq B - A$.

68. True. For all scalars a and all matrices B and C, $a(B + C) = aB + aC$. As an example,
Let $a = 2$, $B = [1 \quad 3]$, and $C = [2 \quad 1]$.   Then $a(B + C) = 2([1 \quad 3] + [2 \quad 1]) = 2[3 \quad 4] = [6 \quad 8]$,
and $aB + aC = 2[1 \quad 3] + 2[2 \quad 1] = [2 \quad 6] + [4 \quad 2] = [6 \quad 8] = a(B + C)$.

69. a) $1.4(14) + 0.7(10) + 0.3(7) = \$28.70$

b) $2.7(12) + 2.8(9) + 0.5(5) = \$60.10$

c) $L \times C = \begin{bmatrix} 28.7 & 24.6 \\ 41.3 & 35.7 \\ 69.3 & 60.1 \end{bmatrix} \begin{matrix} \text{small} \\ \text{medium} \\ \text{large} \end{matrix}$

$\qquad \qquad \begin{matrix} \text{Ames} & \text{Bay} \end{matrix}$

This array shows the total cost of each sofa at each plant.

70. $A + B = \begin{bmatrix} 1 & 2 & 3 \\ 3 & 2 & 1 \end{bmatrix} + \begin{bmatrix} 0 & 1 & 2 \\ 4 & 5 & 1 \end{bmatrix} = \begin{bmatrix} 1 & 3 & 5 \\ 7 & 7 & 2 \end{bmatrix}$

A x B cannot be calculated because the # of columns ≠ # of rows.

71. A + B cannot be calculated because the # of columns ≠ # of rows.

$A \times B = \begin{bmatrix} 1 & 2 & 3 \\ 3 & 2 & 1 \end{bmatrix} \begin{bmatrix} 0 & 4 \\ 1 & 5 \\ 2 & 1 \end{bmatrix} = \begin{bmatrix} 1(0)+2(1)+3(2) & 1(4)+2(5)+3(1) \\ 3(0)+2(1)+1(2) & 3(4)+2(5)+1(1) \end{bmatrix} = \begin{bmatrix} 7 & 17 \\ 4 & 23 \end{bmatrix}$

72. Answers will vary.

Exercise Set 7.4

1. a) An augmented matrix is a matrix formed with the coefficients of the variables and the constants. The coefficients are separated from the constants by a vertical bar.

b) $\begin{bmatrix} 1 & 3 & | & 7 \\ 2 & -1 & | & 4 \end{bmatrix}$

2. 1) Rows of a matrix can be interchanged.

2) All values in a row can be multiplied by a nonzero real number.

3) All the values in a row may be added to the corresponding values in another row.

3. If you obtain an augmented matrix in which one row of numbers on the left side of the vertical line are all zeroes but a zero does not appear in the same row on the other side of the vertical line, the system is inconsistent.

4. If you obtain an augmented matrix in which a 0 appears across an entire row, the system of equations is dependent.

5. 1) Multiply the 2nd row by -1/2; 2) multiply the 2nd row by -3 and add to the 1st row; and 3) identify the values of x and y.

1) $\begin{bmatrix} 1 & 3 & | & 5 \\ 0 & 1 & | & (-1/2) \end{bmatrix}$ 2) $\begin{bmatrix} 1+0 & 3+(-3) & | & 5+(3/2) \\ 0 & 1 & | & (-1/2) \end{bmatrix} = \begin{bmatrix} 1 & 0 & | & 13/2 \\ 0 & 1 & | & -1/2 \end{bmatrix}$ 3) $(x, y) = \left(\dfrac{13}{2}, \dfrac{-1}{2} \right)$

6. 1) Multiply the 2nd row by 2 and add to the 1st row, and 2) identify the values of x and y.

$$\begin{bmatrix} 1 & -2 & | & 1 \\ 0 & 1 & | & 3 \end{bmatrix} \quad 1) \begin{bmatrix} 1+0 & -2+2 & | & 1+6 \\ 0 & 1 & | & 3 \end{bmatrix} = \begin{bmatrix} 1 & 0 & | & 7 \\ 0 & 1 & | & 3 \end{bmatrix} \quad 2) \; (x, y) = \left(\frac{13}{2}, \frac{-1}{2} \right)$$

7. $x + 3y = 3 \qquad -x + y = -3$

$$\begin{bmatrix} 1 & 3 & | & 3 \\ -1 & 1 & | & -3 \end{bmatrix} \rightarrow \begin{bmatrix} 1 & 3 & | & 3 \\ -1+1 & 1+3 & | & -3+3 \end{bmatrix} = \begin{bmatrix} 1 & 3 & | & 3 \\ 0 & 4 & | & 0 \end{bmatrix} \rightarrow \begin{bmatrix} 1+0 & 3+(-3) & | & 3+0 \\ 0 & 1 & | & 0 \end{bmatrix} = \begin{bmatrix} 1 & 0 & | & 3 \\ 0 & 1 & | & 0 \end{bmatrix} \rightarrow (3, 0)$$

8. $x - y = 5 \qquad 2x - y = 6$

$$\begin{bmatrix} 1 & -1 & | & 5 \\ 2 & -1 & | & 6 \end{bmatrix} \rightarrow \begin{bmatrix} 1 & -1 & | & 5 \\ 2-2 & -1+2 & | & 6-10 \end{bmatrix} = \begin{bmatrix} 1 & -1 & | & 5 \\ 0 & 1 & | & -4 \end{bmatrix} \rightarrow \begin{bmatrix} 1+0 & -1+1 & | & 5-4 \\ 0 & 1 & | & -4 \end{bmatrix} = \begin{bmatrix} 1 & 0 & | & 1 \\ 0 & 1 & | & -4 \end{bmatrix} \rightarrow (1, -4)$$

9. $x - 2y = -1 \qquad 2x + y = 8$

$$\begin{bmatrix} 1 & -2 & | & -1 \\ 2 & 1 & | & 8 \end{bmatrix} \rightarrow \begin{bmatrix} 1 & -2 & | & -1 \\ 2-2 & 1+4 & | & 8+2 \end{bmatrix} = \begin{bmatrix} 1 & -2 & | & -1 \\ 0 & 5 & | & 10 \end{bmatrix} \rightarrow \begin{bmatrix} 1+0 & -2+2 & | & -1+4 \\ 0 & 5 & | & 10 \end{bmatrix} \rightarrow$$

$$\begin{bmatrix} 1 & 0 & | & 3 \\ 0 & 5 & | & 10 \end{bmatrix} = \begin{bmatrix} 1 & 0 & | & 3 \\ 0 & 1 & | & 2 \end{bmatrix} \rightarrow (3, 2)$$

10. $x + y = -1 \qquad 2x + 3y = -5$

$$\begin{bmatrix} 1 & 1 & | & -1 \\ 2 & 3 & | & -5 \end{bmatrix} \rightarrow \begin{bmatrix} 1 & 1 & | & -1 \\ 2-2 & 3-2 & | & -5+2 \end{bmatrix} = \begin{bmatrix} 1 & 1 & | & -1 \\ 0 & 1 & | & -3 \end{bmatrix} \rightarrow \begin{bmatrix} 1+0 & 1-1 & | & -1+3 \\ 0 & 1 & | & -3 \end{bmatrix} = \begin{bmatrix} 1 & 0 & | & 2 \\ 0 & 1 & | & -3 \end{bmatrix} \rightarrow (2, -3)$$

11. $\begin{bmatrix} 2 & -5 & | & -6 \\ -4 & 10 & | & 12 \end{bmatrix} \underset{(r_2 + 2r_1)}{=} \begin{bmatrix} 2 & -5 & | & -6 \\ 0 & 0 & | & 0 \end{bmatrix} \Rightarrow$ Dependent system

The solution is all points on the line $2x - 5y = -6$.

12. $\begin{bmatrix} 1 & 1 & | & 5 \\ 3 & -1 & | & 3 \end{bmatrix} \underset{(r_2 - 3r_1)}{=} \begin{bmatrix} 1 & 1 & | & 5 \\ 0 & -4 & | & -12 \end{bmatrix} \underset{(r_2 \div (-4))}{=} \begin{bmatrix} 1 & 1 & | & 5 \\ 0 & 1 & | & 3 \end{bmatrix} \underset{=}{(r_1 - r_2)} \begin{bmatrix} 1 & 0 & | & 2 \\ 0 & 1 & | & 3 \end{bmatrix}$ The solution is (2, 3).

13. $\begin{bmatrix} 2 & -3 & | & 10 \\ 2 & 2 & | & 5 \end{bmatrix} \underset{(r_2 - 2r_1)}{(r_1 \div 2)} \begin{bmatrix} 1 & \frac{-3}{2} & | & 5 \\ 0 & 5 & | & -5 \end{bmatrix} \underset{(r_2 \div (5))}{=} \begin{bmatrix} 1 & -\frac{3}{2} & | & 5 \\ 0 & 1 & | & -1 \end{bmatrix} \underset{=}{(r_1 + \frac{3}{2}r_2)} \begin{bmatrix} 1 & 0 & | & \frac{7}{2} \\ 0 & 1 & | & -1 \end{bmatrix}$ The solution is (7/2, -1).

14. $\begin{bmatrix} 1 & 3 & | & 1 \\ -2 & 1 & | & 5 \end{bmatrix} \underset{(r_2 + 2r_1)}{=} \begin{bmatrix} 1 & 3 & | & 1 \\ 0 & 7 & | & 7 \end{bmatrix} \underset{(r_2 \div 7)}{=} \begin{bmatrix} 1 & 3 & | & 1 \\ 0 & 1 & | & 1 \end{bmatrix} \underset{=}{(r_1 - 3r_2)} \begin{bmatrix} 1 & 0 & | & -2 \\ 0 & 1 & | & 1 \end{bmatrix}$ The solution is (-2, 1).

15. $\begin{bmatrix} 4 & 2 & | & -10 \\ -2 & 1 & | & -7 \end{bmatrix} \begin{matrix} (r_1 \div 2) \\ (r_2 - r_1) \end{matrix} \begin{bmatrix} 4 & 2 & | & -10 \\ 0 & 2 & | & -12 \end{bmatrix} \begin{matrix} \\ (r_2 \div 4) \end{matrix} = \begin{bmatrix} 1 & 1/2 & | & -10/4 \\ 0 & 2 & | & -12 \end{bmatrix} \begin{matrix} \\ (r_2 \div 2) \end{matrix} = \begin{bmatrix} 1 & 1/2 & | & -10/4 \\ 0 & 1 & | & -6 \end{bmatrix}$

$\begin{bmatrix} 1 & 1/2 & | & -10/4 \\ 0 & 1 & | & -6 \end{bmatrix} \begin{matrix} (r_2 \div -2) \\ (r_2 + r_1) \end{matrix} \begin{bmatrix} 1 & 0 & | & 1/2 \\ 0 & 1 & | & -6 \end{bmatrix}$ The solution is (1/2, -6).

16. $\begin{bmatrix} 4 & 2 & | & 6 \\ 5 & 4 & | & 9 \end{bmatrix} \begin{matrix} (r_1 \div 4) \\ = \end{matrix} \begin{bmatrix} 1 & \frac{1}{2} & | & \frac{3}{2} \\ 5 & 4 & | & 9 \end{bmatrix} \begin{matrix} \\ (r_2 - 5r_1) \end{matrix} = \begin{bmatrix} 1 & \frac{1}{2} & | & \frac{3}{2} \\ 0 & \frac{3}{2} & | & \frac{3}{2} \end{bmatrix} \begin{matrix} \\ (\frac{2}{3}r_2) \end{matrix} = \begin{bmatrix} 1 & \frac{1}{2} & | & \frac{3}{2} \\ 0 & 1 & | & 1 \end{bmatrix} \begin{matrix} (r_1 - \frac{1}{2}r_2) \\ = \end{matrix} \begin{bmatrix} 1 & 0 & | & 1 \\ 0 & 1 & | & 1 \end{bmatrix}$

The solution is (1, 1).

17. $\begin{bmatrix} -3 & 6 & | & 5 \\ 2 & -4 & | & 8 \end{bmatrix} \begin{matrix} (r_1 \div (-3)) \\ = \end{matrix} \begin{bmatrix} 1 & -2 & | & \frac{-5}{3} \\ 2 & -4 & | & 8 \end{bmatrix} \begin{matrix} \\ (r_2 - 2r_1) \end{matrix} = \begin{bmatrix} 1 & -2 & | & \frac{-5}{3} \\ 0 & 0 & | & \frac{34}{3} \end{bmatrix} \Rightarrow$ Inconsistent system   No solution.

18. $\begin{bmatrix} 2 & -5 & | & 10 \\ 3 & 1 & | & 15 \end{bmatrix} = \begin{bmatrix} 1 & \frac{-5}{2} & | & 5 \\ 0 & \frac{15}{2} & | & 0 \end{bmatrix} = \begin{bmatrix} 1 & \frac{-5}{2} & | & 5 \\ 0 & 1 & | & 0 \end{bmatrix} = \begin{bmatrix} 1 & 0 & | & 5 \\ 0 & 1 & | & 0 \end{bmatrix}$ The solution is (5, 0).

19. $\begin{bmatrix} 2 & 1 & | & 11 \\ 1 & 3 & | & 18 \end{bmatrix} \begin{matrix} (r_1 \div 2) \\ = \end{matrix} \begin{bmatrix} 1 & \frac{1}{2} & | & \frac{11}{2} \\ 1 & 3 & | & 18 \end{bmatrix} \begin{matrix} \\ (r_2 - r_1) \end{matrix} = \begin{bmatrix} 1 & \frac{1}{2} & | & \frac{11}{2} \\ 0 & \frac{5}{2} & | & \frac{25}{2} \end{bmatrix} \begin{matrix} \\ (\frac{2}{5}r_2) \end{matrix} = \begin{bmatrix} 1 & \frac{1}{2} & | & \frac{11}{2} \\ 0 & 1 & | & 5 \end{bmatrix} \begin{matrix} (r_1 - \frac{1}{2}r_2) \\ = \end{matrix} \begin{bmatrix} 1 & 0 & | & 3 \\ 0 & 1 & | & 5 \end{bmatrix}$

The solution is (3, 5).

20. $\begin{bmatrix} 4 & -3 & | & 7 \\ -2 & 5 & | & 14 \end{bmatrix} \begin{matrix} (r_1 \div 4) \\ = \end{matrix} \begin{bmatrix} 1 & -\frac{3}{4} & | & \frac{7}{4} \\ -2 & 5 & | & 14 \end{bmatrix} \begin{matrix} \\ (r_2 + 2r_1) \end{matrix} = \begin{bmatrix} 1 & -\frac{3}{4} & | & \frac{7}{4} \\ 0 & \frac{7}{2} & | & \frac{35}{2} \end{bmatrix} \begin{matrix} \\ (\frac{2}{7}r_2) \end{matrix} = \begin{bmatrix} 1 & -\frac{3}{4} & | & \frac{7}{4} \\ 0 & 1 & | & 5 \end{bmatrix} \begin{matrix} (r_1 + \frac{3}{4}r_2) \\ = \end{matrix} \begin{bmatrix} 1 & 0 & | & \frac{11}{2} \\ 0 & 1 & | & 5 \end{bmatrix}$

The solution is (11/2, 5).

21. S + L = 55   4S + 6L = 290

$\begin{bmatrix} 1 & 1 & | & 55 \\ 4 & 6 & | & 290 \end{bmatrix} \begin{matrix} (r_1 \bullet -4) \\ (r_2 + r_1) \end{matrix} \begin{bmatrix} 1 & 1 & | & 55 \\ 0 & 2 & | & 70 \end{bmatrix} \begin{matrix} \\ (r_2 \div 2) \end{matrix} = \begin{bmatrix} 1 & 1 & | & 55 \\ 0 & 1 & | & 35 \end{bmatrix} \begin{matrix} (r_2 \bullet -1) \\ (r_1 + r_2) \end{matrix} \begin{bmatrix} 1 & 0 & | & 20 \\ 0 & 1 & | & 35 \end{bmatrix}$ The solution is (20, 35).

22. p = 2H + 2W   2H + 2W = 124   H − W = 8

$\begin{bmatrix} 2 & 2 & | & 124 \\ 1 & -1 & | & 8 \end{bmatrix} \begin{matrix} (r_2 \bullet 2) \\ (r_1 + r_2) \end{matrix} \begin{bmatrix} 4 & 0 & | & 140 \\ 1 & -1 & | & 8 \end{bmatrix} \begin{matrix} \\ (r_1 \div 4) \end{matrix} = \begin{bmatrix} 1 & 0 & | & 35 \\ 1 & -1 & | & 8 \end{bmatrix} \begin{matrix} (r_1 \bullet -1) \\ (r_2 + r_1) \end{matrix} \begin{bmatrix} 1 & 0 & | & 35 \\ 0 & -1 & | & -27 \end{bmatrix} \begin{matrix} (r_2 \bullet -1) \\ \end{matrix} \begin{bmatrix} 1 & 0 & | & 35 \\ 0 & 1 & | & 27 \end{bmatrix}$

The solution is (35, 27).

23. Let T = # of hours for truck driver   L = # of hours for laborer

10T + 8L = 144   L = T + 2 → T = L − 2

$\begin{bmatrix} 10 & 8 & | & 144 \\ 1 & -1 & | & -2 \end{bmatrix} \begin{matrix} (r_2 \bullet 8) \\ (r_1 + r_2) \end{matrix} \begin{bmatrix} 18 & 0 & | & 128 \\ 1 & -1 & | & -2 \end{bmatrix} \begin{matrix} (r_1 \div 18) \\ \end{matrix} \begin{bmatrix} 1 & 0 & | & 64/9 \\ 1 & -1 & | & -2 \end{bmatrix} \begin{matrix} (r_1 \bullet -1) \\ (r_2 + r_1) \end{matrix} \begin{bmatrix} 1 & 0 & | & 64/9 \\ 0 & -1 & | & -82/9 \end{bmatrix}$

$\begin{bmatrix} 1 & 0 & | & 64/9 \\ 0 & -1 & | & -82/9 \end{bmatrix} \begin{matrix} (r_2 \bullet -1) \\ \end{matrix} \begin{bmatrix} 1 & 0 & | & 64/9 \\ 0 & 1 & | & 82/9 \end{bmatrix}$ (64/9, 82/9)

7 1/9 hours for the truck driver and 9 1/9 hours for the laborer.

24. Let x = cost per pound of cherries   y = cost per pound of mints

$2x + 3y = 23$ $1x + 2y = 14$

$$\begin{bmatrix} 2 & 3 & | & 23 \\ 1 & 2 & | & 14 \end{bmatrix} = \begin{bmatrix} 1 & \frac{3}{2} & | & \frac{23}{2} \\ 1 & 2 & | & 14 \end{bmatrix} = \begin{bmatrix} 1 & \frac{3}{2} & | & \frac{23}{2} \\ 0 & \frac{1}{2} & | & \frac{5}{2} \end{bmatrix} = \begin{bmatrix} 1 & \frac{3}{2} & | & \frac{23}{2} \\ 0 & 1 & | & 5 \end{bmatrix} = \begin{bmatrix} 1 & 0 & | & 4 \\ 0 & 1 & | & 5 \end{bmatrix}$$

The cherries are $4 per pound and the mints are $5 per pound.

25. $1.5x + 2y = 337.5$ $x + y = 200$

$$\begin{bmatrix} 1.5 & 2 & | & 337.5 \\ 1 & 1 & | & 200 \end{bmatrix} = \begin{bmatrix} 1 & 1.\overline{33} & | & 225 \\ 1 & 1 & | & 200 \end{bmatrix} = \begin{bmatrix} 1 & 1.\overline{33} & | & 225 \\ 1 & 1 & | & 200 \end{bmatrix} = \begin{bmatrix} 1 & 1.\overline{33} & | & 225 \\ 0 & -.\overline{33} & | & -25 \end{bmatrix} = \begin{bmatrix} 1 & 1.\overline{33} & | & 225 \\ 0 & 1 & | & 75 \end{bmatrix} = \begin{bmatrix} 1 & 0 & | & 125 \\ 0 & 1 & | & 75 \end{bmatrix}$$

The solution is 125 non-refillable pencils @ $1.50 and 75 refillable pencils @ $2.00.

Exercise Set 7.5

1. The solution set of a system of linear inequalities is the set of points that satisfy all inequalities in the system.

2. Graph and shade the solution set to each of the inequalities. The intersection of the shaded areas and any solid lines common to both inequalities is the solution set.

3.

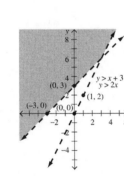

4.

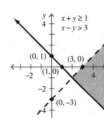

5.

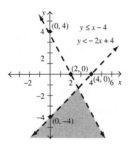

6.

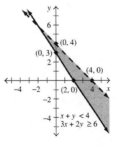

7.

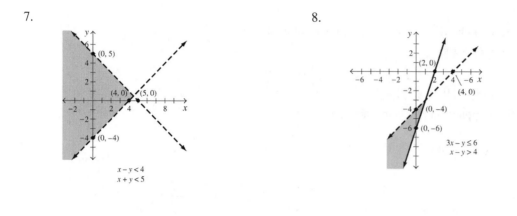

$x - y < 4$
$x + y < 5$

8.

$3x - y \le 6$
$x - y > 4$

9.

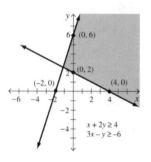

$x + 2y \ge 4$
$3x - y \ge -6$

10.

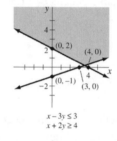

$x - 3y \le 3$
$x + 2y \ge 4$

11.

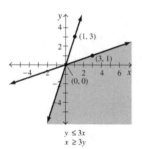

$y \le 3x$
$x \ge 3y$

12.

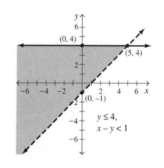

$y \le 4,$
$x - y < 1$

13.

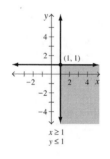

$x \ge 1$
$y \le 1$

14.

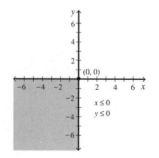

$x \le 0$
$y \le 0$

15.

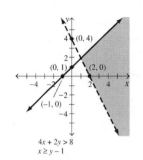

$4x + 2y > 8$
$x \geq y - 1$

16.

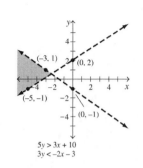

$5y > 3x + 10$
$3y < -2x - 3$

17.

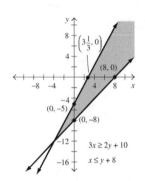

$3x \geq 2y + 10$
$x \leq y + 8$

18.

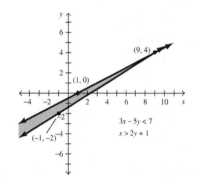

$3x - 5y < 7$
$x > 2y + 1$

19. a) Let P = Panasonic,   S = Sony

$600P + 900S \leq 18000$

$P \geq 2S$ $P \geq 10$

$S \geq 5$

c) (15, 6) means 15 Panasonic models and 6
   Sony models.
   $600 (15) + 900 (6) = 9000 + 5400$ or
   $\$ 14,400$

b)

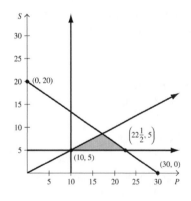

20. $x < 0,\ y > 0$

21. a) No, if the lines are parallel there may not be a solution to the system.

   b) Example: $y \geq x$ $y \leq x$ 2
     This system has no solution.

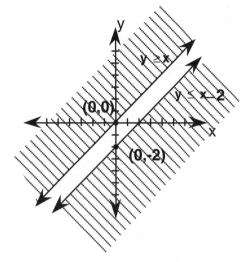

22. Yes. One example is $x < 0$, $y < 0$, $x > 0$, $y > 0$.

23. No. Every line divides the plane into two halves only one of which can be part of the solution. Therefore, the points in the other half cannot satisfy both inequalities and so do not solve the system.

   Example: $y \geq x$ $x \geq 2$

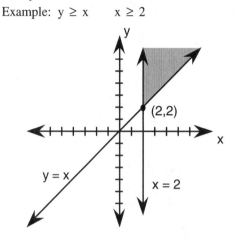

Exercise Set 7.6

1. Constraints are restrictions that are represented as linear inequalities.

2. The feasible region is formed by graphing the system of inequalities.

3. Vertices

4. Objective function: $K = Ax + By$

5. If a linear equation of the form $K = Ax + By$ is evaluated at each point in a closed polygonal region, the maximum and minimum values of the equation occur at a corner.

6. At $(1, 1)$, $P = 4(1) + 6(1) = 10$
   At $(1, 4)$, $P = 4(1) + 6(4) = 28$
   At $(5, 1)$, $P = 4(5) + 6(1) = 26$
   At $(7, 1)$, $P = 4(7) + 6(1) = 34$
   The maximum profit is 34. Determine the value of the profit function at each vertex; the largest profit value is the maximum.

7. At $(0, 0)$, $K = 6(0) + 4(0) = 0$
   At $(0, 4)$, $K = 6(0) + 4(4) = 16$
   At $(2, 3)$, $K = 6(2) + 4(3) = 24$
   At $(5, 0)$, $K = 6(5) + 4(0) = 30$
   The maximum value is 30 at $(5, 0)$; the minimum value is 0 at $(0, 0)$.

8. At (10, 20), K = 2(10) + 3(20) = 80

At (10, 40), K = 2(10) + 3(40) = 140

At (50, 30), K = 2(50) + 3(30) = 190

At (50, 10), K = 2(50) + 3(10) = 130

At (20, 10), K = 2(20) + 3(10) = 70

The maximum value is 190 at (50, 30); the minimum value is 70 at (20, 10).

9. a)

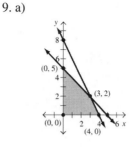

b) $x + y \leq 5$ $2x + y \leq 8$ $x \geq 0$ $y \geq 0$
    $P = 5x + 4y$
    At (0,0), P = 5(0) + 4(0) = 0   min. at (0, 0)
    At (0,4), P = 5(0) + 4(4) = 16
    At (3, 2, P = 5(3) + 4(2) = 23   max. at (2, 3)
    At (0,5), P = 5(0) + 4(5) = 20

10. a)

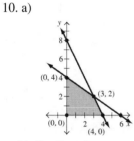

b) P = 2x + 4y
    At (0,0), P= 2(0) + 4(0) = 0   min. at (0, 0)
    At (3,2), P= 2(3) + 4(2) = 14
    At (4,0), P= 2(4) + 4(0) = 8
    At (0,4), P= 2(0) + 4(4) = 16   max. at (0, 4)

11. a)

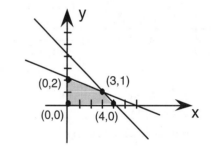

b) P = 7x + 6y
    At (0,0), P = 7(0) + 6(0) = 0   min. at (0, 0)
    At (0,2), P = 7(0) + 6(2) = 12
    At (3,1), P = 7(3) + 6(1) = 27
    At (4,0), P = 7(4) + 6(0) = 28   max. at (4, 0)

12. a)

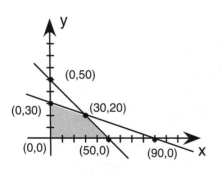

13. a)

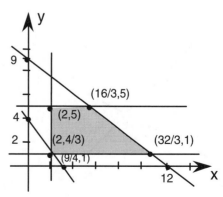

12. b) P = 20x + 40y
    At (0, 0), P = 20(0)+40(0) = 0
    At (0,30), P = 20(0+40(30) = 120
    At (30,20), P = 20(30)+40(20) = 1400
    At (50, 0), P = 20(50)+40(0) = 1000

    Min. (0, 0) and Max. at (20, 30)

13 b) P = 2.20x + 1.65y
    At (2,4/3), P= 2.20(2)+1.65(4/3) = 6.60
    At (2, 5), P= 2.20(2)+1.65(5) = 12.65
    At (16/3,5), P=2.20(16/3)+1.65(5) = 19.98
    At (32/3,1), P=2.20(32/3)+1.65(1) = 25.12
    At (9/4,1), P=2.20(9/4)+1.65(1) = 6.60

    Max. at (32/3,1) and Min. at (2,4/3), (9/4,1)

14. a)

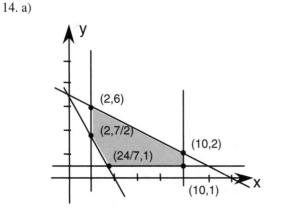

  b) P = 15.13x + 9.35y
      Max. profit is 170 at (10,2)
      Min. profit is 61.22 at (24/7,1)

15. a) Let x = profit on Eastman Kodak film
            y = profit on Fuji film
        x + y ≤ 24   x ≥ 8   x ≤ 24
        y ≥ 4   y ≤ 12
    b) P = .35x + .50y
    c)

    d) At (8, 4), P = .35(8) + .50(4) = .35
        At (16, 8), P = .35(16) + .50(8) = 9.6 max. at
        (16, 8)
        At (20, 4), P = .35(20) + .50(4) = 9
    e) 16 rolls of Kodak film and 8 rolls of Fuji film
    f) Max. profit = $9.60

16. a) Let x = number of skateboards
            y = number of in-line skates
        x + y ≤ 20   x ≥ 3   x ≤ 6   y ≥ 2
    b) P = 25x + 20y
    c)

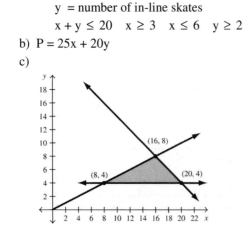

    d) (3,2) (3,17) (6,14) (6,2)

17. Let x = gallons of indoor paint
         y = gallons of outdoor paint
        x ≥ 60   y ≥ 100
    c)

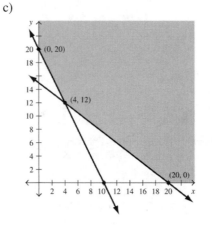

16. e) At (3,2), P = 25(3) + 20(2) = 115
  At (3,17), P = 25(3) + 20(17) = 415
  At (6,14), P = 25(6) + 20(14) = 430
  At (6,2), P = 25(6) + 20(2) = 190
  Six skateboards and 14 pairs of in-line
  skates.

  f) Max. profit = $430

17. a) $3x + 4y \geq 60 \quad x \geq 0$
  $10x + 5y \geq 100 \quad y \geq 0$

  b) C = 28x + 33y

  d) At (0, 20), C = 28(0) + 33(20) = 660
  At (20, 0), C = 28(20) + 33(0) = 560
  At (4, 12), C = 28(4) + 33(12) = 508

  e) 4 hours on Mach. 1 and 12 hours on Mach. 2

  f) Max. profit = $ 660.00

18. Let x = pounds of all-beef hot dogs
  y = pounds of regular hot dogs
  $x + (1/2)y \leq 200$
  $(1/2)y \leq 150 \qquad x \geq 0 \quad y \geq 0$

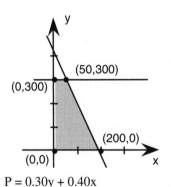

  P = 0.30y + 0.40x
  Maximum profit occurs at (50,300).
  Thus the manufacturer should make 50 lb.
  of the all-beef hot dogs and 300 lb. of the
  regular hot dogs for a profit of $110.

19. Let x = # of car seats
  y = # of strollers
  $x \geq 60 \quad y \geq 100$
  $x + 3y \leq 24 \qquad 2x + y \leq 16 \qquad x + y \leq 10$

  P = 25x + 35y

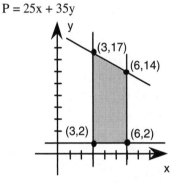

  At (0, 8), P = 25(0) + 35(8) = 280
  At (3, 7), P = 25(3) + 35(7) = 320
  At (4, 6), P = 25(4) + 35(6) = 310
  At (8, 0), P = 25(8) + 35(0) = 200
  3 car seats and 7 strollers
  Max. profit = $ 320.00

Review Exercises

1.

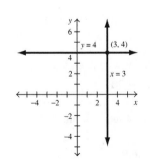

2.

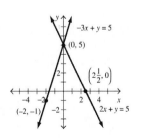

3.

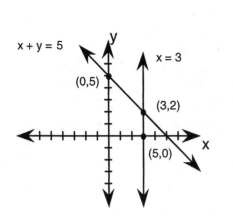

The solution is (3,2).

4.

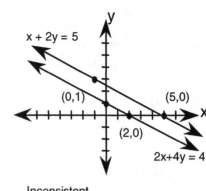

Inconsistent

5. $y = (2/3)x + 5$
    $y = (2/3)x + 5$
    Same slope and y-intercept. Infinite # of solutions.

6. $y = 2x + 6$
    $y = 2x + 7.5$
    Same slope but different y-intercepts. No solution.

7. $6y - 2x = 20$ becomes $y = (1/3)x + 10/3$
    $4y + 2x = 10$ becomes $y = -(1/2)x + 5/2$
    Different slopes. One solution.

8. $y = (1/2)x - 2$
    $y = 2x + 6$
    Different slopes. One solution.

9. (1) $-x + y = 12$
    (2) $\underline{x + 2y = -3}$ (add)
    $\qquad\quad 3y = 9 \qquad\quad y = 3$
    Substitute 3 in place of y in the first equation.
    $-x + 3 = 12$
    $-x = 9 \qquad x = -9$
    The solution is $(-9, 3)$.

10. $x - 2y = 9$
    $y = 2x - 3$
    Substitute $(2x - 3)$ in place of y in the 1st equation.
    $x - 2(2x - 3) = -11$ (solve for x)
    $x - 4x - 6 = -11$
    $5x - 6 = -1$
    $5x = -5 \qquad x = -1$
    Substitute (-1) in place of x in the 2nd equation.
    $y = 2(-1) - 3 = -2 - 3 = -5$
    The solution is $(-1, -5)$.

11. $2x - y = 4 \qquad y = 2x - 4$
    $3x - y = 2$
    Substitute $2x - 4$ for y in the second equation.
    $3x - (2x - 4) = 2$ (solve for x)
    $3x - 2x + 4 = 2$
    $x + 4 = 2 \qquad x = -2$
    Substitute -2 for x in an equation.
    $2(-2) - y = 4$
    $-4 - y = 4 \qquad y = -8$
    The solution is $(-2, -8)$.

12. $3x + y = 1 \qquad y = -3x + 1$
    $3y = -9x - 4$
    Substitute $-3x + 1$ for y in the second equation.
    $3(-3x + 1) = -9x - 4$ (solve for x)
    $-9x + 3 = -9x - 4$
    $3 \neq 4$ False   There is no solution to this system.
    The equations are inconsistent.

13. $x - 2y = 1$ $x = 2y + 1$
    $2x + y = 7$
    Substitute $(2y + 1)$ for x in the 2nd equation.
    $2(2y + 1) + y = 7$ (solve for y)
    $4y + 2 + y = 7$
    $5y + 2 = 7$
    $5y = 5$ $y = 1$
    Substitute 1 in place of y in the equation.
    $x = 2y + 1$.
    $x = 2(1) + 1 = 2 + 1 = 3$
    The solution is (3,1).

14. (1) $2x + y = 2$
    (2) $\underline{-3x - y = 5}$ (add)
        $-x = 7$ $x = -7$
    Substitute (-7) in place of x in the 1st equation.
    $2(-7) + y = 2$
    $-14 y = 2$ $y = 16$
    The solution is $(-7, 16)$.

15. (1) $x + y = 2$
    (2) $x + 3y = -2$
    Multiply the first equation by -1.
    $-x - y = -2$
    $\underline{x + 3y = -2}$ (add)
    $2y = -4$ $y = -2$
    Substitute (-2) for y in equation (2).
    $x + 3(-2) = -2$
    $x - 6 = -2$ $x = 4$
    The solution is (4,–2).

16. (1) $4x - 8y = 16$
    (2) $x - 2y = 4$ $x = 2y + 4$
    Substitute $2y + 4$ for x in the first equation.
    $4(2y + 4) - 8y = 16$
    $8y + 16 - 8y = 16$
    $16 = 16$ True
    There are an infinite number of solutions.
    The system is dependent.

17. (1) $3x + 5y = 15$
    (2) $2x + 4y = 0$
    Multiply the first equation by 2, and the 2nd
    equation by (-3).
    $6x + 10y = 30$
    $\underline{-6x - 12y = 0}$ (add)
    $-2y = 30$ $y = -15$
    Substitute (-15) for y in the second equation.
    $2x + 4(-15) = 0$
    $2x - 60 = 0$ $x = 30$
    The solution is $(30, -5)$.

18. (1) $3x + 4y = 6$
    (2) $2x - 3y = 4$
    Multiply the first equation by 2, and the second
    equation by -3.
    $6x + 8y = 12$
    $\underline{-6x + 9y = -12}$ (add)
    $17y = 0$ $y = 0$
    Substitute 0 for y in the first equation.
    $3x + 4(0) = 6$
    $3x = 6$ $x = 2$
    The solution is (2,0).

19. $A + B = \begin{bmatrix} 1 & -3 \\ 2 & 4 \end{bmatrix} + \begin{bmatrix} -2 & -5 \\ 6 & 3 \end{bmatrix} = \begin{bmatrix} 1+(-2) & -3+(-5) \\ 2+6 & 4+3 \end{bmatrix} = \begin{bmatrix} -1 & -8 \\ 8 & 7 \end{bmatrix}$

20. $A - B = \begin{bmatrix} 1 & -3 \\ 2 & 4 \end{bmatrix} - \begin{bmatrix} -2 & -5 \\ 6 & 3 \end{bmatrix} = \begin{bmatrix} 1-(-2) & -3-(-5) \\ 2-6 & 4-3 \end{bmatrix} = \begin{bmatrix} 3 & 2 \\ -4 & 1 \end{bmatrix}$

21. $2A = 2 \begin{bmatrix} 1 & -3 \\ 2 & 4 \end{bmatrix} = \begin{bmatrix} 2(1) & 2(-3) \\ 2(2) & 2(4) \end{bmatrix} = \begin{bmatrix} 2 & -6 \\ 4 & 8 \end{bmatrix}$

22. $2A - 3B = 2\begin{bmatrix} 1 & -3 \\ 2 & 4 \end{bmatrix} - 3\begin{bmatrix} -2 & -5 \\ 6 & 3 \end{bmatrix} = \begin{bmatrix} 2 & -6 \\ 4 & 8 \end{bmatrix} + \begin{bmatrix} 6 & 15 \\ -18 & -9 \end{bmatrix} = \begin{bmatrix} 2+6 & -6+15 \\ 4-18 & 8-9 \end{bmatrix} = \begin{bmatrix} 8 & 9 \\ -14 & -1 \end{bmatrix}$

23. $A \times B = \begin{bmatrix} 1 & -3 \\ 2 & 4 \end{bmatrix} \times \begin{bmatrix} -2 & -5 \\ 6 & 3 \end{bmatrix} = \begin{bmatrix} 1(-2)+(-3)6 & 1(-5)+(-3)3 \\ 2(-2)+4(6) & 2(-5)+4(3) \end{bmatrix} = \begin{bmatrix} -20 & -14 \\ 20 & 2 \end{bmatrix}$

24. $B \times A = \begin{bmatrix} -2 & -5 \\ 6 & 3 \end{bmatrix} \times \begin{bmatrix} 1 & -3 \\ 2 & 4 \end{bmatrix} = \begin{bmatrix} (-2)1+(-5)2 & (-2)(-3)+(-5)4 \\ 6(1)+3(2) & 6(-3)+3(4) \end{bmatrix} = \begin{bmatrix} -12 & -14 \\ 12 & -6 \end{bmatrix}$

25. $\begin{bmatrix} 1 & 2 & | & 4 \\ 1 & 1 & | & 2 \end{bmatrix}\begin{matrix} \\ (r_2 - r_1) \end{matrix} = \begin{bmatrix} 1 & 2 & | & 6 \\ 0 & -1 & | & -2 \end{bmatrix}\begin{matrix} (r_1 - 2r_2) \\ 2r_2 \end{matrix} = \begin{bmatrix} 1 & 0 & | & -2 \\ 0 & 1 & | & 2 \end{bmatrix}$ The solution is $(2, 2)$.

26. $\begin{bmatrix} -1 & 1 & | & 4 \\ 1 & 2 & | & 2 \end{bmatrix} = \begin{bmatrix} 1 & -1 & | & -4 \\ 0 & 3 & | & 6 \end{bmatrix} = \begin{bmatrix} 1 & 0 & | & -2 \\ 0 & 1 & | & 2 \end{bmatrix}$ The solution is $(-2, 2)$.

27. $\begin{bmatrix} 2 & 1 & | & 3 \\ 3 & -1 & | & 12 \end{bmatrix}\begin{matrix} (r_1 \div 2) \\ = \end{matrix}\begin{bmatrix} 1 & \frac{1}{2} & | & \frac{3}{2} \\ 3 & -1 & | & 12 \end{bmatrix}\begin{matrix} \\ (r_2 - 3r_1) \end{matrix} = \begin{bmatrix} 1 & \frac{1}{2} & | & \frac{3}{2} \\ 0 & -\frac{5}{2} & | & \frac{15}{2} \end{bmatrix}\begin{matrix} \\ (-\frac{2}{5}r_2) \end{matrix} = \begin{bmatrix} 1 & \frac{1}{2} & | & \frac{3}{2} \\ 0 & 1 & | & -3 \end{bmatrix}\begin{matrix} (-\frac{1}{2}r_2 + r_1) \\ = \end{matrix}\begin{bmatrix} 1 & 0 & | & 3 \\ 0 & 1 & | & -3 \end{bmatrix}$

    The solution is $(3, -3)$.

28. $\begin{bmatrix} 2 & 3 & | & 2 \\ 4 & -9 & | & 4 \end{bmatrix} = \begin{bmatrix} 1 & \frac{3}{2} & | & 1 \\ 0 & -15 & | & 0 \end{bmatrix} = \begin{bmatrix} 1 & \frac{3}{2} & | & 1 \\ 0 & 1 & | & 0 \end{bmatrix} = \begin{bmatrix} 1 & 0 & | & 1 \\ 0 & 1 & | & 0 \end{bmatrix}$ The solution is $(1, 0)$

29. $\begin{bmatrix} 1 & 3 & | & 3 \\ 3 & -2 & | & 2 \end{bmatrix} = \begin{bmatrix} 1 & 3 & | & 3 \\ 0 & -11 & | & -7 \end{bmatrix} = \begin{bmatrix} 1 & 3 & | & 3 \\ 0 & 1 & | & \frac{7}{11} \end{bmatrix} = \begin{bmatrix} 1 & 0 & | & \frac{12}{11} \\ 0 & 1 & | & \frac{7}{11} \end{bmatrix}$ The solution is $\left(\frac{12}{11}, \frac{7}{11} \right)$

30. $\begin{bmatrix} 3 & -6 & | & -9 \\ 4 & 5 & | & 14 \end{bmatrix}\begin{matrix} (r_1 \bullet -1) \\ \end{matrix}\begin{bmatrix} -3 & 6 & | & 9 \\ 4 & 5 & | & 14 \end{bmatrix}\begin{matrix} (r_2 + r_1) \\ \end{matrix}\begin{bmatrix} 1 & 11 & | & 23 \\ 4 & 5 & | & 14 \end{bmatrix}\begin{matrix} (r_1 \bullet -4) \\ (r_2 + r_1) \end{matrix}\begin{bmatrix} 1 & 11 & | & 23 \\ 0 & -39 & | & -78 \end{bmatrix}$

    $\begin{bmatrix} 1 & 11 & | & 23 \\ 0 & -39 & | & -78 \end{bmatrix}\begin{matrix} (r_1 \bullet 11/39) \\ (r_2 + r_1) \end{matrix}\begin{bmatrix} 1 & 0 & | & 1 \\ 0 & -39 & | & -78 \end{bmatrix}\begin{matrix} (r_2 \div -2) \\ \end{matrix}\begin{bmatrix} 1 & 0 & | & 1 \\ 0 & 1 & | & 2 \end{bmatrix}$ The solution is $(1, 2)$.

31. Let x = amount borrowed at 8%   y = amount borrowed at 10%

    $.08x + .10y = 53000$ $x + y = 600000$

    $\begin{bmatrix} .08 & .10 & | & 53000 \\ 1 & 1 & | & 600000 \end{bmatrix}\begin{matrix} (r_2 \bullet -10) \\ (r_2 + r) \end{matrix}\begin{bmatrix} -2 & 0 & | & -700000 \\ 1 & 1 & | & 600000 \end{bmatrix}\begin{matrix} (r_1 \bullet -1) \\ (r_2 + r) \end{matrix}\begin{bmatrix} 1 & 0 & | & 350000 \\ 0 & 1 & | & 250000 \end{bmatrix}$

    x = \$350,000 and y = \$ 250,000

32. Let s = liters of 80% acid solution
    w = liters of 50% acid solution
s + w = 100
0.80s + 0.50w = 100(0.75)
0.80s + 0.50w = 75
s = 100 − w
0.80(100 − w) + 0.50w = 75
80 − 0.80w + 0.50w = 75
0.30w = −5
w = −5/(−0.30) = 16 2/3 liters
s = 100 − 16 2/3 = 83 1/3 liters

33. Let s = salary   r = commission rate
    (1) s + 4000r = 660
    (2) s + 6000r = 740 (subtract 1 from 2)
        2000r = 80
    r = 80/2000 = 0.04
Substitute 0.04 for r in eq n. 1.
s = 660 − 4000(.04)   s = 500
His salary is 500 per week and his commission
rate is 4%.

34. Let c = total cost   x = no. of months to
operate
    a) model 1600A: $c_A = 950 + 32x$
       model 6070B: $c_B = 1275 + 22x$
       950 + 32x = 1275 + 22x
       10x = 325   x = 32.5 months
       After 32.5 months of operation the total cost
       of the units will be equal.
    b) After 32.5 months or 2.7 years, the most cost
       effective unit is the unit with the lower per
       month to operate cost. Thus, model 6070B is
       the better deal in the long run.

35. a) Let C = total cost for parking
       x = number of additional hours
       All-Day: C = 5 + 0.50x
       Sav-A-Lot: C = 4.25 + 0.75x
       5 + 0.50x = 4.25 + 0.75x
       0.75 = 0.25x   3 = x
       The total cost will be the same after 3
       additional hours or 4 hours total.
    b) After 5 hours or x = 4 additional hours:
       All-Day: C = 5 + 0.50(4) = $7.00
       Sav-A-Lot: C = 4.25 + 0.75(4) = $7.25
       All-Day would be less expensive.

36.

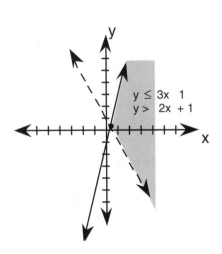

$y \leq 3x - 1$
$y > -2x + 1$

37.

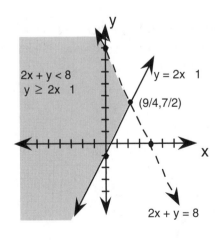

2x + y < 8
$y \geq 2x - 1$

y = 2x − 1

(9/4, 7/2)

2x + y = 8

38.

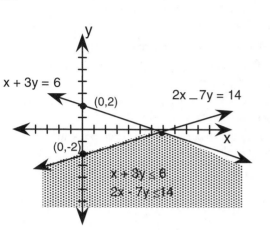

39.

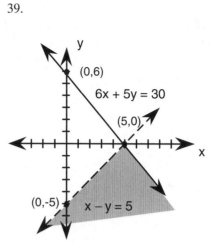

40. P = 6x + 3y

At (0,0), P = 6(0) + 3(0) = 0

At (0,10), P = 6(0) + 3(10) = 30

At (9,0), P = 6(9) + 3(0) = 54

The maximum profit is $54 at (9,0).

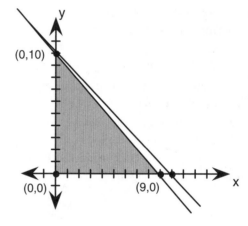

Chapter Test

1. If the lines do not intersect (parallel) the system of equations is inconsistent. The system of equations is consistent if the lines intersect only once. If both equations represent the same line then the system of equations is dependent.

2.

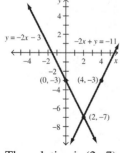

The solution is (2, 7).

3. Write each equation in slope intercept form, then

   compare slopes and intercepts.

   $$4x + 5y = 6 \qquad\qquad 3x + 5y = 13$$
   $$5y = 4x + 6 \qquad\qquad 5y = 3x + 13$$
   $$y = (4/5)x + 6/5$$

   $$y = (3/5)x + 13/5$$

   The slopes are different so there is only one solution.

4. $x - y = 5 \qquad x = y + 5$

$2x + 3y \qquad\qquad = -5$

Substitute $(y + 5)$ for x in the second equation.

$2(y + 5) + 3y = -5$ (solve for y)

$2y + 10 + 3y = -5$

$5y + 10 = -5$

$5y = -15 \qquad y = -3$

Substitute (-3) for y in the equation $x = y + 5$.

$x = -3 + 5 = 2 \qquad$ The solution is $(2, -3)$.

5. $y = 5x + 7 \qquad y = 2x + 1$

Substitute $(5x + 7)$ for y in the second equation.

$5x + 7 = 2x + 1$ (solve for x)

$3x = -6 \qquad x = -2$

Substitute -2 for x in the first equation.

$y = 5(-2) + 7 = -10 + 7 = -3$

The solution is $(-2, -3)$.

6. $x - y = 4$

$\underline{2x + y = 5}$ (add)

$3x = 9 \qquad x = 3$

Substitute 3 for x in the 2nd equation.

$2(3) + y = 5$

$6 + y = 5 \qquad y = -1 \qquad$ The solution is $(3, -1)$.

7. $4x + 3y = 5$

$2x + 4y = 10$

Multiply the second equation by (-2).

$4x + 3y = 5$

$\underline{-4x - 8y = -20}$ (add)

$-5y = -15 \qquad\qquad y = 3$

Substitute 3 for y in the first equation.

$4x + 3(3) = 5$

$4x + 9 = 5$

$4x = -4 \qquad x = -1$

The solution is $(-1, 3)$.

8. $3x + 4y = 6$

$2x - 3y = 4$

Multiply the 1st eq'n. by 3 and the 2nd eq'n. by 4.

$9x + 12y = 18$

$\underline{8x - 12y = 16}$

$17x = 34 \qquad\qquad x = 2$

8. Substitute 2 for x in an equation.

$2(2) - 3y = 4$ (solve for y)

$-3y = 0 \qquad\qquad y = 0$

The solution is $(2, 0)$.

9. $\begin{bmatrix} 1 & 3 & | & 4 \\ 5 & 7 & | & 4 \end{bmatrix}(-5r_1 + r_2) \overset{=}{} \begin{bmatrix} 1 & 3 & | & 4 \\ 0 & -8 & | & -16 \end{bmatrix}(r_2 \div (-8)) \overset{=}{} \begin{bmatrix} 1 & 3 & | & 4 \\ 0 & 1 & | & 2 \end{bmatrix}(r_1 - 3r_2) \begin{bmatrix} 1 & 0 & | & -2 \\ 0 & 1 & | & 2 \end{bmatrix}$

The solution is $(-2, 2)$.

10. $A + B = \begin{bmatrix} 2 & -5 \\ 1 & 3 \end{bmatrix} + \begin{bmatrix} -1 & -3 \\ 5 & 2 \end{bmatrix} = \begin{bmatrix} 2 + (-1) & -5 - 3 \\ 1 + 5 & 3 + 2 \end{bmatrix} = \begin{bmatrix} 1 & -8 \\ 6 & 5 \end{bmatrix}$

11. $3A - B = 3\begin{bmatrix} 2 & -5 \\ 1 & 3 \end{bmatrix} - \begin{bmatrix} -1 & -3 \\ 5 & 2 \end{bmatrix} = \begin{bmatrix} 3(2) - (-1) & 3(-5) - (-3) \\ 3(1) - 5 & 3(3) - 2 \end{bmatrix} = \begin{bmatrix} 7 & -12 \\ -2 & 7 \end{bmatrix}$

12. $A \times B = \begin{bmatrix} 2 & -5 \\ 1 & 3 \end{bmatrix}\begin{bmatrix} -1 & -3 \\ 5 & 2 \end{bmatrix} = \begin{bmatrix} 2(-1) + (-5)(5) & 2(-3) + (-5)(2) \\ 1(-1) + (3)(5) & 1(-3) + 3(2) \end{bmatrix} = \begin{bmatrix} -27 & -16 \\ 14 & 3 \end{bmatrix}$

13. $y < -2x + 2$ $y > 3x + 2$

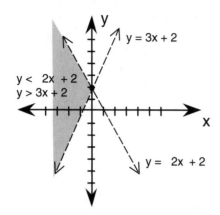

14. Let x = lb of $6.00 coffee
        y = lb of $7.50 coffee
     $x + y = 30$ $y = 30 - x$
     $6x + 7.5y = 7.00(30)$
     Substitute (30 – x) for y in the 2nd equation.
     $6x + 7.5(30 – x) = 210$
     $6x + 225 – 7.5x = 210$
       $-1.5x = -15$ $x = 10$
     Substitute 10 for x in the equation
      $y = 30 – x$.
      $y = 30 – 10 = 20$
     Mix 10 lb of the $6.00 coffee with 20 lb of
     the $7.50 coffee.

15. Let x = no. of one bedroom units
        y = no. of two bedroom units
     $x + y = 20$ $x = 20 - y$
     $425x + 500y = 9100$
     Substitute (20 – y) for x in the second equation.
     $425(20 – y) + 500y = 9100$
     $75y = 600$ $y = 8$
     Substitute 8 for y in the first equation.
     $x + 8 = 20$ $x = 12$
     The building has 12 one bedroom and 8 two
     bedroom apartments.

16. a)

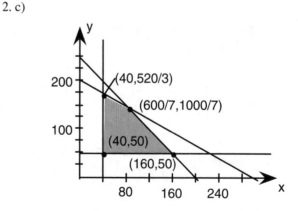

b) $P = 5x + 3y$
    At (0, 0) $P = 5(0) + 3(0) = 0$
    At (0, 2) $P = 5(0) + 3(2) = 6$
    At (3, 1) $P = 5(3) + 3(1) = 18$
    At (3.75, 0) $P = 5(3.75) + 3(0) = 18.75$
    Max. at (3.75, 0) and Min. at (0, 0)

Group Projects

1. Answers will vary.

3. Answers will vary.

2. Let x = # of board feet of oak
       y = # of board feet of walnut
    a) $5x + 2y = 75$ $4x + 3y = 125$
        $x \geq 40$ $y \geq 50$
    b) $P = 75x + 125y$
    d) Determine the maximum profit.

    At (40,173) $P = 75(40) + 125(173) = \$24,625$

    At (40,50) $P = 75(40) + 125(50) = \$9,250$

    At (160,50) $P = 75(160) + 125(50) = \$18,250$

    At (86,143) $P = 75(86) + 125(143) = \$24,325$

    Maximum profit occurs at 40 of model 01 and

    173 of model 02.

    e) Maximum profit = $24,625

2. c)

CHAPTER EIGHT

THE METRIC SYSTEM

Exercise Set 8.1

1. The metric system.   2. The U.S. customary system.

3. It is the worldwide accepted standard of measurement. There is only 1 basic unit of measurement for each quantity. It is based on the number 10 which makes many calculations easier.

4. a) meter   b) kilogram   c) liter   d) celsius

5. a) Move the decimal point one place for each change in unit of measure.

   b) $714.6 \text{ cm} = \dfrac{714.6}{10^5} \text{ km} = 714.6 \times 10^{-5} \text{ km} = 0.007146 \text{ km}$

   c) 30.8 hm = (30.8)(1000) dm = 30800 dm

6. a) mega   b) micro   7. kilo   1000 times the base unit   k
                                          hecto   100 times the base unit   h
                                          deka   10 times the base unit   da
                                          dec   1/10 times the base unit   d
                                          centi   1/100 times the base unit   c
                                          milli   1/1000 times the base unit m

8. a) 10,000 times greater   9. a) 100 times greater   10. a) 0° C
   b) 1 h = 10,000 cm   b) 1 dam = 100 dm   b) 100° C
   c) 1 cm = 0.0001 hm   c) 1dm = 0.01 dam   c) 37° C

11. 2 pounds   12. 1 yard   13. 5 grams   14. 30° C
15. 22° C   16. 2 m.   17. kilo d   18. milli b
19. hector c   20. deka e   21. deci f   22. centi a

23. a) 10 liters   b) 1/100 liter   c) 1/1000 liter   d) 1/10 liter   e) 1000 liters   f) 100 liters
24. a) 100 grams   b) 0.001 gram   c) 1000 grams   d) 0.01 gram   e) 10 grams   f) 0.1 gram

25. mg   1/1000 gm   26. cg   1/100 gm   27. dg   1/10 gm
28. dag   10 gm   29. hg   100 gm   30. kg   1000 gm

31. Max. load 320 kg = (320 x 1,000) g = 320 000 g   32. Max. load 320 kg = (320 x 1,000,000) mg
                                                            = 320,000,000 mg

33. 2 m = (2 x 100) cm = 200 cm   34. 35.7 hg = (35.7 x 100) g = 3,570 g

35. 0.095 hl = (0.095)(100) = 9.5 l

36. 7 dam = (7 x 10) m = 70 m

37. 242.6 cm = (242.6)(0.0001) hm = 0.02426 hm

38. 1.34 ml = (1.34)(0.001) l = 0.00134 l

39. 4036 mg = (4036)(0.00001) hg = 0.04036 hg

40. 14.27 kl = (14.27)(1000) l = 14 270 l

41. 1.34 hm = (1.34)(10000) cm = 13,400 cm

42. 0.000062 kg = 62 mg

43. 92.5 kg = 92,500 g

44. 7.3 m = 7300 mm

45. 895 l = 895,000 ml

46. 24 dm = 0.0024 km

47. 240 cm = 0.0240 hm

48. 6,049 mm = 6.049 m

49. 40,302 ml = 4.0302 dal

50. 0.034 ml = 0.00034 l

51. 590 cm, 5.1 dam, 0.47 km

52. 680 m, 514 hm, 62 km

53. 2.42 kg, 2,400 g, 24,300 dg

54. 420 cl, 4.3 l, 0.045 kl

55. 203,000 mm, 2.6 km, 52.6 hm

56. 0.032 kl, 460 dl, 48,000 cl

57. Jim, since a meter is longer than a yard.

58. 1 hectometer in 10 min. 1 hm > 1 dm

59. The pump that removes 1 dal of water per min.
     1 dekaliter > 1 deciliter

60. The side with the 15 lb. weight would go down.
     5 kg = 5(2.2 lbs.) = 11 lbs.

61. a) Perimeter= 2l + 2w= 2(74) + 2(99)= 346 cm
     b) 346 cm = (346 x 10) mm = 3,460 mm

62. a) (2)(250)(7) = 3,500 mg / week
     b) 3,500 mg / week = 3.5 g / week

63. a) (4)(27 m) = 108 m   b) 108 m = 0.108 km
     c) 108 m = 108 000 mm

64. a) 1,200 km / 187 l = 6.417 km/l
     b) 1,200,00 / 187 l = 6,417 m/l

65. 8 (400) m = 3,200 m;   3,200 m = 3.2 km

66. 360 l = (360 l) $\left(\dfrac{1000 \text{ ml}}{1 \text{ l}} \right) \left(\dfrac{1 \text{ min}}{360 \text{ ml}} \right)$

     = 83.333...min. or 1 hr 23.333...min.

67. a) 6(360) ml = 2,160 ml
     b) 2160(1000) = 2.16 l
     c) 2.45 / 2.16 = $1.13 per liter

68. ($ 1.03)(37.7 l) = $ 38.83

69. a) (6.9)(1000) = 6,900 gm
     b) 6,900 / 3 = 2300 gm   2300 gm = 23,000 dg

70. 750 km = 7500 hm   750 − 32.5 = 717.5 km

71. 1 gigameter = 1000 megameters

72. 1 nanogram = .001 microgram

73. 1 teraliter = 1 x 10^{24} picoliters

74. 1 megagram = 1 x 10^{15} nanogms

75. 0.8/.027 = 29.6   30 eggs

76. 0.8/.288 = 2.777...   2.8 cups

77. 195 mg = 0.195 g
     0.8/0.195 = 4.1 cups

78. 1.6 mg = 0.0016 g
     0.8/0.016 = 500
     500(49) = 24,500 g

79. 5000 cm = 5 dam

80. 2000 mm = 2 m

81. 0.00006 hg = 6 mg

82. 3000 dm = 3 hm

83. 0.02 kl = 2 dal

84. 500 cm = 5 m

85. magr   gram

86. migradec   decigram

87. rteli   liter

88. raktileed   dekaliter

89. terem   meter

90. leritililm   milliliter

91. reketolim   kilometer

92. timenceret   centimeter

93. greeed sulesic   degree celsius

94. togmeharc   hectogram

Exercise Set 8.2

1. volume
2. length
3. area
4. length
5. volume
6. volume
7. volume
8. volume
9. area
10. volume
11. length
12. area
13. Answers will vary. (AWV)
14. Answers will vary.
15. Answers will vary.
16. Answers will vary.
17. Answers will vary.
18. Answers will vary.
19. 1 cubic decimeter
20. 1000 l = 1 kiloliter
21. 1 cubic centimeter
22. square kilometers
23. area
24. 2.5 acres
25. centimeters
26. kilometers
27. cm or mm
28. centimeters
29. centimeters
30. millimeters
31. millimeters
32. meters
33. cm or mm
34. cm or mm
35. kilometers
36. cm or m
37. c   27 m
38. a   2 cm x 3 cm
39. c   5 km
40. a   160 cm
41. a   2 cm
42. b   8 cm
43. a   93 dam
44. c   375 m
45. mm   AWV
46. cm   AWV
47. cm or m   AWV
48. mm or cm   AWV
49. mm or cm   AWV
50. mm   AWV
51. cm, km
52. km
53. m
54. m
55. cm
56. km
57. sq. mm. or sq. cm.
58. hectares or sq. km.
59. sq. m.
60. sq. mm. or sq. cm.
61. sq. m. or hectares
62. sq. m. or hectares
63. sq. mm. or sq. cm.
64. sq. cm. or sq. m.
65. hectares or sq. km.
66. sq. m.
67. b   2.2 sq .m.
68. a   5 sq. cm.
69. a   800 sq. m.
70. b   1/8 ha
71. c   360 sq. cm.
72. a   2.5 sq. cm.
73. c   1200 sq. mm.
74. c   4900 sq. km.
75. AWV
76. AWV
77. AWV
78. AWV
79. AWV
80. AWV
81. kiloliters
82. liters
83. milliliters
84. cubic centimeters
85. liters
86. cubic meters
87. cubic meters
88. cubic meters
89. liters
90. cubic meters
91. c   7780 cu. cm.
92. a   0.5 cu. m.
93. c   55 kl
94. b   355 ml
95. a   550 cu. m.
96. b   120 ml
97. a   24 cu. m.
98. b   14,000 cu. cm.
99. a) 144,000 cc
99. b) 152,561 cc

100. a) AWV   b) $(2)(1.5)(.25) = .75 \ m^3$
101. a) AWV   b) $v \approx (3.14)(0.25)^2 (1) = 0.20 \ m^3$
102. a) AWV   b) $v = \pi r^2 h \approx (3.14)(0.20)^2(2) = 0.25 m^3$
103. a) AWV   b) $A = lw = (4)(2.2) = 8.8 \ cm^2$
104. $A = \pi r^2 = (3.14)(1.2)^2 = 4.5216 \ cm^2$
105. $(82)(62) - (50)(42) = 5084 - 2100 = 2984 \ cm^2$

106. a) $(73)(53) = 3869 \ m^2$
  b) $3869 - (70)(50) = 3869 - 3500 = 869 \ m^2$
107. a) $(3.75)(1.4) = 5.25 \ km$
  b) $(5.25)(100 \ ha) = 525 \ ha$
108. a) $(22.5)(18.3) = 411.75 \ m^2$
  b) $(411.75)(0.0001 \ ha) = 0.041175 \ ha$
109. a) $(18)(10)(2.5) = 450 \ m^3$
  b) $450 \ m^3 = 450 \ kl$

110. Total Surface Area of 4 walls = 2lh + 2wh = 2(20)(6) + 2(12)(6) = 384 m^2

Liters for first coat = $(384 \ m^2)\left(\dfrac{1 \ l}{10 \ m^2}\right) = 38.4 \ l$ Liters for second coat = $(384 \ m^2)\left(\dfrac{1 \ l}{15 \ m^2}\right) = 25.6 \ l$

Total liters = 38.4 + 25.6 = 64 l   Total cost = (64)($4.75) = $304

111. a) $V = lwh = (70)(40)(20) = 56,000 \text{ cm}^3$ b) $56,000 \text{ cm}^3 = 56,000 \text{ ml}$ c) c) $56\,000 \text{ ml} = \left(\dfrac{56000}{1000}\right) 1 = 56 \ 1$

112. $V = \pi r^2 h \approx (3.14)(4.0)^2(12.5)$
        628 cm^3

113. $10^2 = 100$ times larger

114. $100^2 = 10,000$ times larger

115. $10^3 = 1000$ times larger

116. $10^3 = 1000$ times larger

117. $1,000,000 \text{ mm}^2$

118. $1,000,000,000 \text{ cm}^2$

119. 100 hm^2

120. 0.0001 m^2

121. $0.000,000,000,1 \text{ hm}^2$

122. $10,000 \text{ mm}^2$

123. $1,000,000 \text{ cm}^3$

124. $1 \text{ hm}^3 = 0.001 \text{ km}^3$

125. $435 \text{ cm}^3 = 435 \text{ ml}$

126. $435 \text{ cm}^3 = 0.435 \ 1$

127. $76 \text{ kl} = 76 \text{ m}^3$

128. $4.2 \ 1 = 4,200 \text{ cm}^3$

129. $(6.0 \times 10^4)(10) = 600,000 \text{ dl}$

130. $(600,000)(100) =$
        $60,000,000 \text{ ml} = 60,000,000 \text{ cc}$

131. AWV

132. AWV

133. $6.7 \text{ kl} = 6.7 \text{ m}^3 = (6.7 \times 10^3) \text{ dm}^3 = 6,700 \text{ dm}^3$

134. $1.4 \text{ ha} = 14,000 \text{ m}^2 = (14000 \times 100^2) \text{ cm}^2 =$
        $140,000,000 \text{ cm}^2$

135. a) $1 \text{ sq mi} = (1 \text{ mi}^2)(5280)^2 \dfrac{\text{ft}^2}{\text{mi}^2} = 27,878,400 \text{ ft}^2$

        $27,878,400 \text{ ft}^2 \times (12)^2 \dfrac{\text{in}^2}{\text{ft}^2} = 4,014,489,600 \text{ in}^2$

    b) It is easier to convert in the metric system
        because it is a base 10 system.

136. a) $(1 \text{ yd}^3) = (36 \text{ in})^3 = 46,656 \text{ in}^3$
        $(46,656 \text{ in}^3)(6) = 279,936 \text{ in}^3$

    b) It is easier to convert in the metric system
        because it is a base 10 system.

137. Answers will vary.

138. a) $1.5 \text{ m} = 150 \text{ cm}$
        $150 - 50 = 100 \text{ cm}.$
    b) $150/50 = 3$ 3 times larger
    c) No.

139. a) 5150.7 liters / day
    b) 493.2 liters / day

Exercise Set 8.3
1. kilogram
2. 5 gm
3. 2 lb
4. metric tonne
5. approx. 35°C AWV
6. approx. -15° C AWV
7. Answers will vary AWV
8. 8.a) Yes; mass is a measure of the amount of matter in an object.
    b) No; weight is a measure of gravitational force.
9. kilograms or grams
10. kilograms
11. grams
12. kilograms or grams
13. grams
14. metric tonnes
15. metric tonnes
16. milligrams
17. grams
18. grams
19. b   2.26 kg
20. a   9.1 mg
21. b   1.4 kg
22. c   0.45 kg
23. b   2800 kg
24. c   1.6 t
25. AWV
26. AWV
27. AWV
28. AWV
29. c 0° C
30. c 90° C
31. b 27°C
32. b   Dress warmly
       and walk.

33. b 5° C

34. c bathing suit

35. c 177° C

36. c 1260° C

37. b -7° C

38. c 40° C

39. $F = \frac{9}{5}(30) + 32 = 54 + 32 = 86^\circ F$

40. $F = \frac{9}{5}(-5) + 32 = -9 + 32 = 23^\circ F$

41. $C = \frac{5}{9}(92 - 32) = \frac{5}{9}(60) = 33.3^\circ C$

42. $C = \frac{5}{9}(-10 - 32) = \frac{5}{9}(-42) = -23.3^\circ C$

43. $C = \frac{5}{9}(180 - 32) = \frac{5}{9}(148) = 82.2^\circ C$

44. $C = \frac{5}{9}(98 - 32) = \frac{5}{9}(66) = 36.7^\circ C$

45. $F = \frac{9}{5}(37) + 32 = 66.6 + 32 = 98.6^\circ F$

46. $F = \frac{9}{5}(-4) + 32 = -7.2 + 32 = 24.8^\circ F$

47. $C = \frac{5}{9}(13 - 32) = \frac{5}{9}(-19) = -10.6^\circ C$

48. $C = \frac{5}{9}(75 - 32) = \frac{5}{9}(43) = 23.9^\circ C$

49. $F = \frac{9}{5}(45) + 32 = 81 + 32 = 113^\circ F$

50. $F = \frac{9}{5}(60) + 32 = 108 + 32 = 140^\circ F$

51. $C = \frac{5}{9}(-20 - 32) = \frac{5}{9}(-52) = -28.9^\circ C$

52. $C = \frac{5}{9}(425 - 32) = \frac{5}{9}(393) = 218.3^\circ C$

53. $F = \frac{9}{5}(22) + 32 = 39.6 + 32 = 71.6^\circ F$

54. $F = \frac{9}{5}(15.6) + 32 = 28.1 + 32 = 60.1^\circ F$

55. $F = \frac{9}{5}(35.1) + 32 = 63.2 + 32 = 95.2^\circ F$

56. $F = \frac{9}{5}(32.3) + 32 = 58.1 + 32 = 90.1^\circ F$

57. low: $F = \frac{9}{5}(17.8) + 32 = 32 + 32 = 64.04^\circ F$

 high: $F = \frac{9}{5}(23.5) + 32 = 42.3 + 32 = 74.3^\circ F$

 Range $= 74.30 - 64.04 = 10.26^\circ F$

58. low: $F = \frac{9}{5}(22) + 32 = 39.6 + 32 = 71.6^\circ F$

 high: $F = \frac{9}{5}(34) + 32 = 61.2 + 32 = 93.2^\circ F$

 Range $= 93.2 - 71.6 = 21.6^\circ F$

59. cost $= (6.2)(.70) = \$ 4.34$

60. cost $= (1.3)(.80) = \$ 1.04$

61. total mass $= 45$ g $+ 29$ g $+ 370$ ml $=$
 45 g $+ 29$ g $+ 370$ g $= 444$ g

62. fuel used $= (4320)(17) = 73,440$ kg

 $73,440 \text{ kg} \sqcup \dfrac{1 \text{ t}}{1000 \text{ kg}} = 73.44$ t

63. a) $V = lwh$, $l = 16$ m, $w = 12$ m, $h = 12$ m
   $V = (16)(12)(12) = 2304 \text{ m}^3$
 b) $2304 \text{ m}^3 = 2304$ kl
 c) 2304 kl $= 2304$ t

64. a) $V = \pi r^2 h$ $r = 50$ cm $= 0.50$ m
   $h = 150$ cm $= 1.5$ m
   $V = (3.14)(0.50)^2(1.50) = 1.1775 \text{ m}^3$
 b) $1.1775 \text{ m}^3 = 1.1775$ kl $= 1177.5$ l
 c) 1177.5 l $= 1177.5$ kg

65. $4.2 \text{ kg} = (4.2 \text{ kg})\left(\dfrac{1 \text{ t}}{1000 \text{ kg}}\right) = 0.0042$ t

66. $9.52 \text{ t} = (9.52 \text{ t})\left(\dfrac{1000 \text{ kg}}{1 \text{ t}}\right) = 9520$ kg

67. $17.4 \text{ t} = (17.4 \text{ t})\left(\dfrac{1000 \text{ kg}}{1 \text{ t}}\right) = 17,400 \text{ kg} =$
 $17,400,000$ g

68. $1,460,000 \text{ mg} = 1.46 \text{ kg} = (1.46 \text{ kg})\left(\dfrac{1 \text{ t}}{1000 \text{ kg}}\right) =$
 0.00146 t

69. Yes, 78° F $= \dfrac{5}{9}(78 - 32) \approx 25.6^\circ$ C, not 20° C

70. Normal body temperature is 98.6° F or 37° C. Maria's temperature is 38.2° C which is above normal. She should take an aspirin.

71. $1.2\,l = 1200$ ml   a) 1200 gm   b) 1200 cm^3

72. -40° C $= \dfrac{9}{5}(-40) + 32 = -72 + 32 \approx -40^\circ$ F

73. a) $V = lwh$ $l = 1$ yd $= 3$ ft $w = 15$ in $= 1.25$ ft
         $h = 1.5$ ft
    $V = (3)(1.25)(1.5) = 5.625$ cubic feet

    b) $(5.625 \text{ ft}^3)\left(62.5\,\dfrac{\text{lbs}}{\text{ft}^3}\right) = 351.6$ lb

    c) $(351.6 \text{ lb})\left(\dfrac{1 \text{ gal}}{8.3 \text{ lb}}\right) = 42.4$ gal

74. 3 kg _____ x

    $(3)(2) = 6 = 4x$
    $x = 6/4 = 3/2 = 1.5$ 1.5 kg
    1.5 kg $= 1500$ g

75. a) -62.11° C $F = \dfrac{9}{5}(-62.11) + 32 = -111.798 + 32 = -79.798^\circ$ F

    b) 2.5° C $F = \dfrac{9}{5}(2.5) + 32 = 4.5 + 32 = 36.5^\circ$ F

    c) $918,000,000^\circ$ F $C = \dfrac{5}{9}(918,000,000 - 32) = \dfrac{5}{9}(917,999,968) = 509,999,982.2 \approx 510,000,000^\circ$ C

Exercise Set 8.4

1. **Dimensional analysis** is a procedure used to convert from one unit of measurement to a different unit of measurement.

2. A **unit fraction** is a fraction in which the numerator and denominator contain different units and the value of the fraction is 1.

3. $\dfrac{60 \text{ seconds}}{1 \text{ minute}}$ or $\dfrac{1 \text{ minute}}{60 \text{ seconds}}$ because $60 \text{ seconds} = 1 \text{ minute}$

4. $\dfrac{3 \text{ ft}}{1 \text{ yd}}$ or $\dfrac{1 \text{ yd}}{3 \text{ ft}}$ because $3 \text{ ft} = 1 \text{ yd}$

5. $\dfrac{1 \text{ ft}}{30 \text{ cm}}$ Since we need to eliminate centimeters, cm must appear in the denominator. Since we need to convert to feet, ft must appear in the numerator.

6. $\dfrac{1 \text{ lb}}{0.45 \text{ kg}}$ Since we need to eliminate kilograms, kg must appear in the denominator. Since we need to convert to pounds, lb must appear in the numerator.

7. $\dfrac{3.8\, \text{l}}{1 \text{ gal}}$ Since we need to eliminate gallons, gal must appear in the denominator. Since we need to convert to liters, l must appear in the numerator.

8. $\dfrac{0.8 \text{ m}^2}{1 \text{ yd}^2}$ Since we need to eliminate square yards, yd^2 must appear in the denominator. Since we need to convert to square meters, m^2 must appear in the numerator.

9. $52 \text{ in.} = (52 \text{ in.})\left(\dfrac{2.54 \text{ cm}}{1 \text{ in.}}\right) = 132.08 \text{ cm}$

10. $9 \text{ lb} = (9 \text{ lb})\left(\dfrac{0.45 \text{ kg}}{1 \text{ lb}}\right) = 4.05 \text{ kg}$

11. $4.2 \text{ ft} = (4.2 \text{ ft})\left(\dfrac{30 \text{ cm}}{1 \text{ ft}}\right)\left(\dfrac{1 \text{ m}}{100 \text{ cm}}\right) = 1.26 \text{ m}$

12. $427 \text{ g} = (427 \text{ g})\left(\dfrac{1 \text{ oz}}{28 \text{ g}}\right) = 15.25 \text{ oz}$

13. $15 \text{ yd}^2 = (15 \text{ yd}^2)\left(\dfrac{0.8 \text{ m}^2}{1 \text{ yd}^2}\right) = 12 \text{ m}^2$

14. $160 \text{ kg} = (160 \text{ kg})\left(\dfrac{1 \text{ lb}}{0.45 \text{ kg}}\right) = 355.\overline{5} \approx 355.6 \text{ lb}$

15. $39 \text{ mi} = (39 \text{ mi})\left(\dfrac{1.6 \text{ km}}{1 \text{ mi}}\right) = 62.4 \text{ km}$

16. $765 \text{ mm} = (765 \text{ mm})\left(\dfrac{1 \text{ cm}}{10 \text{ mm}}\right)\left(\dfrac{1 \text{ in.}}{2.54 \text{ cm}}\right) = 30.11811024 \approx 30.12 \text{ in.}$

17. $675 \text{ ha} = (675 \text{ ha})\left(\dfrac{1 \text{ acre}}{0.4 \text{ ha}}\right) = 1687.5 \text{ acres}$

18. $192 \text{ oz} = (192 \text{ oz})\left(\dfrac{28 \text{ g}}{1 \text{ oz}}\right) = 5376 \text{ g}$

19. $15.6 \ \text{l} = \left(15.6 \ \text{l}\right)\left(\dfrac{1 \ \text{pt}}{0.47 \ \text{l}}\right) = 33.19148936 \approx 33.19 \ \text{pints}$

20. $4 \ \text{T} = \left(4 \ \text{T}\right)\left(\dfrac{0.9 \ \text{t}}{1 \ \text{T}}\right) = 3.6 \ \text{t}$

21. $45.6 \ \text{ml} = \left(45.6 \ \text{ml}\right)\left(\dfrac{1 \ \text{fl oz}}{30 \ \text{ml}}\right) = 1.52 \ \text{fl oz}$

22. $1.6 \ \text{km}^2 = \left(1.6 \ \text{km}^2\right)\left(\dfrac{1 \ \text{mi}^2}{2.6 \ \text{km}^2}\right) = 0.615384615 \approx 0.62 \ \text{mi}^2$

23. $120 \ \text{lb} = \left(120 \ \text{lb}\right)\left(\dfrac{0.45 \ \text{kg}}{1 \ \text{lb}}\right) = 54 \ \text{kg}$

24. $6.2 \ \text{acres} = \left(6.2 \ \text{acres}\right)\left(\dfrac{0.4 \ \text{ha}}{1 \ \text{acre}}\right) = 2.48 \ \text{ha}$

25. 28 grams

26. 28 grams, 0.45 kilogram

27. 0.45 kilogram

28. $5 \ \text{ft} = \left(5 \ \text{ft}\right)\left(\dfrac{12 \ \text{in.}}{1 \ \text{ft}}\right)\left(\dfrac{2.54 \ \text{cm}}{1 \ \text{in.}}\right) = 152.4 \ \text{cm}$

    $2 \ \text{in.} = \left(2 \ \text{in.}\right)\left(\dfrac{2.54 \ \text{cm}}{1 \ \text{in.}}\right) = 5.08 \ \text{cm}$

    $152.4 \ \text{cm} + 5.08 \ \text{cm} = 157.48 \ \text{centimeters}$

    $157.48 \ \text{cm} = \left(157.48 \ \text{cm}\right)\left(\dfrac{1 \ \text{m}}{100 \ \text{cm}}\right) = 1.5748 \approx 1.57 \ \text{meters}$

29. 2.54 centimeters, 1.6 kilometers

30. 1.6 kilometers

31. $10 \ \text{yd} = \left(10 \ \text{yd}\right)\left(\dfrac{0.9 \ \text{m}}{1 \ \text{yd}}\right) = 9 \ \text{meters}$

32. 0.9 meter

33. $505 \ \text{m} = \left(505 \ \text{m}\right)\left(\dfrac{1 \ \text{yd}}{0.9 \ \text{m}}\right) = 561.\overline{1} \approx 561.11 \ \text{yd}$

34. $175 \ \text{m} = \left(175 \ \text{m}\right)\left(\dfrac{1 \ \text{yd}}{0.9 \ \text{m}}\right) = 194.\overline{4} \approx 194.44 \ \text{yd}$

35. $344 \ \text{m} = \left(344 \ \text{m}\right)\left(\dfrac{100 \ \text{cm}}{1 \ \text{m}}\right)\left(\dfrac{1 \ \text{ft}}{30 \ \text{cm}}\right) = 1146.\overline{6} \approx 1146.67 \ \text{ft}$

36. $303 \ \text{m} = \left(303 \ \text{m}\right)\left(\dfrac{100 \ \text{cm}}{1 \ \text{m}}\right)\left(\dfrac{1 \ \text{ft}}{30 \ \text{cm}}\right) = 1010 \ \text{ft}$

37. $85 \ \text{km} = \left(85 \ \text{km}\right)\left(\dfrac{1 \ \text{mi}}{1.6 \ \text{km}}\right) = 53.125 \approx 53.13 \ \text{mph}$

38. $105 \ \text{mi} = \left(105 \ \text{mi}\right)\left(\dfrac{1.6 \ \text{km}}{1 \ \text{mi}}\right) = 168 \ \text{km}$

39. $\left(6 \ \text{yd}\right)\left(9 \ \text{yd}\right) = 54 \ \text{yd}^2$

    $54 \ \text{yd}^2 = \left(54 \ \text{yd}^2\right)\left(\dfrac{0.8 \ \text{m}^2}{1 \ \text{yd}^2}\right) = 43.2 \ \text{m}^2$

40. $110 \text{ mi} = (110 \text{ mi})\left(\dfrac{1.6 \text{ km}}{1 \text{ mi}}\right) = 176 \text{ km}$

41. $400 \text{ g} = (400 \text{ g})\left(\dfrac{1 \text{ oz}}{28 \text{ g}}\right) = 14.28571429 \approx 14.29 \text{ oz}$

42. $80 \text{ km} = (80 \text{ km})\left(\dfrac{1 \text{ mi}}{1.6 \text{ km}}\right) = 50 \text{ mph}$

43. $8 \text{ fl oz} = (8 \text{ fl oz})\left(\dfrac{30 \text{ ml}}{1 \text{ fl oz}}\right) = 240 \text{ ml}$

44. $12{,}500 \text{ gal} = (12{,}500 \text{ gal})\left(\dfrac{3.8 \text{ l}}{1 \text{ gal}}\right)\left(\dfrac{1 \text{ kl}}{1000 \text{ l}}\right) = 47.5 \text{ kl}$

45. $(50 \text{ ft})(30 \text{ ft})(8 \text{ ft}) = 12{,}000 \text{ ft}^3$

    $12{,}000 \text{ ft}^3 = (12{,}000 \text{ ft}^3)\left(\dfrac{0.03 \text{ m}^3}{1 \text{ ft}^3}\right) = 360 \text{ m}^3$

46. $1189 \text{ mi}^2 = (1189 \text{ mi}^2)\left(\dfrac{2.6 \text{ km}^2}{1 \text{ mi}^2}\right) = 3091.4 \text{ km}^2$

47. $1 \text{ kg} = (1 \text{ kg})\left(\dfrac{1 \text{ lb}}{0.45 \text{ kg}}\right) = 2.\overline{2} \text{ lb}$

    $\dfrac{\$1.10}{2.\overline{2}} = \$0.495 \text{ per pound}$

48. a) $1.3 \text{ t} = (1.3 \text{ t})\left(\dfrac{1 \text{ T}}{0.9 \text{ t}}\right) = 1.\overline{4} \approx 1.44 \text{ T}$

    b) $1.\overline{4} \text{ T} = (1.\overline{4} \text{ T})\left(\dfrac{2000 \text{ lb}}{1 \text{ T}}\right) = 2888.\overline{8} \approx 2888.9 \text{ lb}$

49. $34.5 \text{ kl} = (34.5 \text{ kl})\left(\dfrac{1000 \text{ l}}{1 \text{ kl}}\right)\left(\dfrac{1 \text{ gal}}{3.8 \text{ l}}\right) = 9078.947368 \approx 9078.95 \text{ gal}$

50. $0.25 \text{ oz} = (0.25 \text{ oz})\left(\dfrac{28 \text{ g}}{1 \text{ oz}}\right) = 7 \text{ g}$

    $\dfrac{\$80}{7} = 11.42857143 \approx \$11.43 \text{ per gram}$

51. a) $8 \text{ stones} = (8 \text{ stones})\left(\dfrac{70 \text{ kg}}{11 \text{ stones}}\right) = 50.\overline{90} \approx 50.91 \text{ kg}$

    b) $50.\overline{90} \text{ kg} = (50.\overline{90} \text{ kg})\left(\dfrac{1 \text{ lb}}{0.45 \text{ kg}}\right) = 113.\overline{13} \approx 113.13 \text{ lb}$

52. $\dfrac{1}{8} \text{ carat} = (0.125 \text{ carat})\left(\dfrac{1 \text{ g}}{5 \text{ carat}}\right) = 0.025 \text{ g}$

53. a) $-282 \text{ ft} = (-282 \text{ ft})\left(\dfrac{30 \text{ cm}}{1 \text{ ft}}\right) = -8460 \text{ cm}$

    b) $-8460 \text{ cm} = (-8460 \text{ cm})\left(\dfrac{1 \text{ m}}{100 \text{ cm}}\right) = -84.6 \text{ m}$

54. $5.7 \, \mathrm{l} = (5.7 \, \mathrm{l})\left(\dfrac{1 \, \mathrm{qt}}{0.95 \, \mathrm{l}}\right) = 6 \, \mathrm{qt}$

55. a) $1 \, \mathrm{m}^2 = (1 \, \mathrm{m}^2)\left(\dfrac{(3.3)^2 \, \mathrm{ft}^2}{1 \, \mathrm{m}^2}\right) = 10.89 \, \mathrm{ft}^2$

    b) $1 \, \mathrm{m}^3 = (1 \, \mathrm{m}^3)\left(\dfrac{(3.3)^3 \, \mathrm{ft}^3}{1 \, \mathrm{m}^3}\right) = 35.937 \, \mathrm{ft}^3$

56. a) $1 \, \mathrm{ft}^2 = (1 \, \mathrm{ft}^2)\left(\dfrac{(30)^2 \, \mathrm{cm}^2}{1 \, \mathrm{ft}^2}\right) = 900 \, \mathrm{cm}^2$

    b) $1 \, \mathrm{ft}^3 = (1 \, \mathrm{ft}^3)\left(\dfrac{(30)^3 \, \mathrm{cm}^3}{1 \, \mathrm{ft}^3}\right) = 27\,000 \, \mathrm{cm}^3$

57. $56 \, \mathrm{lb} = (56 \, \mathrm{lb})\left(\dfrac{0.45 \, \mathrm{kg}}{1 \, \mathrm{lb}}\right)\left(\dfrac{1 \, \mathrm{mg}}{1 \, \mathrm{kg}}\right) = 25.2 \, \mathrm{mg}$

58. $170 \, \mathrm{lb} = (170 \, \mathrm{lb})\left(\dfrac{0.45 \, \mathrm{kg}}{1 \, \mathrm{lb}}\right)\left(\dfrac{1.5 \, \mathrm{mg}}{1 \, \mathrm{kg}}\right) = 114.75 \, \mathrm{mg}$

59. $76 \, \mathrm{lb} = (76 \, \mathrm{lb})\left(\dfrac{0.45 \, \mathrm{kg}}{1 \, \mathrm{lb}}\right)\left(\dfrac{200 \, \mathrm{mg}}{1 \, \mathrm{kg}}\right) = 6840 \, \mathrm{mg}$

    $6840 \, \mathrm{mg} = (6840 \, \mathrm{mg})\left(\dfrac{1 \, \mathrm{g}}{1000 \, \mathrm{mg}}\right) = 6.84 \, \mathrm{g}$

60. $82 \, \mathrm{lb} = (82 \, \mathrm{lb})\left(\dfrac{0.45 \, \mathrm{kg}}{1 \, \mathrm{lb}}\right)\left(\dfrac{5 \, \mathrm{mg}}{1 \, \mathrm{kg}}\right) = 184.5 \, \mathrm{mg}$

61. a) $2 \text{ teaspoons} = (2 \text{ teaspoons})\left(\dfrac{12.5 \, \mathrm{mg}}{1 \text{ teaspoon}}\right) = 25 \, \mathrm{mg}$

    b) $12 \, \mathrm{fl\,oz} = (12 \, \mathrm{fl\,oz})\left(\dfrac{30 \, \mathrm{ml}}{1 \, \mathrm{fl\,oz}}\right)\left(\dfrac{12.5 \, \mathrm{mg}}{5 \, \mathrm{ml}}\right) = 900 \, \mathrm{mg}$

62. a) $2 \text{ tablespoons} = (2 \text{ tablespoons})\left(\dfrac{236 \, \mathrm{mg}}{1 \text{ tablespoon}}\right) = 472 \, \mathrm{mg}$

    b) $8 \, \mathrm{fl\,oz} = (8 \, \mathrm{fl\,oz})\left(\dfrac{30 \, \mathrm{ml}}{1 \, \mathrm{fl\,oz}}\right)\left(\dfrac{1 \text{ tablespoon}}{15 \, \mathrm{ml}}\right)\left(\dfrac{236 \, \mathrm{mg}}{1 \text{ tablespoon}}\right) = 3776 \, \mathrm{mg}$

63. a) $964 \, \mathrm{ft} = (964 \, \mathrm{ft})\left(\dfrac{30 \, \mathrm{cm}}{1 \, \mathrm{ft}}\right)\left(\dfrac{1 \, \mathrm{m}}{100 \, \mathrm{cm}}\right) = 289.2 \, \mathrm{m}$

    b) $85,000 \text{ tons} = (85,000 \text{ tons})\left(\dfrac{0.9 \text{ tonne}}{1 \text{ ton}}\right) = 76\,500 \, \mathrm{t}$

    c) $28 \, \mathrm{mi} = (28 \, \mathrm{mi})\left(\dfrac{1.6 \, \mathrm{km}}{1 \, \mathrm{mi}}\right) = 44.8 \, \mathrm{kph}$

64. $(0.5 \text{ c})\left(\dfrac{0.24 \text{ l}}{1 \text{ c}}\right) = 0.12 \text{ l}$ graham cracker crumbs

$(12 \text{ oz})\left(\dfrac{28 \text{ g}}{1 \text{ oz}}\right) = 336 \text{ g}$ nuts

$(8 \text{ oz})\left(\dfrac{28 \text{ g}}{1 \text{ oz}}\right) = 224 \text{ g}$ chocolate pieces

$\left(\dfrac{4}{3} \text{ c}\right)\left(\dfrac{0.24 \text{ l}}{1 \text{ c}}\right) = 0.32 \text{ l}$ flaked coconut

$\left(\dfrac{4}{3} \text{ c}\right)\left(\dfrac{0.24 \text{ l}}{1 \text{ c}}\right) = 0.32 \text{ l}$ condensed milk

$(9 \text{ in.})\left(\dfrac{2.54 \text{ cm}}{1 \text{ in.}}\right) \times (13 \text{ in.})\left(\dfrac{2.54 \text{ cm}}{1 \text{ in.}}\right) = 22.86 \text{ cm} \times 33.02 \text{ cm}$ baking pan

$350° \text{ F} = \dfrac{5}{9}(350 - 32) = 176.\overline{6} \approx 176.7° \text{ C}$

$(1.5 \text{ in.})\left(\dfrac{2.54 \text{ cm}}{1 \text{ in.}}\right) \times (3 \text{ in.})\left(\dfrac{2.54 \text{ cm}}{1 \text{ in.}}\right) = 3.81 \text{ cm} \times 7.62 \text{ cm}$ bars

65. a) $(37 \text{ m})\left(\dfrac{1 \text{ yd}}{0.9 \text{ m}}\right) = 41.\overline{1} \approx 41.1 \text{ yd}$

   b) $(370\ 140 \text{ km})\left(\dfrac{1 \text{ mi}}{1.6 \text{ km}}\right) = 231{,}337.5 \text{ mi}$

   c) $(44 \text{ km})\left(\dfrac{1 \text{ mi}}{1.6 \text{ km}}\right) = 27.5 \text{ mi}$

   d) $1260° \text{ C} = \dfrac{9}{5}(1260) + 32 = 2300° \text{ F}$

   e) $(335 \text{ km})\left(\dfrac{1 \text{ mi}}{1.6 \text{ km}}\right) = 209.375 \text{ mph}$

   f) $(29\ 484 \text{ kg})\left(\dfrac{1 \text{ lb}}{0.45 \text{ kg}}\right) = 65{,}520 \text{ lb}$

   g) $(4.5 \text{ m})\left(\dfrac{1 \text{ yd}}{0.9 \text{ m}}\right) \times (18 \text{ m})\left(\dfrac{1 \text{ yd}}{0.9 \text{ m}}\right) = 5 \text{ yd} \times 20 \text{ yd}$

   h) $(171\ 396 \text{ l})\left(\dfrac{1 \text{ gal}}{3.8 \text{ l}}\right) = 45{,}104.21053 \approx 45{,}104.21 \text{ gal/min}$

   i) $(63\ 588 \text{ l})\left(\dfrac{1 \text{ gal}}{3.8 \text{ l}}\right) = 16{,}733.68421 \approx 16{,}733.68 \text{ gal/min}$

   j) $(46.89 \text{ m})\left(\dfrac{1 \text{ yd}}{0.9 \text{ m}}\right) = 52.1 \text{ yd}$

   k) $(8.4 \text{ m})\left(\dfrac{1 \text{ yd}}{0.9 \text{ m}}\right) = 9.\overline{3} \approx 9.33 \text{ yd}$

   l) $(632\ 772 \text{ kg})\left(\dfrac{1 \text{ lb}}{0.45 \text{ kg}}\right) = 1{,}406{,}160 \text{ lb}$

65.   m) $(106\ 142\ \text{kg})\left(\dfrac{1\ \text{lb}}{0.45\ \text{kg}}\right) = 235{,}871.\overline{1} \approx 235{,}871.11\ \text{lb}$

   n) $-251°\,\text{C} = \dfrac{9}{5}(-251) + 32 = -419.8°\,\text{F}$

66. $(0.2\ \text{mg})\left(\dfrac{1\ \text{grain}}{60\ \text{mg}}\right)\left(\dfrac{1\ \text{ml}}{\frac{1}{300}\ \text{grain}}\right) = 1.0\ \text{cc, or b)}$

67. $15(130\ \text{lb}) = 1950\ \text{lb}$

   $(1950\ \text{lb})\left(\dfrac{0.18\ \text{kg}}{100\ \text{lb}}\right)\left(\dfrac{1\ \text{lb}}{0.45\ \text{kg}}\right) = 7.8\ \text{lb}$

68.   a) $(4.0\ \text{l})\left(\dfrac{1000\ \text{ml}}{1\ \text{l}}\right)\left(\dfrac{1\ \text{cm}^3}{1\ \text{ml}}\right) = 4000\ \text{cc}$

   b) $(4000\ \text{cm}^3)\left(\dfrac{1\ \text{in.}^3}{(2.54)^3\ \text{cm}^3}\right) = \dfrac{4000}{16.387064} = 244.0949764 \approx 244.09\ \text{in.}^3$

69.   A meter
70.   A kilogram
71.   A hectare
72.   A liter
73.   A tonne
74.   A decimeter
75.   wonton
76.   1 microscope

77.   1 kilohurtz
78.   1 pound cake $\left(1\ \text{lb} = 16\ \text{oz}; 16\ \text{oz}\left(\dfrac{28\ \text{g}}{1\ \text{oz}}\right) = 448\ \text{g}\right)$

79.   1 megaphone
80.   2 megacycles
81.   2 kilomockingbird
82.   1 decacards
83.   1 decoration
84.   1 microfiche

Review Exercises

1. $\dfrac{1}{100}$ of base unit
2. $1000\times$ base unit
3. $\dfrac{1}{1000}$ of base unit

4. $100\times$ base unit
5.   10 times base unit
6. $\dfrac{1}{10}$ of base unit

7.   20 cg = 0.20 g
8.   3.2 l = 320 cl
9.   0.0004 cm = 0.004 mm
10.   1 000 000 mg = 1 kg
11.   4.62 kl = 4620 l
12.   192.6 dag = 19 260 dg
13.   2.67 kl = 2 670 000 ml
14.   0.047 km = 47 m
15.   Centimeters

   14 630 cl = 146 300 ml
   47 000 cm = 470 m

   3000 ml, 14 630 cl, 2.67 kl
   0.047 km, 47 000 cm,

   4700 m

16.   Grams
17.   Degrees Celsius
18.   Millimeters or centimeters
19.   Square meters
20.   Milliliters or cubic centimeters
21.   Millimeters
22.   Kilograms or tonnes
23.   Kilometers
24.   Meters or centimeters
25.   a) and b) Answers will vary.
26.   a) and b) Answers will vary.
27.   c
28.   b
29.   c
30.   a
31.   a
32.   b

33. $2500 \text{ kg} = (2500 \text{ kg})\left(\dfrac{1 \text{ lb}}{0.45 \text{ kg}}\right)\left(\dfrac{1 \text{ T}}{2000 \text{ lb}}\right)\left(\dfrac{0.9 \text{ t}}{1 \text{ T}}\right) = 2.5 \text{ t}$

34. $6.3 \text{ t} = (6.3 \text{ t})\left(\dfrac{1 \text{ T}}{0.9 \text{ t}}\right)\left(\dfrac{2000 \text{ lb}}{1 \text{ T}}\right)\left(\dfrac{0.45 \text{ kg}}{1 \text{ lb}}\right)\left(\dfrac{1000 \text{ g}}{1 \text{ kg}}\right) = 6\ 300\ 000 \text{ g}$

35. $18°\text{ C} = \dfrac{9}{5}(18) + 32 = 64.4°\text{ F}$

36. $68°\text{ F} = \dfrac{5}{9}(68 - 32) = 20°\text{ C}$

37. $-6°\text{ F} = \dfrac{5}{9}(-6 - 32) = -21.\overline{1} \approx -21.1°\text{ C}$

38. $39°\text{ C} = \dfrac{9}{5}(39) + 32 = 102.2°\text{ F}$

39. $l = 4 \text{ cm}, \quad w = 1.6 \text{ cm}$

 $A = lw = 4(1.6) = 6.4 \text{ cm}^2$

40. $r = 1.5 \text{ cm}$

 $A = \pi r^2 \approx 3.14(1.5)^2 = 7.065 \approx 7.07 \text{ cm}^2$

41. a) $V = lwh = (10)(4)(2) = 80 \text{ m}^3$

 b) $(80 \text{ m}^3)\left(\dfrac{1 \text{ kl}}{1 \text{ m}^3}\right)\left(\dfrac{1000 \text{ l}}{1 \text{ kl}}\right)\left(\dfrac{1 \text{ kg}}{1 \text{ l}}\right) = 80\ 000 \text{ kg}$

42. a) $A = lw = 30(22) = 660 \text{ m}^2$

 b) $660 \text{ m}^2 = (660 \text{ m}^2)\left(\dfrac{1 \text{ km}^2}{(1000)^2 \text{ m}^2}\right) = 0.000\ 66 \text{ km}^2$

43. a) $V = lwh = (80)(40)(30) = 96\ 000 \text{ cm}^3$

 b) $96\ 000 \text{ cm}^3 = (96\ 000 \text{ cm}^3)\left(\dfrac{1 \text{ m}^3}{(100)^3 \text{ cm}^3}\right) = 0.096 \text{ m}^3$

 c) $96\ 000 \text{ cm}^3 = (96\ 000 \text{ cm}^3)\left(\dfrac{1 \text{ ml}}{1 \text{ cm}^3}\right) = 96\ 000 \text{ ml}$

 d) $0.096 \text{ m}^3 = (0.096 \text{ m}^3)\left(\dfrac{1 \text{ kl}}{1 \text{ m}^3}\right) = 0.096 \text{ kl}$

44. Since $1 \text{ km} = 100 \times 1 \text{ dam}$, $1 \text{ km}^2 = 100^2 \times 1 \text{ dam}^2 = 10\ 000 \text{ dam}^2$.

 Thus, 1 square kilometer is 10,000 times larger than a square dekameter.

45. $(20 \text{ cm})\left(\dfrac{1 \text{ in.}}{2.54 \text{ cm}}\right) = 7.874015748 \approx 7.87 \text{ in.}$

46. $(105 \text{ kg})\left(\dfrac{1 \text{ lb}}{0.45 \text{ kg}}\right) = 233.\overline{3} \approx 233.33 \text{ lb}$

47. $(83 \text{ yd})\left(\dfrac{0.9 \text{ m}}{1 \text{ yd}}\right) = 74.7 \text{ m}$

48. $(100 \text{ m})\left(\dfrac{1 \text{ yd}}{0.9 \text{ m}}\right) = 111.\overline{1} \approx 111.11 \text{ yd}$

49. $(45 \text{ mi})\left(\dfrac{1.6 \text{ km}}{1 \text{ mi}}\right) = 72 \text{ kph}$

50. $(40 \text{ l})\left(\dfrac{1 \text{ qt}}{0.95 \text{ l}}\right) = 42.10526316 \approx 42.11 \text{ qt}$

51. $(15 \text{ gal})\left(\dfrac{3.8 \text{ l}}{1 \text{ gal}}\right) = 57 \text{ l}$

52. $(40 \text{ m}^3)\left(\dfrac{1 \text{ yd}^3}{0.76 \text{ m}^3}\right) = 52.63157895 \approx 52.63 \text{ yd}^3$

53. $(83 \text{ cm}^2)\left(\dfrac{1 \text{ in.}^2}{6.5 \text{ cm}^2}\right) = 12.76923077 \approx 12.77 \text{ in.}^2$

54. $(4 \text{ qt})\left(\dfrac{0.95 \text{ l}}{1 \text{ qt}}\right) = 3.8 \text{ l}$

55. $\left(15 \text{ yd}^3\right)\left(\dfrac{0.76 \text{ m}^3}{1 \text{ yd}^3}\right) = 11.4 \text{ m}^3$

56. $\left(62 \text{ mi}\right)\left(\dfrac{1.6 \text{ km}}{1 \text{ mi}}\right) = 99.2 \text{ km}$

57. $\left(27 \text{ cm}\right)\left(\dfrac{1 \text{ ft}}{30 \text{ cm}}\right) = 0.9 \text{ ft}$

58. $\left(3.25 \text{ in.}\right)\left(\dfrac{2.54 \text{ cm}}{1 \text{ in.}}\right)\left(\dfrac{10 \text{ mm}}{1 \text{ cm}}\right) = 82.55 \text{ mm}$

59. a) $700\left(1.5 \text{ kg}\right) = 1050 \text{ kg}$

    b) $1050 \text{ kg} = \left(1050 \text{ kg}\right)\left(\dfrac{1 \text{ lb}}{0.45 \text{ kg}}\right) = 2333.\overline{3} \approx 2333.33 \text{ lb}$

60. $A = lw = (24)(15) = 360 \text{ ft}^2$

    $360 \text{ ft}^2 = \left(360 \text{ ft}^2\right)\left(\dfrac{0.09 \text{ m}^2}{1 \text{ ft}^2}\right) = 32.4 \text{ m}^2$

61. a) $\left(50{,}000 \text{ gal}\right)\left(\dfrac{3.8 \text{ l}}{1 \text{ gal}}\right)\left(\dfrac{1 \text{ kl}}{1000 \text{ l}}\right) = 190 \text{ kl}$

    b) $\left(190 \text{ kl}\right)\left(\dfrac{1000 \text{ l}}{1 \text{ kl}}\right)\left(\dfrac{1 \text{ kg}}{1 \text{ l}}\right) = 190\,000 \text{ kg}$

62. a) $35 \text{ mi} = \left(35 \text{ mi}\right)\left(\dfrac{1.6 \text{ km}}{1 \text{ mi}}\right) = 56 \text{ kph}$

    b) $56 \text{ km} = \left(56 \text{ km}\right)\left(\dfrac{1000 \text{ m}}{1 \text{ km}}\right) = 56\,000 \text{ meters per hour}$

63. a) $V = lwh = (90)(70)(40) = 252\,000 \text{ cm}^3$

    $252\,000 \text{ cm}^3 = \left(252\,000 \text{ cm}^3\right)\left(\dfrac{1 \text{ ml}}{1 \text{ cm}^3}\right)\left(\dfrac{1 \text{ l}}{1000 \text{ ml}}\right) = 252 \text{ l}$

    b) $252 \text{ l} = \left(252 \text{ l}\right)\left(\dfrac{1 \text{ kg}}{1 \text{ l}}\right) = 252 \text{ kg}$

64. $1 \text{ kg} = \left(1 \text{ kg}\right)\left(\dfrac{1 \text{ lb}}{0.45 \text{ kg}}\right) = 2.\overline{2} \text{ lb}$

    $\dfrac{\$3.50}{2.\overline{2}} = \$1.575 \approx \$1.58 \text{ per pound}$

Chapter Test

1. $204 \text{ cl} = 0.204 \text{ dal}$

2. $123 \text{ km} = 123\,000\,000 \text{ mm}$

3. $1 \text{ km} = \left(1 \text{ km}\right)\left(\dfrac{100 \text{ dam}}{1 \text{ km}}\right) = 100 \text{ dam}$ or 100 times greater

4. $400(6) = 2400 \text{ m}$

   $\left(2400 \text{ m}\right)\left(\dfrac{1 \text{ km}}{1000 \text{ m}}\right) = 2.4 \text{ km}$

5. b

6. a

7. c

8. c

9. b

10. $1 \, m^2 = \left(1 \, m^2\right)\left(\dfrac{100^2 \, cm^2}{1 \, m^2}\right) = 10 \; 000 \, cm^2$ or 10,000 times greater

11. $1 \, m^3 = \left(1 \, m^3\right)\left(\dfrac{1000^3 \, mm^3}{1 \, m^3}\right) = 1 \; 000 \; 000 \; 000 \, mm^3$ or 1,000,000,000 times greater

12. $452 \, in. = \left(452 \, in.\right)\left(\dfrac{2.54 \, cm}{1 \, in.}\right) = 1148.08 \, cm$

13. $150 \, m = \left(150 \, m\right)\left(\dfrac{1 \, yd}{0.9 \, m}\right) = 166.\overline{6} \approx 166.67 \, yd$

14. $-10° \, F = \dfrac{5}{9}\left(-10-32\right) = -23.\overline{3} \approx -23.33° \, C$　　　　15. $20° \, C = \dfrac{9}{5}\left(20\right)+32 = 68° \, F$

16. $12 \, ft = \left(12 \, ft\right)\left(\dfrac{30 \, cm}{1 \, ft}\right) = 360 \, cm$ or $12 \, ft = \left(12 \, ft\right)\left(\dfrac{12 \, in.}{1 \, ft}\right)\left(\dfrac{2.54 \, cm}{1 \, in.}\right) = 365.76 \, cm$

17. a) $V = lwh = 20(20)(8) = 3200 \, m^3$

    b) $3200 \, m^3 = \left(3200 \, m^3\right)\left(\dfrac{1000 \, l}{1 \, m^3}\right) = 3 \; 200 \; 000 \, l$ or $3 \; 200 \; 000 \, l = \left(3 \; 200 \; 000 \, l\right)\left(\dfrac{1 \, kl}{1000 \, l}\right) = 3200 \, kl$

    c) $3 \; 200 \; 000 \, l = \left(3 \; 200 \; 000 \, l\right)\left(\dfrac{1 \, kg}{1 \, l}\right) = 3 \; 200 \; 000 \, kg$

18. Total surface area: $2lh + 2wh = 2(20)(6) + 2(15)(6) = 420 \, m^2$

    Liters needed for first coat: $\left(420 \, m^2\right)\left(\dfrac{1 \, l}{10 \, m^2}\right) = 42 \, l$

    Liters needed for second coat: $\left(420 \, m^2\right)\left(\dfrac{1 \, l}{15 \, m^2}\right) = 28 \, l$

    Total liters needed: $42 + 28 = 70 \, l$

    Total cost: $\left(70 \, l\right)\left(\dfrac{\$3.50}{1 \, l}\right) = \$245$

Group Projects

1. a) $\left(196 \, lb\right)\left(\dfrac{0.45 \, kg}{1 \, lb}\right)\left(\dfrac{20 \, mg}{1 \, kg}\right) = 1764 \, mg$

   b) $\left(\dfrac{250 \, cc}{1 \, hr}\right)\left(\dfrac{1 \, hr}{60 \, min}\right) = 4.1\overline{6} \approx 4.17 \, cc/min$

2. a) $(60 \text{ lb})\left(\dfrac{0.45 \text{ kg}}{1 \text{ lb}}\right) = 27 \text{ kg}$

   Child's dose: $\dfrac{27 \text{ kg}}{67.5 \text{ kg}}(70 \text{ mg}) = 28 \text{ mg}$

   b) $\dfrac{\text{child's weight in kg}}{67.5 \text{ kg}} \times 70 \text{ mg} = 70 \text{ mg}$

   $\dfrac{\text{child's weight in kg}}{67.5 \text{ kg}} = 1$

   Child's weight: $67.5 \text{ kg} = (67.5 \text{ kg})\left(\dfrac{1 \text{ lb}}{0.45 \text{ kg}}\right) = 150 \text{ lb}$

3. a) $5 \text{ ft } 2 \text{ in.} = 62 \text{ in.}$

   $62 \text{ in.} = (62 \text{ in.})\left(\dfrac{2.54 \text{ cm}}{1 \text{ in.}}\right) = 157.48 \text{ cm}$

   b) $8695.5 \text{ yen} = (8695.5 \text{ yen})\left(\dfrac{\$1 \text{ U.S.}}{118.25 \text{ yen}}\right) = \$73.53488372 \text{ U.S.} \approx \$73.53 \text{ U.S.}$

   c) $6 \text{ lb} = (6 \text{ lb})\left(\dfrac{16 \text{ oz}}{1 \text{ lb}}\right)\left(\dfrac{28 \text{ g}}{1 \text{ oz}}\right) = 2688 \text{ g}$

   $2688 \text{ g} = (2688 \text{ g})\left(\dfrac{10 \text{ pesos}}{100 \text{ g}}\right)\left(\dfrac{\$0.095 \text{ U.S.}}{1 \text{ peso}}\right) = 25.536 \approx \$25.54$

   Note: If you use different conversion factors, your answer will be slightly different because the conversion factors are rounded values.

   d) To fill the tank in New Zealand dollars:

   $53 \text{ l}\left(\dfrac{\$0.929 \text{ New Zealand}}{1 \text{ l}}\right) = \$49.237 \text{ New Zealand} \approx \$49.24 \text{ New Zealand}$

   To fill the tank in U.S. dollars:

   $\$49.237 \text{ New Zealand} = (\$49.237 \text{ New Zealand})\left(\dfrac{\$0.584 \text{ U.S.}}{\$1 \text{ New Zealand}}\right) = \$28.754408 \text{ U.S.} \approx \$28.75 \text{ U.S.}$

   $\$28.754408 \text{ U.S. for } 53 \text{ l}$

   $\dfrac{\$28.754408 \text{ U.S.}}{53 \text{ l}} = \$0.542536 \text{ U.S. per l}$

   $\left(\dfrac{\$0.542536 \text{ U.S.}}{1 \text{ l}}\right)\left(\dfrac{3.8 \text{ l}}{1 \text{ gal}}\right) = \$2.0616368 \text{ U.S. per gal} \approx \$2.06 \text{ U.S. per gal}$

CHAPTER NINE

GEOMETRY

Exercise Set 9.1

1. a) Undefined terms, definitions, postulates (axioms), and theorems

   b) First, Euclid introduced **undefined terms**. Second, he introduced certain **definitions**. Third, he stated primitive propositions called **postulates (axioms)** about the undefined terms and definitions. Fourth, he proved, using deductive reasoning, other propositions called **theorems**.

2. An **axiom (postulate)** is a statement that is accepted as being true on the basis of its "obviousness" and its relation to the physical world. A **theorem** is a statement that has been proven using undefined terms, definitions, and axioms.

3. Two lines in the same plane that do not intersect are **parallel lines**.

4. Two lines that do not lie in the same plane and do not intersect are called **skewed lines**.

5. Two angles in the same plane are **adjacent angles** when they have a common vertex and a common side but no common interior points.

6. Two angles the sum of whose measure is 180° are called **supplementary angles**.

7. Two angles the sum of whose measure is 90° are called **complementary angles**.

8. An angle whose measure is 180° is a **straight angle**.

9. An angle whose measure is greater than 90° but less than 180° is an **obtuse angle**.

10. An angle whose measure is less than 90° is an **acute angle**.

11. An angle whose measure is 90° is a **right angle**.

12. In the pair of intersecting lines below, ∡ 1 and ∡ 3 are vertical angles as are ∡ 2 and ∡ 4.

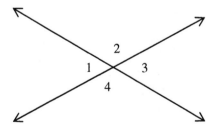

13. Half line, $\overset{\circ}{AB}\!\!\rightarrow$

14. Half open line segment, $\overset{\circ}{AB}$

15. Line segment, $\overline{AB}$

16. Ray, $\overrightarrow{AB}$

17. Line, $\overleftrightarrow{AB}$

18. Half line, $\overset{\circ}{BA}\!\!\rightarrow$

19. Open line segment, $\overset{\circ\circ}{AB}$

20. Ray, $\overrightarrow{BA}$

21. $\overline{BD}$

22. $\overline{EG}$

23. $\overset{\circ}{BD}$

24. $\overrightarrow{AD}$

25. $\{B,F\}$

26. $\{C\}$

27. $\{C\}$

28. $\overset{\circ\circ}{BC}$

29. $\overline{BC}$ 30. $\triangle BCF$ 31. $\overrightarrow{BC}$ 32. $\varnothing$

33. $\varnothing$ 34. $\overset{\circ\;\circ}{ED}$ 35. $\overline{BC}$ 36. $\overline{DE}$

37. $\angle ABE$ 38. $\angle FBE$ 39. $\angle EBC$ 40. $\{B\}$

41. $\overset{\circ\;\circ}{AC}$ 42. $\varnothing$ 43. $\overset{\circ}{BE}$ 44. $\{B\}$

45. Obtuse   46. Straight   47. Straight   48. Acute

49. Right   50. None of these   51. None of these   52. Right

53. $90° - 19° = 71°$ 54. $90° - 89° = 1°$

55. $90° - 32\frac{3}{4}° = 57\frac{1}{4}°$ 56. $90° - 43\frac{1}{3}° = 46\frac{2}{3}°$

57. $90° - 64.7° = 25.3°$ 58. $90° - 0.01° = 89.99°$

59. $180° - 91° = 89°$ 60. $180° - 8° = 172°$

61. $180° - 20.5° = 159.5°$ 62. $180° - 179.99° = 0.01°$

63. $180° - 43\frac{5}{7}° = 136\frac{2}{7}°$ 64. $180° - 64\frac{7}{16}° = 115\frac{9}{16}°$

65. d   66. b   67. c

68. f   69. e   70. a

71. Let x = measure of $\angle 2$
$x+4$ = measure of $\angle 1$
$$x + x + 4 = 90$$
$$2x + 4 = 90$$
$$2x = 86$$
$$x = \frac{86}{2} = 43°, m\angle 2$$
$$x + 4 = 43 + 4 = 47°, m\angle 1$$

72. Let x = measure of $\angle 1$
$90 - x$ = measure of $\angle 2$
$$x - (90 - x) = 62$$
$$x - 90 + x = 62$$
$$2x - 90 = 62$$
$$2x = 152$$
$$x = \frac{152}{2} = 76°, m\angle 1$$
$$90 - x = 90 - 76 = 14°, m\angle 2$$

73. Let x = measure of $\angle 1$
$180 - x$ = measure of $\angle 2$
$$x - (180 - x) = 88$$
$$x - 180 + x = 88$$
$$2x - 180 = 88$$
$$2x = 268$$
$$x = \frac{268}{2} = 134°, m\angle 1$$
$$180 - x = 180 - 134 = 46°, m\angle 2$$

74. Let x = measure of $\angle 1$
$17x$ = measure of $\angle 2$
$$x + 17x = 180$$
$$18x = 180$$
$$x = \frac{180}{18} = 10°, m\angle 1$$
$$17x = 17(10) = 170°, m\angle 2$$

75. $m\angle 1 + 125° = 180°$
    $m\angle 1 = 55°$
    $m\angle 2 = m\angle 1$ (vertical angles)
    $m\angle 3 = 125°$ (vertical angles)
    $m\angle 5 = m\angle 2$ (alternate interior angles)
    $m\angle 4 = m\angle 3$ (alternate interior angles)
    $m\angle 7 = m\angle 4$ (vertical angles)
    $m\angle 6 = m\angle 5$ (vertical angles)
    Measures of angles 3, 4, and 7 are each 125°.
    Measures of angles 1, 2, 5, and 6 are each 55°.

76. $m\angle 3 + 30° = 180°$
    $m\angle 3 = 150°$
    $m\angle 1 = 30°$ (vertical angles)
    $m\angle 2 = m\angle 3$ (vertical angles)
    $m\angle 4 = m\angle 1$ (corresponding angles)
    $m\angle 7 = m\angle 4$ (vertical angles)
    $m\angle 6 = m\angle 3$ (alternate interior angles)
    $m\angle 5 = m\angle 6$ (vertical angles)
    Measures of angles 1, 4, and 7 are each 30°.
    Measures of angles 2, 3, 5, and 6 are each 150°.

77. $m\angle 1 + 25° = 180°$
    $m\angle 1 = 155°$
    $m\angle 3 = m\angle 1$ (vertical angles)
    $m\angle 2 = 25°$ (vertical angles)
    $m\angle 4 = m\angle 3$ (alternate interior angles)
    $m\angle 7 = m\angle 4$ (vertical angles)
    $m\angle 5 = m\angle 2$ (corresponding angles)
    $m\angle 6 = m\angle 5$ (vertical angles)
    Measures of angles 2, 5, and 6 are each 25°.
    Measures of angles 1, 3, 4, and 7 are each 155°.

78. $m\angle 3 + 120° = 180°$
    $m\angle 3 = 60°$
    $m\angle 4 = 120°$ (vertical angles)
    $m\angle 7 = m\angle 3$ (vertical angles)
    $m\angle 6 = m\angle 3$ (alternate interior angles)
    $m\angle 1 = m\angle 6$ (vertical angles)
    $m\angle 5 = m\angle 4$ (alternate exterior angles)
    $m\angle 2 = m\angle 5$ (vertical angles)
    Measures of angles 2, 4, and 5 are each 120°.
    Measures of angles 1, 3, 6, and 7 are each 60°.

79.
$$x + 3x + 10 = 90$$
$$4x + 10 = 90$$
$$4x = 80$$
$$x = \frac{80}{4} = 20°, \; m\angle 2$$
$$3x + 10 = 3(20) + 10 = 70°, \; m\angle 1$$

80.
$$x + 7x + 2 = 90$$
$$8x + 2 = 90$$
$$8x = 88$$
$$x = \frac{88}{8} = 11°, \; m\angle 1$$
$$7x + 2 = 7(11) + 2 = 79°, \; m\angle 2$$

81.
$$x + 2x - 9 = 90$$
$$3x - 9 = 90$$
$$3x = 99$$
$$x = \frac{99}{3} = 33°, \; m\angle 1$$
$$2x - 9 = 2(33) - 9 = 57°, \; m\angle 2$$

82.
$$x + 8x - 9 = 90$$
$$9x - 9 = 90$$
$$9x = 99$$
$$x = \frac{99}{9} = 11°, \; m\angle 2$$
$$8x - 9 = 8(11) - 9 = 79°, \; m\angle 1$$

83.
$$x + 2x - 15 = 180$$
$$3x - 15 = 180$$
$$3x = 195$$
$$x = \frac{195}{3} = 65°, \; m\angle 2$$
$$2x - 15 = 2(65) - 15 = 115°, \; m\angle 1$$

84.
$$x + 4x + 10 = 180$$
$$5x + 10 = 180$$
$$5x = 170$$
$$x = \frac{170}{5} = 34°, \; m\angle 2$$
$$4x + 10 = 4(34) + 10 = 146°, \; m\angle 1$$

85. $$x + 5x + 6 = 180$$
$$6x + 6 = 180$$
$$6x = 174$$
$$x = \frac{174}{6} = 29°, \ m\measuredangle \ 1$$
$$5x + 6 = 5(29) + 6 = 151°, \ m\measuredangle \ 2$$

86. $$x + 6x + 5 = 180$$
$$7x + 5 = 180$$
$$7x = 175$$
$$x = \frac{175}{7} = 25°, \ m\measuredangle \ 1$$
$$6x + 5 = 6(25) + 5 = 155°, \ m\measuredangle \ 2$$

87. a) An infinite number of lines can be drawn through a given point.

b) An infinite number of planes can be drawn through a given point.

88. If the two planes are not parallel, the intersection is a straight line.

89. An infinite number of planes can be drawn through a given line.

90. a) Yes, any three noncollinear points always determine a plane.

b) No, the plane determined is unique.

c) An infinite number of planes can be drawn through three collinear points.

For Exercises 91 - 98, the answers given are one of many possible answers.

91. Plane ABG and plane JCD

92. $\overleftrightarrow{EF}$ and $\overleftrightarrow{DG}$

93. $\overrightarrow{BG}$ and $\overrightarrow{DG}$

94. Plane ABG and plane BCD

95. Plane $AGB \cap$ plane $ABC \cap$ plane $BCD = \{B\}$

96. Plane $HGD \cap$ plane $FGD \cap$ plane $BGD = \overleftrightarrow{GD}$

97. $\overleftrightarrow{BC} \cap$ plane $ABG = \{B\}$

98. $\overleftrightarrow{AB} \cap$ plane $ABG = \overleftrightarrow{AB}$

99. Always true. If any two lines are parallel to a third line, then they must be parallel to each other.

100. Sometimes true. A triangle must always contain at least two acute angles. Some triangles contain three acute angles.

101. Sometimes true. Vertical angles are only complementary when each is equal to 45°.

102. Sometimes true. Alternate exterior angles are only supplementary when each is equal to 90°.

103. Sometimes true. Alternate interior angles are only complementary when each is equal to 45°.

104. Never true. The sum of two obtuse angles is greater than 180°.

105. No. Line m and line n may intersect.

106. No. Line l and line n may be parallel or skewed.

107.

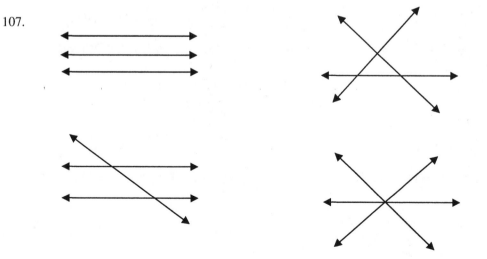

108.

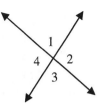

$$m\angle 1 + m\angle 2 = 180°$$
$$m\angle 3 + m\angle 4 = 180°$$
$$180° + 180° = 360°$$

109. a)

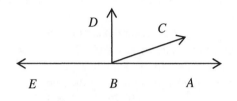

Other answers are possible.

b) Let $m\angle ABC = x$ and $m\angle CBD = y$.

$x + y = 90°$ and $y = 2x$

Substitute $y = 2x$ into $x + y = 90°$.

$$x + 2x = 90°$$
$$3x = 90°$$
$$\frac{3x}{3} = \frac{90°}{3}$$
$$x = 30° = m\angle ABC$$

c) $m\angle CBD = y$

$$y = 2x = 2(30°) = 60°$$

d) $m\angle ABD + m\angle DBE = 180°$

$m\angle ABD = x + y = 30° + 60° = 90°$.

$$90° + m\angle DBE = 180°$$
$$m\angle DBE = 180° - 90° = 90°.$$

Exercise Set 9.2

1. A **polygon** is a closed figure in a plane determined by three or more straight line segments.
2. A **regular polygon** is one whose sides are all the same length and whose interior angles all have the same measure; other polygons may have sides of different length and interior angles with different meaures.
3. The different types of triangles are acute, obtuse, right, isosceles, equilateral, and scalene. Descriptions will vary.
4. The different types of quadrilaterals are trapezoid, parallelogram, rhombus, rectangle, and square. Descriptions will vary.
5. If the corresponding sides of two similar figures are the same length, the figures are **congruent figures**.
6. Figures that have the same shape but may be of different sizes are **similar figures**.

7. a) Rectangle
   b) Not regular
8. a) Triangle
   b) Regular
9. a) Hexagon
   b) Regular
10. a) Octagon
    b) Not regular
11. a) Rhombus
    b) Not regular
12. a) Pentagon
    b) Regular
13. a) Octagon
    b) Not regular
14. a) Dodecagon
    b) Not regular

15. a) Scalene
    b) Right
16. a) Isosceles
    b) Acute
17. a) Isosceles
    b) Obtuse
18. a) Isosceles
    b) Right
19. a) Equilateral
    b) Acute
20. a) Scalene
    b) Acute
21. a) Scalene
    b) Obtuse
22. a) Scalene
    b) Right
23. Parallelogram
24. Rectangle
25. Rhombus
26. Trapezoid
27. Trapezoid
28. Square

29. The measures of the other two angles of the triangle are 138° and 25° (by vertical angles). Therefore, the measure of angle x is 180° - 138° - 25° = 17°.

30. The measure of one angle of the triangle is 75° (by vertical angles). The measure of another angle of the triangle is 180° - 133° = 47°. The measure of the third angle of the triangle is 180° - 75° - 47° = 58°. Since angle x is a vertical angle with the 58° angle, the measure of angle x is 58°.

31. The measure of one angle of the triangle is 27° (by vertical angles). The measure of another angle of the triangle is 180° - 57° = 123°. The measure of the third angle of the triangle is 180° - 27° - 123° = 30°. The measure of angle x is 180° - 30° = 150° (The 30° angle and angle x form a straight angle.).

32. The given measure of one angle of the triangle is 35°. The measure of another angle of the triangle is 30° (by vertical angles). The measure of the third angle of the triangle is 180° - 35° - 30° = 115°. The measure of angle x is 180° - 115° = 65° (The 115° angle and angle x form a straight angle.).

33.

Angle	Measure	Reason
1	50°	∡ 1 and ∡ 5 are vertical angles
2	63°	Vertical angle with the given 63° angle
3	67°	∡ 1, ∡ 2, and ∡ 3 form a straight angle
4	67°	∡ 3 and ∡ 4 are vertical angles
5	50°	∡ 5 and ∡ 12 are corresponding angles
6	113°	∡ 6 and the given 67° angle form a straight angle
7	50°	The sum of the measures of the interior angles of a triangle is 180°
8	130°	∡ 8 and ∡ 12 form a straight angle
9	67°	∡ 4 and ∡ 9 are corresponding angles
10	113°	∡ 6 and ∡ 10 are vertical angles
11	130°	∡ 8 and ∡ 11 are vertical angles
12	50°	∡ 7 and ∡ 12 are vertical angles

34.

Angle	Measure	Reason
1	90°	∡ 1 and ∡ 7 are vertical angles
2	50°	∡ 2 and ∡ 4 are corresponding angles
3	130°	∡ 3 and ∡ 4 form a straight angle
4	50°	Vertical angle with the given 50° angle
5	50°	∡ 2 and ∡ 5 are vertical angles
6	40°	Vertical angle with the given 40° angle
7	90°	∡ 2, ∡ 6, and ∡ 7 form a straight angle
8	130°	∡ 3 and ∡ 8 are vertical angles
9	140°	∡ 9 and ∡ 10 form a straight angle
10	40°	∡ 10 and ∡ 12 are vertical angles
11	140°	∡ 9 and ∡ 11 are vertical angles
12	40°	∡ 6 and ∡ 12 are corresponding angles

35. $n = 5$
    $(5 - 2) \times 180° = 3 \times 180° = 540°$

36. $n = 9$
    $(9 - 2) \times 180° = 7 \times 180° = 1260°$

37. $n = 6$

    $(6 - 2) \times 180° = 4 \times 180° = 720°$

38. $n = 10$

    $(10 - 2) \times 180° = 8 \times 180° = 1440°$

39. $n = 20$

    $(20 - 2) \times 180° = 18 \times 180° = 3240°$

40. $n = 12$

    $(12 - 2) \times 180° = 10 \times 180° = 1800°$

41. a) The sum of the measures of the interior angles of a triangle is 180°. Dividing by 3, the number of angles, each interior angle measures 60°.

    b) Each exterior angle measures 180° - 60° = 120°.

42. a) The sum of the measures of the interior angles of a quadrilateral is $(4 - 2) \times 180° = 2 \times 180° = 360°$. Dividing by 4, the number of angles, each interior angle measures 90°.

    b) Each exterior angle measures 180° - 90° = 90°.

43. a) The sum of the measures of the interior angles of an octagon is $(8 - 2) \times 180° = 6 \times 180° = 1080°$. Dividing by 8, the number of angles, each interior angle measures 135°.

    b) Each exterior angle measures 180° - 135° = 45°.

44. a) The sum of the measures of the interior angles of a nonagon is $(9 - 2) \times 180° = 7 \times 180° = 1260°$. Dividing by 9, the number of angles, each interior angle measures 140°.

    b) Each exterior angle measures 180° - 140° = 40°.

45. a) The sum of the measures of the interior angles of a dodecagon is $(12 - 2) \times 180° = 10 \times 180° = 1800°$. Dividing by 12, the number of angles, each interior angle measures 150°.

    b) Each exterior angle measures 180° - 150° = 30°.

46. a) The sum of the measures of the interior angles of an icosagon is $(20 - 2) \times 180° = 18 \times 180° = 3240°$. Dividing by 20, the number of angles, each interior angle measures 162°.

    b) Each exterior angle measures 180° - 162° = 18°.

47. Let $x = BC$

$$\frac{BC}{B'C'} = \frac{AB}{A'B'}$$

$$\frac{x}{2.4} = \frac{10}{4}$$

$$4x = 24$$

$$x = 6$$

Let $y = A'C'$

$$\frac{A'C'}{AC} = \frac{A'B'}{AB}$$

$$\frac{y}{8} = \frac{4}{10}$$

$$10y = 32$$

$$y = \frac{32}{10} = \frac{16}{5}$$

48. Let $x = A'C'$

$$\frac{A'C'}{AC} = \frac{A'B'}{AB}$$

$$\frac{x}{10} = \frac{2}{5}$$

$$5x = 20$$

$$x = 4$$

Let $y = B'C'$

$$\frac{B'C'}{BC} = \frac{A'B'}{AB}$$

$$\frac{y}{8} = \frac{2}{5}$$

$$5y = 16$$

$$y = \frac{16}{5}$$

49. Let $x = DC$

$$\frac{DC}{D'C'} = \frac{AB}{A'B'}$$

$$\frac{x}{6} = \frac{4}{10}$$

$$10x = 24$$

$$x = \frac{24}{10} = \frac{12}{5}$$

Let $y = B'C'$

$$\frac{B'C'}{BC} = \frac{A'B'}{AB}$$

$$\frac{y}{3} = \frac{10}{4}$$

$$4y = 30$$

$$y = \frac{30}{4} = \frac{15}{2}$$

50. Let $x = AB$

$$\frac{AB}{A'B'} = \frac{AD}{A'D'}$$

$$\frac{x}{5} = \frac{5}{12}$$

$$12x = 25$$

$$x = \frac{25}{12}$$

Let $y = C'D'$

$$\frac{C'D'}{CD} = \frac{A'D'}{AD}$$

$$\frac{y}{1} = \frac{12}{5}$$

$$5y = 12$$

$$y = \frac{12}{5}$$

51. Let $x = AC$

$$\frac{AC}{A'C'} = \frac{BC}{B'C'}$$

$$\frac{x}{0.75} = \frac{2}{1.25}$$

$$1.25x = 1.5$$

$$x = 1.2$$

Let $y = A'B'$

$$\frac{A'B'}{AB} = \frac{B'C'}{BC}$$

$$\frac{y}{1} = \frac{1.25}{2}$$

$$2y = 1.25$$

$$y = 0.625$$

52. Let $x = BC$

$$\frac{BC}{B'C'} = \frac{DC}{D'C'}$$

$$\frac{x}{0.875} = \frac{1}{1.75}$$

$$1.75x = 0.875$$

$$x = 0.5$$

Let $y = DE$

$$\frac{DE}{DC} = \frac{D'E'}{D'C'}$$

$$\frac{y}{1} = \frac{0.7}{1.75}$$

$$1.75y = 0.7$$

$$y = 0.4$$

53. Let $x = BC$

$$\frac{BC}{EC} = \frac{AB}{DE}$$

$$\frac{x}{2} = \frac{6}{2}$$

$$2x = 12$$

$$x = 6$$

54. Let $x = DC$

$$\frac{DC}{AC} = \frac{DE}{AB}$$

$$\frac{x}{10} = \frac{2}{6}$$

$$6x = 20$$

$$x = \frac{20}{6} = \frac{10}{3}$$

55. $AD = AC - DC = 10 - \dfrac{10}{3} = \dfrac{30}{3} - \dfrac{10}{3} = \dfrac{20}{3}$

56. $BE = BC - EC = 6 - 2 = 4$

57. $A'B' = AB = 14$

58. $B'C' = BC = 30$

59. $AC = A'C' = 28$

60. $m\angle B'A'C' = m\angle BAC = 84°$

61. $m\angle ACB = m\angle A'C'B' = 28°$

62. $m\angle ABC = m\angle A'B'C' = 180° - 84° - 28° = 68°$

63. $A'B' = AB = 8$

64. $AD = A'D' = 6$

65. $B'C' = BC = 16$

66. $m\angle BCD = m\angle B'C'D' = 50°$

67. $m\angle A'D'C' = m\angle ADC = 70°$

68. $m\angle DAB = m\angle D'A'B'$
$= 360° - 130° - 70° - 50° = 110°$

69. $180° - 125° = 55°$

70. $55°$

71. $180° - 90° - 55° = 35°$

72. $90° + 35° = 125°$

73. Let x = height of silo
$$\dfrac{x}{6} = \dfrac{105}{9}$$
$$9x = 630$$
$$x = 70 \text{ ft}$$

74. $m\angle BAC + m\angle BCA + 80° = 180°$
$m\angle BAC + m\angle BCA = 100°$
$m\angle BAC = m\angle BCA$
$m\angle BAC = 50°, m\angle BCA = 50°$
$m\angle x = 50°$ since $\angle x$ and $\angle BAC$ are
alternate interior angles.
The measure of the angle adjacent to $\angle y$
is $180° - 50° - 80° = 50°$.
$m\angle y = 180° - 50° = 130°$

75. a) $197 \text{ mi} = (197 \text{ mi}) \left(\dfrac{5280 \text{ ft}}{1 \text{ mi}} \right) \left(\dfrac{12 \text{ in.}}{1 \text{ ft}} \right)$

$= 12,481,920 \text{ in.}$

Let x = the actual distance from Dallas to Houston

$$\dfrac{x}{3.75} = \dfrac{12,481,920}{3}$$
$$3x = 46,807,200$$
$$x = 15,602,400 \text{ in.}$$

$15,602,400 \text{ in.} = (15,602,400 \text{ in.}) \left(\dfrac{1 \text{ ft}}{12 \text{ in.}} \right) \left(\dfrac{1 \text{ mi}}{5280 \text{ ft}} \right)$

$= 246.25 \text{ mi}$

b) Let x = the actual distance from Dallas to San Antonio

$$\dfrac{x}{4.125} = \dfrac{12,481,920}{3}$$
$$3x = 51,487,920$$
$$x = 17,162,640 \text{ in.}$$

$17,162,640 \text{ in.} = (17,162,640 \text{ in.}) \left(\dfrac{1 \text{ ft}}{12 \text{ in.}} \right) \left(\dfrac{1 \text{ mi}}{5280 \text{ ft}} \right)$

$= 270.875 \text{ mi}$

76. a) $44 \text{ mi} = (44 \text{ mi}) \left(\dfrac{5280 \text{ ft}}{1 \text{ mi}} \right) \left(\dfrac{12 \text{ in.}}{1 \text{ ft}} \right)$

$= 2,787,840 \text{ in.}$

Let x = the actual distance from St. Paul to Austin

$$\dfrac{x}{2.25} = \dfrac{2,787,840}{0.875}$$
$$0.875x = 6,272,640$$
$$x = 7,168,731.429 \text{ in.}$$

$7,168,731.429 \text{ in.}$

$= (7,168,731.429 \text{ in.}) \left(\dfrac{1 \text{ ft}}{12 \text{ in.}} \right) \left(\dfrac{1 \text{ mi}}{5280 \text{ ft}} \right)$

$= 113.1428571 \approx 113.14 \text{ mi}$

b) Let x = the actual distance from St. Paul to Rochester

$$\dfrac{x}{1.5} = \dfrac{2,787,840}{0.875}$$
$$0.875x = 4,181,760$$
$$x = 4,779,154.286 \text{ in.}$$

$4,779,154.286 \text{ in.}$

$= (4,779,154.286 \text{ in.}) \left(\dfrac{1 \text{ ft}}{12 \text{ in.}} \right) \left(\dfrac{1 \text{ mi}}{5280 \text{ ft}} \right)$

$= 75.42857143 \approx 75.43 \text{ mi}$

77.

$$\dfrac{DE}{D'E'} = 3$$

$$\dfrac{12}{D'E'} = 3$$

$$3D'E' = 12$$

$$\overline{D'E'} = 4$$

$$\dfrac{EF}{E'F'} = 3$$

$$\dfrac{15}{E'F'} = 3$$

$$3E'F' = 15$$

$$\overline{E'F'} = 5$$

$$\dfrac{DF}{D'F'} = 3$$

$$\dfrac{9}{D'F'} = 3$$

$$3D'F' = 9$$

$$\overline{D'F'} = 3$$

78.

$$\dfrac{E'F'}{EF} = \dfrac{1}{3}$$

$$\dfrac{E'F'}{21} = \dfrac{1}{3}$$

$$3E'F' = 21$$

$$\overline{E'F'} = 7$$

$$\dfrac{F'G'}{FG} = \dfrac{1}{3}$$

$$\dfrac{F'G'}{9} = \dfrac{1}{3}$$

$$3F'G' = 9$$

$$\overline{F'G'} = 3$$

$$\dfrac{G'H'}{GH} = \dfrac{1}{3}$$

$$\dfrac{G'H'}{9} = \dfrac{1}{3}$$

$$3G'H' = 9$$

$$\overline{G'H'} = 3$$

$$\dfrac{E'H'}{EH} = \dfrac{1}{3}$$

$$\dfrac{E'H'}{12} = \dfrac{1}{3}$$

$$3E'H' = 12$$

$$\overline{E'H'} = 4$$

79. a) $m\angle HMF = m\angle TMB$, $m\angle HFM = m\angle TBM$, $m\angle MHF = m\angle MTB$

b) Let $x = $ height of the wall

$$\dfrac{x}{20} = \dfrac{5.5}{2.5}$$

$$2.5x = 110$$

$$x = \dfrac{110}{2.5} = 44 \text{ ft}$$

80. a) $m\angle CED = m\angle ABC$; $m\angle ACB = m\angle DCE$ (vertical angles); $m\angle BAC = m\angle CDE$ (alternate interior angles)

b) Let $x = DE$

$$\dfrac{x}{AB} = \dfrac{CE}{BC}$$

$$\dfrac{x}{543} = \dfrac{1404}{356}$$

$$356x = 762,372$$

$$x = 2141.494382 \approx 2141.49 \text{ ft}$$

Exercise Set 9.3

Throughout this section, on exercises involving π, we used the π key on a scientific calculator to determine the answer. If you use 3.14 for π, your answers may vary slightly.

1. a) The **perimeter** of a two-dimensional figure is the sum of the lengths of the sides of the figure.

b) The **area** of a two-dimensional figure is the region within the boundaries of the figure.

c)

$A = lw = 6(2) = 12$ square units

$P = 2l + 2w = 2(6) + 2(2) = 12 + 4 = 16$ units

2. The **radius** of a circle is half the **diameter** or the **diameter** of a circle is twice the **radius**.

3. a) To determine the number of square inches, multiply the number of square feet by $12 \times 12 = 144$.

   b) To determine the number of square feet, divide the number of square inches by $12 \times 12 = 144$.

4. a) To determine the number of square feet, multiply the number of square yards by $3 \times 3 = 9$.

   b) To determine the number of square yards, divide the number of square feet by $3 \times 3 = 9$.

5. $A = \dfrac{1}{2}bh = \dfrac{1}{2}(10)(7) = 35$ in.^{2}

6. $3 \text{ yd} = 3(3) = 9$ ft

   $A = \dfrac{1}{2}bh = \dfrac{1}{2}(1)(9) = 4.5 \text{ ft}^2 = \dfrac{4.5}{9} = 0.5 \text{ yd}^2$

7. $A = \dfrac{1}{2}bh = \dfrac{1}{2}(7)(5) = 17.5 \text{ cm}^2$

8. $A = \dfrac{1}{2}bh = \dfrac{1}{2}(2)\left(\sqrt{3}\right) = \sqrt{3} \text{ m}^2$

9. $A = lw = (15)(7) = 105 \text{ ft}^2$

   $P = 2l + 2w = 2(15) + 2(7) = 44$ ft

10. $A = bh = (7)(5) = 35$ in.^{2}

    $P = 2b + 2w = 2(7) + 2(6) = 26$ in.

11. $3 \text{ m} = 3(100) = 300$ cm

    $A = bh = 300(20) = 6000 \text{ cm}^2$

    $P = 2b + 2w = 2(300) + 2(27) = 654$ cm

12. $2 \text{ yd} = 2(3) = 6$ ft

    $A = s^2 = (6)^2 = 36 \text{ ft}^2$

    $P = 4s = 4(6) = 24$ ft

13. $2 \text{ ft} = 2(12) = 24$ in.

    $A = \dfrac{1}{2}h(b_1 + b_2) = \dfrac{1}{2}(24)(5 + 19)$

    $= \dfrac{1}{2}(24)(24) = 288$ in.^{2}

    $P = s_1 + s_2 + b_1 + b_2 = 25 + 25 + 5 + 19 = 74$ in.

14. $A = \dfrac{1}{2}h(b_1 + b_2) = \dfrac{1}{2}(12)(6 + 16)$

    $= \dfrac{1}{2}(12)(22) = 132$ in.^{2}

    $P = s_1 + s_2 + b_1 + b_2 = 13 + 13 + 6 + 16 = 48$ in.

15. $A = \pi r^2 = \pi(7)^2 = 49\pi = 153.93804 \approx 153.94$ in.^{2}

    $C = 2\pi r = 2\pi(7) = 14\pi = 43.98229715 \approx 43.98$ in.

16. $r = \dfrac{100}{2} = 50$ cm

    $A = \pi r^2 = \pi(50)^2 = 2500\pi = 7853.981634$

    $\approx 7853.98 \text{ cm}^2$

    $C = 2\pi r = 2\pi(50) = 100\pi = 314.1592654$

    ≈ 314.16 cm

17. $r = \dfrac{9}{2} = 4.5$ ft

    $A = \pi r^2 = \pi(4.5)^2 = 20.25\pi = 63.61725124$

    $\approx 63.62 \text{ ft}^2$

    $C = 2\pi r = 2\pi(4.5) = 9\pi = 28.27433388$

    ≈ 28.27 ft

18. $A = \pi r^2 = \pi(13)^2 = 169\pi = 530.9291585$

    $\approx 530.93 \text{ mm}^2$

    $C = 2\pi r = 2\pi(13) = 26\pi = 81.68140899$

    ≈ 81.68 mm

19. a) $a^2 + 12^2 = 15^2$

    $a^2 + 144 = 225$

    $a^2 = 81$

    $a = \sqrt{81} = 9$ in.

    b) $P = s_1 + s_2 + s_3 = 9 + 12 + 15 = 36$ in.

    c) $A = \dfrac{1}{2}bh = \dfrac{1}{2}(9)(12) = 54$ in.^{2}

20. a) $c^2 = 5^2 + 12^2$

    $c^2 = 25 + 144$

    $c^2 = 169$

    $c = \sqrt{169} = 13$ ft

    b) $P = s_1 + s_2 + s_3 = 5 + 12 + 13 = 30$ ft

    c) $A = \dfrac{1}{2}bh = \dfrac{1}{2}(5)(12) = 30 \text{ ft}^2$

21. a) $c^2 = 10^2 + 24^2$
    $c^2 = 100 + 576$
    $c^2 = 676$
    $c = \sqrt{676} = 26$ cm

    b) $P = s_1 + s_2 + s_3 = 10 + 24 + 26 = 60$ cm

    c) $A = \frac{1}{2}bh = \frac{1}{2}(10)(24) = 120$ cm^2

22. a) $b^2 + 15^2 = 39^2$
    $b^2 + 225 = 1521$
    $b^2 = 1296$
    $c = \sqrt{1296} = 36$ m

    b) $P = s_1 + s_2 + s_3 = 15 + 36 + 39 = 90$ m

    c) $A = \frac{1}{2}bh = \frac{1}{2}(36)(15) = 270$ m^2

23. Area of larger circle:
    $\pi(4)^2 = 16\pi = 50.265\ 482\ 46$ cm^2
    Area of smaller circle:
    $\pi(3)^2 = 9\pi = 28.274\ 333\ 88$ cm^2
    Shaded area:
    $50.26548246 - 28.27433388 = 21.99114858$
    ≈ 21.99 cm^2

24. Area of square: $(10)^2 = 100$ m^2
    Area of circle: $\pi(5)^2 = 25\pi = 78.539\ 816\ 34$ m^2
    Shaded area:
    $100 - 78.53981634 = 21.46018366 \approx 21.46$ m^2

25. Use the Pythagorean Theorem to find the length of a side of the shaded square.
    $x^2 = 2^2 + 2^2$
    $x^2 = 4 + 4$
    $x^2 = 8$
    $x = \sqrt{8}$
    Shaded area: $\sqrt{8}\left(\sqrt{8}\right) = 8$ in.^{2}

26. Area of rectangle: $7(4) = 28$ ft^2
    Area of trapezoid: $\frac{1}{2}(4)(3+7) = \frac{1}{2}(4)(10) = 20$ ft^2
    Shaded area: $28 - 20 = 8$ ft^2

27. Area of trapezoid:
    $\frac{1}{2}(8)(9+20) = \frac{1}{2}(8)(29) = 116$ in.^{2}
    Area of circle: $\pi(4)^2 = 16\pi = 50.26548246$ in.^{2}
    Shaded area:
    $116 - 50.26548246 = 65.73451754 \approx 65.73$ in.^{2}

28. Area of circle: $\pi(5)^2 = 25\pi = 78.539\ 816\ 34$ m^2
    Area of rectangle: $8(6) = 48$ m^2
    Shaded area:
    $78.53981634 - 48 = 30.53981634 \approx 30.54$ m^2

29. Area of small rectangle on the right side:
    $12(6) = 72$ ft^2
    Area of semi-circle on the right side:
    $\frac{1}{2}\pi(6)^2 = 18\pi = 56.54866776$ ft^2
    Area of shaded region on the right side:
    $72 - 56.54866776 = 15.45133224$ ft^2
    Area of shaded region on the left side:
    15.45133224 ft^2
    Area of triangle: $\frac{1}{2}(14)(12) = 84$ ft^2
    Shaded area:
    $15.45133224 + 15.45133224 + 84$
    $= 114.9026645 \approx 114.90$ ft^2

30. Radius of larger circle: $\frac{28}{2} = 14$ cm
    Area of large circle:
    $\pi(14)^2 = 196\pi = 615.7521601$ cm^2
    Radius of each smaller circle: $\frac{14}{2} = 7$ cm
    Area of each smaller circle:
    $\pi(7)^2 = 49\pi = 153.93804$ cm^2
    Shaded area:
    $615.7521601 - 153.93804 - 153.93804$
    $= 307.8760801 \approx 307.88$ cm^2

31. Length of rectangle: $3(8) = 24$ in.

    Area of rectangle: $24(8) = 192$ in.^{2}

    Radius of each circle: $\dfrac{8}{2} = 4$ in.

    Area of each circle: $\pi(4)^2 = 16\pi = 50.26548246$

    Shaded area:

    $192 - 50.26548246 - 50.26548246 - 50.26548246$

    $= 41.20355262 \approx 41.20$ in.^{2}

32. Area of each outer rectangle: $2(4) = 8$ cm^2

    Area of four outer rectangles: $4(8) = 32$ cm^2

    Area of inner square: $4(4) = 16$ cm^2

    Radius of circle: $\dfrac{4}{2} = 2$ cm

    Area of circle: $\pi(2)^2 = 4\pi = 12.56637061$

    Shaded area: $32 + 16 - 12.56637061$

    $= 35.43362939 \approx 35.43$ cm^2

33. $\dfrac{1}{x} = \dfrac{9}{107}$

    $9x = 107$

    $x = \dfrac{107}{9} = 11.\overline{8} \approx 11.89$ yd^2

34. $\dfrac{1}{x} = \dfrac{9}{15.2}$

    $9x = 15.2$

    $x = \dfrac{15.2}{9} = 1.6\overline{8} \approx 1.69$ yd^2

35. $\dfrac{1}{14.7} = \dfrac{9}{x}$

    $x = 14.7(9) = 132.3$ ft^2

36. $\dfrac{1}{18.3} = \dfrac{9}{x}$

    $x = 18.3(9) = 164.7$ ft^2

37. $\dfrac{1}{23.4} = \dfrac{10,000}{x}$

    $x = 23.4(10,000) = 234,000$ cm^2

38. $\dfrac{1}{14.7} = \dfrac{10,000}{x}$

    $x = 14.7(10,000) = 147,000$ cm^2

39. $\dfrac{1}{x} = \dfrac{10,000}{1075}$

    $10,000x = 1075$

    $x = \dfrac{1075}{10,000} = 0.1075$ m^2

40. $\dfrac{1}{x} = \dfrac{10,000}{608}$

    $10,000x = 608$

    $x = \dfrac{608}{10,000} = 0.0608$ m^2

41. Area of living/dining room: $25(22) = 550$ ft^2

    a) $550(5.89) = \$3239.50$

    b) $550(8.89) = \$4889.50$

42. Area of living/dining room: $25(22) = 550$ ft^2

    a) $550(10.86) = \$5973$

    b) $550(13.86) = \$7623$

43. Area of kitchen: $12(14) = 168$ ft^2

    Area of first floor bathroom: $6(10) = 60$ ft^2

    Area of second floor bathroom: $8(14) = 112$ ft^2

    Total area: $168 + 60 + 112 = 340$ ft^2

    Cost: $340(\$5) = \$1700$

44. Area of kitchen and both bathrooms: 340 ft^2
    (See Exercise 43.)

    Cost: $340(\$8.50) = \$2890$

45. Area of bedroom 1: $10(14) = 140$ ft^2

    Area of bedroom 2: $10(20) = 200$ ft^2

    Area of bedroom 3: $10(14) = 140$ ft^2

    Total area: $140 + 200 + 140 = 480$ ft^2

    Cost: $480(\$6.06) = \$2908.80$

46. Area of all three bedrooms: 480 ft^2
    (See Exercise 45.)

    Cost: $480(\$5.56) = \$2668.80$

47. Area of entire lawn if all grass:
$200(100) = 20,000 \text{ ft}^2$

Area of patio: $40(10) = 400 \text{ ft}^2$

Area of shed: $10(8) = 80 \text{ ft}^2$

Area of house: $50(25) = 1250 \text{ ft}^2$

Area of drive: $30(10) = 300 \text{ ft}^2$

Area of pool: $\pi(12)^2 = 144\pi = 452.3893421 \text{ ft}^2$

Area of lawn:
$20,000 - 400 - 80 - 1250 - 300 - 452.3893421$

$= 17,517.61066 \text{ ft}^2 = \dfrac{17,517.61066}{9}$

$= 1946.401184 \text{ yd}^2$

Cost:
$1946.401184(\$0.02) = \$38.92802368 \approx \$38.93$

48. Area of entire lawn if all grass:
$400(300) = 120,000 \text{ ft}^2$

Area of house: $\dfrac{1}{2}(50)(100 + 150) = 6250 \text{ ft}^2$

Area of goldfish pond:
$\pi(20)^2 = 400\pi = 1256.637061 \text{ ft}^2$

Area of privacy hedge: $200(20) = 4000 \text{ ft}^2$

Area of garage: $70(30) = 2100 \text{ ft}^2$

Area of driveway: $40(25) = 1000 \text{ ft}^2$

Area of lawn:
$120,000 - 6250 - 1256.637061 - 4000 - 2100 - 1000$

$= 105,393.3629 \text{ ft}^2 = \dfrac{105,393.3629}{9}$

$= 11,710.37366 \text{ yd}^2$

Cost:
$11,710.37366(\$0.02) = \$234.2074732 \approx \$234.21$

49. a) $A = 11.5(15.4) = 177.1 \text{ m}^2$

b) $\dfrac{1}{x} = \dfrac{10,000}{177.1}$

$10,000x = 177.1$

$x = \dfrac{177.1}{10,000} = 0.01771 \text{ hectare}$

50. Wendy's hamburger: $A = 3(3) = 9 \text{ in.}^2$

Burger King hamburger:

$A = \pi\left(\dfrac{3.5}{2}\right)^2 = \pi(1.75)^2 = 3.0625\pi$

$= 9.621127502 \approx 9.62 \text{ in.}^2$

Burger King's hamburger is larger by
$\approx 9.62 - 9 = 0.62 \text{ in.}^2$

51. Let c = length of guy wire
$90^2 + 52^2 = c^2$

$8100 + 2704 = c^2$

$10,804 = c^2$

$c = \sqrt{10,804} = 103.9422917 \approx 103.94 \text{ ft}$

52. Let a = height on the wall that the ladder reaches
$a^2 + 20^2 = 29^2$

$a^2 + 400 = 841$

$a^2 = 441$

$a = \sqrt{441} = 21 \text{ ft}$

53. Let a = horizontal distance from dock to boat
$a^2 + 9^2 = 41^2$

$a^2 + 81 = 1681$

$a^2 = 1600$

$a = \sqrt{1600} = 40 \text{ ft}$

54.

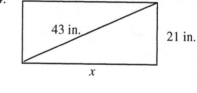

$x^2 + 21^2 = 43^2$

$x^2 + 441 = 1849$

$x^2 = 1408$

$x = \sqrt{1408} = 37.52332608 \approx 37.52 \text{ in.}$

55. a) $A = s^2$

    b) $A = (2s)^2 = 4s^2$

    c) The area of the square in part b) is four times larger than the area of the square in part a).

57. $s = \dfrac{1}{2}(a+b+c) = \dfrac{1}{2}(8+6+10) = 12$

    $A = \sqrt{12(12-8)(12-6)(12-10)}$

    $\quad = \sqrt{12(4)(6)(2)} = \sqrt{576} = 24 \text{ cm}^2$

59. Answers will vary.

56. a) $A = bh$

    b) $A = 2b(2h) = 4bh$

    c) The area of the parallelogram in part b) is four times larger than the area of the parallelogram in part a).

58. a) $A = a^2$

    b) $A = ab$

    c) $A = ab$

    d) $A = b^2$

    e) $(a+b)^2 = a^2 + ab + ab + b^2 = a^2 + 2ab + b^2$

Exercise Set 9.4

In this section, we use the π key on the calculator to determine answers in calculations involving π.
If you use 3.14 for π, your answers may vary slightly.

1. **Volume** is a measure of the capacity of a figure.
2. **Solid geometry** is the study of three-dimensional solid figures.
3. A **polyhedron** is a closed surface formed by the union of polygonal regions.
   A **regular polyhedron** is one whose faces are all regular polygons of the same size and shape.
4. A **prism** is a polyhedron whose bases are congruent polygons and whose sides are parallelograms.
   A **right prism** is one in which all of the lateral faces are rectangles.
5. A **prism** and a **pyramid** are both polyhedrons, but a prism has a top and a bottom base while a pyramid only has one base.
6. For any polyhedron, the number of vertices minus the number of edges plus the number of faces equals two.

7. $V = s^3 = (3)^3 = 27 \text{ ft}^3$

8. $V = lwh = 8(3)(3) = 72 \text{ ft}^3$

9. $1 \text{ ft} = 12 \text{ in.}$

   $V = \pi r^2 h = \pi(2)^2(12) = 48\pi$

   $\quad = 150.7964474 \approx 150.80 \text{ in.}^3$

10. $2 \text{ ft} = 2(12) = 24 \text{ in.}$

    $V = \pi r^2 h = \pi(6)^2(24) = 864\pi$

    $\quad = 2714.336053 \approx 2714.34 \text{ in.}^3$

11. $V = \dfrac{1}{3}\pi r^2 h = \dfrac{1}{3}\pi(3)^2(14) = 42\pi$

    $\quad = 131.9468915 \approx 131.95 \text{ cm}^3$

12. $r = \dfrac{10}{2} = 5 \text{ ft}$

    $V = \dfrac{1}{3}\pi r^2 h = \dfrac{1}{3}\pi(5)^2(24) = 200\pi$

    $\quad = 628.3185307 \approx 628.32 \text{ ft}^3$

13. Area of the base:

    $B = \dfrac{1}{2}h(b_1 + b_2) = \dfrac{1}{2}(10)(8+12) = 100 \text{ in.}^2$

    $V = Bh = 100(24) = 2400 \text{ in.}^3$

14. Area of the base: $B = \dfrac{1}{2}bh = \dfrac{1}{2}(8)(8) = 32 \text{ in.}^2$

    $V = Bh = 32(12) = 384 \text{ in.}^3$

15. $r = \dfrac{9}{2} = 4.5$ cm

$V = \dfrac{4}{3}\pi r^3 = \dfrac{4}{3}\pi (4.5)^3 = 121.5\pi$

$= 381.7035074 \approx 381.70$ cm^3

16. $V = \dfrac{4}{3}\pi r^3 = \dfrac{4}{3}\pi (7)^3 = 457.\overline{3}\pi$

$= 1436.75504 \approx 1436.76$ cm^3

17. Area of the base: $B = s^2 = (11)^2 = 121$ cm^2

$V = \dfrac{1}{3}Bh = \dfrac{1}{3}(121)(13) = 524.\overline{3} \approx 524.33$ cm^3

18. Area of the base: $B = \dfrac{1}{2}bh = \dfrac{1}{2}(9)(15) = 67.5$ ft^2

$V = \dfrac{1}{3}Bh = \dfrac{1}{3}(67.5)(13) = 292.5$ ft^3

19. Area of the base:

$B = \dfrac{1}{2}h(b_1 + b_2) = \dfrac{1}{2}(5)(7+9) = 40$ in.^{2}

$V = \dfrac{1}{3}Bh = \dfrac{1}{3}(40)(8) = 106.\overline{6} \approx 106.67$ in.^{3}

20. Area of the base: $B = lw = 18(15) = 270$ in.^{2}

$V = \dfrac{1}{3}Bh = \dfrac{1}{3}(270)(10) = 900$ in.^{3}

21. V = volume of rect. solid - volume of cylinder

$= 6(4)(3) - \pi(1)^2(4) = 72 - 4\pi$

$= 72 - 12.56637061$

$= 59.43362939 \approx 59.43$ m^3

22. V = volume of cylinder - volume of cone

$= \pi(2)^2(9) - \dfrac{1}{3}\pi(2)^2(9) = 36\pi - 12\pi = 24\pi$

$= 75.39822369 \approx 75.40$ cm^3

23. V = volume of rect. solid - volume of sphere

$= 4(4)(4) - \dfrac{4}{3}\pi(2)^3 = 64 - 33.51032164$

$= 30.48967836 \approx 30.49$ ft^3

24. V = vol. of large sphere - vol. of small sphere

$= \dfrac{4}{3}\pi(6)^3 - \dfrac{4}{3}\pi(3)^3 = 288\pi - 36\pi = 252\pi$

$= 791.6813487 \approx 791.68$ cm^3

25. V = vol. of large cylinder - vol. of small cylinder

$= \pi(1.5)^2(5) - \pi(0.5)^2(5) = 11.25\pi - 1.25\pi = 10\pi$

$= 31.41592654 \approx 31.42$ m^3

26. V = volume of cylinder - volume of 3 spheres

$= \pi(3.5)^2(20.8) - 3\left[\dfrac{4}{3}\pi(3.45)^3\right]$

$= 254.8\pi - 164.2545\pi = 90.5455\pi$

$= 284.4570776 \approx 284.46$ cm^3

27. V = volume of rect. solid - volume of pyramid

$= 3(3)(4) - \dfrac{1}{3}(3)^2(4) = 36 - 12 = 24$ ft^3

28. V = volume of prism $-$ volume of rectangular solid

$= \dfrac{1}{2}(6)(8)(11) - 3(4)(11)$

$= 264 - 132 = 132$ in.^{3}

29. 7 yd$^3 = 7(27) = 189$ ft^3

30. 3.8 yd$^3 = 3.8(27) = 102.6$ ft^3

31. 153 ft$^3 = \dfrac{153}{27} = 5.\overline{6} \approx 5.67$ yd^3

32. 2457 ft$^3 = \dfrac{2457}{27} = 91$ yd^3

33. 5.9 m$^3 = 5.9(1,000,000) = 5,900,000$ cm^3

34. 17.6 m$^3 = 17.6(1,000,000) = 17,600,000$ cm^3

35. $3,000,000$ cm$^3 = \dfrac{3,000,000}{1,000,000} = 3$ m^3

36. $7,300,000$ cm$^3 = \dfrac{7,300,000}{1,000,000} = 7.3$ m^3

37. a) $V = 46(25)(25) = 28,750$ in.^{3}

b) $(1\text{ ft})^3 = (12\text{ in.})(12\text{ in.})(12\text{ in.}) = 1728$ in.^{3}

$28,750$ in.$^3 = \dfrac{28,750}{1728} = 16.63773148 \approx 16.64$ ft^3

38. Tubs: $V = \pi r^2 h = \pi(3)^2(5) = 45\pi$

$= 141.3716694 \approx 141.37$ in.^{3}

Boxes: $V = s^3 = (5)^3 = 125$ in.^{3}

39. $V = 12(4)(3) = 144$ in.^{3}

    144 in.^{3} = 144(0.01736) = 2.49984 ≈ 2.50 qt

40. Wendy's Volume: $4(4)\left(\dfrac{3}{16}\right) = 3$ in.^{3}

    Magic Burger's Volume:

$$\pi\left(\dfrac{4.5}{2}\right)^2 (0.25) = \pi(2.25)^2(0.25)$$

$$= 3.976078202 ≈ 3.98 \text{ in.}^3$$

    The Magic Burger has the greater volume
by ≈ 0.98 in.^{3}

41. a) Cylinder 1:

$$V = \pi\left(\dfrac{10}{2}\right)^2 (12) = 300\pi = 942.4777961 ≈ 942.48 \text{ in.}^3$$

    Cylinder 2:

$$V = \pi\left(\dfrac{12}{2}\right)^2 (10) = 360\pi$$

$$= 1130.973355 ≈ 1130.97 \text{ in.}^3$$

    The container with the larger diameter holds
more.

    b) $1130.97 - 942.48 = 188.49 ≈ 188.50$ in.^{3}

42. a) $V = 15(9)(2) = 270$ m^3

    b) 270 kl

43. $V = \dfrac{1}{3}Bh = \dfrac{1}{3}(720)^2(480) = 82{,}944{,}000$ ft^3

44. a) $V = 80(50)(30) = 120{,}000$ cm^3

    b) 120,000 ml

    c) $120{,}000 \text{ ml} = \dfrac{120{,}000}{1000} = 120$ l

45. $r = \dfrac{3.875}{2} = 1.9375$ in.

    Volume of each cylinder:

    $\pi r^2 h = \pi(1.9375)^2(3)$

    $= 11.26171875\pi = 35.37973289$

    Total volume:

    $8(35.37973289) = 283.0378631 ≈ 283.04$ in.^{3}

46. a) $4 \text{ in.} = \dfrac{4}{12} = \dfrac{1}{3}$ ft

    $V = lwh = 9(18)\left(\dfrac{1}{3}\right) = 54$ ft^3

    $\dfrac{54}{27} = 2$ yd^3

    b) $2(\$32.95) = \$65.90$

47. a) $5.5 \text{ ft} = 5.5(12) = 66$ in.

    $r = \dfrac{2.5}{2} = 1.25$ in.

    $V = \pi r^2 h = \pi(1.25)^2(66) = 103.125\pi$

    $= 323.9767424 ≈ 323.98$ in.^{3}

    b) $\dfrac{323.98}{1728} = 0.187488425 ≈ 0.19$ ft^3

48. a) Volume of water needed to fill the pool to a height
of $\dfrac{1}{2}$ ft: $\pi r^2 h = \pi(2)^2\left(\dfrac{1}{2}\right) = 2\pi = 6.283185307$ ft^3

    Radius of bucket of water: $\dfrac{1}{2}$ ft

    Volume of bucket of water:

    $\pi r^2 h = \pi\left(\dfrac{1}{2}\right)^2(1) = \dfrac{1}{4}\pi = 0.785398163$ ft^3

    $\dfrac{6.283185307}{0.785398163} = 8.000000004 ≈ 8$ bucketsful

    b) $6.283185307(62.5) = 392.6990817 ≈ 392.70$ lb

    c) $6.283185307(7.5) = 47.1238898 ≈ 47.12$ gal

49. a) Round pan:

$$A = \pi r^2 = \pi \left(\frac{9}{2}\right)^2 = 20.25\pi$$

$$= 63.61725124 \approx 63.62 \text{ in.}^2$$

Rectangular pan: $A = lw = 7(9) = 63 \text{ in.}^2$

b) Round pan:

$$V = \pi r^2 h \approx 63.62(2) = 127.24 \text{ in.}^3$$

Rectangular pan: $V = lwh = 7(9)(2) = 126 \text{ in.}^3$

c) Round pan

51. a) B = area of trapezoid = $\frac{1}{2}(9)(8+12) = 90 \text{ in.}^2$

$4 \text{ ft} = 4(12) = 48 \text{ in.}$

$V = Bh = 90(48) = 4320 \text{ in.}^3$

b) $1 \text{ ft}^3 = (12)(12)(12) = 1728 \text{ in.}^3$

$$4320 \text{ in.}^3 = \frac{4320}{1728} = 2.5 \text{ ft}^3$$

53. $8 - x + 3 = 2$

$11 - x = 2$

$-x = -9$

$x = 9$ edges

55. $x - 8 + 4 = 2$

$x - 4 = 2$

$x = 6$ vertices

57. $11 - x + 5 = 2$

$16 - x = 2$

$-x = -14$

$x = 14$ edges

50. $V = \frac{1}{3}\pi r^2 h = \frac{1}{3}\pi \left(\frac{3}{2}\right)^2 (6) = 4.5\pi$

$$= 14.13716694 \approx 14.14 \text{ in.}^3$$

52. a) $C = 2\pi r = 2\pi \left(\frac{19.6}{2}\right) = 19.6\pi$

$$= 61.57521601 \approx 61.58 \text{ m}$$

b) $V = \pi r^2 h = \pi \left(\frac{19.6}{2}\right)^2 (60) = 5762.4\pi$

$$= 18{,}103.11351 \approx 18{,}103.11 \text{ m}^3$$

54. $12 - 16 + x = 2$

$-4 + x = 2$

$x = 6$ faces

56. $7 - 12 + x = 2$

$-5 + x = 2$

$x = 7$ faces

58. $x - 10 + 4 = 2$

$x - 6 = 2$

$x = 8$ vertices

59. Let $r =$ the radius of one of the cans of orange juice

The length of the box $= 6r$ and the width of the box $= 4r$

Volume of box - volume of cans:

$$lwh - 6(\pi r^2 h) = (6r)(4r)h - 6\pi r^2 h = 24r^2 h - 6\pi r^2 h = 6r^2 h(4 - \pi)$$

Percent of the volume of the interior of the box that is not occupied by the cans:

$$\frac{6r^2 h(4 - \pi)}{lwh} = \frac{6r^2(4 - \pi)}{(6r)(4r)} = \frac{4 - \pi}{4} = 0.2146018366 \approx 21.46\%$$

60. a) – e) Answers will vary.

f) If we double the length of each edge of a cube, the new volume will be eight times the original volume.

61. a) – e) Answers will vary.

f) If we double the radius of a sphere, the new volume will be eight times the original volume.

62. a) $42(60)(24)(365) = 22,075,200$ drops

    b) $\dfrac{22,075,200}{20} = 1,103,760$ ml

    $\dfrac{1,103,760}{1000} = 1103.76$ l

    $\dfrac{1103.76}{3.79} = 291.2295515 \approx 291.23$ gal

    c) $291.23(\$0.11) = 32.0353 \approx \$32.04$

63. a) Find the volume of each numbered region. Since the length of each side is $a+b$, the sum of the volumes of each region will equal $(a+b)^3$.

    b) $V_1 = a(a)(a) = a^3 \qquad V_2 = a(a)(b) = a^2 b \qquad V_3 = a(a)(b) = a^2 b \qquad V_4 = a(b)(b) = ab^2$

    $V_5 = a(a)(b) = a^2 b \qquad V_6 = a(b)(b) = ab^2 \qquad V_7 = b(b)(b) = b^3$

    c) The volume of the piece not shown is ab^2.

64. a) 5.5 ft $= 5.5(12) = 66$ in.

    $V = Bh = 5(66) = 330$ in.^{3}

    b) Radius of cylinder: $\dfrac{0.75}{2} = 0.375$ in.

    Volume of cylinder: $\pi r^2 h = \pi(0.375)^2(66) = 9.28125\pi = 29.15790682$ in.^{3}

    Volume of hollow noodle: $330 - 29.15790682 = 300.8420932 \approx 300.84$ in.^{3}

Exercise Set 9.5

1. The act of moving a geometric figure from some starting position to some ending position without altering its shape or size is called **rigid motion**. The four main rigid motions studied in this section are reflections, translations, rotations, and glide reflections.

2. **Transformational geometry** is a type of geometry in which we study how to use a geometric figure to obtain other geometric figures by conducting one of several changes, called rigid motions, to the figure.

3. A **reflection** is a rigid motion that moves a figure to a new position that is a mirror image of the figure in the starting position.

4. Answers will vary.

5. A **translation** is a rigid motion that moves a figure by sliding it along a straight line segment in the plane.

6. Answers will vary.

7. A **rotation** is a rigid motion performed by rotating a figure in the plane about a specific point.

8. Answers will vary.

9. A **glide reflection** is a rigid motion formed by performing a translation (or glide) followed by a reflection.

10. Answers will vary.

11. A geometric figure is said to have **reflective symmetry** if the positions of a figure before and after a reflection are identical (except for vertex labels).

12. A geometric figure is said to have **rotational symmetry** if the positions of a figure before and after a rotation are identical (except for vertex labels).

13. A **tessellation** is a pattern consisting of the repeated use of the same geometric figures to entirely cover a plane, leaving no gaps.

14. Answers will vary.

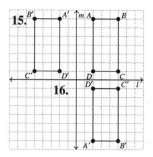

15.

16.

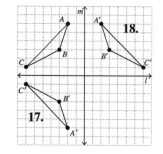

18.

17.

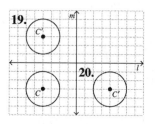

19.

20.

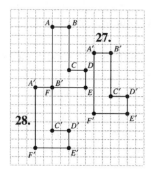

22.

21.

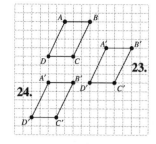

24.

23.

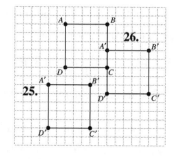

26.

25.

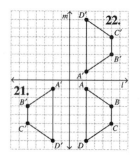

27.

28.

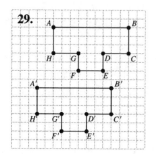

29.

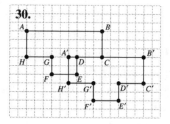

30.

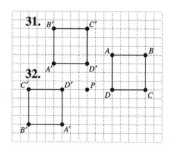

31.

32.

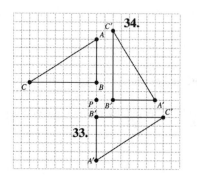

34.

33.

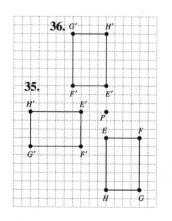

36.

35.

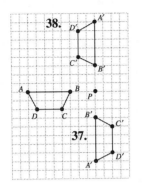

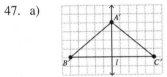

37.

38.

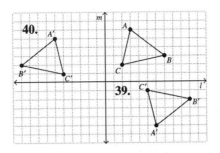

39.

40.

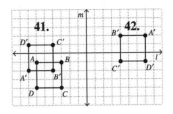

41.

42.

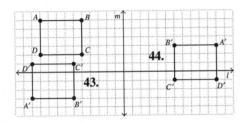

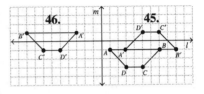

43.

44.

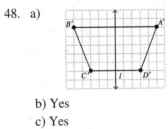

45.

46.

47. a)

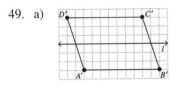

   b) Yes
   c) Yes

48. a)

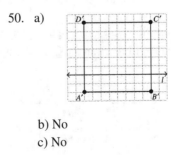

   b) Yes
   c) Yes

49. a)

   b) No
   c) No

50. a)

   b) No
   c) No

51. a)

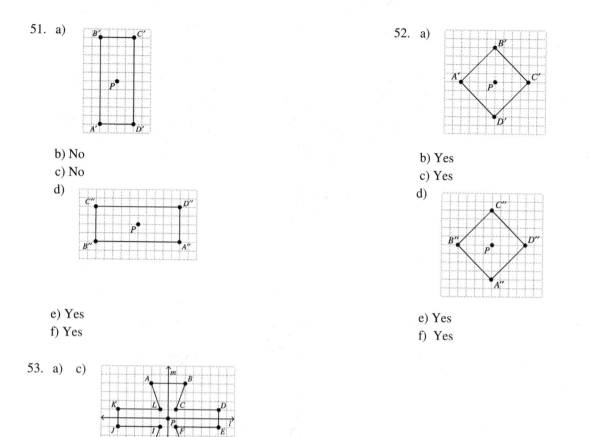

   b) No
   c) No
   d)

   e) Yes
   f) Yes

52. a)

   b) Yes
   c) Yes
   d)

   e) Yes
   f) Yes

53. a) c)

   d) No. Any 90° rotation will result in the figure being in a different position than the starting position.

54. a)

   b) No. Any reflection about any horizontal line will result in the figure being in a different position than the starting position.
   c) No. Any 90° rotation will result in the figure being in a different position than the starting position.
   d) No. Any 180° rotation will result in the figure being in a different position than the starting position.

55. a) b)

   c) No
   d) The order in which the translation and the reflection are performed is important. The figure obtained in part b) is the glide reflection.

56. Answers will vary.

57. Answers will vary.

58. a) Answers will vary.

    b) A regular octagon cannot be used as a tessellating shape.

59. a) Answers will vary.

    b) A regular pentagon cannot be used as a tessellating shape.

60. Although answers will vary depending on the font, the following capital letters have reflective symmetry about a horizontal line drawn through the center of the letter: B, C, D, E, H, I, K, O, X.

61. Although answers will vary depending on the font, the following capital letters have reflective symmetry about a vertical line drawn through the center of the letter: A, H, I, M, O, T, U, V, W, X, Y.

62. Although answers will vary depending on the font, the following capital letters have 180° rotational symmetry about a point in the center of the letter: H, I, O, S, X, Z.

Exercise Set 9.6

1. Topology is sometimes referred to as "rubber sheet geometry" because it deals with bending and stretching of geometric figures.

2. A **Möbius strip** is a one-sided, one-edged surface.

3. You can construct a Möbius strip by taking a strip of paper, giving one end a half twist, and taping the ends together.

4. A **Klein bottle** is a topological object that resembles a bottle but has only one side.

5. Four

6. a) Six

   b) Seven

7. A **Jordan curve** is a topological object that can be thought of as a circle twisted out of shape.

8. Since you must cross the curve to get from outside to inside, two crosses puts you back where you started. Thus, if you cross the curve twice (or any even number of times) to get outside, you must have started outside. Also, if you cross the curve once (or any odd number of times) to get outside, you must have started inside.

9. The number of holes in the object determines the **genus** of an object.

10. Two figures are **topologically equivalent** if one figure can be elastically twisted, stretched, bent, or shrunk into the other figure without ripping or puncturing the original figure.

11. 1, 4, 6 – Red; 2,3 – Yellow; 7 – Green; 5 – Blue

12. 1, 3, 7 – Red; 2, 6, 8 – Blue; 4,5 – Green

13. 1, 4, 6 – Red; 2, 5, 8 – Blue; 3, 7, 9 – Yellow

14. 1 – Red; 2, 5 – Yellow; 3, 6 – Blue; 4, 7 – Green

15. 1 – Red; 2, 5 – Yellow; 3, 6 – Blue; 4, 7 – Green

16. 1 – Red; 2, 5 – Yellow; 3, 6 – Blue; 4, 7 – Green

17. YT, NU, AB, ON, NS – Red

    NT, QC – Blue

    BC, SK, NB, NF – Green

    MB, PE – Yellow

18. BCS, SON, DGO, NLE – Red

    BCA, CHH, ZAC, TMP – Blue

    SIN, COA – Green

    NAY, SLP – Yellow

19. TX, KS, MS, KY, SC, FL – Red

    OK, LA, TN – Green

    MO, GA, VA – Blue

    AR, AL, NC – Yellow

20. CA, WA, MT, UT – Red

    OR, WY, AZ – Green

    ID, NM – Blue

    NV, CO – Yellow

21. Outside; a straight line from point *A* to a point clearly outside the curve crosses the curve an even number of times.

22. Inside; a straight line from point *B* to a point clearly outside the curve crosses the curve an odd number of times.

23. Outside; a straight line from point *A* to a point clearly outside the curve crosses the curve an even number of times.

24. Outside; a straight line from point *B* to a point clearly outside the curve crosses the curve an even number of times.

25. Outside; a straight line from point *C* to a point clearly outside the curve crosses the curve an even number of times.

26. Outside; a straight line from point *D* to a point clearly outside the curve crosses the curve an even number of times.

27. Inside; a straight line from point *E* to a point clearly outside the curve crosses the curve an odd number of times.

28. Inside; a straight line from point *F* to a point clearly outside the curve crosses the curve an odd number of times.

29. 1

30. 1

31. 1

32. 5

33. Larger than 5

34. 0

35. 5

36. 1

37. 0

38. Larger than 5

39. 5

40. 1

41. a) - d) Answers will vary.

42. One

43. One

44. One

45. Two

46. a) No, it has an inside and an outside.
    b) Two
    c) Two
    d) Two strips, one inside the other

47. The smaller one is a Möbius strip; the larger one is not.

48. No, it does not.

49. Yes. "Both sides" of the belt experience wear.

50. Answers will vary.

51. Ecuador, Brazil, Chile – Red
    Colombia, Guyana, French Guiana, Bolivia – Green
    Peru, Venezuela, Suriname, Paraguay, Uruguay – Yellow
    Argentina - Blue

52. Answers will vary.

53. a) 1
    b) 1
    c) Answers will vary.

Exercise Set 9.7

1. Girolamo Saccheri - proved many theorems of hyperbolic geometry

2. Janos Bolyai - discovered hyperbolic geometry

3. Carl Friedrich Gauss - discovered hyperbolic geometry

4. Nikolay Ivanovich Lobachevsky - discovered hyperbolic geometry

5. G.F. Bernhard Riemann - discovered elliptical geometry

6. Benoi Mandelbrot – first to use the word fractal to describe shapes that had several common characteristics, including some form of "self-similarity"

7. a) Euclidean - Given a line and a point not on the line, one and only one line can be drawn parallel to the given line through the given point.

   b) Hyperbolic - Given a line and a point not on the line, two or more lines can be drawn through the given point parallel to the given line.

   c) Elliptical - Given a line and a point not on the line, no line can be drawn through the given point parallel to the given line.

8. a) Euclidean - The sum of the measures of the angles of a triangle is 180°.

   b) Hyperbolic - The sum of the measures of the angles of a triangle is less than 180°.

   c) Elliptical - The sum of the measures of the angles of a triangle is greater than 180°.

9. A plane

10. A sphere

11. A pseudosphere

12. Each type of geometry can be used in its own frame of reference.

13. Spherical - elliptical geometry; flat - Euclidean geometry; saddle-shaped - hyperbolic geometry

14. Coastlines, trees, mountains, galaxies, polymers, rivers, weather patterns, brains, lungs, blood supply

15.

16.

17.

18.

19. a)

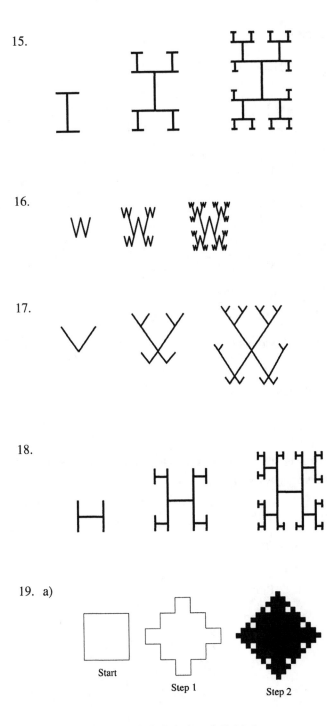

Start

Step 1

Step 2

b) Infinite since it is infinitely subdivided.
c) Finite since it covers a finite or closed area.

20. a)

Step	Perimeter
1	$3\left(\dfrac{4}{3}\right)^0 = 3(1) = 3$
2	$3\left(\dfrac{4}{3}\right)^1 = 3\left(\dfrac{4}{3}\right) = 4$
3	$3\left(\dfrac{4}{3}\right)^2 = 3\left(\dfrac{16}{9}\right) = \dfrac{16}{3}$
4	$3\left(\dfrac{4}{3}\right)^3 = 3\left(\dfrac{64}{27}\right) = \dfrac{64}{9}$
5	$3\left(\dfrac{4}{3}\right)^4 = 3\left(\dfrac{256}{81}\right) = \dfrac{256}{27}$
6	$3\left(\dfrac{4}{3}\right)^5 = 3\left(\dfrac{1024}{243}\right) = \dfrac{1024}{81}$

b) At each stage, the perimeter is $\dfrac{4}{3}$ multiplied by the previous perimeter.

c) The area is finite because it encloses a finite region.
The perimeter is infinite because it consists of an infinite number of pieces.

Review Exercises

In the Review Exercises and Chapter Test questions, the π key on the calculator is used to determine answers in calculations involving π. If you use 3.14 for π, your answers may vary slightly.

1. $\{F\}$

2. $\triangle BFC$

3. $\overline{BC}$

4. $\overleftrightarrow{BH}$

5. $\{F\}$

6. $\{\ \}$

7. $90° - 51.2° = 38.8°$

8. $180° - 124.7° = 55.3°$

9. Let $x = BC$

$$\frac{BC}{B'C} = \frac{AC}{A'C}$$

$$\frac{x}{3.4} = \frac{12}{4}$$

$$4x = 40.8$$

$$x = \frac{40.8}{4} = 10.2 \text{ in.}$$

10. Let $x = A'B'$

$$\frac{A'B'}{AB} = \frac{A'C}{AC}$$

$$\frac{x}{6} = \frac{4}{12}$$

$$12x = 24$$

$$x = \frac{24}{12} = 2 \text{ in.}$$

11. $m\angle ABC = m\angle A'B'C$
$m\angle A'B'C = 180° - 88° = 92°$
Thus, $m\angle ABC = 92°$
$m\angle BAC = 180° - 30° - 92° = 58°$

12. $m\angle ABC = m\angle A'B'C$
$m\angle A'B'C = 180° - 88° = 92°$
Thus, $m\angle ABC = 92°$

13. $m\angle 1 = 180° - 110° = 70°$

$m\angle 6 = 70°$ (angle 1 and angle 6 are vertical angles)

The measure of the top angle of the triangle is $50°$, by vertical angles. The measure of the angle on the bottom right of the triangle is $180° - 70° - 50° = 60°$.

$m\angle 2 = 60°$ (angle 2 and the angle on the bottom right of the triangle are vertical angles)

The measure of the alternate interior angle of angle 2 is $60°$. Thus, $m\angle 3 = 180° - 60° = 120°$.

The measure of the alternate interior angle of angle 6 is $70°$. Thus, $m\angle 5 = 180° - 70° = 110°$.

$m\angle 4 = 180° - 110° = 70°$

14. $n = 6$

$(n-2)180° = (6-2)180° = 4(180°) = 720°$

15. $A = lw = 9(7) = 63 \text{ cm}^2$

16. $A = \dfrac{1}{2}bh = \dfrac{1}{2}(14)(5) = 35 \text{ in.}^2$

17. $A = \dfrac{1}{2}h(b_1 + b_2) = \dfrac{1}{2}(2)(4+9) = 13 \text{ in.}^2$

18. $A = bh = 12(7) = 84 \text{ in.}^2$

19. $A = \pi r^2 = \pi(13)^2 = 169\pi$

$= 530.9291585 \approx 530.93 \text{ cm}^2$

20. $A = lw = 14(16) = 224 \text{ ft}^2$

Cost: $224(\$2.75) = \$616$

21. $V = \pi r^2 h = \pi(5)^2(15) = 375\pi$

$= 1178.097245 \approx 1178.10 \text{ in.}^3$

22. $V = lwh = 10(3)(4) = 120 \text{ cm}^3$

23. If h represents the height of the triangle which is the base of the pyramid, then

$$
\begin{aligned}
h^2 + 3^2 &= 5^2 \\
h^2 + 9 &= 25 \\
h^2 &= 16 \\
h &= \sqrt{16} = 4 \text{ ft}
\end{aligned}
$$

$B = \dfrac{1}{2}bh = \dfrac{1}{2}(6)(4) = 12 \text{ ft}^2$

$V = \dfrac{1}{3}Bh = \dfrac{1}{3}(12)(7) = 28 \text{ ft}^3$

24. $B = \dfrac{1}{2}bh = \dfrac{1}{2}(9)(12) = 54 \text{ m}^2$

$V = Bh = 54(8) = 432 \text{ m}^3$

25. $r = \dfrac{12}{2} = 6 \text{ mm}$

$V = \dfrac{1}{3}\pi r^2 h = \dfrac{1}{3}\pi(6)^2(16) = 192\pi$

$= 603.1857895 \approx 603.19 \text{ mm}^3$

26. $V = \frac{4}{3}\pi r^3 = \frac{4}{3}\pi(7)^3 = 457.\overline{3}\pi$

$= 1436.75504 \approx 1436.76 \text{ ft}^3$

27.

$h^2 + 1^2 = 3^2$

$h^2 + 1 = 9$

$h^2 = 8$

$h = \sqrt{8}$

$A = \frac{1}{2}h(b_1 + b_2) = \frac{1}{2}(\sqrt{8})(2+4) = 8.485281374 \text{ ft}^2$

a) $V = Bh = 8.485281374(8)$

$= 67.88225099 \approx 67.88 \text{ ft}^3$

b) Weight:

$67.88(62.5) + 375 = 4617.5 \text{ lb}$

Yes, it will support the trough filled with water.

c) $(4617.5 - 375) = 4242.5 \text{ lb of water}$

$\frac{4242.5}{8.3} = 511.1445783 \approx 511.14 \text{ gal}$

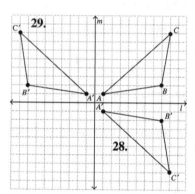

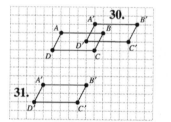

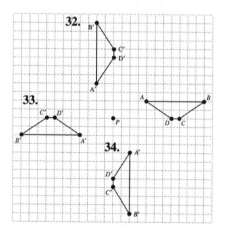

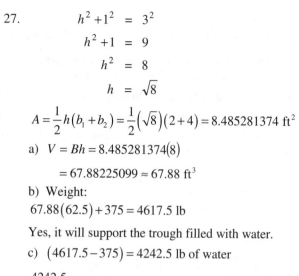

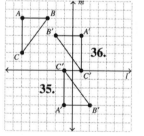

37. Yes   38. No   39. No   40. Yes

41. 1

42. Saarland, North Rhine-Westphalia, Bremen, Mecklenburg-Western Pomerania, Berlin, Thuringia, Baden-W r ttemberg, Hamburg   Red

Rhineland-Palatinate, Lower Saxony, Saxony   Green

Schleswig-Holstein, Hesse, Brandenburg   Yellow

Bavaria, Saxony-Anhalt - Blue

43. Outside; a straight line from point A to a point clearly outside the curve crosses the curve an even number of times.

44. Euclidean: Given a line and a point not on the line, one and only one line can be drawn parallel to the given line through the given point.

Elliptical: Given a line and a point not on the line, no line can be drawn through the given point parallel to the given line.

Hyperbolic: Given a line and a point not on the line, two or more lines can be drawn through the given point parallel to the given line.

45.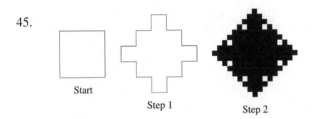

Chapter Test

1. $\overset{\circ\circ}{EF}$

2. $\triangle BCD$

3. $\{D\}$

4. $\overrightarrow{AC}$

5. $90° - 36.9° = 53.1°$

6. $180° - 101.5° = 78.5°$

7. The other two angles of the triangle are 48 (by vertical angles) and $180 - 112 = 68$. Thus, the measure of angle $x = 180 - 48 - 68 = 64$.

8. $n = 8$

$$(n-2)180° = (8-2)180° = 6(180°) = 1080°$$

9. Let $x = B'C'$

$$\frac{B'C'}{BC} = \frac{A'C'}{AC}$$

$$\frac{x}{7} = \frac{5}{13}$$

$$13x = 35$$

$$x = \frac{35}{13} = 2.692307692 \approx 2.69 \text{ cm}$$

10. a)

$$x^2 + 5^2 = 13^2$$
$$x^2 + 25 = 169$$
$$x^2 = 144$$
$$x = \sqrt{144} = 12 \text{ in.}$$

b) $P = 5 + 13 + 12 = 30$ in.

c) $A = \frac{1}{2}bh = \frac{1}{2}(5)(12) = 30 \text{ in.}^2$

11. $r = \frac{16}{2} = 8$ cm

$$V = \frac{4}{3}\pi r^3 = \frac{4}{3}\pi(8)^3 = 682.\overline{6}\pi$$

$$= 2144.660585 \approx 2144.66 \text{ cm}^3$$

12. $B = 9(14) + \pi(4.5)^2 = 126 + 20.25\pi = 189.6172512$

$$V = Bh = 189.6172512(6) = 1137.703507 \text{ ft}^3$$

$$1137.703507 \text{ ft}^3 = \frac{1137.703507}{27}$$

$$= 42.13716694 \approx 42.14 \text{ yd}^3$$

13. $B = lw = 4(7) = 28 \text{ ft}^2$

$V = \dfrac{1}{3}Bh = \dfrac{1}{3}(28)(12) = 112 \text{ ft}^3$

14.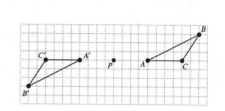

15.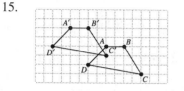

16.

17.

18. a) No
    b) Yes

19. A **Möbius strip** is a surface with one side and one edge.

20. a) and b) Answers will vary.

21. Euclidean: Given a line and a point not on the line, one and only one line can be drawn parallel to the given line through the given point.

Elliptical: Given a line and a point not on the line, no line can be drawn through the given point parallel to the given line.

Hyperbolic: Given a line and a point not on the line, two or more lines can be drawn through the given point parallel to the given line.

Group Projects

1. a) $B = \pi r^2 = \pi\left(\dfrac{12}{2}\right)^2 = 36\pi = 113.0973355$

$V = Bh = 113.0973355(4) = 452.3893421 \approx 452 \text{ ft}^3$

b) $452.3893421(7.5) = 3392.920066 \approx 3393 \text{ gal}$

c) $452.3893421(52.4) = 23{,}705.20153 \approx 23{,}705 \text{ lb}$

d) Weight of Jacuzzi and water: $475 + 23{,}705.20153 = 24{,}180.20153 \text{ lb}$
Yes

e) Weight of Jacuzzi, water, and four people: $24{,}180.20153 + 4(115) = 24{,}640.20153 \text{ lb}$
Yes

2. a) 12 ft

   b) 4 in.×3 ft 6 in.×12 ft 6 in.

   c) $V = \dfrac{4}{12} \text{ ft} \times 3.5 \text{ ft} \times 12.5 \text{ ft} = 14.58\overline{3} \text{ ft}^3$

      $14.58\overline{3} \text{ ft}^3 = \dfrac{14.58\overline{3}}{27} = 0.5401234568 \text{ yd}^3$

   d) $0.5401234568(45) = 24.30\overline{5} \approx \$24.31$

   e) 1 sheet

   f) \$18.95

   g) Five 8 ft $2 \times 4's$

   h) $5(\$2.14) = \$10.70$

   i) $B = \dfrac{1}{2}bh = \dfrac{1}{2}(2)(12) = 12 \text{ ft}^2$

     $V = Bh = 12(3) = 36 \text{ ft}^3$

     $36 \text{ ft}^3 = \dfrac{36}{27} = 1.\overline{3} \text{ yd}^3 \approx 1.33 \text{ yd}^3$

   j) $1.\overline{3}(\$45) = \$60$

   k) $\$24.31 + \$18.95 + \$10.70 + \$60 = \$113.96$

   l) $2^2 + 12^2 = x^2$

     $4 + 144 = x^2$

     $x^2 = 148$

     $x = \sqrt{148} = 12.16552506 \approx 12.17 \text{ ft}$

  m) 8 boards

   n) $8(\$6.47) = \$51.76$

   o) $10(\$2.44) = \$24.40$

   p) $\$24.31 + \$51.76 + \$24.40 = \$100.47$

   q) The materials are less expensive for the wooden ramp.

CHAPTER TEN

MATHEMATICAL SYSTEMS

Exercise Set 10.1

1. A binary operation is an operation that is performed on two elements, and the result is a single element.

2. A set of elements and at least one binary operation.

3. Each of these operations can be performed on only two elements at a time and the result is always a single element.   a) $2 + 3 = 5$ b) $5 - 3 = 2$ c) $2 \times 3 = 6$ d) $6 \div 3 = 2$

4. Closure, identity, each element must have a unique inverse, associative property.

5. Closure, identity, each element must have a unique inverse, associative property, commutative property.

6. Abelian group

7. If a binary operation is performed on any two elements of a set and the result is an element of the set, then that set is closed under the given binary operation. For all integers a and b, $a + b$ is an integer. Therefore, the set of integers is closed under the operation of addition.

8. An identity element is an element in a set such that when a binary operation is performed on it and any given element in the set, the result is the given element. The additive identity element is 0, and the multiplicative identity element is 1.   Examples: $5 + 0 = 5$, $5 \times 1 = 5$

9. When a binary operation is performed on two elements in a set and the result is the identity element for the binary operation, then each element is said to be the inverse of the other. The additive inverse of 2 is (-2) since $2 + (-2) = 0$, and the multiplicative inverse of 2 is $(1/2)$ since $2 \times 1/2 = 1$.

10. A specific example illustrating that a specific property is not true is called a counterexample.

11. No. Every commutative group is also a group.

12. Yes. For a group, the Commutative property need not apply.

13. d   The Commutative property need not apply.

14. Squaring, finding square roots, finding the reciprocal, finding the absolute value

15. The associative property of addition states that $(a + b) + c = a + (b + c)$, for any elements a, b, and c. Example: $(3 + 4) + 5 = 3 + (4 + 5)$

16. The associative property of multiplication states that $(a \times b) \times c = a \times (b \times c)$, for any real numbers a, b, and c.   Example: $(3 \times 4) \times 5 = 3 \times (4 \times 5)$

17. The commutative property of multiplication stated that $a \times b = b \times a$, for any real numbers a, b, and c. Example: $3 \times 4 = 4 \times 3$

18. The commutative property of addition stated that $a + b = b + a$, for any elements a, b, and c. Example: $3 + 4 = 4 + 3$

19. $8 \div 4 = 2$, but $4 \div 8 = \frac{1}{2}$

20. $7 - 3 = 4$, BUT $3 - 7 = -4$

21. $(6 - 3) - 2 = 3 - 2 = 1$, but $6 - (3 - 2) = 6 - 1 = 5$

22. $(16 \div 4) \div 2 = 4 \div 2, = 2$ but $16 \div (4 \div 2) = 16 \div 2 = 8$

23. No. No inverse element

24. No. No inverse element

25. Yes. Satisfies 5 properties needed

26. Yes. Satisfies 4 properties needed

27. No. Not closed

28. No. Not closed

29. No. No identity or inverse elements

30. No. Not all elements have inverses

31. No. Not closed

32. No. Not all elements have inverses

33. Yes. Satisfies 4 properties needed

34. No. Not all elements have inverses

35. No. Not closed   ie.: 1/0 is undefined

36. No. Does not satisfy Associative property

37. No. Does not satisfy Associative property

38. No. Not closed

39. No; the system is not closed, $\pi + (-\pi) = 0$ which is not an irrational number.

40. No; $\pi \square (1/\pi) = 1$ which is not an irrational number.

41. Yes. Closure: The sum of any two real numbers is a real number. The identity element is zero. Example: $5 + 0 = 0 + 5 = 5$
Each element has a unique inverse.
Example: $6 + (-6) = 0$
The associative property holds:
Example: $(2 + 3) + 4 = 2 + (3 + 4)$

42. No. Closure: The product of any two real numbers is a real number. The identity element is one. Example: $5 \bullet 0 = 0 \bullet 5 = 5$
Not every element has an inverse.
Example: $2 \bullet ? = 1$
The associative property holds:
Example: $(2 \bullet 3) \bullet 4 = 2 \bullet (3 \bullet 4)$

43. Answers will vary.

44. 999

45. 9/19/29/39/49/59/69/79/89   9 } 20
    90/91/92/93/94/95/96/97/98/99   11 }

Exercise Set 10.2

1. The clock addition table is formed by adding all pairs of integers between 1 and 12 using the 12 hour clock to determine the result. Example: If the clock is at 7 and we add 8, then the clock will read 3. Thus, $7 + 8 = 3$ in clock arithmetic.

2. $12 + 12 = 12$. Start at 12 move clockwise 12 hours, the result is 12.

3. a) First add $(6 + 9)$ on the clock, then add that result to 5 on the clock to obtain the final answer.
   b) $(4 + 10) + 3 = 2$ $(2) + 3 = 5$

4. a) Start at the first number on the face of the clock, then count counterclockwise the number being subtracted. The number you end at is the difference.
   b) $4 - 7 = 9$

5. a) $5 - 9$ $5 + 12 = 17$ $17 - 9$
   b) $17 - 9 = 8$

6. The system is commutative if the elements in the table are symmetric about the main diagonal.

7. If a binary operation is performed on any two elements of a set and the result is an element of the set, then that set is closed under the given binary operation. For all integers a and b, $a + b$ is an integer. Therefore, the set of integers is closed under the operation of addition.

8. Yes. 12

9. Yes. One and 11 are inverses, 2 and 10 are inverses, 3 and 9 are inverses, 4 and 8 are inverses, 5 and 7 are inverses, 6 is its own inverse, and 12 is its own inverse.

10. $(2 + 3) + 8 = 2 + (3 + 8)$
    $5 + 8 = 2 + 11$
    $1 = 1$

11. Yes. $6 + 9 = 3$ and $9 + 6 = 3$

12. Yes, the five properties are met.

    1) The system is closed. All results are from the set {1, 2, 3, 4, 5, 6, 7, 8, 9, 10, 11, 12}

    2) The identity element is 12.

    3) Each element has an inverse.

    4) The associative property holds true.

    5) The system is commutative.

13. a) Identity element = 5

    b) Add inverse of 2, which is 3. $2 + 3 = 5$

14. a) Identity element = 8

    b) Add inverse of 3, which is 5. $3 + 5 = 8$

15. Yes. Commutative, symmetrical around main diagonal

16. No. Not commutative, Non-symmetrical around main diagonal

17. Identity element = C, Row 3 is identical to top row and column 3 is identical to left column

18. There is no identity. While the top row = 3rd row, the left column $\neq$ any other column.

19. The inverse of A is B, because A operate B = C and B operate A = C.

20. The inverse of A is A, because A operate A = A.

21. $4 + 7 = 11$

22. $8 + 7 = 3$

23. $9 + 8 = 5$

24. $10 + 4 = 2$

25. $4 + 12 = 4$

26. $12 + 12 = 12$

27. $3 + (8 + 9) = 3 + 5 = 8$

28. $(8 + 7) + 6 = 3 + 6 = 9$

29. $(6 + 4) + 8 = 10 + 8 = 6$

30. $(6 + 10) + 12 = 4 + 12 = 4$

31. $(7 + 8) + (9 + 6) = 3 + 3 = 6$

32. $(7 + 11) + (9 + 5) = 6 + 2 = 8$

33. $7 - 4 = 3$

34. $11 - 8 = 3$

35. $4 - 12 = 4$

36. $3 - 9 = 6$

37. $5 - 10 = 7$

38. $3 - 10 = 5$

39. $1 - 12 = 1$

40. $6 - 10 = 8$

41. $5 - 5 = 12$

42. $8 - 8 = 12$

43. $12 - 12 = 12$

44. $5 - 8 = 9$

45.

+	1	2	3	4	5	6
1	2	3	4	5	6	1
2	3	4	5	6	1	2
3	4	5	6	1	2	3
4	5	6	1	2	3	4
5	6	1	2	3	4	5
6	1	2	3	4	5	6

45.

+	1	2	3	4	5	6	7
1	2	3	4	5	6	7	1
2	3	4	5	6	7	1	2
3	4	5	6	7	1	2	3
4	5	6	7	1	2	3	4
5	6	7	1	2	3	4	5
6	7	1	2	3	4	5	6
7	1	2	3	4	5	6	7

46. $4 + 5 = 3$

47. $1 + 6 = 1$

48. $6 + 4 = 4$

49. $5 - 2 = 3$

50. $4 - 5 = 5$

51. $2 - 6 = 2$

52. $3 - 4 = 5$

53. $4 - 6 = 4$

54. $2 + (1 - 3) = 2 + 4 = 6$

55. See above.

56. $5 + 4 = 2$

57. $6 + 5 = 4$

58. $4 + 4 = 1$

59. $7 + 6 = 6$

60. $2 - 3 = 6$

61. $3 - 6 = 4$

62. $2 - 4 = 5$

63. $(4 - 5) - 6 = 6 - 6 = 7$

64. $3 - (2 - 6) = 3 - 3 = 7$

65. Yes. Satisfies 5 required properties

66. No, not necessarily. It may not have an inverse, identity element, or satisfy the Commutative or Associative properties.

67. a) $\{0, 1, 2, 3\}$
    b) ✈ airplane
    c) Yes. All solutions are members of the original set.
    d) Identity element is 0.
    e) Yes; $0 ✈ 0 = 0$, $1 ✈ 3 = 0$, $2 ✈ 2 = 0$, $3 ✈ 1 = 0$
    f) $(1 ✈ 2) ✈ 3 = 3 ✈ 3 = 2$ and $1 ✈ (2 ✈ 3) = 1 ✈ 1 = 2$
    g) Yes; $3 ✈ 2 = 1 = 2 ✈ 3$
    h) Yes, system satisfies five properties needed.

68. a) $\{*, 5, L\}$ b)
    b) 🐎
    c) Yes. All solutions are members of the original set.
    d) Identity element is L.
    e) Yes; $* 🐎 5 = L$, $5 🐎 * = L$, $L 🐎 L = L$
    f) $(* 🐎 5) 🐎 5 = L 🐎 5 = 5$ and $* 🐎 (5 🐎 5) = * 🐎 * = 5$
    g) Yes; $L 🐎 * = *$ and $* 🐎 L = *$
    h) Yes, system satisfies five properties needed.

69. a) $\{r, s, t, u\}$ b) 🛸
    c) Yes. All solutions are members of the original set.
    d) Yes, the identity element is t.
    e) Yes; $r 🛸 r = t$, $s 🛸 u = t$, $t 🛸 t = t$, $u 🛸 s = t$
    f) $(r 🛸 s) 🛸 u = u 🛸 u = r$ and $r 🛸 (s 🛸 u) = r 🛸 t = r$
    g) Yes; $s 🛸 r = u$ and $r 🛸 s = u$
    h) Yes, system satisfies five properties needed.

70. a) $\{3, 5, 8, 4\}$ b) 🧍
    c) Yes. All solutions are members of the original set.
    d) Identity element is 4.
    e) Yes. $3 🧍 8 = 4$, $5 🧍 5 = 4$, $8 🧍 3 = 4$, $4 🧍 4 = 4$
    f) $(5 🧍 8) 🧍 4 = 3 🧍 4 = 3$ and $5 🧍 (8 🧍 4) = 5 🧍 8 = 3$
    g) Yes. $8 🧍 5 = 3 = 5 🧍 8$
    h) Yes, system satisfies five properties needed.

71. a) $\{f, r, o, m\}$ b) 🐕
    c) The system is closed. All elements in the table are elements of the set.
    d) $(r 🐕 o) 🐕 f = m 🐕 f = m$
    e) $(f 🐕 r) 🐕 m) = r 🐕 m = f$
    f) Identity element is f.
    g) Inverse of r is m since $m 🐕 r = f$.
    h) Inverse of m is r since $r 🐕 m = f$.

72. a) No, there is no identity element.
    b) $(1\ w\ 3)\ w\ 4 \neq 1\ w\ (3\ w\ 4)$
       $4\ w\ 4 \neq 1\ w\ 3$

73. a) Is closed; all solutions are members of the original set.   b) Identity $= ⊡$
    c) Inverse: of $⊡$ is $⊡$; of M is M; of $⊿$ is $⊿$
    d) $(M ⊗ ⊿) ⊗ M = ⊿ ⊗ M = M$
       $M ⊗ (⊿ ⊗ M) = M ⊗ M = ⊡$
       Not associative since M $⊡$
    e) $⊿ ⊗ M = M$ $M ⊗ ⊿ = ⊿$
       Not commutative since $M \neq ⊿$

74. Not closed: y ^ x = a and a is not a member of
the set {w, x, y}
No identity element, and therefore no inverses.
(x ^ w) ^ x = y ^ x = a
x ^ (w ^ x) = x ^ y = w
Not associative since a   w
y ^ x = a and x ^ y = w
Not commutative since a ≠ w

75. No inverses for ⊙ and *
(* ⊗ *) ⊗ T = ⊙ ⊗ T = *
* ⊗ (* ⊗ T) = * ⊗ T = ⊙
Not associative since * ≠ ⊙

76. (a 😊 a) 😊 Δ = Δ 😊 Δ = a
a 😊 (a 😊 Δ) = a 😊 0 = ≺
Not associative since a   ≺
Δ 😊 ≺ = ≺   ≺ 😊 Δ = a
Not commutative since ≺ ≠ a

77. No identity element and therefore no inverses.
(d ⇔ e) ⇔ d = d ⇔ d = e
d ⇔ (e ⇔ d) = d ⇔ e = d
Not associative since e   d
e ⇔ d = e   d ⇔ e = d
Not commutative since e ≠ d

78. No inverses for 0, 2, 3, and 4

79. a)

+	E	O
E	E	O
O	O	E

b) The system is closed, the identity element is E,
each element is its own inverse, and the system
is commutative since the table is symmetric
about the main diagonal. Since the system has
fewer than 6 elements satisfying the above
properties, it is a commutative group.

80. a)

□	E	O
E	E	E
O	E	O

b) The identity is 0, but since E has no inverse,
the system is not a group.

81. Student activity - Answers will vary.

82. Student activity - Answers will vary.

83. a) All elements in the table are in the set
{1, 2, 3, 4, 5, 6} so the system is closed.
The identity is 6. 5 and 1 are inverses of each
other, and 2, 3, 4, and 6 are their own inverses.
Thus, if the associative property is assumed,
the system is a group.

b) 4 ∞ 5 = 2, but 5 ∞ 4 = 3

83. Examples of associativity
(2 ∞ 3) ∞ 4 = 5 ∞ 4 = 3 and
2 ∞ (3 ∞ 4) = 2 ∞ 5 = 3
(1 ∞ 3) ∞ 5 = 4 ∞ 5= 2 and
1 ∞ (3 ∞ 5) = 1 ∞ 4 = 2

84. a) Is closed   Identity = F $\odot$

  $(C \odot D) \odot A = E \odot A = F$

  $C \odot (D \odot A) = C \odot C = F$

  Is Associative since F = F

  Inverses of: $A \odot E = F$, $B \odot B = F$,

  $C \odot C = F$, $D \odot D = F$,

  $E \odot A = F$, $F \odot F = F$

  $B \odot C = E$ $C \odot B = A$

  Not Commutative since $E \neq A$

86. $4^3 = 64$ ways

87.

+	0	1	2	3	4
0	0	1	2	3	4
1	1	2	3	4	0
2	2	3	4	0	1
3	3	4	0	1	2
4	4	0	1	2	3

89.

+	0	1	2	3
0	0	1	2	3
1	1	2	3	4
2	2	3	4	0
3	3	4	0	1

85. a)

*	R	S	T	U	V	I
R	V	T	U	S	I	R
S	U	I	V	R	T	S
T	S	R	I	V	U	T
U	T	V	R	I	S	U
V	I	U	S	T	R	V
I	R	S	T	U	V	I

  $R * (T * V) = R * U = S$

  $(R * T) * V = U * V = S$

  Is Associative since S = S

b) Is closed

c) $R * S = T$ $S * R = U$

  Not Commutative since $T \neq U$

  $R. (S. V) = R. T = U$

88.

+	0	1	2	3	4	5
0	0	1	2	3	4	5
1	1	2	3	4	5	0
2	2	3	4	5	0	1
3	3	4	5	0	1	2
4	4	5	0	1	2	3
5	5	0	1	2	3	4

89. 1) Add # in left column to # in top row

  2) Divide by 4

  3) Replace remainder in table

Exercise Set 10.3

1. A modulo m system consists of m elements, 0 through m – 1, and a binary operation.

2. a) a is congruent to b modulo m, written $a \cong b$ (mod m), means a and b have the same remainder when divided by m.

  b) 13 and 3 have the same remainder, 3, when divided by 5.

3. In a modulo 5 system there will be 5 modulo classes. When a number is divided by 5 the remainder will be a number from 0 – 4.

0	1	2	3	4
0	1	2	3	4
5	6	7	8	9
10	11	12	13	14

4. In any modulo system, modulo classes are developed by placing all numbers with the same remainder in the same modulo class.

5. In a modulo 12 system there will be 12 modulo classes. When a number is divided by 12 the remainder will be a number 0 – 11.

6. In a modulo n system there will be n modulo classes. When a number is divided by n the remainder will be a number from 0 – (n–1).

7. $27 \cong ?$ (mod 5) c or d 27, 12, and 107 have the same remainder, 2, when divided by 5.

8. $167 \cong ?$ (mod 7) b or d 106, 71, and 22 have the same remainder, 1, when divided by 7.

9. Thursday = Day 4 $30 \cong 2$ (mod 7) Saturday

10. 4 + 161 = 165 and 165 ÷ 7 = 23, remainder 4
    Day 4 = Thursday

11. 4 + 366 = 370 and 370 ÷ 7 = 52, remainder 6
    Day 6 = Saturday

12. 5 years = (5 • 365) days = 1825 days
    4 + 1825 = 1829 and 1829 ÷ 7 = 261, remainder 2
    Day 2 = Tuesday

13. 3 years, 34 days = (3)(365 + 34) days = 1129 days
    4 + 1129 = 1133 and 1133 ÷ 7 = 161, remainder 6
    Day 6 = Saturday

14. 4 + 463 = 467 and 467 ÷ 7 = 66, remainder 5
    Day 5 = Friday

15. 728 days / 7 = 104 remainder 0 Thursday

16. 3 yrs. 27 days = 1122 days
    1122 / 7 = 160 remainder 2 Saturday

17. Answers will vary. 18. Answers will vary. 19. Answers will vary. 20. Answers will vary.
21. Answers will vary. 22. Answers will vary. 23. Answers will vary. 24. Answers will vary.

25. 8 + 6 = 14 $14 \cong 4$ (mod 5)

26. 5 + 10 = 15 $15 \cong 0$ (mod 5)

27. 1 + 9 + 12 = 22 $22 \cong 2$ (mod 5)

28. 9 - 3 = 6 $6 \cong 1$ (mod 5)

29. 5 - 12 = 3 $3 \cong 3$ (mod 5)

30. 7 • 4 = 28 $28 \cong 3$ (mod 5)

31. 8 • 9 = 72 $72 = 2$ (mod 5)

32. 10 - 15 = 0 $0 \cong 0$ (mod 5)

33. 4 - 8 = 1 $1 \cong 1$ (mod 5)

34. 3 - 7 = 1 $1 \cong 1$ (mod 5)

35. (15 • 4) - 8 = 60 – 8 = 52 $52 \cong 2$ (mod 5)

36. (4 – 9)7 = (-5)7 = 5(7) = 35 $35 \cong 0$ (mod 5)

37. 15 (mod 5) $\cong$ 0 38. 23 (mod 7) $\cong$ 2 39. 84 (mod 12) $\cong$ 0 40. 43 (mod 6) $\cong$ 1
41. 60 (mod 9) $\cong$ 6 42. 75 (mod 8) $\cong$ 3 43. 30 (mod 7) $\cong$ 2 44. 53 (mod 4) $\cong$ 1

45. -5 (mod 7) $\cong$ 2 46. -7 (mod 4) $\cong$ 1 47. -13 (mod 11) $\cong$ 9 48. -11 (mod 13) $\cong$ 2
49. 135 (mod 10) $\cong$ 5 50. -12 (mod 4) $\cong$ 0 51. 3 + 4 = 7 $\cong$ 1 (mod 6) 52. 6 + 5 $\cong$ 3 (mod 8)
53. 2 + 2 $\cong$ 4 (mod 5) 54. 4 + 5 $\cong$ 3 (mod 6) 55. 4 - 5 $\cong$ 5 (mod 6) 56. 4 • 5 $\cong$ 6 (mod 7)
57. 5 • 5 $\cong$ 7 (mod 9) 58. 3 • { } $\cong$ 5 (mod 6) 59. 3 • { } $\cong$ 1 (mod 6) 60. 3 • { } $\cong$ 3 (mod 12)
  No solution No solution {1, 5, 9}

61. $4 \bullet \{ \ \} \cong 4 \pmod{10}$
    $\{1, 6\}$

62. $2 - 6 \cong 4 \pmod 8$

63. $4 - 7 \cong 9 \pmod{12}$

64. $6 - 7 \cong 8 \pmod 9$

65. $3 \bullet 0 \cong 05 \pmod{10}$

66. $4 \bullet \{ \ \} \cong 5 \pmod 8$
    No solution

67. a) 2016, 2020, 2024,
       20,28, 2032
    b) 3004

67. c) 2552, 2556, 2560,
       2564, 2568, 2572

68. a) flying   7 R 4
    c) resting 30 R 0
    e) flying
    b) flying   11 R 2
    d) flying $7 - 6 = 1$
    f) $7 - 20 = 3$

69. a) $28/8 = 3$ R 4   resting 2nd day
    b) $60/8 = 7$ R 4   resting 2nd day
    c) $127/8 = 15$ R 7   am/pm practice
    d) no am practice

70. a) $20/10 = 2$ R 0   twice a day
    b) $49/10 = 4$ R 9   twice a day
    c) $103/10 = 10$ R 3   twice a day
    d) $78/10 = 7$ R 8   yes, rest

71. The manager's schedule is repeated every seven weeks. If this is week two of her schedule, then this is her second weekend that she works, or week 1 in a mod 7 system. Her schedule in mod 7 on any given weekend is shown in the following table:
    Weekend (mod 7):

Work/off	0	1	2	3	4	5	6
	w	w	w	w	w	w	o

    a) If this is weekend 1, then in 5 more weeks $(1 + 5 = 6)$ she will have the weekend off.
    b) $25 \cong 7 = 3$, remainder 4. Thus $25 \cong \pmod 7$ and 4 weeks from weekend 1 will be weekend 5. She will not have off.
    c) $50 \cong 7 = 7$, remainder 1. One week from weekend 1 will be weekend 2. It will be 4 more weeks before she has off. Thus, in 54 weeks she will have the weekend off.

72. a) $6 \cong 1 \pmod 5$
       If this is week 3, then $3 + 1 \cong 4 \pmod 5$ indicates the 3 P.M. - 11 P.M. shift.
    b) $7 \cong 2 \pmod 5$
       If this is week 4, then $4 + 2 \cong 1 \pmod 5$ indicates the 7 A.M. - 3 P.M. shift.
    c) $11 \cong 1 \pmod 5$
       If this is week 1, then $1 + 1 \cong 2 \pmod 5$ indicates the 7 A.M. - 3 P.M. shift.

73. The waiter's schedule in a mod 14 system is given in the following table:

Day:	0	1	2	3	4	5	6	7	8	9	10	11	12	13
shift:	d	d	d	d	d	e	e	e	d	d	d	d	e	e

    Note: This is his second day shift which is day 1 in the mod 14 system.
    a) $20 \cong 14 = 1$, remainder 6. Six days from day 1 is day 7 which is the evening shift.
    b) $52 \cong 14 = 3$, remainder 10. Ten days from day 1 is day 11, which is the day shift.
    c) $365 \cong 14 = 26$, remainder 1. One day from day 1 is day 2, which is the day shift.

74. The truck driver's schedule is repeated every 17 days as indicated by the following table:

Days	Activity
0 - 2	N.Y. - Chicago
3	Rest in Chicago
4 - 6	Chicago - L.A.
7 - 8	Rest in L.A.
9 - 13	L.A. - N.Y.
14 – 16	

    a) $30 \cong 13 \pmod{17}$ indicates that he will be driving from L.A. to N.Y.
    b) $70 \cong 2 \pmod{17}$ indicates that he will be driving from N.Y. to Chicago.
    c) 2 years = 730 days $\cong 16 \pmod{17}$

75. a)

+	0	1	2	3
0	0	1	2	3
1	1	2	3	0
2	2	3	0	1
3	3	0	1	2

b) Yes. All the numbers in the table are from the set {0, 1, 2, 3}.

c) The identity element is 0.

d) Yes. element + inverse = identity
$0 + 0 = 0$ $1 + 3 = 0$ $2 + 2 = 0$ $3 + 1 = 0$

e) $(1 + 3) + 2\ 0 + 2 = 2$ $1 + (3 + 2) = 1 + 1 = 2$
Associative since $2 = 2$.

f) Yes, the table is symmetric about the main diagonal. $1 + 3 = 0 = 3 + 1$

76. a)

+	0	1	2	3	4	5	6	7
0	0	1	2	3	4	5	6	7
1	1	2	3	4	5	6	7	0
2	2	3	4	5	6	7	0	1
3	3	4	5	6	7	0	1	2
4	4	5	6	7	0	1	2	3
5	5	6	7	0	1	2	3	4
6	6	7	0	1	2	3	4	5
7	7	0	1	2	3	4	5	6

b) Yes. All the numbers in the table are from the set {0, 1, 2, 3, 4, 5, 6, 7}.

c) The identity element is 0.

d) elem. + inverse = identity
$0 + 0 = 0$ $1 + 7 = 0$ $2 + 6 = 0$ $3 + 5 = 0$
$4 + 4 = 0$ $5 + 3 = 0$ $6 + 2 = 0$ $7 + 1 = 0$

e) $(1 + 2) + 5 = 3 + 5 = 0$
$1 + (2 + 5) = 1 + 7 = 0$ Yes, Associative

f) Yes. $2 + 4 = 6 = 4 + 2$

g) Yes. All five properties are satisfied.

h) Same answer as problem 63 part h.

77. a)

□	0	1	2	3
0	0	1	2	3
1	1	2	3	0
2	2	3	0	1
3	3	0	1	2

b) Yes. All the elements in the table are from the set {0, 1, 2, 3}.

c) Yes. The identity element is 1.

d) elem. □ inverse = identity
$0\ □$ none $= 1$ $1\ □\ 1 = 1$ $2\ □$ none $= 1$
$3\ □\ 3 = 1$ Elements 0 and 2 do not have inverses.

e) $(1\ □\ 3)\ □\ 0 = 3\ □\ 0 = 0$
$1\ □\ (3\ □\ 0) = 1\ □\ 0 = 0$ Yes, Associative

f) Yes. $2\ □\ 3 = 2 = 3\ □\ 2$

g) No. Not all elements have inverses.

78. a)

□	0	1	2	3	4	5	6
0	0	0	0	0	0	0	0
1	0	1	2	3	4	5	6
2	0	2	4	6	1	3	5
3	0	3	6	2	5	1	4
4	0	4	1	5	2	6	3
5	0	5	3	1	6	4	2
6	0	6	5	4	3	2	1

b) Yes. All the elements in the table are from the set {0, 1, 2, 3, 4, 5, 6}.

c) Yes. The identity element is 1.

d) No. elem. → inverse $0 →$ none $1 → 1$
$2 → 4$ $3 → 5$ $4 → 2$ $5 → 3$ $6 → 6$
The element 0 does not have an inverse.

e) $(1\ □\ 2)\ □\ 4 = 2\ □\ 4 = 1$
$1\ □\ (2\ □\ 4) = 1\ □\ 1 = 1$ Yes, Associative

f) Yes. $2\ □\ 3 = 6 = 3\ □\ 2$

g) No. 0 does not have an inverse.

For the operation of division in modular systems, we define $n \div d = n \bullet i$, where i is the multiplicative inverse of d.

79. $5 \div 7 \cong ?$ (mod 9)

Since $7 \bullet 4 = 28 \cong 1$ (mod 9), 4 is the inverse of 7.

Thus, $5 \div 7 = 0$ R 2 $5\ \ 7 \cong 2$ (mod 9) $? = 2$

80. $? \div 5 \cong 5$ (mod 9)

Since 5 $5 \cong 5$ (mod 9), $1 \cong 5$ (mod 5) $? = 7$

81. $? \div ? \cong 1$ (mod 4) $0 \div 0$ is undefined.

    $1 \div 1 \cong 1$)mod 4) $2 \div 2 \cong 1$ (mod 4)

    $3 \div 3 \cong 1$ (mod 4) $? = \{1, 2, 3\}$

82. $1 \div 2 \cong ?$ (mod 5)

    $2(1/2) \cong 3?$ $1 \cong 6$ (mod 5) $1 = 1$

    $? = 3$

83. $5k \cong x$ (mod 5) $5(1) \cong 0$ (mod 5)

    $5(2) = 10 \cong 0$ (mod 5) $x = 0$

84. $5k + 4 \cong x$ (mod 5) $5(1) + 4 = 9 \cong 4$ (mod 5)

    $5(2) + 4 = 14 \cong 4$ (mod 5) $x = 4$

85. $4k - 2 \cong x$ (mod 4) $4(0) - 2 = -2 \cong 2$ (mod 4)

    $4(1) - 2 = 2 \cong 2$ (mod 4) $4(2) - 2 = 6 \cong 2$ (mod 4)

    $x = 2$

86. Check the numbers divisible by 5 until you find
    one that is also congruent to 2 in modulo 6.
    $20 \cong 2$ (mod 6) and 20 is also divisible by 5.

87. (365 days)(24 hrs./day)(60 min./hr.) = 525,600 hrs.

    $(525,600)/(4) = 131,400$ rolls $131400 \cong 0$ (mod 4)

88. 1 yr. 21 days = 365 + 21 = 386 days

    $386/5 = 77$ R 1   Halfway up the mountain

89. If 10 is subtracted from each number on the wheel,

    23 11   3 18 10 19   2 10 16   4 24   becomes

    13   1 20   8   0   9 19   0   6 21 14   which is equivalent to

    M   A   T   H   I S   F   U   N

Review Exercises

1. A set of elements and at least one binary operation.
2. A binary operation is an operation that can be performed on two and only two elements of a set. The result is a single element.
3. Yes. The sum of any two integers is always an integer.
4. No. Example: $2 - 3 = -1$, but -1 is not a natural number.
5. $9 + 10 = 19 \cong 7$ (mod 12)   6. $5 + 12 = 17 \cong 5$ (mod 12)   7. $8 - 10 = -2 \cong 10$ (mod 12)
8. $4 + 7 + 9 = 20 \cong 8$ (mod 12)   9. $7 - 4 + 6 = 9 \cong 9$ (mod 12)   10. $2 - 8 - 7 = -13 \cong 11$ (mod 12)
11. a) The system is closed. If the binary operation is $\boxdot$ then for any elements a and b in the set, a ⊡ b is a member of the set.

    b) There exists an identity element in the set. For any element a in the set, if a $\boxdot$ i = i $\boxdot$ a = a, then i is called the identity element.

    c) Every element in the set has a unique inverse. For any element a in the set, there exists an element b such that a $\boxdot$ b = b $\boxdot$ a = i. Then b is the inverse of a, and a is the inverse of b.

    d) The set is associative under the operation For elements a, b, and c in the set, (a $\boxdot$ b) $\boxdot$ c = a $\boxdot$ (b $\boxdot$ c).

12. An Abelian group is a group in which the operation has the commutative property.

13. Yes. Closure: The sum of any two integers is an   Yes, Associative   Example: $(2 + 3) + 4 = 2 + (3 + 4)$
    integer. The identity element is zero.   Each element has a unique inverse.

14. The set of integers with the operation of multiplication does not form a group since not all elements have an inverse. $4 \bullet \underline{?} = 1$

15. Yes. Closure: The sum of any two rational #s is a rational number.
    The identity element is zero.   Ex.: $5 + 0 = 0 + 5 = 5$

    Yes, Associative   Example: $(2 + 3) + 4 = 2 + (3 + 4)$
    Each element has a unique inverse. Ex. ; $6 + (-6) = 0$

16. The set of rational numbers with the operation of multiplication does not form a group since zero does not have an inverse. $0 \bullet \underline{?} = 1$

17. There is no identity element. Therefore the system does not form a group.

18. Not Associative   Example: $(! \square p) \square ? = p \square ? = !$ $! \square (p \square ?) = ! \square ! = \triangle$ $! \neq \triangle$

19. Not Associative   Example: $(p ? p) ? 4 = L ? 4 = \#$ $p ? (p ? 4) = p ? L = 4$ $\# \neq 4$

20. a) { ⊢, ☉, ?, △ }
    b) ⌐⌐
    c) Yes. All the elements in the table are from the set { ⊢, ☉, ?, △ }.

    d) The identity element is ⊢ .
    e) Yes. elem. ⊢ inverse = identity
       ⊢ ⌐⌐ ⊢ = ⊢   ☉ ⌐⌐ △ = ⊢
       ? ⌐⌐ ? = ⊢   △ ⌐⌐ ☉ = ⊢

    f) Yes, Associative
       (⊢ ⌐⌐ ?) ⌐⌐ △ = ? ⌐⌐ △ = ☉
       ⊢ ⌐⌐ (? ⌐⌐ △) = ⊢ ⌐⌐ ☉ = ☉
    g) Yes. △ ⌐⌐ ? = ☉ = ? ⌐⌐ △
    h) Yes, all five properties are satisfied.

21. $21 \div 3 = 7$, remainder 0 $21 \cong 0 \pmod 3$

22. $31 \div 8 = 3$, remainder 7 $31 \cong 7 \pmod 8$

23. $31 \div 6 = 5$, remainder 1 $31 \cong 1 \pmod 6$

24. $59 \div 8 = 7$, remainder 3 $59 \cong 3 \pmod 8$

25. $82 \div 13 = 6$, remainder 4 $82 \cong 4 \pmod{13}$

26. $54 \div 4 = 13$, remainder 2 $54 \cong 2 \pmod 4$

27. $52 \div 12 = 4$, remainder 4 $52 \cong 4 \pmod{12}$

28. $54 \div 14 = 3$, remainder 12 $54 \cong 12 \pmod{14}$

29. $97 \div 11 = 8$, remainder 9 $97 \cong 9 \pmod{11}$

30. $42 \div 11 = 3$, remainder 9 $42 \cong 9 \pmod{11}$

31. $5 + 8 = 13 \cong 4 \pmod 9$
    Thus, replace ? with 4.

32. $? - 3 \cong 0 \pmod 5$
    $0 - 3 \cong 2 \pmod 5$ $1 - 3 \cong 3 \pmod 5$
    $2 - 3 \cong 4 \pmod 5$ $3 - 3 \cong 0 \pmod 5$
    Replace ? with 3.

33. $4 \bullet ? \cong 3 \pmod 6$
    $4 \bullet 0 \cong 0 \pmod 6$ $4 \bullet 1 \cong 4 \pmod 6$
    $4 \bullet 2 = 8 \cong 2 \pmod 6$ $4 \bullet 3 = 12 \cong 0 \pmod 6$
    $4 \bullet 4 = 16 \cong 4 \pmod 6$ $4 \bullet 5 = 20 \cong 2 \pmod 6$
    There is no solution. $? = \{ \ \}$

34. $6 - ? \cong 5 \pmod 7$

    $6 - 0 \cong 6 \pmod 7$ $6 - 1 \cong 5 \pmod 7$

    $6 - 2 \cong 4 \pmod 7$ $6 - 3 \cong 3 \pmod 7$

    $6 - 4 \cong 2 \pmod 7$ $6 - 5 \cong 1 \pmod 7$

    Replace ? with 1.

35. $? \bullet 4 \cong 0 \pmod 8$

    $0 \bullet 4 \cong 0 \pmod 8$ $1 \bullet 4 \cong 4 \pmod 8$

    $2 \bullet 4 = 8 \cong 0 \pmod 8$ $3 \bullet 4 = 12 \cong 4 \pmod 8$

    $4 \bullet 4 = 16 \cong 0 \pmod 8$ $5 \bullet 4 = 20 \cong 4 \pmod 8$

    $6 \bullet 4 = 24 \cong 0 \pmod 8$ $7 \bullet 4 = 28 \cong 4 \pmod 8$

    Replace ? with $\{0, 2, 4, 6\}$.

36. $10 \bullet 7 \cong ? \pmod{12}$

    $10 \bullet 7 = 70$;   70 $12 \cong 5$, remainder 10

    Thus, $10 \bullet 7 \cong 10 \pmod{12}$.

    Replace ? with 10.

37. $3 - 5 \cong ? \pmod 7$

    $3 - 5 = (3+7) - 5 = 5 \cong 5 \pmod 7$

    Replace ? with 5.

38. $? \bullet 7 \cong 3 \pmod{10}$

    $0 \bullet 7 \cong 0 \pmod{10}$ $1 \bullet 7 \cong 7 \pmod{10}$

    $2 \bullet 7 = 14 \cong 4 \pmod{10}$ $3 \bullet 7 = 21 \cong 1 \pmod{10}$

    $4 \bullet 7 = 28 \cong 8 \pmod{10}$ $5 \bullet 7 = 35 \cong 5 \pmod{10}$

    $6 \bullet 7 = 42 \cong 2 \pmod{10}$ $7 \bullet 7 = 49 \cong 9 \pmod{10}$

    $8 \bullet 7 = 56 \cong 6 \pmod{10}$ $9 \bullet 7 = 63 \cong 3 \pmod{10}$

    $10 \bullet 7 = 70 \cong 0 \pmod{10}$

    Replace ? with 9.

39. $5 \bullet ? \cong 3 \pmod 8$

    $5 \bullet 0 \cong 0 \pmod 8$ $5 \bullet 1 \cong 5 \pmod 8$

    $5 \bullet 2 = 10 \cong 2 \pmod 8$ $5 \bullet 3 = 15 \cong 7 \pmod 8$

    $5 \bullet 4 = 20 \cong 4 \pmod 8$ $5 \bullet 5 = 25 \cong 1 \pmod 8$

    $5 \bullet 6 = 30 \cong 6 \pmod 8$ $5 \bullet 7 = 35 \cong 3 \pmod 8$

    Replace ? with 7.

40. $7 \bullet ? \square 2 \pmod 9$

    $7 \bullet 0 \cong 0 \pmod 9$ $7 \bullet 1 \cong 7 \pmod 9$

    $7 \bullet 2 = 14 \cong 5 \pmod 9$ $7 \bullet 3 = 21 \cong 3 \pmod 9$

    $7 \bullet 4 = 28 \cong 1 \pmod 9$ $7 \bullet 5 = 35 \cong 7 \pmod 9$

    $7 \bullet 6 = 42 \cong 6 \pmod 9$ $7 \bullet 7 = 49 \cong 4 \pmod 9$

    $7 \bullet 8 = 56 \cong 2 \pmod 9$

    Replace ? with 8.

41. a)

+	0	1	2	3	4	5
0	0	1	2	3	4	5
1	1	2	3	4	5	0
2	2	3	4	5	0	1
3	3	4	5	0	1	2
4	4	5	0	1	2	3
5	5	0	1	2	3	4

41. b) Since all the numbers in the table are elements of
    $\{0, 1, 2, 3, 4, 5\}$, the system has the closure
    property.

    c) The commutative property holds since the
    elements are symmetric about the main
    diagonal.

    d) The identity element is 0 and the inverses of
    each element are $0 - 0$, $1 - 5$, $2 - 4$, $3 - 3$,
    $4 - 2$, $5 - 1$

    e) If it is assumed the associative property holds as
    illustrated by the example: $(2 + 3) + 5 = 4 =$
    $2 + (3 + 5)$, then the system is a commutative
    group.

42. a)

$\square$	0	1	2	3
0	0	1	2	3
1	1	2	3	0
2	2	3	0	1
3	3	0	1	2

   b) The identity element is 1, but because 0 and 2
   have no inverses, the system does not form a
   group.

43. Day (mod 10):   0   1   2   3   4   5   6   7   8   9
    Work/off :   w   w   w   o   o   w   o   o   o   o

    a) If today is the first day of her work pattern, day 0, then $18 \cong 8 \pmod{10}$ indicates Toni will not be working
    in 18 days.

    b) $38 \cong 8 \pmod{10}$ indicates that Toni will have the evening off in 38 days.

Chapter Test

1. A mathematical system consists of a set of elements and at least one binary operation.

2. Closure, identity element, inverses, associative property, and commutative property.

3. No, the numbers greater than 0 do not have inverses.

4.

+	1	2	3	4	5
1	2	3	4	5	1
2	3	4	5	1	2
3	4	5	1	2	3
4	5	1	2	3	4
5	1	2	3	4	5

5. Yes. It is closed since the only elements in the table
   are from the set $\{1, 2, 3, 4, 5\}$. The identity element
   is 5. The inverses are $1 - 4$, $2 - 3$, $3 - 2$, $4 - 1$, and
   $5 - 5$. The system is associative. The system is
   commutative since the table is symmetric about the
   main diagonal. Thus, all five properties are satisfied.

6. $9 + 3 + 2 = 14 \cong 4 \bmod 5$

7. $5 - 18 = (15 + 5) - 18 = 20 - 18 = 2 \cong 2 \bmod 5$

8. a) The binary operation is Δ.

   b) Yes. All elements in the table are from the set $\{W, S, T, R\}$.

   c) The identity element is T, since $T \Delta x = x = x \Delta T$, where x is any member of the set $\{W, S, T, R\}$.

   d) The inverse of R is S, since $R \Delta S = T$

   e) $(T \Delta R) \Delta W = R \Delta W = S$

9. The system is not a group. It does not have the closure property since $c \cdot c = d$, and d is not a member of $\{a, b, c\}$.

10. Since all the numbers in the table are elements of $\{1, 2, 3\}$, the system is closed. The commutative property
    holds since the elements are symmetric about the main diagonal. The identity element is 2 and the inverses are
    $1 - 3$, $2 - 2$, $3 - 1$. If it is assumed the associative property holds as illustrated by the example:
    $(1 \; ? \; 2) \; ? \; 1 = 2 = 1 \; ? \; (2 \; ? \; 3)$, then the system is a commutative group.

11. Since all the numbers in the table are elements of { @, $, &, % }, the system is closed. The commutative property holds since the elements are symmetric about the main diagonal. The identity element is $ and the inverses are @ – &, $ – $, & – @, % – %. It is assumed the associative property holds as illustrated by the example: (@ O $) O % = & = @ O ($ O %), then the system is a commutative group.

12. $64 \div 9 = 7$, remainder 1 $64 \cong 1 \pmod 9$

13. $58 \div 11 = 5$, remainder 3 $3 \cong 1 \pmod{11}$

14. $7 + 7 = 6 \bmod 8$

15. $2 - 3 = (5 + 2) - 3 = 4 \cong 4 \bmod 5$

16. $3 - 5 \cong 7 \pmod 9$
    $3 - 5 = (3 + 9) - 5 = 12 - 5 \cong 7 \pmod 9$
    $12 - 5 \cong 7 \pmod 9$
    Replace ? with 5.

17. $4 \bullet 2 = 8$ and $8 \div 6 = 1$, remainder 2
    $4 \bullet 2 \cong 2 \pmod 6$
    Replace ? with 2.

18. $3 \bullet ? \bullet \cong 2 \pmod 6$
    $3 \bullet 0 \cong 0 \pmod 6$ $3 \bullet 1 \cong 3 \pmod 6$
    $3 \bullet 2 \cong 0 \pmod 6$ $3 \bullet 3 \cong 3 \pmod 6$
    $3 \bullet 4 \cong 0 \pmod 6$ $3 \bullet 5 \cong 3 \pmod 6$
    There is no solution for ?   The answer is { }.

19. $103 \div 7 = 14$, remainder 5
    $103 \cong 5 \pmod 7$
    Replace ? with 5.

20. a)

□	0	1	2	3	4
0	0	0	0	0	0
1	0	1	2	3	4
2	0	2	4	1	3
3	0	3	1	4	2
4	0	4	3	2	1

   b) The system is closed. The identity is 1. However, 0 does not have an inverse, so the system is <u>not</u> a group.

Group Projects

1. a)

♣	A	B	C	D
A	B	C	D	A
B	C	D	A	B
C	D	A	B	C
D	A	B	C	D

b) The system is closed. The identity is D.

c) (A ♣ B) ♣ C = C ♣ C = B

A ♣ (B ♣ C) = A ♣ A = B

Yes, Associative

d) A ♣ C = D   B ♣ B = D   C ♣ A = D

D ♣ D = D   All elements have inverses.

e) A ♣ B = C = B ♣ A

Yes, Commutative, symmetrical around the main diagonal

Therefore, the system is a group.

2. a) Yes, see Group Project exercise 3. a).

b) Product = 0 when factors   0

mod 4, mod 6, mod 8, mod 9

c) Product = 0 when at least 1 factor = 0

mod 3, mod 5, mod 7

d) The systems in which the modulo is a composite number system have factors   0.

3. a)

mod 3

•	0	1	2
0	0	0	0
1	0	1	2
2	0	2	1

mod 4

•	0	1	2	3
0	0	0	0	0
1	0	1	2	3
2	0	2	0	2
3	0	3	2	1

3. a)

mod 5

•	0	1	2	3	4
0	0	0	0	0	0
1	0	1	2	3	4
2	0	2	4	1	3
3	0	3	1	4	2
4	0	4	3	2	1

mod 6

•	0	1	2	3	4	5
0	0	0	0	0	0	0
1	0	1	2	3	4	5
2	0	2	4	0	2	4
3	0	3	0	3	0	3
4	0	4	2	0	4	2
5	0	5	4	3	2	1

3. a)

mod 7

•	0	1	2	3	4	5	6
0	0	0	0	0	0	0	0
1	0	1	2	3	4	5	6
2	0	2	4	6	1	3	5
3	0	3	6	2	5	1	4
4	0	4	1	5	2	6	3
5	0	5	3	1	6	4	2
6	0	6	5	4	3	2	1

mod 8

•	0	1	2	3	4	5	6	7
0	0	0	0	0	0	0	0	0
1	0	1	2	3	4	5	6	7
2	0	2	4	6	0	2	4	6
3	0	3	6	1	4	7	2	5
4	0	4	0	4	0	4	0	4
5	0	5	2	7	4	1	6	3
6	0	6	4	2	0	6	4	2
7	0	7	6	5	4	3	2	1

mod 9

•	0	1	2	3	4	5	6	7	8
0	0	0	0	0	0	0	0	0	0
1	0	1	2	3	4	5	6	7	8
2	0	2	4	6	8	1	3	5	7
3	0	3	6	0	3	6	0	3	6
4	0	4	8	3	7	2	6	1	5
5	0	5	1	6	2	7	3	8	4
6	0	6	3	0	6	3	0	6	3
7	0	7	5	3	1	8	6	4	2
8	0	8	7	6	5	4	3	2	1

3. b) mod 3, mod 5, mod 7

c) mod 4, mod 6, mod 8, mod 9

d) Modulo systems that have composite numbers have multiplicative inverses for all nonzero numbers.

CHAPTER ELEVEN

CONSUMER MATHEMATICS

Exercise Set 11.1

1. A percent is a ratio of some number to 100.
2. (i) Divide the number by 100.　　　(ii) Remove the percent sign.
3. (i) Divide the number by the denominator.
   (ii) Multiply the quotient by 100 (which has the effect of moving the decimal point two places to the right).
   (iii) Add a percent sign.
4. Multiply the decimal number by 100 and add a percent sign.

5. Percent change = $\dfrac{(\text{Amount in latest period}) - (\text{Amount in previous period})}{\text{Amount in previous period}} \times 100$

6. Percent markup on cost = $\dfrac{\text{Selling Price} - \text{Dealer's Cost}}{\text{Dealer's Cost}} \times 100$

7. $\dfrac{1}{2} = 0.500 = (0.500)(100)\% = 50.0\%$

8. $\dfrac{1}{4} = 0.25 = (0.25)(100)\% = 25.0\%$

9. $\dfrac{2}{5} = 0.400 = (0.400)(100)\% = 40.0\%$

10. $\dfrac{7}{8} = 0.875 = (0.875)(100)\% = 87.5\%$

11. $0.007654 = (0.007654)(100)\% = 0.8\%$

12. $0.5688 = (0.5688)(100)\% = 56.9\%$

13. $3.78 = (3.78)(100)\% = 378.0\%$

14. $13.678 = (13.678)(100)\% = 1367.8\%$

15. $4\% = \dfrac{4}{100} = 0.04$

16. $6.9\% = \dfrac{6.9}{100} = 0.069$

17. $1.34\% = \dfrac{1.34}{100} = 0.0134$

18. $0.0005\% = \dfrac{0.0005}{100} = 0.000005$

19. $\dfrac{1}{4}\% = 0.25\% = \dfrac{0.25}{100} = 0.0025$

20. $\dfrac{3}{8}\% = 0.375\% = \dfrac{0.375}{100} = 0.00375$

21. $\dfrac{1}{5}\% = 0.2\% = \dfrac{0.2}{100} = 0.002$

22. $135.9\% = \dfrac{135.9}{100} = 1.359$

23. $1\% = \dfrac{1}{100} = 0.01$

24. $0.50\% = \dfrac{0.50}{100} = 0.005$

25. $\dfrac{95}{3500} \approx .0271428571 = (0.0271428571 \times 100)\% = $ 2.714 %

26. $50 + 50 + 40 + 40 = 180 \qquad (180)(.06) = \$\,10.80$

27. $8(.4125) = 3.3 \qquad 8.0 - 3.3 = 4.7 \text{ g}$

28. $(693,905)(.36) = 249,805.8 \text{ miles}$

29. $(591,000)(.08) = \$\,47,280,000 = \$\,47.28 \text{ M}$

30. $(591 \text{ M})(.06) = \$\,35.46 \text{ M}$

31. $(591 \text{ M})(.27) = \$\,159.57 \text{ M}$

32. $(591 \text{ M})(.59) = \$\,348.69 \text{ M}$

33. $(32.3 \text{ B})(.176) = \$\,5.6848 \text{ B}$

34. $(32.3 \text{ B})(.058) = \$\,1.8734 \text{ B}$

35. $(32.3 \text{ B})(.306) = \$\,9.8838 \text{ B}$

36. $(32.3 \text{ B})(.46) = \$\,14.858 \text{ B}$

37. $\dfrac{1,553 \text{ M}}{8,105 \text{ M}} = .1916 \qquad (.1916)(100) = 19.2\%$

38. $\dfrac{1,392 \text{ M}}{8,105 \text{ M}} = .1717 \qquad (.1717)(100) = 17.2\%$

39. $\dfrac{1,592 \text{ M}}{8,105 \text{ M}} = .1964 \qquad (.1964)(100) = 19.6\%$

40. $\dfrac{2,188 \text{ M}}{8,105 \text{ M}} = .2699 \qquad (.2699)(100) = 27.0\%$

41. $\dfrac{45,793}{48,622} = .942 \qquad 100 - 94.2 \approx 5.8\%$

42. a) $\dfrac{288.4 \text{ M}}{248.7 \text{ M}} = 1.1596 \qquad 115.96 - 100.0 \approx 16\%$

42. b) $(288.4 \text{ M})(1.1596) = 334.4 \text{ M}$

43. a) $\dfrac{9,457 \text{ M}}{8,059 \text{ M}} = 1.1735 \qquad 117.35 - 100.0 = 17.3\%$

    b) $\dfrac{13,577 \text{ M}}{9,457 \text{ M}} = 1.4357 \qquad 143.57 - 100.0 = 43.6\%$

    c) $\dfrac{20,947 \text{ M}}{13,577 \text{ M}} = 1.5428 \qquad 154.28 - 100.0 = 54.3\%$

    d) $\dfrac{32,240 \text{ M}}{20,947 \text{ M}} = 1.5391 \qquad 153.91 - 100.0 = 53.9\%$

44. a) $\dfrac{9,457 \text{ M}}{8,059 \text{ M}} = 1.1735 \qquad 117.35 - 100.0 = 17.3\%$

    b) $\dfrac{13,577 \text{ M}}{9,457 \text{ M}} = 1.4357 \qquad 143.57 - 100.0 = 43.6\%$

    c) $\dfrac{9,457 \text{ M}}{8,059 \text{ M}} = 1.1735 \qquad 117.35 - 100.0 = 17.3\%$

    d) $\dfrac{13,577 \text{ M}}{9,457 \text{ M}} = 1.4357 \qquad 143.57 - 100.0 = 43.6\%$

45. a) $\dfrac{10,403.94}{9,920} = 1.0488 \qquad 104.88 - 100.0 = 4.9\%$

    b) $\dfrac{7,591.93}{10,403.94} = 0.7297 \qquad 100.0 - 72.97 = 27.0\%$

    c) $\dfrac{7,591.93}{9,920} = 0.7653 \qquad 100.0 - 76.53 = 23.5\%$

    d) $\dfrac{8,397.03}{7,591.93} = 1.1060 \qquad 110.60 - 100.0 = 10.6\%$

46. a) $\dfrac{50,000}{25,000} = 2.00 \qquad 200.0 - 100.0 = 100\%$

    b) $\dfrac{100,000}{75,000} = 1.3333 \qquad 133.33 - 100.0 = 33.3\%$

    c) $\dfrac{400,000}{200,000} = 2.00 \qquad 200.0 - 100.0 = 100\%$

    d) $\dfrac{400,000}{25,000} = 16.00 \qquad 1600.0 - 100.0 = 1500\%$

47. $(.15)(45) = \$6.75$

48. $(.065)(150) = \$9.75$

49. $24/96 = .25$ $(.25)(100) = 25.0\%$

50. $15/75 = .20$ $(.20)(100) = 20.0\%$

51. $.05x = 75$ $x = 75/.05 = 300$

52. $10x = 75$ $x = 75/.10 = 750$

53. a) tax = 6% of $43.50 = $(0.06)(43.50) = \$2.61$
    b) total bill before tip = $\$43.50 + \$2.61 = \$46.11$
    c) tip = 15% of $46.11 = 0.15(46.11) = \$6.92$
    d) total cost = $46.11 + 6.92 = \$53.03$

54. 25% of what number is 10?

$$0.25x = 10 \qquad x = \frac{10}{0.25} = 40$$

original number of crew is 40.

55. $1.50(x) = 18$ $x = \dfrac{18}{1.50} = 12$

12 students got an A on the 2nd test.

56. $0.30(x) = 57$ $x = \dfrac{57}{0.30} = 190$

The original number of employees was 190.

57. Mr. Browns' increase was $0.07(36,500) = \$2,555$
His new salary = $\$36,500 + \$2,555 = \$39,055$

58. $(0.17)(300) = x = 51$ 51 prefer Ranch.

59. Percent change = $\left(\dfrac{407 - 430}{430}\right)(100) =$

$\left(\dfrac{-23}{430}\right)(100) = -5.3\%$

There was a 5.3% decrease in the # of units sold.

60.

Percent markup = $\left(\dfrac{699 - 320}{320}\right)(100) = \dfrac{379}{320} = 1.184$

$(1.184)(100) = 118.4\%$

61. $\dfrac{31.1\,\text{M}}{39.3\,\text{M}} = .7913$ $100.0 - 79.13 = 20.9\%$

62. Percent increase in great grandchildren =

$\left(\dfrac{12 - 8}{8}\right)(100) = 0.50$ $\rightarrow$ $(0.50(100) = 50.0\%$

63. Percent decrease from regular price =

$\left(\dfrac{\$439 - 539.62}{539.62}\right)(100) = \left(\dfrac{-100.62}{539.62}\right)(100) =$

- 18.6 %
The sale price is 18.6% lower than the regular price.

64. Percent markup = $\left(\dfrac{11.95 - 7.95}{7.95}\right)(100) =$

$(0.5031)(100) = 50.3\%$

65. $(0.18)(\text{sale price}) = \$675$

sale price = $\dfrac{675}{0.18} = \$3,750$

66. No, 15% of $115 is $(0.15)(\$115) = \$17.25$
The sale price should be $115 - 17.25 = \$97.75$
not $100.

67. $1000 increased by 10% is $\$1000 + 0.10(\$1000) = \$1000 + \$100 = \$1,100.$
$1,100 decreased by 10% is $\$1,100 - 0.10(\$1,100) = \$1,100 - \$110 = \$990.$
Therefore if he sells the car at the reduced price he will lose $10.

68. a) No, the 25% discount is greater. (see part b)
    b) $189.99 - 0.10(189.99) = 189.99 - 19.00 = 170.99$ $170.99 - 0.15(170.99) = 170.99 - 25.65 = \$145.34$
    c) $189.99 - 0.25(189.99) = 189.99 - 47.50 = \$142.49$
    d) Yes

69. Total profit must = $0.40(\$5901.79) = \$2,360.72$ Total revenue must = $\$5901.79 + 2360.72 = \$8,262.51$
Revenue from first sale = $100 \times \$9.00 = \$900$ Revenue from second sale = $150 \times \$12.50 = \$1,875.00$

Exercise Set 11.2

1. Interest is the money the borrower pays for the use of the lender's money.
2. The amount of money that a bank is willing to lend to a person is called credit.
3. Security is anything of value pledged by the borrower that the lender may sell or keep if the borrower does not repay the loan.
4. A cosigner is a person, other than the person who received the loan, who guarantees that a loan will be repaid.
5. i = interest, p = principal, r = interest rate, t = time
   The rate and time must be expressed for the same period of time, i.e. days, months or years.
6. A personal note is an agreement that states the conditions of the loan.
7. The difference between ordinary interest and interest calculated using the Banker's rule is the way in which time is used in the simple interest formula. Ordinary interest: a month is 30 days and year is 360 days. Banker's rule: any fractional part of a year is the exact number of days, and a year is 360 days.
8. The United States Rule states that if a partial payment is made on a loan, interest is computed on the principal from the first day of the loan (or previous partial payment) up to the date of the partial payment. For each partial payment, the partial payment is used to pay the interest first, then the remainder of the payment is applied to the principle. On the due date of the loan the interest is calculated from the date of the last partial payment.

9. $i = prt = (300)(.04)(5) = \$60.00$

10. $(450)(.055)(2) = \$49.50$

11. $(900)(.0375)(30/360) = \$2.81$

12. $i = (365.45)(0.115)\left(\dfrac{8}{12}\right) = \$28.02$

13. $i = prt = (587)(0.00045)(60) = \$15.85$

14. $i = (6,742.75)(0.0605)\left(\dfrac{90}{360}\right) = \$101.98$

15. $i = (2,756.78)(0.1015)\left(\dfrac{103}{360}\right) = \$80.06$

16. $i = (550.31)(0.089)\left(\dfrac{67}{360}\right) = \$9.12$

17. $i = (1372.11)(.01375)(12)(.5) = \$113.20$

18. $i = (41864)(.000375)(360)\left(\dfrac{60}{360}\right) = \$941.94$

19. $(1500)(r)(3 = 450 \qquad r = \left(\dfrac{450}{4500}\right)(100) = 10.0\%$

20. $p(.03)\left(\dfrac{90}{360}\right) = 600 \qquad p = \left(\dfrac{600}{.0075}\right) = \$80,000.00$

21. $12.00 = p(0.08)\left(\dfrac{3}{12}\right) = p(0.02) \quad p = \dfrac{12.00}{0.02} = \$600$

22. $64.00 = (800)(0.06)(t) = 48t$
    $t = \dfrac{64.00}{48} = 1.\overline{33}$ years, or 1 yr. 4 months

23. $124.49 = (957.62)(0.065)(t) = 62.2453t$
    $\dfrac{124.49}{62.2453} = t \qquad t = 2$ years

24. $343.20 = (1650.00)(r)(6.5) = 10725r$
    $\dfrac{343.20}{10725} = r \qquad r = 0.032$ or 3.2% per year

25. $i = (1000)(.03)(1) = \$30.00 \qquad 15 + 1000 = \$1015$

26. a) $(4500)(.0475)(3) = \$641.25$
    b) $4500 + 641.25 = \$5,141.25$

27. a) i= prt   i = (3500)(0.075)(6/12) = \$131.25
    b) A= p + i   A = 3500 + 131.25 = \$3,631.25

28. a) i= prt   I = (2500)(0.08)(5/12) = \$83.33
    b) 2500.00 – 83.33 = \$2416.67
    c) i= prt   83.33 = (2416.67)(5/12) = 1006.95r

$$\frac{83.33}{1006.95} = r = 0.08275 \text{ or } 8.3\%$$

29. a) i= prt   I = (3650)(0.075)(8/12) = \$182.50
    b) 3650.00 – 182.50 = \$3467.50, which is the amount Julie received.
    c) i= prt   182.50 = (3467.50)(r)(8/12) = 2311.67r

$$\frac{182.50}{2311.67} = r = 0.0789 \text{ or } 7.9\%$$

30. a) 0.80x = 350   x = 350/0.80 = \$437.50
    \$437.50 is needed in savings
    b) $3\frac{1}{4}\% + 2\% = 5\frac{1}{4}\%$
    c) i = prt   i = (350)(0.0525)(0.5) = \$9.19
    A = p + i = 350 + 9.19 = \$359.19

31. Amt. collected = (470)(4500/2) = \$1,057,500
    i = prt = (1,057,500)(0.054)(5/12) = \$23,793.75

32. i = 80.25 – 75.00 = 5.25
    5.25 = (75.00)(r)(14/360) = 2.92r

$$r = \frac{5.25}{2.92} = 1.80 \text{ or } 180\%$$

33. [Jan 17 – July 4] = 185 – 17 = 168 days

34. [06/19 – 02/12] = 170 – 43 = 127 days
    Because of Leap Year, 127 + 1 = 128 days

35. [12/08 – 03/17] = 342 – 76 = 266 days

36. [06/14 – 01/24] = (365 – 165) + 24 = 200 + 24 = 224 days

37. [08/24 – 05/15] = (365 – 236) + 135 = 129 + 135 = 264 days

38. [12/21 – 04/28] = (365 – 355) + 118 = 10 + 118 = 128 days

39. [04/15] for 60 days   105 + 60 = 165, which is June 14

40. [05/18] for 180 days   138 + 180 = 318, which is November 14

41. [11/25] for 120 days   329 + 120 = 449;
    449 – 365 = 84   84 – 1 leap year day = day 83, which is March 24

42. July 5 for 210 days   186 + 210 = 396;
    396 – 365 = day 31, which is January 31

43. [03/01 to 05/01] = 91 – 60 or 30 days

    (2000)(.05)(31/360) = 8.61   400.00 - 8.61 = 391.39
    2000.00 – 391.39 = \$1608.61

    (1608.61)(.05)(31/360) = 6.70
    1608.61 + 6.70 = \$1615.31

44. [01/15 to 03/01] = 60 – 15 or 45 days

    (4500)(.03)(45/360) = 16.875
    2000.00 – 16.875 = \$1983.125
    4500.00 – 1983.125 = \$2516.875

    (2516.875)(.03)(5/360) = 9.44
    2516.875 + 9.44 = \$2526.32

45. [08/01 to 11/15] = 319 − 213 or 106 days,
    to [12/15] = 30 days

    (7000)(.0575)(106/360) = 118.51
    3500.00 − 118.51 = $3381.49
    7000.00 − 3381.49 = $3618.51

    (3618.51)(.0575)(30/360) = 17.34
    3618.51 + 17.34 = $3635.85

46. [04/15 to 08/01] = 213 − 105 or 108 days,
    to [10/01] = 61 days

    (7500)(.12)(108/360) = 270.00
    1000.00 − 270.00 = $730.00
    7500.00 − 730.00 = $6770.00

    (6770)(.12)(61/360) = 137.66
    6770.00 + 137.66 = $6907.66

47. [07/15 to 12/27] = 361 − 196 or 165 days,
    to [02/01] = 4 + 32 = 36 days

    (9000)(.06)(165/360) = 247.50
    4000.00 − 247.50 = $3752.50
    9000.00 − 3752.50 = $5247.50

    (5247.50)(.06)(36/360) = 31.485
    5247.50 + 31.49 = $5278.99

48. [01/01 to 01/15] = 14 days,
    to [02/15] = 31 days

    (1000)(.125)(14/360) = 4.86
    300.00 − 4.86 = $295.14
    1000.00 − 295.14 = $704.86

    (704.86)(.125)(31/360) = 7.59
    704.86 + 7.59 = $712.45

49. [08/01 to 09/01] = 31 days,
    to [10/01] = 30 days
    to [11/01] = 31 days

    (1800)(.15)(31/360) = 23.25
    500.00 − 23.25 = $476.75
    1800.00 − 476.75 = $1323.25

    (1323.25)(.15)(30/360) = 16.54
    500.00 − 16.54 = 483.46
    1323.25 − 483.46 = $839.79

    (839.79)(.15)(31/360) = 10.85
    839.79 + 10.85 = $850.64

50. [10/15 to 11/15] = 31 days,
    to [12/15] = 30 days
    to [01/01] = 16 days

    (5000)(.14)(31/360) = 60.28
    800.00 − 60.28 = $739.72
    5000.00 − 739.72 = $4260.28

    (4260.28)(.14)(30/360) = 49.70
    800.00 − 49.70 = 750.30
    4260.28 − 750.30 = $3509.98

    (3509.98)(.14)(16/360) = 21.84
    3509.98 + 21.84 = $3531.82

51. [03/01 to 08/01] = 153 days,
    to [11/15] = 106 days
    to [12/01] = 16 days

    (11600)(.06)(153/360) = 295.80
    2000.00 − 295.80 = $1704.20
    11600.00 − 1704.20 = $9895.80

    (9895.80)(.06)(106/360) = 174.83
    4000.00 − 174.83 = 3825.17
    9895.8 − 3825.17 = $6070.63

    (6070.63)(.06)(16/360) = 16.19
    6070.63 + 16.19 = $6086.82

52. [07/12 to 10/10] = 90 days,
    to [12/08] = 59 days
    to [01/30] = 53 days

    (21000)(.04375)(90/360) = 229.69
    8000.00 − 229.69 = $7770.31
    21000.00 − 7770.31 = $13229.69

    (13229.69)(.04375)(59/360) = 94.86
    6000.00 − 94.86 = 5905.14
    13229.69 − 5905.14 = $7324.55

    (7324.55)(.04375)(53/360) = 47.18
    7324.55 + 47.18 = $7371.73

53. [03/01 to 05/01] = 61 days,
    to [07/01] = 61 days
    to [08/28] = 58 days

    (6500)(.105)(61/360) = 115.65
    1750.00 – 115.65 = $1634.35
    6500.00 – 1634.35 = $4865.65

    (4865.65)(.105)(61/360) = 86.57
    2350.00 – 86.57 = 2263.43
    4865.65 – 2263.43 = $2602.22

    (2602.22)(.105)(58/360) = 44.02
    2602.22 + 44.02 = $2646.24

54. [05/15 to 06/15] = 31 days,
    to [08/01] = 47 days
    to [09/01] = 31 days

    (3000)(.11)(31/360) = 28.42
    875.00 – 28.42 = $846.58
    3000.00 – 846.58 = $2153.42

    (2153.42)(.11)(47/360) = 30.93
    940.00 – 30.93 = 909.07
    2153.42 – 909.07 = $1244.35

    (1244.35)(.11)(31/360) = 11.79
    1244.35 + 11.79 = $1256.14

55. a) May 5 is day 125   125 + 182 = 307
       day 307 is Nov. 3
    b) i = (1000)(0.0434)(182/360) = $21.94
       Amt. paid = 1000 – 21.94 = $978.06
    c) interest = $21.94
    d) $r = \dfrac{i}{pt} = \dfrac{21.94}{978.06\left(\frac{182}{360}\right)} = 0.0444$ or 4.44%

56. a) Aug. 31 is day 243   243 + 364 = 607
       (607 – 1) – 365 = 241   day 241 is Aug. 29
    b) i = (6000)(0.044)(364/360) = $266.93
       Amt. paid = 6000 – 266.93 = $5,733.07
    c) interest = $266.93
    d) $r = \dfrac{266.93}{5733.07\left(\frac{364}{360}\right)} = 0.0460$ or 4.6%

57. a) Amt. received = 743.21 – 39.95 = $703.26
       i = prt
       39.95 = (703.26)(r)(5/360)
       39.95 = (9.7675)(r)
       r = 39.95/9.7675 = 4.09 or 409%
    b) 39.95 = (703.26)(r)10/360)
       39.95 = (19.535)(r)
       r = 39.95/19.535 = 2.045 or 204.5%
    c) 39.95 = (703.26)(r)(20/360)
       39.95 = (39.07)r
       r = 39.95/39.07 = 1.023 or 102.3%

58. a) (600)(.0675)(30/360) = 3.38
       200.00 + 3.38 = $203.38

       (400)(.07)(30/360) = 2.33
       200.00 + 2.33 = $202.33

       (200)(.0725)(30/360) = 1.21
       200.00 + 1.21 = $201.21

    b) 3.38 + 2.33 + 1.21 = $6.92= total interest

59. a) $\dfrac{93337}{100000} = 0.93337$
       1.00000 – 0.93337 = .06663 or 6.663 %
    b) 100000 – 93337 = $6663.00
    c) $\dfrac{100000}{93337} = 1.071386$
       1.071386 – 1.000000 = .071386 or 7.139 %
    d) (6663)(.05)(1) = 33.15
       6663.00 + 33.15 = $6696.15

60. a) [08/03/1492 to 12/01/1620]
       1492 to 1620 = 127 years = 45720 days
       08/03 to 12/31 = 365 – 215 = 150 days
       01/01 to 12/01 = 335 days
       45720 + 150 + 335 = 46205 days
       (1)(.05)(46205/360) = 6.417361 = $6.42
    b) [07/04/1776 to 08/03/1492]
       284 yrs. minus 30 days = 102,210
       (1)(.05)(102,210/360) = 14.1958 = $14.20
    c) [08/03/1492 to 12/07/1941]
       449 yrs plus 126 days = 161,766 days
       (1)(.05)(161,766/360) = $22.47
    d) Answers will vary.

Exercise Set 11.3

1. An investment is the use of money or capital for income or profit.
2. With a fixed investment the amount invested as principal is guaranteed and interest is computed at a fixed rate.
3. For a variable investment neither the principal nor the interest is guaranteed.
4. Interest that is computed on the principal and any accumulated interest is called compound interest.
5. The effective annual yield on an investment is the simple interest rate that gives the same amount of interest as a compound rate over the same period of time.
6. The principal that would have to be invested today to have a fixed amount of money in the future.

7. a) $n = 1, r = 2.0\%, t = 3, p = \$2000$

$$A = 2000\left(1+\frac{0.02}{1}\right)^{1\cdot3} = \$2122.42$$

   b) $i = \$2122.42 - \$2000 = \$122.42$

8. a) $n = 2, r = 2.0\%, t = 3, p = \$2000$

$$A = 2000\left(1+\frac{0.02}{2}\right)^{2\cdot3} = \$2123.04$$

   b) $i = \$2123.04 - \$2000 = \$123.04$

9. a) $n = 2, r = 3.0\%, t = 4, p = \$3500$

$$A = 3500\left(1+\frac{0.03}{2}\right)^{2\cdot4} = \$3942.72$$

   b) $i = \$3942.72 - \$3500 = \$442.72$

10. a) $n = 1, r = 3.0\%, t = 4, p = \$3500$

$$A = 3000\,3500\left(1+\frac{0.03}{1}\right)^{1\cdot4} = \$3939.28$$

   b) $i = \$3939.28 - \$3500 = \$439.28$

11. a) $n = 4, r = 4.75\%, t = 3, p = \$1500$

$$A = 1500\left(1+\frac{0.0475}{4}\right)^{4\cdot3} = \$1728.28$$

   b) $i = \$1728.28 - \$1500 = \$228.28$

12. a) $n = 4, r = 4.75\%, t = 4, p = \$1500$

$$A = 1500\left(1+\frac{0.0475}{4}\right)^{4\cdot4} = \$1811.85$$

   b) $i = \$1811.85 - \$1500 = \$311.85$

13. a) $n = 12, r = 6.25\%, t = 2, p = \$2500$

$$A = 2500\left(1+\frac{0.0625}{12}\right)^{12\cdot2} = \$2831.95$$

   b) $i = \$2831.95 - \$2500 = \$331.95$

14. a) $n = 12, r = 6.25\%, t = 2, p = \$3000$

$$A = 3000\left(1+\frac{0.0625}{12}\right)^{12\cdot2} = \$3398.34$$

   b) $i = \$3398.34 - \$3000 = \$398.34$

15. a) $n = 360, r = 4.59\%, t = 4 \text{ yr.}, p = \$4000$

$$A = 4000\left(1+\frac{0.0459}{360}\right)^{360\cdot4} = \$4806.08$$

   b) $i = \$4806.08 - \$4000 = \$806.08$

16. a) $n = 360, r = 4.59\%, t = 8 \text{ yr.}, p = \$4000$

$$A = 4000\left(1+\frac{0.0459}{360}\right)^{360\cdot8} = \$5774.61$$

   b) $i = \$5774.61 - \$4000 = \$1774.61$

17. $A = 7500\left(1+\frac{0.0266}{2}\right)^{2\cdot4} = \$8336.15$

18. $A = 9500\left(1+\frac{0.0412}{4}\right)^{4\cdot3} = \$10743.06$

19. $A = 1500\left(1+\frac{0.039}{12}\right)^{12\cdot2.5} = \$1653.36$

20. $p = 250,000 - 10,000 = 240,000$

$$A = 240,000\left(1+\frac{0.015}{12}\right)^{12\cdot10} = \$278814.00$$

21. $p = 800 + 150 + 300 + 1000 = \$2250$

$A = 2250\left(1+\dfrac{0.02}{360}\right)^{360\cdot 2} = \$2,341.82$

22. $A = 5000\left(1+\dfrac{0.0335}{4}\right)^{4\cdot 5} = \$5907.60$

23. a) $A = 2000\left(1+\dfrac{0.05}{2}\right)^{2\cdot 15} = \$4,195.14$

b) $A = 2000\left(1+\dfrac{0.05}{4}\right)^{2\cdot 15} = \$4,214.36$

24. a) $A = 2000\left(1+\dfrac{0.06}{2}\right)^{2\cdot 10} = \$3612.22$ - 1st 10 yrs.

b) $A = 3612.22\left(1+\dfrac{0.06}{4}\right)^{4\cdot 8} = \$5,816.85$ - 18 yrs.

25. $A = 3000\left(1+\dfrac{0.08}{4}\right)^{8} = \$3514.98$

26. $A = 6000\left(1+\dfrac{0.0525}{12}\right)^{24} = \$6,662.74$

$i = \$6662.74 - \$6000 = \$662.74$

27. $A = 6000\left(1+\dfrac{0.08}{4}\right)^{12} = \$7,609.45$

28. Let $p = 1.00$. Then

$A = 1\left(1+\dfrac{0.056}{360}\right)^{360} = \$1.0576$

$i = 1.0576 - 1.00 = 0.0576$

The effective annual yield is 5.76%

29. a) $A = 1000\left(1+\dfrac{0.02}{2}\right)^{4} = \$1,040.60$

$i = \$1040.60 - \$1000 = \$40.60$

b) $A = 1000\left(1+\dfrac{0.04}{2}\right)^{4} = \$1,082.43$

$i = \$1082.43 - \$1000 = \$82.43$

c) $A = 1000\left(1+\dfrac{0.08}{2}\right)^{4} = \$1,169.86$

$i = \$1169.86 - \$1000 = \$169.86$

d) No predictable outcome.

30. a) $A = 100\left(1+\dfrac{0.12}{12}\right)^{24} = \$126.97$

$i = \$126.97 - \$100 = \$26.97$

b) $A = 200\left(1+\dfrac{0.12}{12}\right)^{24} = \$253.95$

$i = \$253.95 - \$200 = \$53.95$

c) $A = 400\left(1+\dfrac{0.12}{12}\right)^{24} = \$507.89$

$i = \$507.89 - \$400 = \$107.89$

d) The interest doubles also.

31. a) $A = 1000\left(1+\dfrac{0.06}{2}\right)^{4} = \$1,125.51$

$i = \$1125.51 - \$1000 = \$125.51$

b) $A = 1000\left(1+\dfrac{0.06}{2}\right)^{8} = \$1,266.77$

$i = \$1266.77 - \$1000 = \$266.77$

c) $A = 1000\left(1+\dfrac{0.06}{2}\right)^{16} = \$1,604.71$

$i = \$1604.71 - \$1000 = \$604.71$

d) New amount $= \dfrac{(\text{old amount})^2}{1000}$

32. a) $A = 1000\left(1+\dfrac{0.04}{2}\right)^{1\cdot 1} = \$1,004.00$

$i = \$1040.00 - \$1000 = \$40.00$

b) $A = 1000\left(1+\dfrac{0.04}{2}\right)^{1\cdot 2} = \$1,040.40$

$i = \$1040.04 - \$1000 = \$40.04$

c) $A = 1000\left(1+\dfrac{0.04}{2}\right)^{4\cdot 1} = \$1,040.60$

$i = \$1040.60 - \$1000 = \$40.60$

d) No

33. $A = 1\left(1 + \dfrac{0.035}{12}\right)^{12 \bullet 1} = 1.03536$ or 3.54%

34. $A = 1\left(1 + \dfrac{0.0475}{12}\right)^{12 \bullet 1} = 1.04854$ or 4.85%

35. $A = 1\left(1 + \dfrac{0.024}{12}\right)^{12 \bullet 1} = 1.02426$

Yes, $APY = 2.43\%$, not 2.6%

36. $A = 1\left(1 + \dfrac{0.045}{12}\right)^{4 \bullet 1} = 1.045765$ Yes, 4.85%

37. The effective rate of the 4.75% account is:

$A = 1\left(1 + \dfrac{0.0475}{12}\right)^{12} = 1.0485$

$1.0485 - 1.00 = 0.0485$ or 4.85%

Therefore the 5% simple interest account pays more interest.

38. The amount Troy owes the bank after two years is:

$A = 1500\left(1 + \dfrac{0.10}{4}\right)^{4 \bullet 2} = \$1,827.60$

Bank's interest charge:
$i = 1827.60 - 1500 = \$327.60$
Grandfather's interest charge:
$i = prt = (1500)(0.07)(2) = \$210.00$
Troy will save $327.60 - 210.00 = \$117.60$

39. a) $\dfrac{A}{\left(1 + i/n\right)^{n \bullet t}} = \dfrac{290000}{\left(1 + 0.0825/2\right)^{20}} = \$129,210.47$

b) surcharge $= \dfrac{129210.47}{958} = \$134.88$

40. a) $\dfrac{A}{\left(1 + i/n\right)^{n \bullet t}} = \dfrac{783000}{\left(1 + 0.09/12\right)^{180}} = \$204,010.21$

b) surcharge $= \dfrac{204010.21 - 50000}{2682} = \$57.42$

41. $p = \dfrac{A}{\left(1 + i/n\right)^{n \bullet t}} = \dfrac{30000}{\left(1 + 0.0515/12\right)^{60}} = \$23,202.23$

42. Present value $= \dfrac{200000}{\left(1 + \dfrac{0.075}{4}\right)^{80}} = \$45,250.17$

43. Present value $= \dfrac{50000}{\left(1 + \dfrac{0.08}{4}\right)^{72}} = \$12,015.94$

44. Present value $= \dfrac{20000}{\left(1 + \dfrac{0.07}{4}\right)^{60}} = \$7,062.61$

45. $p = 1.35$, $r = 0.025$, $t = 10$, $n = 1$
$A = 1.35(1 + 0.025)^5 = \$1.53$

46. $p = 2000$, $A = 3586.58$, $n = 12$, $t = 5$

$3586.58 = 2000\left(1 + \dfrac{r}{12}\right)^{60}$

$\dfrac{3586.58}{2000} = \left(1 + \dfrac{r}{12}\right)^{60}$

$(1.79329)^{1/60} = 1 + \dfrac{r}{12} = 1.00978$

$0.00978 = \dfrac{r}{12}$ $r = 0.00978(12) = .117$ or 11.7%

47. a) $72/3 = 24$ years   b) $72/6 = 12$ years
    c) $72/8 = 9$ years   d) $72/12 = 6$ years
    e) $72/r = 22$ $72 = 22r$ $r = 72/22 = 0.0327$
       $r = 3.27\%$

48. $A = 2000 [1 + (.08/2)]^6 = 2000 (1.04)^6 = \$2530.64$
$i = \$2530.64 - 2500 = \$530.64$
Simple interest: $i = prt = 530.64 = 2000(r)(3)$

$530.64 = 6000r \quad r = \dfrac{530.64}{6000} = 0.0884$ or 8.84 %

49. $R = \$500, r = 5.5\%, n = 2, t = 17$

$$S = 500 \ \dfrac{\left[\left(1+\dfrac{0.055}{2}\right)^{34} -1\right]}{\dfrac{0.055}{2}} =$$

$$500 [1.51526] \left(\dfrac{2}{0.055}\right) = \$27,550.11$$

50.

$$S = 50 \ \dfrac{\left[\left(1+\dfrac{0.08}{4}\right)^{120} -1\right]}{\dfrac{0.08}{4}} = \$24,412.91$$

51. Use the formula given in exercise 45.
  a) $R = 150, r = 0.056, n = 12, t = 18$
      ans. $S = \$55,726.01$
  b) $R = 900, r = 0.058, n = 2, t = 18$
      ans. $S = \$55,821.15$

Exercise Set 11.4

1. An open-end installment loan is a loan on which you can make different payment amounts each month. A fixed installment loan is one in which you pay a fixed amount each month for a set number of months.

2. With an installment plan, the borrower repays the principal plus the interest with weekly or monthly payments that usually begin shortly after the loan is made. With a personal note, the borrower repays the principal plus the interest as a single payment at the end of the specified time period.

3. The APR is the true rate of interest charged on a loan.

4. The finance charge is the total amount of money the borrower must pay for the use of the money borrowed.

5. The total installment price is the sum of all the monthly payments and the down payment, if any.

6. The Actuarial method and the Rule of 78's.

7. The unpaid balance method and the average daily balance method.

8. A cash advance is a loan obtained through a credit card.

9. a) Amount financed $= 43000 - 0.15(43000) = \$36,550.00$
      From table 11.2 the finance charge per \$100 at 5.5 % for 60 payments is 14.61.

      Total finance charge $= (14.61) \left(\dfrac{36550}{100}\right) = \$5340.00$

  b) Total amount due after down payment $= 36550.00 + 5340.00 = \$41889.96$

      Monthly payment $= \dfrac{41889.96}{60} = \$698.17$

10. a) Amount financed $= 2900 - 0.20(2900) = \$2,320.00$
      From table 11.2, the finance charge per \$100 financed at 8.5% for 24 months is 9.09.

      Total finance charge $= (9.09)\left(\dfrac{2320}{100}\right) = \$210.89$

  b) Total amount due after down payment $= 2320 + 210.89 = \$2,530.89$

      Monthly payment $= \dfrac{2530.89}{24} = \$105.45$

11. a) From table 11.2, the finance charge per $100 financed at 7.5% for 60 months is $20.23.

Total finance charge is $(20.23)\left(\dfrac{4000}{100}\right) = \$809.20$

b) Total amount due = 4000 + 809.20 = $4,809.20

Monthly payment = $\dfrac{4809.20}{60} = \$80.15$

12. a) From table 11.2, the finance charge per $100 financed at 4.5% for 48 months is $9.46.

Total finance charge = $(9.46)\left(\dfrac{2500}{100}\right) = \$236.50$

b) Total amount due = 2500 + 236.50 = $2736.50

Monthly payment = $\dfrac{2736.50}{48} = \$57.01$

13. a) Down payment = 0.20(3200) = $640
Total installment price = 640 + (60 • 53.14) = $3828.40
Finance charge = 3828.40 − 3200 = $628.40

b) $\left(\dfrac{\text{finance charge}}{\text{amt. financed}}\right)(100) = \left(\dfrac{628.40}{2560}\right)(100) = 24.55$

From Table 11.2 for 60 payments, the value of 24.55 corresponds with an APR of 9.0 %.

14. a) Total installment price = (64)(24) = $1536.00
Finance charge = 1536.00 − 1420.25 = $115.75

b) $\left(\dfrac{\text{finance charge}}{\text{amt. financed}}\right)(100) = \left(\dfrac{115.75}{1420.25}\right)(100) = 8.15$

From Table 11.2, for 24 payments, the value of $8.15 is closest to $8.00 which corresponds with an APR of 7.5 %.

15. a) Total installment price = (224)(48) = $10752.00
Finance charge = 10752.00 − 9000.00 = $1752.00

b) $\left(\dfrac{\text{finance charge}}{\text{amt. financed}}\right)(100) = \left(\dfrac{1752.00}{9000}\right)(100) = 19.47$

From Table 11.2, for 48 payments, the value of $19.47 is closest to $19.45 which corresponds with an APR of 9.0 %.

16. Down payment = (1/4)(3450) = $862.50   Amount financed = (3/4)(3450) = $2587.50

a) Installment price = (6)(437) = 2622
Finance charge = $2622.00 − $2587.50 = $34.50

b) $\left(\dfrac{\text{finance charge}}{\text{amt. financed}}\right)(100) = \left(\dfrac{34.50}{2587.50}\right)(100) = 1.33$

From Table 11.2, for 6 payments, the value of $1.33 is closest to $1.32 which corresponds with an APR of 4.5 %.

17. Down payment = $0.00   Amount financed = $12000.00

   a) Installment price = (60)(232) = 13920

      Finance charge = $13920.00 − $12000.00 = $1920.00

$$\left(\frac{\text{finance charge}}{\text{amt. financed}}\right)(100) = \left(\frac{1920.00}{12000}\right)(100) = 16.00$$

      From Table 11.2, for 6 payments, the value of $16.00 corresponds with an APR of 6.0 %.

   b) $u = \dfrac{npv}{100+v} = \dfrac{(36)(232)(9.52)}{(100+9.52)} = \dfrac{79511.04}{109.52} = 725.9956 = \$726.00$

   c) (232)(23) = 5336   5336 + 726 = 6062   13920 − 6062 = $7858.00

18. (167.67)(48) = 8048.16   8048.16 − 7500.00 = 548.16

   a) $\left(\dfrac{548.16}{7500}\right)(100) = \$7.31$ per $100   From Table 11.2, $7.31 corresponds with an APR of 3.5 %.

   b) $u = \dfrac{npv}{100+v} = \dfrac{(30)(167.67)(4.58)}{(100+4.58)} = \dfrac{23037.86}{104.58} = \$220.29$)

   c) (167.67)(17) = 2850.39   2850.39 + 220.29 = 3070.68   8048.16 − 3070.68 = $4977.48

19. a) Amount financed = 32000 − 10000 = $22000

      From table 11.2, the finance charge per 100 financed at 8 % for 36 payments is 12.81.

      Total finance charge = $(12.81)\left(\dfrac{22000}{100}\right) = 2818.20$

   b) Total amt. due = 22000 + 2818.20 = $24,818.20

      Monthly payment = $\dfrac{24818.20}{36} = \$689.39$

   c) $u = \dfrac{(12)(689.89)(4.39)}{100+4.39} = \dfrac{36317.07}{104.39} = \$347.90$

   d) (23)(689.39) = 15855.97   15855.97 − 347.90 = 16203.87   24818.20 − 16203.87 = $8614.17

20. a) Amount financed = (110.52)(24) = $2652.48   2652.48 − 2558.00 = 94.48 $\left(\dfrac{94.48}{2558}\right)(100) = 3.69$

      From table 11.2, the interest rate that would generate a finance charge of $3.69 is 3.5 % for 24 payments.

   b) $u = \dfrac{(110.52)(12)(1.91)}{100+1.91} = \dfrac{2533.12}{101.91} = 24.86$

   c) (110.52)(11) = 1215.72   1215.72 + 24.86 = 1240.58   2652.48 − 1240.58 = $1411.90

21. a) Amount financed = $7345.00 with no down payment.
       From table 11.2, the finance charge per 100 financed at 8.5 % for 48 payments is 18.31.

   $$\text{Total finance charge} = (18.31)\left(\frac{7345}{100}\right) = 1344.87$$

   b) Total amt. due = 7345.00 + 1344.87 = $8,689.87

   $$\text{Monthly payment} = \frac{8689.87}{48} = \$181.04$$

   c) $u = \dfrac{(1344.87)(36)(36+1)}{48(48+1)} = \dfrac{1791366.84}{2352} = \$761.64$

   d) (181.04)(48) = 8689.92   (11)(181.04) = 1991.44   1991.44 + 761.64 = 2753.08
       8689.92 − 2753.08 = $5936.84

22. a) From table 11.2, at 8.5% for 36 payments the finance charge per 100 is 13.64.

   $$\text{Finance charge} = (13.64)\left(\frac{3600}{100}\right) = \$491.04$$

   b) Total installment price = 3600 + 491.04 = $4091.04

   $$\text{Monthly payment} = \frac{4091.04}{36} = \$113.64$$

   c) K = 24, n = 36, f = 491.04

   $$u = \frac{(491.04)(24)(25)}{(36)(37)} = \$221.19$$

   d) $2727.36 Total of remaining payments   2727.36 − 221.19 = 2506.17
       2506.17 + 113.64 = $2619.81   Total amount due

23. a) Interest = 500 + (151.39)(18) - 3000 = $225.02   k = 6, n = 18, and f = 225.02

   $$u = \frac{(225.02)(6)(6+1)}{18(18+1)} = \frac{9450.84}{342} = \$27.63$$

   b) $908.34 Total of remaining payments   908.34 − 27.63 = 880.71
       880.71 + 151.39 = $1032.10   Total amount due

24. a) Interest = 850 +(134.71)(12) − 2375 = $91.52   k = 6, n = 12, and f = 91.52

   $$u = \frac{(91.52)(6)(6+1)}{12(12+1)} = \frac{3843.84}{156} = \$24.64$$

   b) $808.26   Total of remaining payments   808.26 − 24.64 = $783.62
       783.62 + 134.71 = $918.33   Total amount due

25. a) Balance due = 365 + 180 + 195 + 84 = $824 $\text{min. payment} = \dfrac{\text{bal. due}}{48} = \dfrac{824}{48} \approx 17.17 \approx \$18$

   b) Bal. due after Dec. 1 payment = 824 − 200 = $624   interest for Dec. = (0.011)(624) = $6.86
       Bal. due Jan. 1 = 624 + 6.86 = $630.86

26. a) Bal. due = 425 + 175 + 450 + 125 = \$1175   min. payment = $\dfrac{\text{bal. due}}{36} = \dfrac{1175}{36} \approx 32.64 \approx \$33$

   b) Bal. due after Sept. 1 payment = 1175 − 650 = \$525   interest for Sept. = (0.012)(525) = \$6.30
     Bal. due Oct. 1 = 525 + 6.30 = \$531.30

27. a) Bal. due = 423 + 36 + 145 + 491 = \$1095   min. payment = $\dfrac{\text{bal. due}}{36} = \dfrac{1095}{36} \approx 30.42 \approx \$31$

   b) Bal. due after Mar. 1 payment = 1095 − 548 = \$547   interest for March = (0.011)(547) = \$6.02
     Bal. due Apr. 1 = 547 + 6.02 = \$553.02

28. a) Bal. due = 512 + 172 + 190 + 350 = \$1224   min. payment = $\dfrac{\text{bal. due}}{36} = \dfrac{1224}{48} \approx 25.50 \approx \$26$

   b) Bal. due after July 1 payment = 1224 − 500 = \$724   interest for July = (0.013)(724) = \$9.41
     Bal. due Aug. 1 = 724 + 9.41 = \$733.41

29. a) Finance charge = (1097.86)(0.018)(1) = \$19.76
   b) Bal. due May 5 = (1097.86 + 19.76 + 425.79) − 800 = \$743.41

30. a) Finance charge = (567.20)(0.011)(1) = \$6.24
   b) old balance + finance charge − payment + airline ticket + hotel bill + clothing = new balance
     567.20 + 6.24 − 275.00 + 330.00 + 190.80 + 84.75 = \$903.99

31. a) Finance charge = (124.78)(0.0125)(1) = \$1.56
   b) old balance + finance charge − payment + art supplies + flowers + music CD = new balance
     124.78 + 1.56 − 100.00 + 25.64 + 67.23 + 13.90 = \$133.11

32. a) Finance charge = (57.88)(0.0135(1) = \$0.78
   b) old balance + finance charge − payment + paint + curtains + chair = new balance
     57.88 + 0.78 − 45.00 + 64.75 + 72.85 + 135.50 = \$903.99

33. a)

Date	Balance Due	Number of Days	(Balance)(Days)
May 12	\$378.50	1	(378.50)(1) = \$ 378.50
May 13	\$508.29	2	(508.29)(2) = 1,016.58
May 15	\$458.29	17	(458.29)(17) = 7,790.93
June 01	\$594.14	7	(594.14)(7) = 4,158.98
June 08	\$631.77	4	(631.77)(4) = 2,527.08
		31	sum = \$15,872.07

Average daily balance = $\dfrac{15872.07}{31} =$ \$512

   b) Finance charge = prt =
     (512.00)(0.013)(1) = \$6.66
   c) Balance due = 631.77 + 6.66 =
     \$638.43

34. a)

Date	Balance Due	Number of Days	(Balance)(Days)
Mar. 23	\$1,578.25	3	(1578.25)(3) = \$4,734.75
Mar. 26	\$1,658.23	4	(1658.23)(4) = 6,632.92
Mar. 30	\$1,710.99	4	(1710.99)(4) = 6,843.96
Apr. 03	\$1,460.99	12	(1460.99)(12) = 17,531.88
Apr. 15	\$1,651.51	7	(1651.51)(7) = 11,560.57
Apr. 22	\$1,842.36	1	(1842.36)(1) = 1,842.36
		31	sum = \$49,146.44

Average daily balance = $\dfrac{49146.44}{31} =$ \$1585.37

   b) Finance charge = prt =
     (1585.37)(0.013)(1) = \$20.61
   c) Balance due = 1842.36 + 20.61 =
     \$1,862.97

35. a)

Date	Balance Due	Number of Days	(Balance)(Days)
Feb. 03	$124.78	5	(124.78)(5) = $623.90
Feb. 08	$150.42	4	(150.42)(4) = 601.68
Feb. 12	$ 50.42	2	(50.42)(2) = 100.84
Feb. 14	$117.65	11	(117.65)(11) = 1294.15
Feb. 25	$131.55	6	(131.55)(6) = 789.30
		28	sum = $3,409.87

Average daily balance = $\dfrac{3409.87}{28}$ = $121.78

b) Finance charge = prt =
(121.78)(0.0125)(1) = $1.52

c) Balance due = 131.55 + 1.52 = $133.07

d) The interest charged using the ave. daily balance method is $0.04 less than the interest charged using the unpaid balance method.

36. a)

Date	Balance Due	Number of Days	(Balance)(Days)
Sept. 05	$385.75	3	(385.75)(3) = $1157.25
Sept. 08	$110.75	13	(110.75)(13) = 1439.75
Sept. 21	$440.75	6	(440.75)(6) = $2644.50
Sept. 27	$631.55	5	(631.55)(5) = $3157.75
Oct. 02	$716.30	3	(716.30)(3) = $2148.90
		28	sum = $10,548.15

Average daily balance = $\dfrac{10548.15}{30}$ = $351.61

b) Finance charge = prt =
(351.61)(0.014)(1) = $4.92

c) Balance due = 716.30 + 4.92 = $721.22

d) Smaller finance charge on Oct. 5 using the ave. daily balance method.

37. 0.05477 % per day = 0.0005477   a) (600)(0.0005477)(27) = $8.87   b) 600.00 + 8.87 = $608.87

38. a) i = (875)(0.0004273)(32) = $11.96   b) A = 875 + 11.96 = $886.96

39. $1000.00   5 %   6 payments
a) State National Bank (SNB):   (1000)(.05)(.5) = $25.00
b) Consumers Credit Union (CCU):   (1000)(x)(1) = 35.60   (86.30)(12) = 1035.60
1035.60 − 1000.00 = $35.60

c) $\left(\dfrac{25}{1000}\right)(100) = 2.50$ In Table 11.2, $2.49 is the closest value to $2.50, which corresponds to an APR of 8.5 %.

d) $\left(\dfrac{35.60}{1000}\right)(100) = 3.56$ In Table 11.2, $3.56 corresponds to an APR of 6.5 %.

40. The interest on $890 at 5.25% annually for 1 month is: i = (890)(0.0525)(1/12) = $3.89
She will be saving $3.89 by using her credit card.

41. a) Amount financed = 3450 − 1150 = $2300

Month	Finance charge	Payment	Balance
1	None	$384.00	$1,916.00
2	(1916)(0.013) = $24.91	408.91	1,532.00
3	(1532)(0.013) = $19.92	403.92	1,148.00
4	(1148)(0.013) = $14.92	398.92	764.00
5	(764)(0.013) = $ 9.93	393.93	380.00
6	(380)(0.013) = $ 4.94	384.94	0.00

Total = $74.62 It will take 6 months to repay the loan.

b) The total amount of interest paid is $74.62

c) The finance charge is $13.38 less using the credit card.

42. Let p = amount Ken borrowed

    p + 2500 = purchase price

    Installment price: 2500 + (379.50)(36) = $16,162

    Interest = Installment price – purchase price

    i = 16,162 – (p + 2500) = 16,162 – p – 2500 = 13,662 – p

Since i = prt we have:

13,662 – p = (p)(.06)(3) = 13,662 – p = .18p

13,662 = .18p + p   p = 11,577.97

purchase price = 11,577.97 + 2500 =

    $14,077.97

43. $35,000   15 % down payment   60 month fixed loan   APR = 8.5 %

    (35000)(.15) = 5250   35000 – 5250 = 29750

  a) From Table 11.2, 60 payments at an APR of 8.5 % yields a finance charge of $23.10 per $100.

$$\left(\frac{29750}{100}\right)(23.10) = \$6872.25$$

  b) 29750.00 + 6872.25 = 36622.25 $\dfrac{36622.25}{60} = \$610.37$

  c) In Table 11.2, 36 payments at an APR of 8.5 % yields a finance charge of $13.64 per $100.

$$u = \frac{(36)(610.37)(13.64)}{100+13.64} = \frac{299716.08}{113.64} = \$2637.42$$

  d) $u = \dfrac{f \cdot k\,(k+1)}{n\,(n+1)} = \dfrac{(6872.25)(36)(37)}{60\,(61)} = \dfrac{9153837}{3660} = \$2501.05$

44. $23,000   10 % down payment   48 month fixed loan   APR = 6.0 %

    (23000)(.10) = 2300   23000 – 2300 = 20700

  a) From Table 11.2, 48 payments at an APR of 6.0 % yields a finance charge of $12.73 per $100.

$$\left(\frac{20700}{100}\right)(12.73) = \$2635.11$$

  b) 20700.00 + 2635.11 = 23335.11 $\dfrac{23335.11}{48} = \$486.15$

  c) In Table 11.2, 36 payments at an APR of 6.0 % yields a finance charge of $9.52 per $100.

$$u = \frac{(36)(486.15)(9.52)}{100+9.52} = \frac{166613.33}{109.52} = \$1521.31$$

  d) $u = \dfrac{f \cdot k\,(k+1)}{n\,(n+1)} = \dfrac{(2635.11)(36)(37)}{48\,(49)} = \dfrac{3509966.52}{2352} = \$1492.33$

45. With her billing date on the 25th of the month she can buy the camera during the period of June 26 - June 29 and the purchase will be on the July 25th bill. Purchasing during these dates she can pay the bill on August 5th or later without paying interest.

Exercise Set 11.5

1. A mortgage is a long term loan in which the property is pledged as security for payment of the difference between the down payment and the sale price.
2. The down payment is the amount of cash the buyer must pay the seller before the lending institution will grant the buyer a mortgage.
3. The major difference between these two types of loans is that the interest rate for a conventional loan is fixed for the duration of the loan, whereas the interest rate for a variable-rate loan may change every period, as specified in the loan agreement.

4. a) A point is 1% of the mortgage.   b) For x points multiply the mortgage by 0.01x.
5. A buyer's adjusted monthly income is found by subtracting any fixed monthly payment with more than
   10 months remaining from the gross monthly income.
6. An add on rate, or margin, is the percent added to the interest rate on which the adjustable rate mortgage is based.
7. An amortization schedule is a list of the payment number, interest, principal, and balance remaining on the loan.
8. The FHA insures the loan and a bank provides the money for the loan.
9. Equity is the difference between the appraised value of your home and the loan balance.
10. A home equity loan is a loan in which the equity in your home is used as collateral.

11. a) Down payment = 15% of $250,000
    (0.15)(250000) = $35,700
    b) amt. of mortgage = 250000 − 35700 = 212500
    Table 11.4 yields $7.65 per $1000 of mortgage

    Monthly payment = $\left(\dfrac{212000}{1000}\right)(7.65) = \$1625.63$

12. a) Down payment = 20% of $175,000
    (0.20)(175000) = $35,000
    b) amt. of mortgage = 175000 − 35000 = 140000
    Table 11.4 yields $5.68 per $1000 of mortgage

    Monthly payment = $\left(\dfrac{140000}{1000}\right)(5.68) = \$795.20$

13. a) Down payment = 10% of $210,000
    (0.10)(210000) = $21,000
    b) amt. of mortgage = 210000 − 21000 = 189000
    Table 11.4 yields $6.60 per $1000 of mortgage

    Monthly payment = $\left(\dfrac{189000}{1000}\right)(6.60) = \$1247.40$

14. a) Down payment = 5% of $95,000
    (0.05)(95000) = $4,750
    b) amt. of mortgage = 95000 − 4750 = 90250
    Table 11.4 yields $6.65 per $1000 of mortgage

    Monthly payment = $\left(\dfrac{90250}{1000}\right)(6.65) = \$600.16$

15. a) Down payment = 20% of $195,000
    (0.20)(195000) = $39,000
    b) amt. of mortgage = 195000 − 39000 = 156000
    c) (156000)(.02) = $3120.00

16. a) 245000 − 45000 = $200,000.00
    b) (200000)(.015) = $3000.00

17. $3,200 = monthly income
    a) (25)(335) = $8,375.00   3200 − 335 = $2865
    b) (2865)(.28) = $802.20
    c) Table 11.4 yields $7.91 per $1000 of mortgage
    $\left(\dfrac{150000}{1000}\right)(7.91) = \$1186.50$
    d) No; $1411.50 > $802.20

18. $4,100 = monthly income
    a) 4100 − 505 = $3595.00
    b) (3595)(.28) = $1006.60
    c) Table 11.4 yields $9.00 per $1000 of mortgage
    $\left(\dfrac{275000}{1000}\right)(9.00) = \$2475.00$
    2475 + 425 = $2900.00
    d) No; $2900.00 > $1006.60

19. a) (490.24)(30)(12) = $176,486.40
    176486.40 + 11250.00 = $187,736.40
    b) 187736.40 − 75000 = $112,736.40
    c) i = prt = (63750)(.085)(1/12) = 451.56
    490.24 − 451.56 = $38.68

20. a) Down payment = 160,000 − 110,000 = $50,000
    Total cost of house = 50000 + (1038.40)(12)(25)
    = $361,520
    b) interest = $361,520 − $160,000 = $201,520
    c) interest on first payment   i = prt
    (110,000)(0.105)(1/12) = $962.50
    amount applied to principal =
    1038.40 − 962.50 = $75.90

21. a) down payment = (0.28)(113500) = $31,780

   b) amount of mortgage = 113500 − 31780 = $81,720

      cost of three points = (0.03)(81720) = $2,451.60

   c) 4750 − 420 = $4330.00 adjusted monthly income

   d) maximum monthly payment =

      (0.28)(4330) = $1,212.40

   e) At a rate of 10% for 20 years, Table 11.4

      yields 9.66.

      mortgage payment = $\left(\dfrac{81720}{1000}\right)(9.66) = \$789.42$

   f) 789.42 + 126.67 = $916.09 total mo. Payment

   g) Since $1,212.40 is greater than $916.09, the

      Yakomo's qualify.

   h) interest on first payment = i = prt =

      (81720)(0.10)(1/12) = $681.00

      amount applied to principal = 789.42 − 681.00 =

      $108.42

22. Cost of house = $95,000.00

   a) (95000)(.20) = $19,000.00

   b) 4000 − 135 = $3865   (3865)(.28) = $1082.20

   c) Table 11.4 yields $8.74 per $1000 of mortgage

      $\left(\dfrac{76000}{1000}\right)(8.74) = \$664.24$

   d) 664.24 + 125.00 + 28.00 = $817.24

   e) Yes; $817.24 < $1082.20

   f) i = prt = (76000)(.095)(1/12) = $601.67

   g) (664.24)(25)(12) = 199272

      199272 + 19000 = $218,272.00

   h) 218272 − 95000 = $123,272.00

23. <u>Bank A</u>   Down payment = (0.10)(105000) =

   $10,500

   amount of mortgage   105000 − 10500 = $94,500

   At a rate of 10% for 30 years, Table 11.4 yields

   $8.70 .

   monthly mortgage payment =

   $\left(\dfrac{94500}{1000}\right)(8.70) = \$822.15$

   cost of three points = (0.03)(94500) = $2835

   Total cost of the house =

      10500 + 2835 + (822.15)(12)(30) = $309,309

   <u>Bank B</u>   Down payment = (0.20)(105000) =

   $21,000

   amount of mortgage   105000 − 21000 = $84,000

   At a rate of 11.5% for 25 years, Table 11.4 yields

   $10.16.

   monthly mortgage payment =

   $\left(\dfrac{84000}{1000}\right)(10.16) = \$853.44$

   cost of the house = 21000 + (853.44)(12)(25) =

      $277,032

   The Nagrockis should select Bank B.

24. Condominium $525,000.00

   GCTCU   20% down payment, 7.5%, 15 years,

                  1 point at closing

   SCCU   15% down payment, 8.5%, 20 years,

                  No points

   (525000)(.80) = 420000

   1 pt.: (420000)(.01) = 42000

   At 7.5% for 15 yrs., Table 11.4 yields $9.27.

   $(9.27)\left(\dfrac{378000}{1000}\right) = \$3504.06$

   GCTCU: 105000.00 + 42000.00 + 630730.80 =

                  $777,730.80

   (525000)(.85) = 446250

   At 8.5% for 20 yrs., Table 11.4 yields $8.68.

   $(8.68)\left(\dfrac{446250}{1000}\right) = \$3873.45$

   (3873.45)(20)(12) = $929,628.00

   SCCU: 929628 + 78750 = $1,008,378.00

   Grant County Teacher's Credit Union would

    provide a lower cost.

25. a) Amount of mortgage = 105000 - 5000 = $100000   Initial monthly payment = $\left(\dfrac{100000}{1000}\right)(8.05) = \$805.00$

b)
Payment #	Interest	Principal	Balance
1	$750.00	$55.00	$99,945.00
2	749.59	55.41	99,889.59
3	749.17	55.83	99,833.76

c) effective interest rate = 6.13% + 3.25% = 9.38%.   The new rate is 9.38%.

d)
Payment #	Interest	Principal	Balance
4	$780.37	$24.63	$99,809.13
5	780.17	24.83	99,784.30
6	779.98	25.02	99,759.28

e)   New rate = 6.21% + 3.25% = 9.46%

26. a) amount of mortgage: $95000 – $13000 = $82,000
     At a rate of 8.5% for 30 years, Table 11.4 yields
     $7.69.
     initial monthly payment =

$\left(\dfrac{82000}{1000}\right)(7.69) = \$630.58$

b) effective interest rate: 5.65 + 3.25 = 8.9%
     8.9% is less than 1% above the old rate of 8.5%.
     Thus, the new rate is 8.9%.

c) effective new interest rate: 4.85 + 3.25 = 8.1%.

27. a) $\left(\dfrac{\text{amount of mortgage}}{1000}\right)(8.4) = 950$

     amount of mortgage = $113,095.24

b) (0.75)(total price) = 113,095.24
     total price = $150,793.65

28. a) The variable rate mortgage would be the
     cheapest.

b) By choosing the variable rate plan, they would
     save $2,672.64

Review Exercises

1. 3/5 = 0.60   (0.60)(100) = 60%

2. 2/3 ≈ 0.667   (0.667)(100) = 66.7%

3. 5/8 = 0.625   (0.625)(100) = 62.5%

4. 0.041   (0.041)(100) = 4.1%

5. 0.0098   (0.0098)(100) = 0.98% ≈ 1.0%

6. 3.141   (3.141)(100) = 314.1%

7. 3% $\dfrac{3}{100} = .03$

8. 12.1% $\dfrac{12.1}{100} = 0.121$

9. 123% $\dfrac{123}{100} = 1.23$

10. $\dfrac{1}{4}\% = 0.25\%$ $\dfrac{.25}{100} = .0025$

11. $\dfrac{5}{6} = 0.8\overline{3}\%$ $\dfrac{0.8\overline{3}}{100} = 0.008\overline{3}$

12. 0.00045% $\dfrac{0.00045}{100} = 0.0000045$

13. $\dfrac{71500}{60790} = 1.17618$ (1.17618)(100) ≈ 17.6%

14. $\dfrac{5100}{46200} = 0.11039$ (0.11039)(100) ≈ 11.0%

15. $(x\%)(80) = 25 \qquad x\% = 25/80 = .3125$
$(.3125)(100) = 31.25\%$
Twenty-five is 31.25% of 80.

16. $0.16x = 44 \qquad x = 44/0.16 = 275$
Forty-four is 16% of 275.

17. $(0.17)(540) = x \qquad 91.8 = x$
Seventeen percent of 540 is 91.8.

18. $\text{Tip} = 15\% \text{ of } \$42.79 = (0.15)(42.79) = \$6.42$

19. $0.20(x) = 8 \qquad x = 8/0.20 = 40$
The original number was 40 people.

20. $\dfrac{(95-75)}{75} = \dfrac{20}{75} = .2\overline{6} \qquad (.267)(100) = 26.7$
The increase was 26.7%.

21. $i = (2500)(.04)(60/360) = \$16.67$

22. $41.56 = (1575)(r)(100/360) = 41.56$
$41.56 = \left(\dfrac{157500}{360}\right)(r) \qquad r = 0.095 \text{ or } 9.5\%$

23. $114.75 = (p)(0.085)(3)$
$114.75 = (p)(0.255) \qquad \$450 = p$

24. $316.25 = (5500)(0.115)(t)$
$316.25 \qquad = (632.50)(t) \quad t = 0.5 \text{ yrs. or } 6 \text{ mos.}$

25. $i = (5300)(.0575)(3) = 914.25$
Total amount due at maturity = 5300 + 914.25 = $6214.25

26. a) $i = (3000)(0.081)(240/360) = \$162$
She paid 3000 + 162 = $3,162

27. a) $i = (6000)(0.115)(24/120 = \$1380.00$
 b) amount received: 6000.00 - 1380.00 = 4,620.00
 c) $i = prt \qquad 1380 = (4620)(r)(24/12) = 9240r$
 $r = (1380)(9240) = .1494 \quad (.1494)(100) = 14.9\%$

28. a) $5\dfrac{1}{2}\% + 2\% = 7\dfrac{1}{2}\%$
 b) $i = (800)(0.75)(6/12) = \$30$
 $A = \$800 + \$30 = \$830.00$
 c) x = amount of money in the account
 85% of x = 800   0.85x = 800   x = $941.18

29. a) $A = 1000\left(1+\dfrac{.10}{1}\right)^{5} = (1.10)^{5} = 1610.51 \qquad 1610.51 - 1000 = \$610.51$

 b) $A = 1000\left(1+\dfrac{.10}{2}\right)^{10} = (1.05)^{10} = 1628.89 \qquad 1628.89 - 1000 = \$628.89$

 c) $A = 1000\left(1+\dfrac{.10}{4}\right)^{20} = (1.025)^{20} = 1638.62 \qquad 1638.62 - 1000 = \$638.62$

 d) $A = 1000\left(1+\dfrac{.10}{12}\right)^{60} = (1.008\overline{3})^{60} = 1645.31 \qquad 1645.31 - 1000 = \$645.31$

 e) $A = 1000\left(1+\dfrac{.10}{360}\right)^{1800} = (1.0002\overline{7})^{1800} = 1648.38 \qquad 1648.38 - 1000 = \$648.38$

30. $A = p\left(1 + \dfrac{r}{n}\right)^{nt}$

$A = 2500\left(1 + \dfrac{0.0475}{4}\right)^{4 \cdot 15} = \$5{,}076.35$

31. Let $p = 1.00$. Then $A = 1\left(1 + \dfrac{0.56}{360}\right)^{360} = 1.05759$

$i = 1.05759 - 1.00 = 0.05759$

The effective annual yield is 5.76% .

32. $p\left(1 + \dfrac{0.055}{4}\right)^{80} = 40000$ $\quad p = \dfrac{40000}{(1.01375)^{80}} = 13415.00$ $\quad$ You need to invest \$13,415.00

33. 48 mo. $176.14/mo. $7500 24 payments

a) $(176.14)(48) = 8454.72$ $\quad 8454.72 - 7500 = \$954.72$ $\quad \left(\dfrac{954.72}{7500}\right)(100) = \$12.73 / \$100$

From Table 11.2, \$12.73 indicates an APR of 6.0%

b) $n = 24$, $p = 176.14$, $v = 6.37$ $\quad u = \dfrac{(24)(176.14)(6.37)}{100 + 6.37} = \dfrac{26928.28}{106.37} = \$253.16$

c) $(176.14)(48) = 8454.72$ $\quad (176.14)(23) = 4051.22$ $\quad 8454.72 - 4051.22 = \$4403.50$

$4403.50 - 253.16 = \$4150.34$

34. a) Amount financed = \$3,500 $\quad$ Finance charge $= (163.33)(24) - 3500 = \$419.92$

$f = 419.92$, $k = 12$, $n = 24$ $\quad u = \dfrac{(419.92)(12)(13)}{(24)(25)} = \$109.18$

b) $1959.96 - 109.18 = \$1850.78$ $\quad 1850.78 + 163.33 = \$2014.11$

35. 24 mo. $111.73/mo. Down payment = \$860 24 payments

a) $3420 - 860 = \$2560.00$ $\quad (111.73)(24) = 2681.52$ $\quad 2681.52 - 2560.00 = \$121.52$

$\left(\dfrac{121.52}{2560}\right)(100) = \$4.75 / \$100$

From Table 11.2, \$4.75 indicates an APR of 4.5%

b) $n = 12$, $p = 111.73$, $v = 2.45$ $\quad u = \dfrac{(12)(111.73)(2.45)}{100 + 2.45} = \dfrac{3284.86}{102.45} = \$32.06$

c) $(111.73)(11) = 1229.03$ $\quad 2681.52 - 1229.03 = 1452.49$ $\quad 1452.49 - 32.06 = \$1420.43$

36. Balance = \$485.75 as of June 01 $\quad i = 1.3\%$

June 04: $485.75 - 375.00) = \$110.75$ $\quad$ June 08: $110.75 + 370.00 = \$480.75$

June 21: $480.75 + 175.80 = \$656.55$ $\quad$ June 28: $656.55 + 184.75 = \$841.30$

a) $(485.75)(.013)(1) = \$6.31$ $\quad$ b) $841.30 + 6.31 = \$847.61$

c) $(485.75)(3) + (110.75)(4) + (480.75)(13) + (656.55)(7) + (841.30)(3) = \$15269.75$

$15269.75 / 30 = \$508.99$

d) $(508.99)(.013)(1) = \$6.62$ $\quad$ e) $841.30 + 6.62 = \$847.92$

37. a) Aug. 01: \$185.72

Aug. 05: $185.72 + 2.60 = \$188.32$

Aug. 08: $188.32 + 85.75 = \$274.07$

Aug. 10: $274.07 - 75.00 = \$199.07$

Aug. 15: $199.07 + 72.85 = \$271.92$

Aug. 21: $271.92 + 275.00 = \$546.92$

b) As of Aug. 31, $544.32 + 2.60 = \$546.92$

c)

Date	Balance	# of Days	Balance-Days
Aug. 01	185.72	4	$(185.72)(4) = $ 742.88
Aug. 05	188.32	3	$(188.32)(3) = $ 564.96
Aug. 08	274.07	2	$(274.07)(2) = $ 548.14
Aug. 10	199.07	5	$(199.07)(5) = $ 995.35
Aug. 15	271.92	6	$(271.92)(6) = $ 1631.52
Aug. 21	546.92	11	$(546.92)(11) = 6016.12$
		31	sum $= \$10{,}498.97$

40. a) down payment = (0.25)(135700) = $33,925
    b) gross monthly income = 64000/12 = $5,333.33
       adjusted monthly income:
       5333.33 - 528.00 = $4,805.33
    c) maximum monthly payment:
       (0.28)(4805.33) = $1,345.49
    d) $\left(\dfrac{101775}{1000}\right)(8.11) = \$825.40$
    e) total monthly payment:
       825.40 + 316.67 = $1,142.07
    f) Yes, $1345.49 is greater than $1142.07.

41. a) down payment = (0.15)(89900) = $13,485
    b) amount of mortgage = 89,900 - 13,485 =
       $76,415
       At 11.5% for 30 years, Table 11.4 yields 9.90.
       monthly mortgage payment:
       $\left(\dfrac{76415}{1000}\right)(9.90) = \$756.51$
    c) i = prt = (76415)(0.115)(1/12) = $732.31
       amount applied to principal:
       756.51 − 732.51 = $24.20
    d) total cost of house: 13485 + (756.51)(12)(30) =
       $285,828.60
    e) total interest paid: 285,828.60 − 89900 =
       $195,928.60

42. a) amount of mortgage: 105,000 − 26,250 = $78,750   First payment = $\left(\dfrac{78750}{1000}\right)(6.99) = \$550.40$

    b) 5.00% + 3.00% = 8.00%   c) 4.75% + 3.00% = 7.75%

Chapter Test

1. i = (2000)(0.04)(1/2) = $40.00

2. 288 = (1200)(0.08)(t)   288 = 96t   t = 3 years

3. i = prt = (5000)(0.085)(18/12) = $637.50

4. Total amount paid to the bank
   5000 + 637.50 = $5,637.50

5. Partial payment on Sept. 15 (45 days)
   i = (5400)(0.125)(45/360) = $84.375
   $3000.00 - 84.375 = $2,915.625
   5400.00 − 2915.625 = $2484.375

   i = (2484.375)(0.125)(45/360) = $38.82
   2484.38 + 38.82 = $2523.20

6. 84.38 + 38.82 = $123.20

7. $A = 7500\left(1+\dfrac{0.03}{4}\right)^8 = \$7961.99$

   interest = 7961.99 − 7500.00 = $461.99

8. $A = 2500\left(1+\dfrac{0.065}{12}\right)^{36} = \$3036.68$

   interest = 3036.68 − 2500.00 = $536.68

9. (2350)(.85) = $1997.50   (2350)(.15) = 352.50

10. (90.79)(24) = 2178.96
    2178.96 − 1997.50 = 181.46

11. $\left(\dfrac{181.46}{1997.50}\right)(100) = \$9.08\ /\$100$

    In Table 11.2, $9.08 is closest to $9.09 which
    yields an APR of 8.5% .

12. $6750 $1550 dp   12 mo.   6750 - 1550 = $5200
    5590.20 − 5200.00 = 390.20
    a) $u = \dfrac{f \bullet k(k+1)}{n(n+1)} = \dfrac{(390.20)(6)(7)}{12(3)} = \$105.05$
    b) (465.85)(5) = $2329.25
       5590.20 − 2329.25 = $3260.95
       3260.95 − 105.05 = $3155.90

13. $7500   36 mo. $223.10 / mo. $(223.10)(36) = 8031.60$ $8031.60 - 7500.00 = 531.60$

 a) $\left(\dfrac{181.46}{1997.50}\right)(100) = \$9.08$ In Table 11.2, $9.08 yields an APR of 4,5% .

 b) $u = \dfrac{(12)(223.10)(2.45)}{100 + 2.45} = \dfrac{6559.14}{102.45} = \$64.02$

 c) $(223.10)(23) = 5131.30$ $8031.60 - 5131.30 = \$2900.30$ $2900.30 - 64.02 = \$2836.28$

14. Mar. 23: $878.25
 Mar. 26: $878.25 + 95.89 = \$974.14$
 Mar. 30: $974.14 + 68.76 = \$1042.90$
 Apr. 03: $1042.90 - 450.00 = \$592.90$
 Apr. 15: $592.90 + 90.52 = \$683.42$
 Apr. 22: $683.42 + 450.85 = \$1134.27$

 a) $i = (878.25)(.014)(1) = \$12.30$
 b) $1134.27 + 12.30 = \$1146.57$

 c)
Date	Balance	# of Days	Balance-Days
Mar. 23	878.25	3	$(878.25)(3) =$ 2634.75
Mar. 26	974.14	4	$(974.14)(4) =$ 3896.56
Mar. 30	1042.90	4	$(1042.90)(4) =$ 4171.60
Apr. 03	592.90	12	$(592.90)(12) =$ 7114.80
Apr. 15	683.42	7	$(683.42)(7) =$ 4783.94
Apr. 22	1134.27	1	$(1134.27)(1) =$ 1134.27
		31	sum = $23,735.92

 d) $(765.67)(.014)(1) = \$10.72$
 e) $1134.27 + 10.72 = \$1144.99$

15. down payment = $(0.15)(144500) = \$21,675.00$

16. gross monthly income = $86500 \div 12 = \$7208.33$
 $7,208.33 - 605.00 = \$6,603.33$ adj. mo. income

17. maximum monthly payment = $(0.28)(6603.33) =$ $1,848.93

18. At 10.5% interest for 30 years, Table 11.4 yields $9.15.
 amount of loan = $144500 - 21675 = \$122,825$
 monthly payments = $\left(\dfrac{122825}{1000}\right)(9.15) = \$1,123.85$

19. $1123.85 + 304.17 = \$1428.02$ total mo. payment

20. Yes, the bank feels he can afford $1,848.93 per month and his payments would be $1,428.02.

21. a) Total cost of the house:
 $21675 + (1123.85)(12)(30) = \$426,261$
 b) interest = $426,261 - 144,500 = \$281,761$

Group Projects

1. a) $340,860.00   b) $308,420.00   c) $23,274.33   d) $174.80   e) $121,135.34
 f) Make a down payment of $20,000 and invest the difference in part (d).

CHAPTER TWELVE

PROBABILITY

Exercise Set 12.1

1. An experiment is a controlled operation that yields a set of results.
2. a) The possible results of an experiment are called its outcomes.
   b) An event is a subcollection of the outcomes of an experiment.
3. Empirical probability is the relative frequency of occurrence of an event. It is determined by actual observation of an experiment.

$$P(E) = \frac{\text{number of times the event occurred}}{\text{number of times the experiment was performed}}$$

4. The equally likely possible outcomes of an experiment.
5. Relative frequency over the long run can accurately be predicted, not individual events or totals.
6. The best way to determine the likelihood of death for a person is to observe others with similar characteristics.
7. Not necessarily, but it does mean that there is a 50:50 chance that 1 flip will land on heads.
8. Not necessarily, but it does mean that each outcome on a die has a chance of 1 in 6 of occurring.
9. Not necessarily, but it does mean that based on expirical data, Mr. Duncan may live until 79.21 years.
10. a) Roll a die 100 times and determine the number of times that a 5 occurs out of 100.
    b) Answers will vary (AWV).   c) AWV
11. AWV   12. AWV   13. AWV   14. AWV

15. Of 30 birds: 14 finches   10 cardinals   6 blue jays
    a) $P(f) = 14/30 = 7/15$ b) $P(c) = 10/30 = 1/3$ c) $P(bj) = 6/30 = 1/5$

16. Of 60 music lovers: 24 like rock   16 like country   8 like classical   12 like other types
    a) $P(r) = 24/60 = 2/5$ b) $P(c) = 16/60 = 4/15$ c) $P(other) = 12/60 = 1/59$

17. Of 95 animals: 40 are dogs.   35 are cats   15 are birds   5 are iguanas
    a) $P(dog) = 40/95 = 8/19$ b) $P(cat) = 35/95 = 7/19$ c) $P(iguana) = 5/95 = 1/19$

18. $5/50000 = 1/10000 = 0.0001$

19. Of 900 people: 19% like bananas   32% like apples   22% like oranges   27% like others
    a) Percents = the relative frequencies of the events occurring.
    b) $P(a) = \frac{32}{100} = 0.32$ c) $P(o) = \frac{22}{100} = 0.22$ d) $P(b) = \frac{19}{100} = 0.19$

20. 40,244 M of a total of 131,100 M   P(filing electronically) = $\dfrac{40,244,000,000}{131,100,000,000} = 0.31$

    131,100 M − 40,244 M = 90,856 M people that filed non-electronically

    P(filing non-electroncally) = $\dfrac{90,856,000,000}{131,100,000,000} = 0.69$

21. a) P(increase) = $\dfrac{\text{freq. of increases}}{\text{no. of observations}} = \dfrac{12}{12} = \dfrac{1}{1} = 1$ 22. a) P(A) = $\dfrac{43}{645} \approx 0.067$

    b) Yes, the answer in part (a) is only an estimate based on observation.

    b) P(C) = $\dfrac{260}{645} \approx 0.403$

    c) P(D or higher) = $\dfrac{90 + 260 + 182 + 43}{645} = 0.891$

23. Of 80 votes: 22 for Austin   18 for Emily   20 for Kimberly   14 for Joshua   6 for others
    a) P(A) = 22/80 = 11/40   b) P(E) = 18/80 = 9/40   c) P(K) = 20/80 = 1/4   d) P(J) = 14/80 = 7/40
    e) P(others) = 6/80 = 3/40

24. Of changes in housing prices across 50 states:
    a) P($\geq 60\%$) = 2/50 = 1/25   b) P(45% to 59.9%) = 4/50 = 2/25   c) P(30% to 44.9%) = 15/50 = 3/10
    d) P(15% to 29.9%) = 27/50   e) P(10% to 14.9%) = 2/50 = 1/25

25. a) P(bulls-eye) = $\dfrac{6}{20} = \dfrac{3}{10}$

    b) P(not bulls-eye) = $\dfrac{14}{20} = \dfrac{7}{10}$

    c) P(at least 20 pts.) = $\dfrac{14}{20} = \dfrac{7}{10}$

    d) P(does not score) = $\dfrac{2}{20} = \dfrac{1}{10}$

26. P(side 4) = $\dfrac{13}{100} = 0.13$

27. a) P(affecting circular) = $\dfrac{0}{150} = 0$

    b) P(affecting elliptical) = $\dfrac{50}{250} = 0.2$

    c) P(affecting irregular) = $\dfrac{100}{100} = 1$

28. Of 4,058,805 babies born, 2,076,960 were male and 1,981,845 were female.

    a) P(m) = $\dfrac{2,076,960}{4,058,805} = 0.51$ b) P(f) = $\dfrac{1,981,845}{4,058,805} = 0.49$

29. a) P(white flowers) = $\dfrac{224}{929} = 0.24$ b) P(purple flowers) = $\dfrac{705}{929} = 0.76$

30. a) P(tall plants) = $\dfrac{787}{1064} = 0.74$ b) P(short plants) = $\dfrac{277}{1064} = 0.26$

31. Answers will vary (AWV).   32. Answers will vary (AWV).

Exercise Set 12.2

1. If each outcome of an experiment has the same chance of occurring as any other outcome, they are said to be equally likely outcomes.

2. $P(\text{event}) = \dfrac{\text{no. of outcomes favorable to the event}}{\text{total number of possible outcomes}}$

3. $P(A) + P(\text{not } A) = 1$

4. $P(\text{event will not occur}) = 1 - \dfrac{4}{9} = \dfrac{9}{9} - \dfrac{4}{9} = \dfrac{5}{9}$

5. $P(\text{event will not occur}) = 1 - 0.3 = 0.7$

6. $P(\text{event will occur}) = 1 - 0.25 = 0.75$

7. $P(\text{event will occur}) = 1 - \dfrac{5}{12} = \dfrac{12}{12} - \dfrac{5}{12} = \dfrac{7}{12}$

8. a) 52 b) 13 c) 26 d) 4 e) 26
   f) 12 g) 4 h) 4

9. None of the possible outcomes is the event in question.

10. The event must include all possible outcomes.

11. All probabilities are between 0 and 1.

12. The sum of the probabilities of all outcomes = 1.

13. a) $P(\text{correct}) = 1/5$ b) $P(\text{correct}) = 1/4$

14. a) $P(\text{channel 3}) = 1/10$
    b) $P(\text{even channel}) = 5/10 = 1/2$
    c) $P(\text{less than 7}) = 7/10$

15. $P(\text{you win}) = \dfrac{\text{one choice}}{48 \text{ possible choices}} = \dfrac{1}{48}$

16. $P(\text{you win}) = \dfrac{\text{one choice}}{52 \text{ possible choices}} = \dfrac{1}{52}$

17. $P(7) = \dfrac{4}{52} = \dfrac{1}{13}$

18. $P(7 \text{ or } 9) = \dfrac{4}{52} + \dfrac{4}{52} = \dfrac{8}{52} = \dfrac{2}{13}$

19. $P(\quad 7) = \dfrac{48}{52} = \dfrac{12}{13}$

20. $P(5 \text{ of diamonds}) = \dfrac{1}{52}$

21. $P(\text{black}) = \dfrac{13}{52} + \dfrac{13}{52} = \dfrac{26}{52} = \dfrac{1}{2}$

22. $P(\text{heart}) = \dfrac{13}{52} = \dfrac{1}{4}$

23. $P(\text{red or black}) = \dfrac{26}{52} + \dfrac{26}{52} = \dfrac{52}{52} = \dfrac{1}{1} = 1$

24. $P(\text{red and black}) = 0$

25. $P(>4 \text{ and } <9) = P(5,6,7,8) = \dfrac{16}{52} = \dfrac{4}{13}$

26. $P(\text{jack of hearts}) = \dfrac{1}{52}$

27. a) $P(\text{red}) = \dfrac{2}{4} = \dfrac{1}{2}$ b) $P(\text{green}) = \dfrac{1}{4}$
    c) $P(\text{yellow}) = \dfrac{1}{4}$ d) $P(\text{blue}) = 0$

28. a) $P(\text{red}) = \dfrac{1}{4}$ b) $P(\text{green}) = \dfrac{1}{2}$
    c) $P(\text{yellow}) = \dfrac{1}{4}$ d) $P(\text{blue}) = 0$

29. a) $P(\text{red}) = \dfrac{2}{4} = \dfrac{1}{2}$ b) $P(\text{green}) = 0$
    c) $P(\text{yellow}) = \dfrac{1}{3}$ d) $P(\text{blue}) = \dfrac{1}{6}$

30. a) $P(\text{red}) = \frac{2}{4} = \frac{1}{2}$ b) $P(\text{green}) = \frac{1}{8}$ c) $P(\text{yellow}) = \frac{1}{8}$ d) $P(\text{blue}) = \frac{1}{4}$

Of 100 cans: 30 are cola (c)   40 are orange (o)   10 are ginger ale (ga)   20 are root beer (rb)

31. $P(o) = \frac{40}{100} = \frac{2}{5}$ 32. $P(c \text{ or } o) = \frac{70}{100} = \frac{7}{10}$ 33. $P(c, rb, o) = \frac{90}{100} = \frac{9}{10}$ 34. $P(ga) = \frac{10}{100} = \frac{1}{10}$

35. $P(600) = \frac{1}{12}$ 36. $P(> 400) = \frac{5}{12}$ 37. $P(\text{lose/bankrupt}) = \frac{2}{12} = \frac{1}{6}$ 38. $P(2500/\text{surprise}) = \frac{2}{12} = \frac{1}{6}$

Of 50 tennis balls: 23 are Wilson (w)   17 are Penn (p)   10 are other (o)

39. $P(W) = \frac{23}{50}$ 40. $P(P) = \frac{17}{50}$ 41. $P(\text{not } P) = \frac{33}{50}$ 42. $P(W \text{ or } P) = \frac{40}{50} = \frac{4}{5}$

For a traffic light: 25 seconds on red (r)   5 seconds on yellow (y)   55 seconds on green (g)

43. $P(g) = \frac{55}{85} = \frac{11}{17}$ 44. $P(y) = \frac{5}{85} = \frac{1}{17}$ 45. $P(\text{not } r) = \frac{60}{85} = \frac{12}{17}$ 46. $P(\text{not } g) = \frac{30}{85} = \frac{6}{17}$

Of 11 letters: 1 = m   4 = i   4 = s   2 = p

47. $P(s) = \frac{4}{11}$ 48. $P(\text{not } s) = \frac{7}{11}$ 49. $P(\text{vowel}) = \frac{4}{11}$ 50. $P(i \text{ or } p) = \frac{6}{11}$ 51. $P(\text{not } v) = 1$ 52. $P(w) = 0$

53. $P(= 60) = \frac{1}{11}$ 54. $P(> 250) = \frac{4}{11}$ 55. $P(> 50 \text{ and } < 250) = \frac{4}{11}$ 56. $P(\leq 40 \text{ and } \geq 163) = \frac{1}{11}$

57. $P(15) =$ $\frac{1}{26}$ 58. $P(\text{orange}) =$ $\frac{13}{26} = \frac{1}{2}$ 59. $P(\geq 22) =$ $\frac{5}{26}$ 60. $P(\leq 6 \text{ and/or } \leq 9) =$ $P(7,8,9) = \frac{3}{26}$

61. $P(\text{male}) =$ $\frac{345}{715} = \frac{69}{143}$ 62. $P(\text{female}) =$ $\frac{370}{715} = \frac{74}{143}$ 63. $P(GM, Ford, C\text{-}D) =$ $\frac{533}{715}$ 64. $P(\text{not } GM, Ford, C\text{-}D)$ $= \frac{533}{715}$

65. $P(\text{female} - \text{other}) =$ $\frac{97}{715}$ 66. $P(\text{male-}GM, Ford, C\text{-}D)$ $= \frac{260}{715} = \frac{52}{143}$ 67. $P(\text{Jiffy}) =$ $\frac{50}{159}$ 68. $P(\text{Skippy}) =$ $\frac{39}{159} = \frac{13}{53}$

69. $P(\text{chunky}) =$ $\frac{66}{159} = \frac{22}{53}$ 70. $P(\text{smooth}) =$ $\frac{93}{159} = \frac{31}{53}$ 71. $P(\text{Peter Pan} - \text{chunky})$ $= \frac{23}{159}$ 72. $P(\text{Jiffy} - \text{smooth}) =$ $\frac{28}{159}$

73. $P(\text{red}) = \frac{2}{18} + \frac{1}{12} + \frac{1}{6} = \frac{4}{36} + \frac{3}{36} + \frac{6}{36} = \frac{13}{36}$ 74. $P(\text{green}) = \frac{1}{18} + \frac{2}{12} + \frac{1}{12} = \frac{1}{18} + \frac{3}{12} = \frac{2}{36} + \frac{9}{36} = \frac{11}{36}$

75. $P(\text{yellow}) = \dfrac{1}{6} + \dfrac{1}{12} + \dfrac{1}{12} = \dfrac{2}{12} + \dfrac{2}{12} = \dfrac{4}{12} = \dfrac{1}{3}$

76. $P(\text{red or green}) = \dfrac{13}{36} + \dfrac{11}{36} = \dfrac{24}{36} = \dfrac{2}{3}$

77. $P(\text{yellow or green}) = \dfrac{1}{3} + \dfrac{11}{36} = \dfrac{23}{36}$

78. $P(\text{red or yellow}) = \left(\dfrac{1}{6} + \dfrac{2}{18} + \dfrac{1}{12}\right) + \dfrac{1}{3} = \dfrac{25}{36}$

79. a) $P(CC) = 0$ b) $P(CC) = 1$

80. a) $P(SCA) = P(S_1S_2) = \dfrac{1}{4}$

   b) $P(SCT) = P(S_1S_2 \text{ or } S_2S_1) = \dfrac{1}{2}$

   c) $P(NSCA \text{ or } NSCT) = 1 - \left(\dfrac{1}{2} + \dfrac{1}{4}\right) = 1 - \dfrac{3}{4} = \dfrac{1}{4}$

81. a) $P(R/R) = \dfrac{2}{4} \cdot \dfrac{2}{4} = \dfrac{4}{16} = \dfrac{1}{4}$

   b) $P(G/G) = \dfrac{2}{4} \cdot \dfrac{2}{4} = \dfrac{4}{16} = \dfrac{1}{4}$

   c) $P(R/G) = \dfrac{2}{4} \cdot \dfrac{2}{4} = \dfrac{4}{16} = \dfrac{1}{4}$

82. a) $P(\text{sparrow w/low attract. to PK}) = \dfrac{2}{7}$

   b) $P(\text{high attract. to CC / low attract to PK}) = 0$

   c) $P(\text{high attract. to BSSS / low attract to PK}) = \dfrac{4}{7}$

83.

+	1	2	3	4	5	6
1	2	3	4	5	6	1
2	3	4	5	6	1	2
3	4	5	6	1	2	3
4	5	6	1	2	3	4
5	6	1	2	3	4	5
6	1	2	3	4	5	6

$5 + 2 + 2 + 5 + 3 + 4 + 4 + 3 + 1 = 29$

Exercise Set 12.3

1. The odds against an event are found by dividing the probability that the event does not occur by the probability that the event does occur. The probabilities used should be expressed in fractional form.

2. The odds in favor of an event are found by dividing the probability that the event does occur by the probability that the event does not occur. The probabilities used should be expressed in fractional form.

3. Odds against are more commonly used.

4. If the odds against an event are a to b, then $P(\text{event occurs}) = \dfrac{b}{a+b}$ and $P(\text{event does not occur}) = \dfrac{a}{a+b}$.

5. $9 : 5$ or 9 to 5

6. $3 : 7$ or 3 to 7

7. a) $P(\text{event occurs}) = \dfrac{1}{1+1} = \dfrac{1}{2}$

   b) $P(\text{event fails to occur}) = \dfrac{1}{1+1} = \dfrac{1}{2}$

8. a) $P(\text{event fails}) = 1 - P(\text{event occurs}) = 1 - \dfrac{1}{2} = \dfrac{1}{2}$

   b) odds against the event $= \dfrac{P(\text{event fails to occur})}{P(\text{event occurs})} =$

   $\dfrac{1/2}{1/2} = \left(\dfrac{1}{2}\right)\left(\dfrac{2}{1}\right) = \dfrac{1}{1}$ or 1:1

   c) odds in favor of the event are 1:1.

9. a) P(tie goes well) = $\dfrac{8}{27}$

b) P(tie does not go well) = $\dfrac{19}{27}$

c) odds against tie going well =

$$\dfrac{\text{P(tie does not go well)}}{\text{P(tie goes well)}} = \dfrac{19/27}{8/27} = \dfrac{19}{27} \cdot \dfrac{27}{8} = \dfrac{19}{8}$$

d) odds in favor of it going well are 8:19 .

10. 14 bills: 7 - \$1; 2 - \$5; 4 - \$10; 1 - \$20

a) P(\$5) = $\dfrac{2}{14} = \dfrac{1}{7}$

b) P(not \$5) = $\dfrac{12}{14} = \dfrac{6}{7}$

c) 1 : 6

d) 6 : 1

11. 5 : 1

12. 3 : 3 or 1 : 1

13. odds against rolling less than 3 = $\dfrac{\text{P(3 or greater)}}{\text{P(less than 3)}} =$

$$\dfrac{4/6}{2/6} = \dfrac{4}{6} \cdot \dfrac{6}{2} = \dfrac{4}{2} = \dfrac{2}{1} \text{ or } 2{:}1$$

14. odds against rolling greater than 4 =

$$\dfrac{\text{P(failure to roll greater than 4)}}{\text{P(roll greater than 4)}} =$$

15. odds against a 6 = $\dfrac{\text{P(failure to pick a 6)}}{\text{P(pick a 6)}} =$

$$\dfrac{48/52}{4/52} = \dfrac{48}{52} \cdot \dfrac{52}{4} = \dfrac{48}{4} = \dfrac{12}{1} \text{ or } 12{:}1$$

Therefore, odds in favor of picking a 6 are 1:12.

16. odds against a heart = $\dfrac{\text{P(failure to pick a heart)}}{\text{P(pick a heart)}} =$

$$\dfrac{39/52}{13/52} = \dfrac{39}{52} \cdot \dfrac{52}{13} = \dfrac{39}{13} = \dfrac{3}{1} \text{ or } 3{:}1$$

Therefore, odds in favor of picking a heart are 1:3.

17. odds against a picture card =

$$\dfrac{\text{P(failure to pick a picture)}}{\text{P(pick a picture)}} = \dfrac{40/52}{12/52} = \dfrac{40}{12} = \dfrac{10}{3}$$

or 10:3

Therefore, odds in favor of picking a picture card are 3:10.

18. odds against card greater than 5 =

$$\dfrac{\text{P(failure to pick a card greater than 5)}}{\text{P(pick a card greater than 5)}} =$$

$$\dfrac{20/52}{32/52} = \dfrac{20}{52} \cdot \dfrac{52}{32} = \dfrac{20}{32} = \dfrac{5}{8} \text{ or } 5{:}8$$

Therefore, odds in favor of picking a card greater than 5 are 8:5.

19. odds against red =

$$\dfrac{\text{P(not red)}}{\text{P(red)}} = \dfrac{1/2}{1/2} = \dfrac{1}{2} \cdot \dfrac{2}{1} = \dfrac{2}{2} = \dfrac{1}{1} \text{ or } 1{:}1$$

20. odds against red =

$$\dfrac{\text{P(not red)}}{\text{P(red)}} = \dfrac{2/3}{1/3} = \dfrac{2}{3} \cdot \dfrac{3}{1} = \dfrac{6}{3} = \dfrac{2}{1} \text{ or } 2{:}1$$

21. odds against red = $\dfrac{\text{P(not red)}}{\text{P(red)}} = \dfrac{5/8}{3/8} = \dfrac{5}{8} \cdot \dfrac{8}{3} = \dfrac{5}{3}$

or 5:3

22. odds against red = $\dfrac{\text{P(not red)}}{\text{P(red)}} = \dfrac{5/8}{3/8} = \dfrac{5}{8} \cdot \dfrac{8}{3} = \dfrac{5}{3}$

or 5:3

23. a) odds against selecting female =

$$\frac{P(\text{failure to select female})}{P(\text{select female})} = \frac{16/30}{14/30} = \frac{16}{14} = \frac{8}{7}$$

   or 8 : 7 .

   b) odds against selecting male =

$$\frac{P(\text{failure to select male})}{P(\text{select male})} = \frac{14/30}{16/30} = \frac{14}{16} = \frac{7}{8}$$

   or 7 : 8 .

24. a) odds against winning = $\dfrac{P(\text{failure to win})}{P(\text{win})} =$

$$\frac{999999/1000000}{1/1000000} = \frac{999999}{1} \quad \text{or } 999,999 : 1$$

   b) odds against winning = $\dfrac{P(\text{failure to win})}{P(\text{win})} =$

$$\frac{999990/1000000}{10/1000000} = \frac{99999}{1} \quad \text{or } 99,999 : 1$$

25. odds against a stripe = $\dfrac{P(\text{not a stripe})}{P(\text{stripe})} =$

$$\frac{8/15}{7/15} = \frac{8}{15} \cdot \frac{15}{7} = \frac{8}{7} \quad \text{or } 8:7$$

26. odds in favor of even are $\dfrac{P(\text{even})}{P(\text{not even})} =$

$$\frac{7/15}{8/15} = \frac{7}{15} \cdot \frac{15}{8} = \frac{7}{8} \quad \text{or } 7:8$$

27. odds in favor of not the 8 ball are

$$\frac{P(\text{not the 8 ball})}{P(\text{the 8 ball})} = \frac{14/15}{1/15} = \frac{14}{15} \cdot \frac{15}{1} = \frac{14}{1} \quad \text{or } 14:1$$

28. odds against a ball with yellow are

$$\frac{P(\text{no yellow})}{P(\text{yellow})} = \frac{13/15}{2/15} = \frac{13}{15} \cdot \frac{15}{2} = \frac{13}{2} \quad \text{or } 13:2$$

29. odds against a ball with 9 or greater are

$$\frac{P(\text{less than 9})}{P(\text{9 or greater})} = \frac{8/15}{7/15} = \frac{8}{15} \cdot \frac{15}{7} = \frac{8}{7} \quad \text{or } 8:7$$

30. The odds in favor of two digits =

$$\frac{P(\text{two digits})}{P(\text{not two digits})} = \frac{6/15}{9/15} = \frac{6}{9} = \frac{2}{3} \quad \text{or } 2:3$$

31. a) $P(> \$5 \text{ M}) = \dfrac{5}{9}$

   b) Odds against payout > $5 M   4 : 5

32. a) $P(2 \text{ dots}) = \dfrac{1}{3}$

   b) Odds against rolling 2 dots   4 : 2 or 2 : 1

33. The odds against testing negative =

$$\frac{P(\text{test positive})}{P(\text{test negative})} = \frac{4/76}{72/76} = \frac{4}{72} = \frac{1}{18} \quad \text{or } 1 : 18$$

34. The odds against red = $\dfrac{P(\text{red})}{P(\text{not red})} = \dfrac{2/11}{9/11} = \dfrac{2}{9}$

   or 2:9

35. a) $P(\text{Carrie wins}) = \dfrac{7}{7+5} = \dfrac{7}{12}$

   b) $P(\text{Carrie loses}) = \dfrac{5}{7+5} = \dfrac{5}{12}$

36. a) $P(\text{Claire wins}) = \dfrac{2}{2+7} = \dfrac{2}{9}$

   b) $P(\text{Carrie loses}) = \dfrac{7}{2+7} = \dfrac{7}{9}$

37. Odds against   4 : 11 $P(\text{promoted}) = \dfrac{11}{4+11} = \dfrac{11}{15}$

38. Odds against   5 : 2   a) $P(\text{wins}) = \dfrac{2}{2+5} = \dfrac{2}{7}$

   b) $P(\text{loses}) = \dfrac{5}{2+5} = \dfrac{5}{7}$

39. $P(G) = \dfrac{15}{75} = \dfrac{1}{5}$

40. $P(\text{not } G) = 1 - \dfrac{1}{5} = \dfrac{4}{5}$

41. Odds in favor of N = $\dfrac{P(N)}{P(\text{not } N)} = \dfrac{1/5}{4/5} = \dfrac{1}{4}$ or 1:4

42. Odds against N are 4:1

43. Odds against I-27 =

$$\dfrac{P(\text{not I-27})}{P(\text{I-27})} = \dfrac{74/75}{1/75} = \left(\dfrac{74}{75}\right)\left(\dfrac{75}{1}\right) = \dfrac{74}{1} \text{ or } 74{:}1$$

44. Odds in favor of I-27 are 1:74

45. P(A+) = $\dfrac{34}{100} = 0.34$

46. P(B-) = $\dfrac{2}{100} = 0.02$

47. $\dfrac{66}{34} = \dfrac{33}{17}$ or 33 : 17

48. $\dfrac{2}{98} = \dfrac{1}{49}$ or 1 : 49

49. P(O or O-) = $\dfrac{43}{100} = \dfrac{43}{43+57}$ or 43 : 57

50. P(A+ or O+) = $\dfrac{71}{100} = \dfrac{71}{71+29}$ or 29 : 71

51. If P(selling out) = $0.9 = \dfrac{9}{10}$, then

P(do not sell your car this week) = $1 - \dfrac{9}{10} = \dfrac{1}{10}$.

The odds against selling out = $\dfrac{1/10}{9/10} = \dfrac{1}{9}$ or 1:9.

52. If P(overtime) = $\dfrac{3}{8}$, then P(no overtime) =

$1 - 1 - \dfrac{3}{8} = \dfrac{5}{8}$　　The odds in favor of being

asked to work overtime = $\dfrac{3/8}{5/8} = \dfrac{3}{5}$ or 3:5

53. If P(all parts are present) = $\dfrac{7}{8}$, then the odds in favor of all parts being present are 7 : 1 .

54. a) P(Mr. Frank is audited) = $\dfrac{1}{42}$

   b) Odds against Mr. Frank being audited are 41 : 1.

55. a) P(Douglas is a male) = $\dfrac{20}{21}$

   b) Odds against being a female are 20 : 1 .

56. P(even or > 3) = $\dfrac{1}{2} + \dfrac{1}{2} - \dfrac{2}{6} = 1 - \dfrac{2}{6} = \dfrac{4}{6} = \dfrac{2}{3}$

Odds against even or > 3 are $\dfrac{1/3}{2/3} = \dfrac{1}{2}$ or 1 : 2 .

57. P(# 1 wins) = $\dfrac{2}{9}$　　P(# 2 wins) = $\dfrac{1}{3}$

   P(# 3 wins) = $\dfrac{1}{16}$　　P(# 4 wins) = $\dfrac{5}{12}$

   P(# 5 wins) = $\dfrac{1}{2}$

58. a) P(R) = $\dfrac{9}{19}$

   b) Odds against red are 10 : 9

   c) P(0 or 00) = $\dfrac{1}{19}$

   d) Odds in favor of 0 or 00 are 1:18.

59. 119648 + 6742 + 506 + 77 = 126,973 multiple births　　.03x = 126973　　x = 4,232,433 total births

Odds against a multiple birth $\dfrac{(4232433 - 126973)}{126973} = \dfrac{4105460}{126973} = \dfrac{97}{3}$ or 97 : 3 .

Exercise Set 12.4

1. Expected value is used to determine the average gain or loss of an experiment over the long run.

2. An expected value of 0 indicates that the individual would break even over the long run.

3. The fair price is the amount charged for the game to be fair and result in an expected value of 0.

4. a) $E = P_1A_1 + P_2A_2$ b) $E = P_1A_1 + P_2A_2 + P_3A_3$

5. To obtain fair price, add the cost to play to the expected value.

6. No, fair price is the price to pay to make the expected value 0. The expected value is the expected outcome of an experiment when the experiment is performed many times

7. $0.50. Since you would lose $1.00 on average for each game you played, the fair price of the game should be $1.00 less. Then the expected value would be 0, and the game would be fair.

8. Fair price $= P_1G_1 + P_2G_2 + P_3G_3$

9. a) A $10 bet is the same as five $2 bets, thus Marty's expected value is $5(-0.40) = -\$2.00$
   b) On average he can expect to lose $2.00

10. a) Paul's expected value on a $5 bet is $5(0.20) = \$1.00$.
    b) If he makes many $5 bets he can expect to win, on average, $1.00 per bet.

11. $E = P_1A_1 + P_2A_2 = 0.70(200) + 0.30(120) = 140 + 36 = 176$ people

12. $E = P_1A_1 + P_2A_2 = 0.60(80000) + 0.40(-20000) = 48000 - 8000 = \$40000$

13. $E = P_1A_1 + P_2A_2 = 0.50(78) + 0.50(62) = 39 + 31 = 70$ points

14. $E = P_1A_1 + P_2A_2 = 0.40(20) + 0.60(12) = 8 + 7.2 = 15.2$ people

15. $E = P_1A_1 + P_2A_2 = 0.40(1.2 \text{ M}) + 0.60(1.6 \text{ M}) = .48 \text{ M} + .96 \text{ M} = 1.44 \text{ M}$ viewers

16. a) $E = P(\text{sunny})(1/2) + P(\text{cloudy})(1/4)$ $E = 0.75(1/2) + 0.25(1/4) = 0.375 + 0.0625 = 0.4375$ inches/day
    b) $(0.4375$ inches per day$)(31$ days$) = 13.5625$ inches of growth during July is expected

17. a) $E = P_1A_1 + P_2A_2 = (.60)(10000) + (.10)(0) + (.30)(7200) = 6000 + 0 + -2160 = \$3840$

18. a) $(.7)(5) + (.3)(10) = .35 + 3 = \$6.50$ b) $100.00 - 6.50 = \$93.50$

19. a) $E = P_1A_1 + P_2A_2 + P_3A_3 = P(\$1 \text{ off})(\$1) + P(\$2 \text{ off})(\$2) + P(\$5 \text{ off})(\$5)$
    $E = (1/10)(1) + (2/10)(2) + (1/10)(5) = 7/10 + 4/10 + 5/10 = 16/10 = \$1.60$

20. a) $(1/4)(5) + (3/4)(-2) = 1.25 - 1.50 = -\$.25$ for Mike
    b) $(1/4)(-5) + (3/4)(2) = -1.25 + 1.50 = \$.50$ for Dave

21. a) $(2/6)(8) + (4/6)(-5) = 8/3 - 20/6 = 16/6 - 20/6 = -4/6 = -\$.67$
    b) $(2/6)(-8) + (4/6)(5) = -8/3 + 20/6 = =16/6 + 20/6 = \$.67$

22. a) $(2/5)(-8) + (3/5)(5) = -16/5 + 3/1 = -3.20 + 3.00 = -\$.20$
    b) $(2/5)(8) + (3/5)(-5) = 16/5 - 3/1 = 3.20 - 3.00 = \$.20$

23. a) $(1/5)(5) + (0)(0) + (4/5)(-1) = 1 - 4/5 = 1/5$
    Yes, positive expectations $= 1/5$
    b) $(1/4)(5) + (0)(0) + (3/4)(-1) = 5/4 - 3/4 = 1/2$
    Yes, positive expectations $= 1/2$

24. a) $(1/4)(5) + (0)(0) + (3/4)(-2) = 5/4 - 6/4 = -1/4$
    No, negative expectations $= -1/4$
    b) $(1/3)(5) + (0)(0) + (2/3)(-2) = 5/3 - 4/3 = 1/3$
    Yes, positive expectations $= 1/3$

25. a) $\left(\dfrac{1}{500}\right)(400)+\left(\dfrac{499}{500}\right)(-2)=\dfrac{400-998}{500}=$

$\dfrac{-598}{500}=\dfrac{-299}{250}=-1.196\approx-\$1.20$

  b) Fair price = -1.20 + 2.00 = $.80

26. a) $\left(\dfrac{1}{1000}\right)(800)+\left(\dfrac{999}{1000}\right)(-1)=\dfrac{800-999}{1000}=$

$\dfrac{800}{1000}-\dfrac{999}{1000}=\dfrac{-199}{1000}=-.199\approx-\$0.20$

  b) Fair price = -0.20 + 1.00 = $.80

27. a) $\left(\dfrac{1}{2000}\right)(1000)+\left(\dfrac{2}{2000}\right)(500)+\left(\dfrac{1997}{2000}\right)(-3)=.50+.50+-2.9955=-\$2.00$

  b) Fair price = -2.00 + 3.00 = $1.00

28. $E = P_1A_1 + P_2A_2 + P_3A_3 + P_4A_4$

$E=\left(\dfrac{1}{10000}\right)(\$9,995)+\left(\dfrac{1}{10000}\right)(\$4,995)+\left(\dfrac{2}{10000}\right)(\$995)+\left(\dfrac{9996}{10000}\right)(-\$5)$

$=\dfrac{9995}{10000}+\dfrac{4995}{10000}+\dfrac{1990}{10000}+\dfrac{49980}{10000}=\dfrac{33000}{10000}=-\$3.30$

29. $\dfrac{1}{2}(1)+\dfrac{1}{2}(10)=\dfrac{1}{2}+5=5.5=\$5.50$

30. $\dfrac{1}{2}(5)+\dfrac{1}{4}(1)+\dfrac{1}{4}(10)=2.50+.25+2.50=\$5.25$

31. $\dfrac{1}{2}(10)+\dfrac{1}{4}(-5)+\dfrac{1}{4}(-20)=5-1.25-5=-\$1.25$

32. $\dfrac{1}{2}(-10)+\dfrac{1}{4}(2)+\dfrac{1}{4}(20)=-5+.50+5=\$0.50$

33. a) $\dfrac{1}{2}(1)+\dfrac{1}{2}(5)=.50+2.50=\$3.00$

  b) Fair price = 3.00 – 2.00 = $1.00

34. a) $\dfrac{1}{2}(10)+\dfrac{1}{4}(1)+\dfrac{1}{4}(5)=5+.25+1.25=\$6.50$

  b) Fair price = 6.50 – 2.00 = $4.50

35. a) $\dfrac{1}{2}(1)+\dfrac{1}{4}(5)+\dfrac{1}{4}(10)=.50+1.25+2.50=\$4.25$

  b) Fair price = 4.25 – 2.00 = $2.25

36. a) $\dfrac{1}{4}(5)+\dfrac{3}{8}(10)+\dfrac{3}{8}(1)=1.25+3.75+.38=\$5.38$

  b) Fair price = 5.38 – 2.00 = $3.38

37. $E = P_1A_1 + P_2A_2 + P_3A_3 + P_4A_4 + P_5A_5 =$
    $0.17(1) + 0.10(2) + 0.02(3) + 0.08(4) + 0.63(0) =$
    0.75 base

38. $E_{company}$ = P(insured lives)(amount gained) +
                  P(insured dies)(amount lost)
   E_{co} = (0.994)(100) + (0.006)(9,900) = 99.4 – 59.4
        = $40, which is the amount the company gains
        on this type of policy.

39. a) $E = P_1A_1 + P_2A_2 + P_3A_3$

   $=\dfrac{3}{10}(4)+\dfrac{5}{10}(3)+\dfrac{2}{10}(1)=1.2+1.5+0.2$

   = 2.9 points

  b) Fair price = 2.9 points

  c) 3(E) = 3(2.9) = 8.7 points

40. a) $E = P_1A_1 + P_2A_2 + P_3A_3$

   $=\dfrac{3}{10}(5)+\dfrac{5}{10}(2)+\dfrac{2}{10}(-3)=1.5+1.0-0.6$

   = 1.9 points

  b) Fair price = 1.9 points

  c) 3(E) = 3(1.9) = 5.7 points

41. $(0.34)(850) + (0.66)(140) = 289 + 92.4 =$
381.4 employees

42. $(.62)(2.3\ M) + (.38)(1.7\ M) = 1.426\ M + .646\ M =$
$\$2.072\ M$

43. $(.11)(10) + (.65)(15) + (.24)(20) = 1.1 + 9.75 + 4.8 =$
15.65 minutes

44. $(.40)(1000) + (.50)(500) + (.10)(0) =$
$400 + 250 = \$650.00$

45. $E = P(1)(1) + P(2)(2) + P(3)(3) + P(4)(4) + P(5)(5)$
$\quad + P(6)(6)$
$\quad = \dfrac{1}{6}(1) + \dfrac{1}{6}(2) + \dfrac{1}{6}(3) + \dfrac{1}{6}(4) + \dfrac{1}{6}(5) + \dfrac{1}{6}(6)$
$\quad = \dfrac{21}{6} = 3.5$ points

46. $E = P_1 A_1 + P_2 A_2 + P_3 A_3$
$\quad = 0.70(40{,}000) + 0.10(0) + 0.20(-30{,}000)$
$\quad = 28{,}000 + 0 - 6{,}000 = \$22{,}000$

47. $E = P_1 A_1 + P_2 A_2 + P_3 A_3$
$\quad = \dfrac{200}{365}(110) + \dfrac{100}{365}(160) + \dfrac{65}{365}(210)$
$\quad = 60.27 + 43.84 + 37.40 = 141.51$ calls/day

48. Profit if Jorge sells the house $= 0.06(100{,}000)$
$\quad = \$6{,}000$
Profit if another Realtor sells the house $=$
$0.03(100{,}000) = \$3{,}00$
$E = P_1 A_1 + P_2 A_2 + P_3 A_3$
$\quad = 0.2(5000) + 0.5(2000) + 0.3(1000)$
$\quad = \$1{,}000 + \$1{,}000 - \$300 = \$1{,}700$ gain
Yes, in the long run if Jorge lists many of these $\$100{,}000$ homes, he can expect to make, on average, $\$1{,}700$ per listing.

49. a) $P(1) = \dfrac{1}{2} + \dfrac{1}{16} = \dfrac{8}{16} + \dfrac{1}{16} = \dfrac{9}{16}$, $P(10) = \dfrac{1}{4} = \dfrac{4}{16}$,
$\quad P(\$20) = \dfrac{1}{8} = \dfrac{2}{16}$, $P(\$100) = \dfrac{1}{16}$

b) $E = P_1 A_1 + P_2 A_2 + P_3 A_3 + P_4 A_4$
$\quad = \dfrac{9}{16}(\$1) + \dfrac{4}{16}(\$10) + \dfrac{2}{16}(\$20) + \dfrac{1}{16}(\$100)$
$\quad = \dfrac{9}{16} + \dfrac{40}{16} + \dfrac{40}{16} + \dfrac{100}{16} = \dfrac{189}{16} = \$11.81$

c) fair price $=$ expected value $-$ cost to play $=$
$\$11.81 - 0 = \$11.81$

50. a) $P(\$1) = \dfrac{1}{6} + \dfrac{1}{4} = \dfrac{2}{12} + \dfrac{3}{12} = \dfrac{5}{12} = \dfrac{10}{24}$,
$\quad P(\$10) = \dfrac{1}{6} = \dfrac{4}{24}$, $P(\$20) = \dfrac{1}{6} + \dfrac{1}{8} = \dfrac{4}{24} + \dfrac{3}{24} = \dfrac{7}{24}$,
$\quad P(\$100) = \dfrac{1}{8} = \dfrac{3}{24}$

c) fair price $=$ expected value $-$ cost to play $=$
$\$20.42 - 0 = \$20.42$

50. b) $E = P_1 A_1 + P_2 A_2 + P_3 A_3 + P_4 A_4$
$\quad = \dfrac{10}{24}(1) + \dfrac{4}{24}(10) + \dfrac{7}{24}(20) + \dfrac{3}{24}(100)$
$\quad = \dfrac{10}{24} + \dfrac{40}{24} + \dfrac{140}{24} + \dfrac{300}{24} = \dfrac{490}{24} = \$20.42$

51. $E = P(\text{insured lives})(\text{cost}) + P(\text{insured dies})(\text{cost} -$
$\quad \$40{,}000)$
$\quad = 0.97(\text{cost}) + 0.03(\text{cost} - 40{,}000)$
$\quad = 0.97(\text{cost}) + 0.03(\text{cost}) - 1200$
$\quad = 1.00(\text{cost}) - 1200$
Thus, in order for the company to make a profit, the cost must exceed $\$1{,}200$

52. No, you don't know how many others are selecting the same numbers that you are selecting.

53. $E = P(\text{win})(\text{amount won}) + P(\text{lose})(\text{amount lost})$
$\quad = \left(\dfrac{1}{38}\right)(35) + \left(\dfrac{37}{38}\right)(-1) = \dfrac{35}{38} - \dfrac{37}{38} = -\dfrac{2}{38}$
$\quad = -\$0.053$

54. $E = P(\text{win})(\text{amount won}) + P(\text{lose})(\text{amount lost}) = \left(\dfrac{18}{38}\right)(1) + \left(\dfrac{20}{38}\right)(-1) = \dfrac{18}{38} - \dfrac{20}{38} = -\dfrac{2}{38} = -\$0.053$

55. a) $E = \dfrac{1}{12}(100) + \dfrac{1}{12}(200) + \dfrac{1}{12}(300) + \dfrac{1}{12}(400) + \dfrac{1}{12}(500) + \dfrac{1}{12}(600) + \dfrac{1}{12}(700) + \dfrac{1}{12}(800) + \dfrac{1}{12}(900)$

$\dfrac{1}{12}(1000) = \left(\dfrac{5500}{12}\right) = \$458.3\overline{3}$

b) $E = \dfrac{1}{12}(5500) + \dfrac{1}{12}(-1800) = \dfrac{3700}{12} = \$308.33$

Exercise Set 12.5

1. If a first experiment can be performed in M distinct way and a second experiment can be performed in N distinct ways, then the two experiments in that specific order can be performed in $M \cdot N$ distinct ways.

2. a) A list of all the possible outcomes of an experiment.

   b) Each individual outcome in a sample space is a sample point.

3. $(2)(7) = 14$ ways. Using the counting principle.

4. Answers will vary.

5. The first selection is made. Then the second selection is made before the first selection is returned to the group of items being selected.

6. $(5)(2) = 10$ ways

7. a) $(50)(50) = 2500$ b) $(50)(49) = 2450$

9. a) $(6)(6)(6) = 216$ b) $(6)(5)(4) = 120$

8. a) $(365)(365) = 133,225$ b) $(365)(364) = 132,860$

10. a) $(10)(10) = 100$ b) $(10)(9) = 90$

11. a) $(2)(2) = 4$ points

    b)

    c) P(no heads) = 1/4

    d) P(exactly one head) = 2/4 = 1/2

    e) P(two heads) = 1/4

12. a) $(2)(2) = 4$ points

    b)

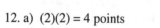

    c) P(two girls) = 1/4

    d) P(at least one girl) = 3/4

    e) P(girl 1st and boy 2nd) = 1/4

13. a) $(3)(3) = 9$ points

    b)

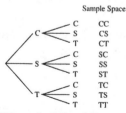

    c) P(two Jacks) = 1/9

    d) P(Jack and then Queen) = 1/9

    e) P(at least one King) = 5/9

14. a) $(3)(2) = 6$ points

    b)

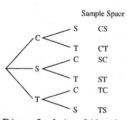

    c) P(two Jacks) = 0/6 = 0

    d) P(Jack and then Queen) = 1/6

    e) P(at least one King) = 4/6 = 2/3

15. a) (4)(3) = 12 points
    b)

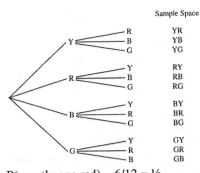

    c) P(exactly one red) = 6/12 = ½
    d) P(at least one is not red) = 12/12 = 1
    e) P(no green) = 6/12 = 1/2

16. a) (2)(2)(2) = 8 points
    b)

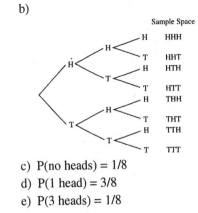

    c) P(no heads) = 1/8
    d) P(1 head) = 3/8
    e) P(3 heads) = 1/8

17. a) (2)(2)(2) = 8 points
    b

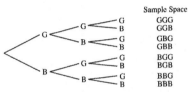

    c) P(no boys) = 1/8
    d) P(at least one girl) = 7/8
    e) P(either exactly 2 boys or 2 girls) = 6/8 = ¾
    f) P(boy 1st and boy 2nd and girl 3rd) = 1/8

18. a) (4)(3) = 12 points
    b)

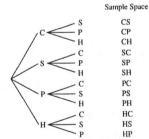

    c) P(Persian cat) = 6/12 = 1/2
    d) P(Persian cat and calico cat) = 2/12 = 1/6
    e) P(not Persiann) = 6/12 = 1/2

19. a) (6)(6) = 36 points
    b)

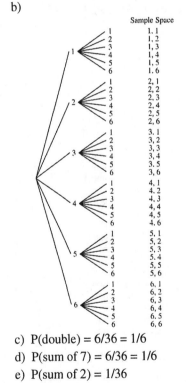

    c) P(double) = 6/36 = 1/6
    d) P(sum of 7) = 6/36 = 1/6
    e) P(sum of 2) = 1/36
    f) No; the P(sum of 2) < P(sum of 7)

20. a) (3)(3)(3) = 27 points
    b)

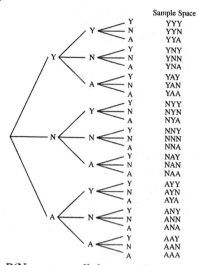

    c) P(No vote on all three motions) = 1/27
    d) P(Yes vote on exactly two motions) = 6/27 = 2/9
    e) P(at least one yes vote) = 19/27

21. a) (3)(2)(1) = 6 points
    b)

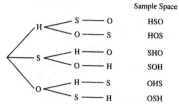

    c) P(Sears – 1st) = 2/6 = 1/3
    d) P(Home Depot – 1st / Outback - last) = 1/6
    e) P(Sears,Outback,Home Depot) = 1/6

22. a) (3)(3)(2) = 18
    b)

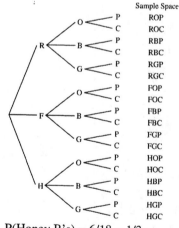

    c) P(Honey B's) = 6/18 = 1/3
    d) P(Rice Krispies and Ginger Ale) = 2/18 = 1/9
    e) P(not black cherry) = 12/18 = 2/3

23. a) (4)(3) = 12 points
    b)

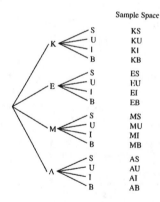

    c) P(M.K. or E.C.) = 6/12 = ½
    d) P(MGM or Univ.) = 6/12 = ½
    e) P(M.K. and (S.W. or B.G.)) = 2/12 = 1/6

24. a) (4)(2)(2) = 16 points
    b)

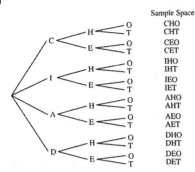

Sample Space

    c) P(Apple) = 1/4
    d) P(H-P) = 1/2
    e) P(Apple and H-P) = 2/16 = 1/8

25. a) (3)(3)(3) = 27
    b)

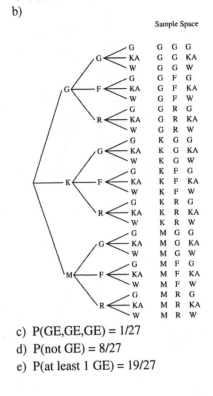

Sample Space

    c) P(GE,GE,GE) = 1/27
    d) P(not GE) = 8/27
    e) P(at least 1 GE) = 19/27

26. a) (2)(3)(4) = 24 points
    b)

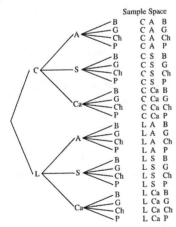

Sample Space

    c) P(geo.) = 1/4
    d) P(geo. or chem.) = 2/4 = ½
    e) P(not calc.) = 16/24 = 2/3

27. a) (2)(4)(3) = 24 sample points
    b)

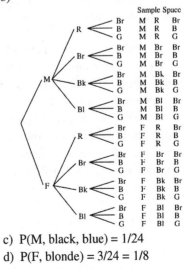

Sample Space

    c) P(M, black, blue) = 1/24
    d) P(F, blonde) = 3/24 = 1/8

28. a) $(2)(2)(2)(2) = 16$ sample points

   b)

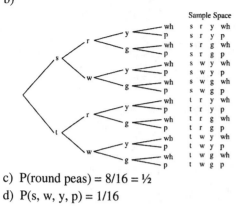

   c) P(round peas) = 8/16 = ½
   d) P(s, w, y, p) = 1/16

29. a) P(white) = 1/3
   b) P(red) = 2/3
   c) No; P(white) < P(red)
   d)

   Sample Space

   e) { ww, wr$_1$,wr$_2$, r$_1$w, r$_1$r$_1$, r$_1$r$_2$, r$_2$w, r$_2$r$_1$, r$_2$r$_2$}
   f) P(2r) = 4/9
   g) P(w, then r) = 2/9
   h) P(w and r, any order) = 4/9
   i) P(at least 1 r) = 8/9

30. a) m or n
   b) 3 or 4
   c) {m3, m4, n3, n4}
   d) No; not unless we know that the outcomes
      are equally likely.

30. e) No; same reason as in d)
   f) Yes; (1/2)(1/2) = 1/4

31. 1 red, 1 blue, and 1 brown

Exercise Set 12.6

1. a) "or" means at least one event A or B must occur.   b) "and" means both events, A and B, must occur.
2. a) P(A or B) = P(A) + P(B) – P(A and B)
3. a) Two events are mutually exclusive if it is impossible for both events to occur simultaneously.
   b) P(A or B) = P(A) + P(B)
4. a) P(A and B) = P(A)⊔P(B, given that A has occurred)
5. We assume that event A has already occurred.

6. Two events are independent if the occurrence of either event in no way affects the probability of occurrence of
   the other event. Ex. toss two coins; find P (tails and tails)
7. Two events are dependent if the occurrence of either event affects the probability of occurrence of the other
   event. Ex. Select two cards from a deck (without replacement); find P (King and King).
8. a) No, it is possible for both to like classical music.
   b) No, if the mother likes classical music the daughter will be more likely to like classical music.
9. a) No, both mother and father may be teachers.
   b) No, studies have shown that if the husband or wife is a teacher there is an increased probability that their
      spouse is also a teacher.

10. a) No, it is possible for an individual to be both happy and healthy at the same time.
    b) No, if you are healthy, you are more likely to be happy.

11. If the events are mutually exclusive, the events   12. Student activity problem.
    cannot happen simultaneously and thus
    P(A and B) = 0.

13. P(A and B) = 0.3
   P(A or B) = P(A) + P(B) – P(A and B)
   $= 0.6 + 0.4 – (0.6)(0.4) = 1.0 – 0.3 = 0.7$

14. P(A or B) = 0.9
   P(A or B) = P(A) + P(B) – P(A and B)
   $0.9 = 0.5 + 0.6 – P(A \text{ and } B)$
   P(A and B) = 0.2

15. P(B) = P(A or B) + P(A and B) – P(A)
   $= 0.8 + 0.1 – 0.4 = 0.5$

16. P(A or B) = P(A) + P(B) – P(A and B)
   $0.6 = P(A) + 0.3 – 0.1$
   $0.6 = P(A) + 0.2 \qquad P(A) = 0.4$

17. P(2 or 5) = 1/6 + 1/6 = 2/6 = 1/3

18. P(odd or greater than 2) = 5/6

19. P(greater than 4 or less than 2) = P(5, 6. or 1) =
   2/6 + 1/6 = 3/6 = 1/2

20. All numbers on the die are either > 3 or < 5.
   P(> 3 or < 5) = 6/6 = 1

21. Since these events are mutually exclusive,
   P(ace or king) = P(ace) + P(king) =
   $= \dfrac{4}{52} + \dfrac{4}{52} = \dfrac{8}{52} = \dfrac{2}{13}$

22. Since it is possible to obtain a card that is both a jack and a diamond when only one card is selected, these events are not mutually exclusive.
   P(jack or diamond) = P(jack) + P(diamond) –
   P(jack and diamond) $= \dfrac{4}{52} + \dfrac{13}{52} - \dfrac{1}{52} = \dfrac{16}{52} = \dfrac{4}{13}$

23. Since it is possible to obtain a card that is a picture card and a red card, these events are not mutually exclusive.
   P(picture or red) = P(pict.) + P(red) – P(pict. & red)
   $= \dfrac{12}{52} + \dfrac{26}{52} - \dfrac{6}{52} = \dfrac{32}{52} = \dfrac{8}{13}$

24. Since it is impossible to obtain a card that is both a heart and a black card, these events are mutually exclusive.
   P(club or red) = P(club) + P(red) =
   $= \dfrac{13}{52} + \dfrac{26}{52} = \dfrac{39}{52} = \dfrac{3}{4}$

25. Since it is possible to obtain a card less than 9 that is a club, these events are not mutually exclusive.
   $P(< 7 \text{ or club}) = \dfrac{24}{52} + \dfrac{13}{52} - \dfrac{6}{52} = \dfrac{31}{52}$

26. Since it is possible to obtain a card greater than 8 that is black, these events are not mutually exclusive.
   $P(> 9 \text{ or black}) = \dfrac{16}{52} + \dfrac{26}{52} - \dfrac{8}{52} = \dfrac{34}{52} = \dfrac{17}{26}$

27. a) P(frog and frog)= $\dfrac{5}{20} \cdot \dfrac{5}{20} = \dfrac{1}{4} \cdot \dfrac{1}{4} = \dfrac{1}{16}$

   b) P(frog and frog)= $\dfrac{5}{20} \cdot \dfrac{4}{19} = \dfrac{1}{4} \cdot \dfrac{4}{19} = \dfrac{1}{19}$

28. a) P(3 and 3) = $\dfrac{4}{20} \cdot \dfrac{4}{20} = \dfrac{1}{5} \cdot \dfrac{1}{5} = \dfrac{1}{25}$

   b) P(3 and 3) = $\dfrac{4}{20} \cdot \dfrac{3}{19} = \dfrac{1}{5} \cdot \dfrac{3}{19} = \dfrac{3}{95}$

29. a) P(lion and bird) = $\dfrac{5}{20} \cdot \dfrac{5}{20} = \dfrac{1}{4} \cdot \dfrac{1}{4} = \dfrac{1}{16}$

   b) P(lion and bird) = $\dfrac{5}{20} \cdot \dfrac{5}{19} = \dfrac{1}{4} \cdot \dfrac{5}{19} = \dfrac{5}{76}$

30. a) P(2 and 4) = $\dfrac{4}{20} \cdot \dfrac{4}{20} = \dfrac{1}{5} \cdot \dfrac{1}{5} = \dfrac{1}{25}$

   b) P(2 and 4) = $\dfrac{4}{20} \cdot \dfrac{4}{19} = \dfrac{1}{5} \cdot \dfrac{4}{19} = \dfrac{4}{95}$

31. a) P(red bird and monkey) = $\dfrac{3}{20} \cdot \dfrac{5}{20} = \dfrac{3}{20} \cdot \dfrac{1}{4} = \dfrac{3}{80}$

    b) P(red bird and monkey) = $\dfrac{3}{20} \cdot \dfrac{5}{19} = \dfrac{15}{380} = \dfrac{3}{76}$

32. a) P(even and even) = $\dfrac{8}{20} \cdot \dfrac{8}{20} = \dfrac{2}{5} \cdot \dfrac{2}{5} = \dfrac{4}{25}$

    b) P(even and even) = $\dfrac{8}{20} \cdot \dfrac{7}{19} = \dfrac{2}{5} \cdot \dfrac{7}{19} = \dfrac{14}{95}$

33. a) P(odd and odd) = $\dfrac{12}{20} \cdot \dfrac{12}{20} = \dfrac{3}{5} \cdot \dfrac{3}{5} = \dfrac{9}{25}$

    b) P(odd and odd) = $\dfrac{12}{20} \cdot \dfrac{11}{19} = \dfrac{3}{5} \cdot \dfrac{11}{19} = \dfrac{33}{95}$

34. a) P(lion and red bird) = $\dfrac{5}{20} \cdot \dfrac{3}{20} = \dfrac{1}{4} \cdot \dfrac{3}{20} = \dfrac{3}{80}$

    b) P(lion and red bird) = $\dfrac{5}{20} \cdot \dfrac{3}{19} = \dfrac{1}{4} \cdot \dfrac{3}{19} = \dfrac{3}{76}$

35. P(monkey or even) = $\dfrac{5}{20} + \dfrac{8}{20} - \dfrac{2}{20} = \dfrac{11}{20}$

36. P(yellow bird or > 4) = $\dfrac{2}{20} + \dfrac{4}{20} = \dfrac{6}{20} = \dfrac{3}{10}$

37. P(lion or a 2) = $\dfrac{5}{20} + \dfrac{4}{20} - \dfrac{1}{20} = \dfrac{8}{20} = \dfrac{2}{5}$

38. P(red bird or even) = $\dfrac{3}{20} + \dfrac{8}{20} - \dfrac{1}{20} = \dfrac{10}{20} = \dfrac{1}{2}$

39. P(2 reds) = $\dfrac{1}{2} \cdot \dfrac{1}{2} = \dfrac{1}{4}$

40. P(red and then yellow) = $\dfrac{1}{2} \cdot \dfrac{1}{2} = \dfrac{1}{4}$

41. P(red and green) = $\dfrac{1}{4} \cdot \dfrac{1}{2} = \dfrac{1}{8}$

42. P(2 reds) = $\dfrac{1}{4} \cdot \dfrac{1}{4} = \dfrac{1}{16}$

43. P(2 yellows) = P(red and red) = $\dfrac{3}{8} \cdot \dfrac{3}{8} = \dfrac{9}{64}$

44. P(both not yellow) = $\dfrac{5}{8} \cdot \dfrac{5}{8} = \dfrac{25}{64}$

45. P(2 reds) = $\dfrac{1}{2} \cdot \dfrac{1}{4} = \dfrac{1}{8}$

46. P(red and yellow) = $\dfrac{1}{2} \cdot \dfrac{1}{4} = \dfrac{1}{8}$

47. P(both not yellow) = $\dfrac{1}{2} \cdot \dfrac{3}{4} = \dfrac{3}{8}$

48. P(yellow and not yellow) = $\dfrac{1}{2} \cdot \dfrac{3}{4} = \dfrac{3}{8}$

49. P(3 girls) = P(1st girl) • P(2nd girl) • P(3rd girl)

    = $\dfrac{1}{2} \cdot \dfrac{1}{2} \cdot \dfrac{1}{2} = \dfrac{1}{8}$

50. P(3 boys) = P(1st boy) • P(2nd boy) • P(3rd boy)

    = $\dfrac{1}{2} \cdot \dfrac{1}{2} \cdot \dfrac{1}{2} = \dfrac{1}{8}$

51. P(G,G,B) = P(1st girl) • P(2nd girl) • P(3rd boy)

    = $\dfrac{1}{2} \cdot \dfrac{1}{2} \cdot \dfrac{1}{2} = \dfrac{1}{8}$

52. P(G,B,G) = P(1st girl) • P(2nd boy) • P(3rd girl)

    = $\dfrac{1}{2} \cdot \dfrac{1}{2} \cdot \dfrac{1}{2} = \dfrac{1}{8}$

53. a) P(5 boys) = P(b) • P(b) • P(b) • P(b) • P(b)

$$= \frac{1}{2} \cdot \frac{1}{2} \cdot \frac{1}{2} \cdot \frac{1}{2} \cdot \frac{1}{2} = \frac{1}{32}$$

b) P(next child is a boy) = $\frac{1}{2}$

54. a) P(7 girls) = $\frac{1}{2} \cdot \frac{1}{2} \cdot \frac{1}{2} \cdot \frac{1}{2} \cdot \frac{1}{2} \cdot \frac{1}{2} \cdot \frac{1}{2} = \frac{1}{128}$

b) P(next child is a girl) = $\frac{1}{2}$

55. a) P(Titleist/Pinnacle) = $\frac{4}{7} \cdot \frac{1}{7} = \frac{4}{49}$

b) P(Titleist/Pinnacle) = $\frac{4}{7} \cdot \frac{1}{6} = \frac{4}{42} = \frac{2}{21}$

56. a) P(Top Flite/ Top Flite) = $\frac{5}{7} \cdot \frac{5}{7} = \frac{25}{49}$

b) P(Top Flite/ Top Flite) = $\frac{5}{7} \cdot \frac{4}{6} = \frac{20}{42} = \frac{10}{21}$

57. a) P(at least 1 Top Flite) = $\frac{2}{7} \cdot \frac{5}{7} + \frac{5}{7} \cdot \frac{2}{7} + \frac{2}{7} \cdot \frac{2}{7} = \frac{24}{49}$

b) P(at least 1 Top Flite) = $\frac{2}{7} \cdot \frac{5}{6} + \frac{5}{7} \cdot \frac{2}{6} + \frac{2}{7} \cdot \frac{1}{6} = \frac{11}{21}$

58. a) P(Pinnacle/Pinnacle) = $\frac{1}{7} \cdot \frac{1}{7} = \frac{1}{49}$

b) P(Pinnacle/Pinnacle) = $\frac{1}{7} \cdot \frac{0}{6} = 0$

59. P(neither had trad. ins.) = $\frac{26}{40} \cdot \frac{25}{39} = \frac{10}{24} = \frac{5}{12}$

60. P(both have managed care) = $\frac{22}{40} \cdot \frac{21}{39} = \frac{77}{260}$

61. P(trad. ins./ trad. ins.) or P(trad. ins./trad. ins.)

$$= \left(\frac{14}{40} \cdot \frac{26}{39} \right) + \left(\frac{14}{40} \cdot \frac{13}{39} \right) = \frac{28}{60} + \frac{7}{60} = \frac{35}{60} = \frac{7}{12}$$

62. P(trad. ins./managed care) = $\frac{14}{40} \cdot \frac{22}{39} = \frac{77}{390}$

63. P(all recommended) = $\frac{19}{30} \cdot \frac{18}{29} \cdot \frac{17}{28} = \frac{969}{4060}$

64. P(no/yes/yes) = $\frac{6}{30} \cdot \frac{19}{29} \cdot \frac{18}{28} = \frac{171}{2030}$

65. P(no/no/not sure) = $\frac{6}{30} \cdot \frac{5}{29} \cdot \frac{5}{28} = \frac{5}{812}$

66. P(yes/no/no) = $\frac{19}{30} \cdot \frac{6}{29} \cdot \frac{5}{28} = \frac{19}{812}$

67. The probability that any individual reacts favorably is 70/100 or 0.7.
P(Mrs. Rivera reacts favorably) = 0.7

68. Since it is assumed the sample is representative of the entire population, it must be assumed this experiment is done with replacement. If done w/o replacement, the number in the population must be known. In addition, since the population is so large, reducing the numerator and/or denominator by 1 has no appreciable effect on the answer.
P(Mr. Rivera and Mrs. Rivera react favorably and Carlos is unaffected) = P(Mr. Rivera reacts favorable) • P(Mrs. Rivera reacts favorable) • P(Carlos is unaffected) = 0.7•0.7•0.2 = 0.098

69. P(all 3 react favorably) = 0.7•0.7•0.7 = 0.343

70. One does not react favorably if the reaction is unfavorable or if it is unaffected.
P(not favorable) = 0.1 + 0.2 = 0.3. Therefore,
P(none reacts favorably) = $(0.3)^3 = 0.027$

71. Since each question has four possible answers of which only one is correct, the probability of guessing correctly on any given question is 1/4.
P(correct answer on any one question) = ¼

72. If you have guessed correctly on only the first question, then you have missed the last four. The probability of missing any given question is 3/4.
P(only the 1st correct) = P(1st corr)•P(2nd incorr)
•P(3rd incorr)•P(4th incorr)•P(5th incorr)
= (1/4)(3/4)(3/4)(3/4)(3/4) = 81/1024

73. P(only the 3rd and 4th questions correct) =

$$\left(\frac{3}{4}\right)\left(\frac{3}{4}\right)\left(\frac{1}{4}\right)\left(\frac{1}{4}\right)\left(\frac{3}{4}\right) = \frac{27}{1024}$$

74. P(all 5 questions correct) =

$$\left(\frac{1}{4}\right)\left(\frac{1}{4}\right)\left(\frac{1}{4}\right)\left(\frac{1}{4}\right)\left(\frac{1}{4}\right) = \frac{1}{1024}$$

75. P(none of the 5 questions correct) =

$$\left(\frac{3}{4}\right)\left(\frac{3}{4}\right)\left(\frac{3}{4}\right)\left(\frac{3}{4}\right)\left(\frac{3}{4}\right) = \frac{243}{1024}$$

76. P(at least one is correct) = 1 − P(none are correct)

$$= 1 - \frac{243}{1024} = \frac{781}{1024}$$

77. P(bell on 1st reel) = 3/22

78. P(orange on all 3 reels) =

P(Or on 1st)P(Or on 2nd)P(Or on 3rd) =

$$\left(\frac{5}{22}\right)\left(\frac{4}{22}\right)\left(\frac{5}{22}\right) = \frac{100}{10648} = \frac{25}{2662}$$

79. P(no bar/no bar/no bar) = $\left(\frac{20}{22}\right)\left(\frac{20}{22}\right)\left(\frac{21}{22}\right) = \frac{1050}{1331}$

80. P(7/7/7) = $\left(\frac{1}{22}\right)\left(\frac{1}{22}\right)\left(\frac{1}{22}\right) = \frac{1}{10648}$

81. P(yellow/yellow) = $\left(\frac{1}{8}\right)\left(\frac{2}{12}\right) = \frac{2}{96} = \frac{1}{48}$

82. P(red on outer and blue on inner) =

$$\frac{4}{12} \cdot \frac{2}{8} = \frac{1}{3} \cdot \frac{1}{4} = \frac{1}{12}$$

83. P(not red on outer and not red on inner) =

$$\frac{8}{12} \cdot \frac{5}{8} = \frac{5}{12}$$

84. P(at least one is red) = 1 − P(neither is red) =

$$1 - \frac{5}{12} = \frac{7}{12}$$

85. P(no hit/no hit) = (0.6)(0.6) = 0.36

86. P(hit/no hit) = (0.4)(0.1) = 0.04

87. P(both hit) = (0.4)(0.9) = 0.36

88. P(1st miss/2nd hit) = (0.6)(0.4) = 0.24

89. a) No; The probability of the 1st depends on the outcome of the first.
    b) P(one afflicted) = .001
    c) P(both afflicted) = (.001)(.04) = .00004
    d) P(not afflicted/afflicted) = (.999)(.001) = .000999
    e) P(not affl/not affl) = (.999)(.999) = .99801

90. a) P(Mrs. Jones # is selected) = (1/10)(1/10(1/10)
       = .001

91. P(audit this year) = .032
92. P(audited next 2 years) = (.032)(.032) = .001024
93. P(audit/no audit) = (.032)(.968) = .030976
94. P(no audit/no audit) = (.968)(.968) = .937024

95. P(2 - same color) = P(2 r) + P(2 b) + P(2 y)

$$= \left(\frac{5}{10}\right)\left(\frac{4}{9}\right) + \left(\frac{3}{10}\right)\left(\frac{2}{9}\right) + \left(\frac{2}{10}\right)\left(\frac{1}{9}\right)$$

$$= \left(\frac{20}{90}\right) + \left(\frac{6}{90}\right) + \left(\frac{2}{90}\right) = \frac{28}{90} = \frac{14}{45}$$

96. P(at least 1 yen) =

$$= \left(\frac{3}{10}\right)\left(\frac{7}{9}\right) + \left(\frac{7}{10}\right)\left(\frac{3}{9}\right) + \left(\frac{3}{10}\right)\left(\frac{2}{9}\right)$$

$$= \left(\frac{21}{90}\right) + \left(\frac{21}{90}\right) + \left(\frac{6}{90}\right) = \frac{48}{90} = \frac{8}{15}$$

97. P(no diamonds) = $\left(\dfrac{39}{52}\right)\left(\dfrac{38}{51}\right)=\dfrac{1482}{2652}=.56$

The game favors the dealer since the probability of no diamonds is greater than 1/2.

98. The other card could be the ace or the queen and it is equally likely that it is either one.

Thus, the probability the card is the queen is 1/2.

99.   P(2/2) = (2/6)(2/6) = 4/36 = 1/9

100. P(3/3) = (3/6)(3/6) = 9/36 = 1/4

101. P(even or < 3) = 2/6 + 3/6 − 2/6 = 3/6 = 1/2

102. P(odd or > 1) = 4/6 + 5/6 -3/6 = 6/6 = 1

Exercise Set 12.7

1. The probability of E_2 given that E_1 has occurred.

2. $P\left(E_2\,|\,E_1\right)=\dfrac{n\left(E_1\text{ and }E_2\right)}{n\left(E_1\right)}$

3. $P(E_2\,|\,E_1)=\dfrac{n(E_1\cap E_2)}{n(E_1)}=\dfrac{4}{12}=\dfrac{1}{3}$

4. $P(E_2\,|\,E_1)=\dfrac{5}{22}$

5. P(5 | orange) = 1

6. P(3 | yellow) = 0

7. P(even | not orange) = 3/5

8. P(> 2 | < 5) = 1/4

9. P(red | orange) = 0

10. P(> 3 | yellow) = 1/2

11. P(circle | odd) = 3/4

12. P(circle | ≥ 5) = 2/3

13. P(red | even) = 2/3

14. P(circle | even) = 0

15. P(circle or square | < 4) = 2/3

16. P(circle | even) = 0

17. P(5 | red) = 1/3

18. P(even | red) = 1/3

19. P(purple | odd) = 2/6 = 1/3

20. P(> 4 | red) = 1

21. P(> 4 | purple) = 3/5

22. P(even | red or purple) = 4/8 = 1/2

23. P(gold | > 5) = 1/7

24. P(gold | > 10) = 0

25. P(1 and 1) = (1/4)(1/4) = 1/16

26. P(1 and 1) = 1/4

27. P(5 | at least a 5) = 1/7

28. P(> 5 | 2nd bill = 10) = 2/4 = ½

29. P(sum = 6) = 5/36

30. P(6 | 1) = 1/6

31. P(6 | 3) = 1/6

32. P(even | 2nd die = 2) = 3/6 = ½

33. P(> 7 | 2nd die = 5) = 4/6 = 2/3

34. P(7 or 11 | 1st die = 5) = 2/6 = 1/3

35. P(Pepsi) = 107/217

36. P(woman) = 112/217

37. P(Coke | woman) = 50/112 = 25/56

38. P(Pepsi | male) = 45/105 = 9/21 = 3/7

39. P(man | prefers Coke) = 60/110 = 12/22 = 6/11

40. P(woman | prefers Pepsi) = 62/107

41. P(girl) = 160/360 = 4/9

42. P(child selected lion) = 165/360 = 33/72

43. P(elephant | boy) = 110/200 = 11/20

44. P(lion | girl) = 75/160 = 15/32

45. P(boy | elephant) = 110/195 = 22/39

46. P(girl | lion) = 75/165 = 13/33 = 5/11

47. P(only tapes) = 133/300

48. P(≥ 30) = 180/300 = 3/5

49. P(DVD | < 30) = 60/120 = 1/2

50. P(both VTs and DVDs | ≥ 30) = 22/180 = 11/90

51. P(> 30 | both VTs and DVDs) = 21/43

52. P(≥ 30 | VTs only) = 94/133

53. P(Air Force) = 8833/27630 = 0.3197

54. P(acquitted) = 1574/27630 = 0.0570

55. P(acquitted | Army) = 434/5458 = 0.0795

56. P(convicted | Navy-MC) = 12866/13339 = 0.9645

57. P(Army | convicted) = 5024/26056 = 0.1928

58. P(Air Force | acquitted) = 667/1574 = 0.4238

59. $P(\text{good}) = \dfrac{300}{330} = \dfrac{10}{11}$

60. $P(\text{good} | 50 \text{ watts}) = \dfrac{100}{105} = \dfrac{20}{21}$

61. $P(\text{defective} | 20 \text{ watts}) = \dfrac{15}{95} = \dfrac{3}{19}$

62. $P(\text{good} | 100 \text{ watts}) = \dfrac{120}{130} = \dfrac{12}{13}$

63. $P(\text{good} | 50 \text{ or } 100 \text{ watts}) = \dfrac{220}{235} = \dfrac{44}{47}$

64. $P(\text{defective} | \text{not } 50 \text{ watts}) = \dfrac{25}{225} = \dfrac{1}{9}$

65. $P(\text{ABC or NBC}) = \dfrac{110}{270} = \dfrac{11}{27}$

66. $P(\text{ABC} | \text{woman}) = \dfrac{50}{125} = \dfrac{2}{5}$

67. $P(\text{ABC or NBC} | \text{man}) = \dfrac{50}{145} = \dfrac{10}{29}$

68. $P(\text{not CBS} | \text{woman}) = \dfrac{105}{125} = \dfrac{21}{25}$

69. $P(\text{ABC,NBC,or CBS} | \text{man}) = \dfrac{11}{29}$

70. $P(\text{NBC or CBS} | \text{Woman}) = \dfrac{30}{125} = \dfrac{6}{25}$

71. P(large company stock) = 93/200

72. P(value stock) = 73/200

73. P(blend | medium co. stock) = 15/52

74. P(large co. stock | blend stock) = 23/50

75. a) n(A) = 140   b) n(B) = 120

c) P(A) = 140/200 = 7/10

d) P(B) = 120/200 = 6/10 = 3/5

e) $P(A \mid B) = \dfrac{n(B \text{ and } A)}{n(B)} = \dfrac{80}{120} = \dfrac{2}{3}$

f) $P(B \mid A) = \dfrac{n(A \text{ and } B)}{n(B)} = \dfrac{80}{140} = \dfrac{4}{7}$

g) $P(A) \bullet P(B) = \left(\dfrac{7}{10}\right)\left(\dfrac{3}{5}\right) = \dfrac{21}{50}$

$P(A \mid B) \qquad P(A) \bullet P(B) \qquad \dfrac{2}{3} \neq \dfrac{21}{50}$

A and B are not independent events.

76. $P(E_2 \mid E_1) = \dfrac{P(E_1 \text{ and } E_2)}{P(E_1)} = \dfrac{\dfrac{n(E_1 \text{ and } E_2)}{n}}{\dfrac{n(E_1)}{n}}$

$= \dfrac{n(E_1 \text{ and } E_2)}{n} \bullet \dfrac{n}{n(E_1)} = \dfrac{n(E_1 \text{ and } E_2)}{n(E_1)}$

$\therefore P(E_2 \mid E_1) = \dfrac{n(E_1 \text{ and } E_2)}{n(E_1)}$

77. a) $P(A \mid B) = \dfrac{n(B \text{ and } A)}{n(B)} = \dfrac{0.12}{0.4} = 0.3$

b) $P(B \mid A) = \dfrac{n(A \text{ and } B)}{n(A)} = \dfrac{0.12}{0.3} = 0.4$

c) Yes, P(A) = P(A | B) and P(B) = P(B | A).

78. P(green circle | +) = 1/3

79. P(+ | orange circle) = 2/3

80. P(yellow circle | -) = 1/3

81. P(green + | +) = 1/3

82. P(green or orange circle | green +) = 1

83. P(orange circle w/green + | +) = 1/3

Exercise Set 12.8

1. Answers will vary.

2. Answers will vary.

3. $n! = n(n-1)(n-2) \cdots 3 \cdot 2 \cdot 1$

4. Multiply the counting numbers from n down to 1.

5. The number of permutations of n items taken r at a time.

6. $\dfrac{n!}{n_1! \, n_2! \cdots n_r!}$

7. $_nP_r = \dfrac{n!}{(n-r)!}$

8. Yes, because 0! = 1! = 1

9. 6! = 720

10. 8! = 40320

11. $_6P_2 = \dfrac{6!}{4!} = 6 \bullet 5 = 30$

12. $_5P_2 = \dfrac{5!}{3!} = 5 \bullet 4 = 20$

13. 0! = 1

14. $_6P_4 = \dfrac{6!}{2!} = 6 \bullet 5 \bullet 4 \bullet 3 = 360$

15. $_8P_0 = \dfrac{8!}{8!} = 1$

16. $_5P_0 = \dfrac{5!}{5!} = 1$

17. $_9P_4 = \dfrac{9!}{5!} = 9 \cdot 8 \cdot 7 \cdot 6 = 3024$

18. $_4P_4 = \dfrac{4!}{0!} = 4! = 24$

19. $_8P_5 = \dfrac{8!}{3!} = 8 \cdot 7 \cdot 6 \cdot 5 \cdot 4 = 6720$

20. $_{10}P_6 = \dfrac{10!}{4!} = 10 \cdot 9 \cdot 8 \cdot 7 \cdot 6 \cdot 5 = 151200$

21. $(10)(10)(10)(10) = 10000$

22. $(8)(7) = 56$

23. a) $(26)(25)(24)(10)(9)(8) = 11,232,000$
    b) $(26)(26)(26)(10)(10)(10) = 17,576,000$

24. a) $(36)(36)(36)(36) = 1,679,616$
    b) $(62)(62)(62)(62) = 14,776,336$

25. a) $5^5 = 3125$

    b) $\dfrac{1}{3125} = 0.00032$

26. $10^9 = 1,000,000,000$

27. $(34)(36)(36)(36) = 57,106,944$

28. $7 \bullet 6 \bullet 5 = 210$

29. $8 \bullet 10 \bullet 9 = 720$ systems

30. a) $5! = 120$ b) $5! = 120$
    c) $4! = 24$ d) $3! = 6$

31. a) $6! = 720$ b) $5! = 120$
    c) $4! = 24$ d) $5! \bullet 5 = 600$

32. $_8P_3 = \dfrac{8!}{(8-3)!} = \dfrac{8!}{5!} = \dfrac{8 \cdot 7 \cdot 6 \cdot 5!}{5!} = 336$

33. $_8P_3 = \dfrac{8!}{(8-3)!} = \dfrac{8!}{5!} = \dfrac{8 \cdot 7 \cdot 6 \cdot 5!}{5!} = 336$

34. $10^{10} = 10,000,000,000$ possible ISBN numbers

35. a) There are 12 individuals and they can be arranged in $12! = 479,001,600$ ways
    b) $10! = 3,628,800$ different ways
    c) $5! \cdot 5! = 14,400$ different ways

36. a) $8! = 40320$
    b) $3 \bullet 6! \bullet 2 = 720 \bullet 6 = 4320$
    c) $5 \bullet 4 \bullet 6! = 14400$

37. $(26)(25)(10)(9)(8)(7) = 650 \bullet 5040 = 3,276,000$

38. $(26)(26)(10)(10)(10)(10) = 6,760,000$

39. $(26)(10)(9)(8)(7) = 131,040$

40. $(4)(25)(10)(9)(8)(7) = 504,000$

41. $(10)(10)(10)(26)(26) = 676,000$

42. $(10)(9)(8)(26)(25) = 468,000$

43. $(5)(4)(8)(26)(25) = 104,000$

44. $(9)(9)(8)(26)(25) = 421,200$

45. a) $(8)(10)(10)(10)(10)(10)(10) = 8,000,000$
    b) $(8)(10)(10)(8,000,000) = 6,400,000,000$
    c) $(8)(10)(10)(8)\left(10^{10}\right) = (64)\left(10^{12}\right)$
       $= 64,000,000,000,000$

46. $_{12}P_3 = \dfrac{12!}{9!} = \dfrac{(12) \cdot (11) \cdot (10) \cdot (9!)}{9!} = 1,320$

47. $_{15}P_6 = \dfrac{15!}{9!} = \dfrac{(15)(14)(13) \cdot (12) \cdot (11)(10)(9!)}{9!}$
       $= 3,603,800$

48. Since the order of the answers is important, this is a permutation problem. $_{10}P_{10} = \dfrac{10!}{(10-10)!} = \dfrac{10!}{0!}$

$= (10)(9)(8)(7)(6)(5)(4)(3)(2)(1) = 3,628,800$

49. $_7P_7 = \dfrac{7!}{0!} = \dfrac{7!}{1} = 7! = 5,040$

50. $(3)(3)(3)(3)(3)(3) = 3^6 = 729$ ways

51. $(5)(4)(7)(2) = 280$ systems

52. $(5)(2)(6) = 60$

53. $_9P_9 = \dfrac{9!}{0!} = 9! = 362,880$

54. $\dfrac{10!}{2!2!} = \dfrac{3628800}{4} = 907,200$

55. $\dfrac{12!}{4!3!2!} = \dfrac{479001600}{(24)(6)(2)} = 1,663,200$

56. $\dfrac{11!}{4!4!2!} = 34,650$

57. $\dfrac{7!}{2!2!2!} = \dfrac{(7)(6)(5)(4)(3)(2)(1)}{(2)(1)(2)(1)(2)(1)} = 630$

58. $\dfrac{7!}{3!2!} = \dfrac{(7)(6)(5)(4)(3!)}{(3!)(2)(1)} = 420$

(There are 3 2's, 2 3's)

59. The order of the flags is important. Thus, it is a permutation problem.

$_8P_5 = \dfrac{8!}{(8-5)!} = \dfrac{8!}{3!} = \dfrac{40320}{6} = 6,720$

60. The order of the flags is important. Thus, it is a permutation problem.

$_8P_5 = 3^{10} = 59,049$

61. a) Since the pitcher must bat last, there is only one possibility for the last position. $\underline{\ \ \ \ \ \ \ \ }\underline{1}$
There are 8 possible batters left for the 1st position. Once the 1st batter has been selected, there are 7 batters left for the 2nd position, 6 for the third, etc. $\underline{(8)}\ \underline{(7)}\ \underline{(6)}\ \underline{(5)}\ \underline{(4)}\ \underline{(3)}\ \underline{(2)}\ \underline{(1)}\ \underline{(1)} = 40,320$

b) $9! = (9)(8)(7)(6)(5)(4)(3)(2)(1) = 362,880$

62. a) Since each arrangement is distinct, this is a permutation. Many problems of this type can be done with both the counting principal and the permutation formula.

counting principal $= (5)(4)(3)(2)(1) = 120$

permutation formula $= _5P_5 = \dfrac{5!}{(5-5)!} = \dfrac{5!}{0!} = 120$

62. b) Consider the possible arrangements as indicated by the dashes. $\underline{\ }\underline{\ }\underline{\ }\underline{\ }\underline{\ }$ There is only one possibility for the middle position. $\underline{\ }\underline{\ }\underline{1}\underline{\ }\underline{\ }$ After the middle one is placed there are 4 possibilities for the 1st position, 3 for the 2nd, 2 for the 4th, and only 1 for the final position. $\underline{(4)}\ \underline{(3)}\ \underline{(1)}\ \underline{(2)}\ \underline{(1)} = 24$

63. a) $5^5 = 3125$ different ways

b) $400,000 \div 3,125 = 128$ cars

c) $\dfrac{128}{400000} = \dfrac{1}{3125} = 0.00032$

64. $_7P_3 + 1 = \dfrac{7!}{4!} + 1 = 7 \cdot 6 \cdot 5 + 1 = 210 + 1 = 211$

65. $_7P_5 = \dfrac{7!}{2!} = \dfrac{7 \cdot 6 \cdot 5 \cdot 4 \cdot 3 \cdot 2!}{2!} = 2,520$ different

letter permutations

66. $\dfrac{7!}{3!2!} = 420$, Time = 420×5 sec. = 2,100 sec.

or 35 min.

67. No, Ex. $_3P_2 \neq {}_3P_{(3-2)}$

$\dfrac{3!}{1!} \neq \dfrac{3!}{2!}$ because $6 \neq 3$

68. A ○ ○ ○ ○ ○ ○ B

1-7 2-6 3-5 4-4 5-3 6-2 7-1

7+6+5+4+3+2+1 = 28

(28)(2) = 56 tickets

69. 25 stops 24+23+22+21+20+19+18 = 172

172-25 = 147

17+16+15+14+13+12+11+10 = 108

9+8+7+6+5+4+3+2+1 = 45

147+108+45 = 300 (300)(2) = 600 tickets

Exercise Set 12.9

1. The selection of a certain number of items without regard to their order.

2. The number of combinations possible when r items are selected from n items.

3. $_nC_r = \dfrac{n!}{(n-r)!\,r!}$

4. $_nC_r = \dfrac{{}_nP_r}{r!}$

5. If the order of the items is important then it is a permutation problem. If order is not important then it is a combination problem.

6. There will be more permutations.

7. $_5C_3 = \dfrac{5!}{(5-3)!3!} = \dfrac{(5)(4)(3)(2)(1)}{(2)(1)(3)(2)(1)} = 10$

8. $_7C_2 = \dfrac{7!}{5!2!} = \dfrac{(7)(6)}{(2)} = 21$

9. a) $_6C_4 = \dfrac{6!}{4!2!} = \dfrac{(6)(5)}{(2)(1)} = 15$

   b) $_6P_4 = \dfrac{6!}{(6-4)!} = \dfrac{6!}{2!} = (6)(5)(4)(3) = 360$

10. a) $_8C_2 = \dfrac{8!}{6!2!} = \dfrac{(8)(7)}{(2)(1)} = 28$

    b) $_8P_2 = \dfrac{8!}{(8-2)!} = \dfrac{8!}{6!} = (8)(7) = 56$

11. a) $_8C_0 = \dfrac{8!}{8!0!} = 1$

    b) $_8P_0 = \dfrac{8!}{(8-0)!} = \dfrac{8!}{8!} = 1$

12. a) $_{12}C_8 = \dfrac{12!}{8!4!} = \dfrac{(12)(11)(10)(9)(8!)}{(8!)(4)(3)(2)(1)} = 495$

    b) $_{12}P_8 = \dfrac{12!}{(12-8)!} = \dfrac{12!}{4!}$

    $= (12)(11)(10)(9)(8)(7)(6)(5) = 19,958,400$

13. a) $_{10}C_3 = \dfrac{10!}{7!3!} = \dfrac{(10)(9)(8)(7!)}{(7!)(3)(2)(1)} = 120$

    b) $_{10}P_3 = \dfrac{10!}{(10-3)!} = \dfrac{(10)(9)(8)(7!)}{7!} = 720$

14. a) $_5C_5 = \dfrac{5!}{0!5!} = \dfrac{5!}{5!} = 1$

    b) $_5P_5 = \dfrac{5!}{(5-5)!} = \dfrac{5!}{1} = 120$

15. $\dfrac{_5C_3}{_5P_3} = \dfrac{\dfrac{5!}{2!3!}}{\dfrac{5!}{2!}} = \left(\dfrac{5!}{2!3!}\right)\left(\dfrac{2!}{5!}\right) = \dfrac{1}{3!} = \dfrac{1}{6}$

16. $\dfrac{_6C_2}{_6P_2} = \dfrac{\dfrac{6!}{4!2!}}{\dfrac{6!}{4!}} = \left(\dfrac{6!}{4!2!}\right)\left(\dfrac{4!}{6!}\right) = \dfrac{1}{2!} = \dfrac{1}{2}$

17. $\dfrac{_8C_5}{_8P_5} = \dfrac{\dfrac{8!}{3!5!}}{\dfrac{8!}{6!2!}} = \left(\dfrac{8!}{3!5!}\right)\left(\dfrac{6!2!}{8!}\right) = \dfrac{6}{3} = 2$

18. $\dfrac{_6C_6}{_8C_0} = \dfrac{\dfrac{6!}{0!6!}}{\dfrac{8!}{8!0!}} = \dfrac{1}{1} = 1$

19. $\dfrac{_9P_5}{_{10}C_4} = \dfrac{\dfrac{9!}{4!}}{\dfrac{10!}{6!4!}} = \dfrac{(9)(8)(7)(6)(5)}{\dfrac{(10)(9)(8)(7)}{(4)(3)(2)(1)}} = \dfrac{144}{2} = 72$

20. $\dfrac{_7P_0}{_7C_0} = \dfrac{\dfrac{7!}{7!}}{\dfrac{7!}{7!0!}} = \dfrac{1}{1} = 1$

21. $_9C_6 = \dfrac{9!}{3!6!} = \dfrac{(9)(8)(7)(6!)}{(3)(2)(1)(6!)} = \dfrac{504}{6} = 84 \text{ ways}$

22. $_{20}C_3 = \dfrac{20!}{17!3!} = \dfrac{(20)(19)(18)(17!)}{(17!)(3)(2)(1)} = 1140$

23. $_5C_4 = \dfrac{5!}{1!4!} = 5$

24. $_8C_3 = \dfrac{8!}{5!3!} = \dfrac{(8)(7)(6)}{(3)(2)(1)} = 56$

25. $_{10}C_4 = \dfrac{10!}{6!4!} = \dfrac{(10)(9)(8)(7)}{(4)(3)(2)(1)} = 210$

26. $_{10}C_3 = \dfrac{10!}{7!3!} = \dfrac{(10)(9)(8)}{(3)(2)(1)} = 120$

27. $_9C_5 = \dfrac{9!}{4!5!} = \dfrac{(9)(8)(7)(6)}{(4)(3)(2)(1)} = 126$

28. $_{24}C_{20} = \dfrac{24!}{4!20!} = \dfrac{(24)(23)(22)(21)}{(4)(3)(2)(1)} = 10,626$

29. $_{12}C_8 = \dfrac{12!}{4!8!} = \dfrac{(12)(11)(10)(9)}{(4)(3)(2)(1)} = 495$

30. $_6C_4 = \dfrac{6!}{2!4!} = \dfrac{(6)(5)}{(2)(1)} = 15$

31. $_{10}C_8 = \dfrac{10!}{2!8!} = \dfrac{(10)(9)}{(2)(1)} = 45$

32. $_9C_3 \bullet {_6}C_2 =$

$\left(\dfrac{9!}{6!3!}\right)\left(\dfrac{6!}{4!2!}\right) = \left(\dfrac{(9)(8)(7)}{(3)(2)(1)}\right)\left(\dfrac{(6)(5)}{(2)(1)}\right) = 1260$

33. $_8C_2 = \dfrac{8!}{6!2!} = \dfrac{(8)(7)}{(2)(1)} = 28 \text{ tickets}$

34. Part I: $_5C_3 = \dfrac{5!}{2!3!} = \dfrac{(5)(4)}{(2)(1)} = 10$

Part II: $_6C_4 = \dfrac{6!}{2!4!} = \dfrac{(6)(5)}{(2)(1)} = 15$

$10 \cdot 15 = 150 \text{ possible combinations}$

35. $_{12}C_3 \bullet {_8}C_2 =$

$$\left(\frac{12!}{9!3!}\right)\left(\frac{8!}{6!2!}\right) = \left(\frac{(12)(11)(10)}{(3)(2)(1)}\right)\left(\frac{(8)(7)}{(2)(1)}\right) = 6160$$

36. $_{10}C_6 \bullet {_9}C_6 =$

$$\left(\frac{10!}{4!6!}\right)\left(\frac{9!}{3!6!}\right) = \left(\frac{(10)(9)(8)(7)}{(4)(3)(2)(1)}\right)\left(\frac{(9)(8)(7)}{(3)(2)(1)}\right)$$
$$= 17,640$$

37. Mathematics: $\quad {_8}C_5 = \dfrac{8!}{3!5!} = \dfrac{(8)(7)(6)}{(3)(2)(1)} = 56$

Computer Sci. $\quad {_5}C_3 = \dfrac{5!}{2!3!} = \dfrac{(5)(4)}{(2)(1)} = 10$

$(56)(10) = 560$ different choices

38. Regular soda: $_{10}C_5 = \dfrac{10!}{5!5!} = \dfrac{(10)(9)(8)(7)(6)}{(5)(4)(3)(2)(1)} = 252$

Diet soda: $\quad {_7}C_3 = \dfrac{7!}{4!3!} = \dfrac{(7)(6)(5)}{(3)(2)(1)} = 35$

$(252)(35) = 8,820$ ways to select the soda

39. Teachers: $\quad {_6}C_2 = \dfrac{6!}{4!2!} = \dfrac{(6)(5)}{(2)(1)} = 15$

Students: $\quad {_{50}}C_3 = \dfrac{50!}{47!3!} = \dfrac{(50)(49)(48)}{(3)(2)(1)} = 19600$

$(1)(19,600) = 294,000$ ways to select the comm.

40. Difficult questions: $\quad {_6}C_3 = \dfrac{6!}{3!3!} = \dfrac{(6)(5)(4)}{(3)(2)(1)} = 20$

Average questions: $\quad {_{10}}C_4 =$

$$\frac{10!}{6!4!} = \frac{(10)(9)(8)(7)}{(4)(3)(2)(1)} = 210$$

Easy questions: $\quad {_{12}}C_3 = \dfrac{12!}{9!3!} = \dfrac{(12)(11)(10)}{(3)(2)(1)} = 220$

Total number of 10-question tests $= (20)(210)(220)$
$= 924,000$

41. $_8C_3 \bullet {_5}C_2 =$

$$\left(\frac{8!}{5!3!}\right)\left(\frac{5!}{3!2!}\right) = \left(\frac{(8)(7)(6)}{(3)(2)(1)}\right)\left(\frac{(5)(4)}{(2)(1)}\right) = 560$$

42. $_6C_3 \bullet {_8}C_3 =$

$$\left(\frac{6!}{3!3!}\right)\left(\frac{8!}{5!3!}\right) = \left(\frac{(6)(5)(4)}{(3)(2)(1)}\right)\left(\frac{(8)(7)(6)}{(3)(2)(1)}\right) = 1120$$

43. $_6C_3 \bullet {_5}C_2 \bullet {_4}C_2 =$

$$\left(\frac{6!}{3!3!}\right)\left(\frac{5!}{3!2!}\right)\left(\frac{4!}{2!2!}\right) =$$

$$\left(\frac{(6)(5)(4)}{(3)(2)(1)}\right)\left(\frac{(5)(4)}{(2)(1)}\right)\left(\frac{(4)(3)}{(2)(1)}\right) = 1200$$

44. $_7C_3 \bullet {_8}C_5 \bullet {_4}C_2 =$

$$\left(\frac{7!}{4!3!}\right)\left(\frac{8!}{3!5!}\right)\left(\frac{4!}{2!2!}\right) =$$

$$\left(\frac{(7)(6)(5)}{(3)(2)(1)}\right)\left(\frac{(8)(7)(6)}{(3)(2)(1)}\right)\left(\frac{(4)(3)}{(2)(1)}\right) = 11,760$$

45. a) $_{10}C_8 = \dfrac{10!}{2!8!} = \dfrac{(10)(9)}{(2)(1)} = 45$

b) $_{10}C_9 = \dfrac{10!}{1!9!} = \dfrac{(10)(9!)}{(1)(9!)} = 10 \qquad {_{10}}C_{10} = \dfrac{10!}{10!} = 1$

$_{10}C_8 + {_{10}}C_9 + {_{10}}C_{10} = 45 + 10 + 1 = 56$

46. a) $_4C_2 = 6$

b) $_5C_2 = 10$

c) $_nC_2$

47. a)
$$
\begin{array}{ccccccccc}
 & & & & 1 & & & & \\
 & & & 1 & & 1 & & & \\
 & & 1 & & 2 & & 1 & & \\
 & 1 & & 3 & & 3 & & 1 & \\
1 & & 4 & & 6 & & 4 & & 1
\end{array}
$$

b)　　1　　5　　10　　10　　5　　1

49. a) $4! = 24$　　　b) $4! = 24$

50. $_nC_{(n-r)} = \dfrac{n!}{(n-(n-r))!(n-r)!} = \dfrac{n!}{(n-n+r)!(n-r)!}$

$= \dfrac{n!}{r!(n-r)!} = \dfrac{n!}{(n-r)!r!} = {}_nC_r$

48. a) $_{46}C_6 = \dfrac{46!}{40!6!} = 9,366,819$

b) $_{47}C_6 = \dfrac{47!}{41!6!} = 10,737,573$

c) $_{48}C_6 = \dfrac{48!}{42!6!} = 12,271,512$

d) $_{49}C_6 = \dfrac{49!}{43!6!} = 13,983,816$

e) No

45. a) The order of the numbers is important. For example: if the combination is 12 - 4 - 23, the lock will not open if 4 - 12 - 23 is used. Since repetition is permitted, it is not a true permutation problem.

b) $(40)(40)(40) = 64,000$

c) $(40)(39)(38) = 59,280$

Exercise Set 12.10

1. P(4 red balls) $= \dfrac{\text{no. of 4 red ball comb.}}{\text{no. of 4 ball comb.}} = \dfrac{_6C_4}{_{10}C_4}$

2. P(all girls) $= \dfrac{\text{no. of girls}}{\text{no. of students}} = \dfrac{_{19}C_{12}}{_{34}C_{12}}$

3. P(3 vowels) $= \dfrac{\text{no. of 3 vowel comb.}}{\text{no. of 3 letter comb.}} = \dfrac{_5C_3}{_{26}C_3}$

4. P(3 aces) $= \dfrac{\text{3 aces of 3 cards}}{\text{no. of 3 letter comb.}} = \dfrac{_3C_3}{_{52}C_3}$

5. P(all 7 are Palaminos) $=$

$\dfrac{\text{no. of 5 Palamino comb.}}{\text{no. of 5 horse comb.}} = \dfrac{_{10}C_5}{_{18}C_5}$

6. P(4 dancers have college degrees) $=$

$\dfrac{\text{no. of 4 college degrees}}{\text{no. of college degs.}} = \dfrac{_{28}C_4}{_{80}C_4}$

7. P(none of the 9 are oak) $=$

$\dfrac{\text{no. of 9 non-oak comb.}}{\text{no. of 9 tree comb.}} = \dfrac{_{14}C_9}{_{30}C_9}$

8. P(none of the 9 are T-I) $=$

$\dfrac{\text{no. of non-October b-days}}{\text{no. of 3 person groups}} = \dfrac{_{12}C_3}{_{16}C_3}$

9. $_5C_3 = \dfrac{5!}{2!3!} = \dfrac{(5)(4)}{(2)(1)} = 10$

$_9C_3 = \dfrac{9!}{6!3!} = \dfrac{(9)(8)(7)}{(3)(2)(1)} = 84$

P(3 reds) $= \dfrac{10}{84} = \dfrac{5}{42}$

10. $_3C_2 = \dfrac{3!}{1!2!} = 3$

$_6C_2 = \dfrac{6!}{4!2!} = \dfrac{(6)(5)}{(2)(1)} = 15$

P(2 evens) $= \dfrac{3}{15} = \dfrac{1}{5}$

11. $_8C_5 = \dfrac{8!}{3!5!} = \dfrac{(8)(7)(6)}{(3)(2)(1)} = 56$

$_{14}C_5 = \dfrac{14!}{5!9!} = \dfrac{(14)(13)(12)(11)(10)}{(5)(4)(3)(2)(1)} = 2002$

P(5 men's names) $= \dfrac{56}{2002} = \dfrac{4}{143}$

12. $_4C_2 = \dfrac{4!}{2!2!} = \dfrac{(4)(3)}{(2)(1)} = 6$

$_8C_2 = \dfrac{8!}{6!2!} = \dfrac{(8)(7)}{(2)(1)} = 28$

P(two $5 bills) $= \dfrac{6}{28} = \dfrac{3}{14}$

13. $_5C_3 = \dfrac{5!}{2!3!} = \dfrac{(5)(4)}{(2)(1)} = 10$

$_{10}C_3 = \dfrac{10!}{7!3!} = \dfrac{(10)(9)(8)}{(3)(2)(1)} = 120$

P(3 greater than 4) $= \dfrac{10}{120} = \dfrac{1}{12}$

14. $_6C_4 = \dfrac{6!}{2!4!} = \dfrac{(6)(5)}{(2)(1)} = 15$

$_{10}C_4 = \dfrac{10!}{6!4!} = \dfrac{(10)(9)(8)(7)}{(4)(3)(2)(1)} = 210$

P(all 4 ride Huffy) $= \dfrac{15}{210} = \dfrac{3}{42} = \dfrac{1}{14}$

15. $_6C_3 = \dfrac{6!}{3!3!} = \dfrac{(6)(5)(4)}{(3)(2)(1)} = 20$

$_{11}C_3 = \dfrac{11!}{8!3!} = \dfrac{(11)(10)(9)}{(3)(2)(1)} = 165$

P(all from manufacturing) $= \dfrac{20}{165} = \dfrac{4}{33}$

16. $_8C_4 = \dfrac{8!}{4!4!} = \dfrac{(8)(7)(6)(5)}{(4)(3)(2)(1)} = 70$

$_{15}C_4 = \dfrac{15!}{11!4!} = \dfrac{(15)(14)(13)(12)}{(4)(3)(2)(1)} = 1365$

P(4 students) $= \dfrac{70}{1365} = \dfrac{2}{39}$

17. $_{46}C_6 = \dfrac{46!}{40!6!} = 9,366,819 \quad _6C_6 = 1$

P(win grand prize) $= \dfrac{1}{9,366,819}$

18. $_{52}C_5 = \dfrac{52!}{47!5!} = \dfrac{(52)(51)(50)(49)(48)}{(5)(4)(3)(2)(1)}$

$= 2,598,960$

$_{26}C_5 = \dfrac{26!}{21!5!} = \dfrac{(26)(25)(24)(23)(22)}{(5)(4)(3)(2)(1)} = 65,700$

P(5 red) $= \dfrac{65700}{2598960} = \dfrac{253}{9996} = 0.253$

19. $_3C_2 = \dfrac{3!}{1!2!} = 3 \quad _5C_2 = \dfrac{5!}{3!2!} = \dfrac{(5)(4)}{(2)(1)} = 10$

P(no cars) $= \dfrac{3}{10}$

20. $_2C_2 = \dfrac{2!}{0!2!} = 1 \quad _5C_2 = \dfrac{5!}{3!2!} = \dfrac{(5)(4)}{(2)(1)} = 10$

P(both cars) $= \dfrac{1}{10}$

21. P(at least 1 car) $= 1 - $ P(no cars) $= 1 - 1 - \dfrac{3}{10} = \dfrac{7}{10}$

22. $_2C_1 = \dfrac{2!}{1!1!} = 2 \quad _3C_1 = \dfrac{3!}{2!1!} = 3 \quad _5C_2 = \dfrac{5!}{3!2!} = 10$

P(exactly on car) $= \dfrac{2 \cdot 3}{10} = \dfrac{6}{10} = \dfrac{3}{5}$

23. $_6C_3 = \dfrac{6!}{3!3!} = \dfrac{(6)(5)(4)}{(3)(2)(1)} = 20$

    $_{25}C_3 = \dfrac{25!}{3!22!} = \dfrac{(25)(24)(23)}{(3)(2)(1)} = 2300$

    P(3 infielders) $= \dfrac{20}{2300} = \dfrac{1}{115}$

24. $_{15}C_3 = \dfrac{15!}{12!\,3!} = \dfrac{(15)(14)(13)}{(3)(2)(1)} = 455$

    P(no pitchers) $= \dfrac{455}{2300} = \dfrac{91}{460}$

25. $_{10}C_2 = \dfrac{10!}{8!2!} = 45 \qquad _6C_1 = \dfrac{6!}{5!1!} = 6$

    P(2 pitchers and 1 infielder) $= \dfrac{(45)(6)}{2300} = \dfrac{27}{230}$

26. $_{10}C_1 = \dfrac{10!}{9!1!} = 10 \qquad _9C_2 = \dfrac{9!}{7!2!} = \dfrac{(9)(8)}{(2)(1)} = 36$

    P(1 pitc. and 2 non-pitch/non-inf) $= \dfrac{(10)(36)}{2300} = \dfrac{18}{115}$

For problems 27 – 30, use the fact that $_{39}C_{12} = \dfrac{39!}{27!12!} = 3,910,797,436$

27. $_{22}C_{12} = \dfrac{22!}{10!12!} = 646,646$

    P(all women) $= \dfrac{646646}{3910797436} = 0.0001653$

28. $_{22}C_8 = \dfrac{22!}{14!8!} = 319,770$

    $_{17}C_4 = \dfrac{17!}{13!4!} = 2,380$

    P(8 women/4 men) $= \dfrac{(319770)(2380)}{3910797436} = 0.1946$

29. $_{17}C_6 = \dfrac{17!}{11!6!} = 12,376$

    $_{22}C_6 = \dfrac{22!}{16!6!} = 74,613$

    P(6 men/6 women) $= \dfrac{(12376)(74613)}{3910797436} = 0.236$

30. P(at least one man) $= 1 - $ P(no men)

    $= 1 - $ P(all women)

    $= 1 - 0.0001653 = 0.9998$

For problems 31 – 34, use the fact that $_{15}C_5 = \dfrac{15!}{10!5!} = \dfrac{(15)(14)(13)(12)(11)}{(5)(4)(3)(2)(1)} = 3003$

31. $_4C_3 = \dfrac{4!}{3!1!} = 4 \qquad _6C_2 = \dfrac{6!}{4!2!} = \dfrac{(6)(5)}{(2)(1)} = 15$

    P(3 in FL/2 in VA) $= \dfrac{(4)(15)}{3003} = \dfrac{60}{3003} = \dfrac{20}{1001}$

32. $_5C_4 = \dfrac{5!}{4!1!} = 5 \qquad _4C_1 = \dfrac{4!}{3!1!} = 4$

    P(4 in KY/1 in FL) $= \dfrac{(5)(4)}{3003} = \dfrac{20}{3003}$

33. $_5C_2 = \dfrac{5!}{2!3!} = \dfrac{(5)(4)}{(2)(1)} = 10 \qquad _4C_1 = \dfrac{4!}{3!1!} = 4$

    $_6C_2 = \dfrac{6!}{2!4!} = \dfrac{(6)(5)}{(2)(1)} = 15$

    P(1 in FL/2 in KY/2 in VA) $= \dfrac{(10)(4)(15)}{3003} = \dfrac{200}{1001}$

34. $_9C_5 = \dfrac{9!}{5!4!} = \dfrac{(9)(8)(7)(6)}{(4)(3)(2)(1)} = 126$

    P(no VA) $= \dfrac{126}{3003}$

    P(≥ 1 VA) $= 1 - $ P(no VA) $=$

    $1 - \dfrac{126}{3003} = \dfrac{3003}{3003} - \dfrac{26}{3003} = \dfrac{2877}{3003} = \dfrac{137}{143}$

For problems 35 – 37, use the fact that $_{11}C_5 = \dfrac{11!}{6!5!} = \dfrac{(11)(10)(9)(8)(7)}{(5)(4)(3)(2)(1)} = 462$

35. $_6C_5 = \dfrac{6!}{1!5!} = 6$

$P(\text{5 women first}) = \dfrac{6}{462} = \dfrac{1}{77}$

36. $_5C_5 = \dfrac{5!}{0!5!} = 1 \qquad P(\text{no women first}) = \dfrac{1}{462}$

$P(\text{at least 1 woman 1st}) =$

$1 - \dfrac{1}{462} = \dfrac{462}{462} - \dfrac{1}{462} = \dfrac{461}{462}$

37. Any one of the 6 women can sit in any one of the five seats - 30 possibilities.

$P(\text{exactly 1 woman}) = \dfrac{30}{462} = \dfrac{5}{77}$

38. $P(\text{3 women and then 2 men})$

$\left(\dfrac{_6C_3}{_{11}C_3}\right)\left(\dfrac{_5C_2}{_8C_2}\right) = \left(\dfrac{20}{165}\right)\left(\dfrac{10}{28}\right) = \dfrac{10}{231}$

39. $_{24}C_6 = \dfrac{24!}{18!6!} = 134,596; \quad _3C_3 = 1$

$_{21}C_3 = \dfrac{21!}{18!3!} = 1,330$

$P(\text{3 brothers}) = \dfrac{_3C_3 \cdot _{21}C_3}{_{24}C_6} = \dfrac{(1)(1330)}{134596} = 0.00988$

40. $_4C_3 = \dfrac{4!}{1!3!} = 4 \qquad _4C_2 = \dfrac{4!}{2!2!} = \dfrac{(4)(3)}{(2)(1)} = 6$

and from problem 9, $_{52}C_5 = 2{,}598{,}960$

$P(\text{3 kings, 2 five's}) = \dfrac{4 \cdot 6}{2598960} = \dfrac{1}{108290}$

41. $_7C_5 = \dfrac{7!}{2!5!} = \dfrac{(7)(6)}{(2)(1)} = 21$ and from problem 9, $_{52}C_5 = 2{,}598{,}960$

a) $P(\text{royal spade flush}) = \dfrac{21}{2598960} = \dfrac{1}{123,760}$

b) $P(\text{any royal flush}) = \dfrac{4}{123760} = \dfrac{1}{30,940}$

42.

$\left(\dfrac{(_8C_3)(_{12}C_4)(_5C_2)}{_{25}C_9}\right) =$

$\dfrac{(8)(7)(6)(12)(11)(10)(9)(5)(4)}{(3)(2)(4)(3)(2)(2)} = \dfrac{277200}{2042975}$

$P(\text{3 waiters/4 waitresses/2 cooks}) = 0.1357$

43. a) $\left(\dfrac{(_4C_2)(_4C_2)(_{44}C_1)}{_{52}C_5}\right) = \dfrac{1584}{2598960} = \dfrac{33}{54,145}$

$P(\text{2 aces/2 8's/other card} \quad \text{ace or 8}) = \dfrac{33}{54,145}$

b) $P(\text{aces of spades and clubs/8's of spades and clubs/9 of diamonds}) =$

$\left(\dfrac{1}{52}\right)\left(\dfrac{1}{51}\right)\left(\dfrac{1}{50}\right)\left(\dfrac{1}{49}\right)\left(\dfrac{1}{48}\right) = \dfrac{1}{2,598,960}$

44. $\left(\dfrac{3}{6}\right)\left(\dfrac{2}{5}\right)\left(\dfrac{1}{4}\right)\left(\dfrac{3}{3}\right)\left(\dfrac{2}{2}\right)\left(\dfrac{1}{1}\right) = \dfrac{1}{20}$

2 ways: $\dfrac{1}{20} + \dfrac{1}{20} = \dfrac{2}{20} = \dfrac{1}{10}$

45. a) $\left(\dfrac{1}{15}\right)\left(\dfrac{1}{14}\right)\left(\dfrac{1}{13}\right)\left(\dfrac{5}{12}\right)\left(\dfrac{4}{11}\right)\left(\dfrac{3}{10}\right)\left(\dfrac{2}{9}\right)\left(\dfrac{1}{8}\right)$

$= \dfrac{120}{259459200} = \dfrac{1}{2,162,160}$

b) $P(\text{any 3 of 8 for officers}) = \dfrac{(8)(7)(6)}{2162160} = \dfrac{1}{6435}$

46. Given any four different numbers, there are $(4)(3)(2)(1) = 24$ different ways they can be arranged. One of these is in ascending order. Thus, the probability of the numbers being in ascending order is 1/24.

47. Since there are more hairs than people, 2 or more people must have the same number of hairs.

Exercise Set 12.11

1. A probability distribution shows the probability associated with each specific outcome of an experiment. In a probability distribution every possible outcome must be listed and the sum of all the probabilities must be 1.

2. Each trial has two possible outcomes, success and failure. There are n repeated independent trials.

3. $P(x) = {}_nC_x\, p^x q^{n-x}$

4. p is the probability of success, $q = 1 - p$ is the probability of failure.

5. $P(2) = {}_4C_2 (0.3)^2 (0.7)^{4-2} = \dfrac{4!}{2!2!}(.09)(.49) = 0.2646$

6. $P(2) = {}_3C_2 (0.6)^2 (0.4)^{3-2} = \dfrac{3!}{2!1!}(.36)(.4) = 0.4320$

7. $P(2) = {}_5C_2 (0.4)^2 (0.6)^{5-2} = \dfrac{5!}{2!3!}(.16)(.216) = 0.3456$

8. $P(3) = {}_3C_3 (0.9)^3 (0.1)^{3-3} = \dfrac{3!}{3!}(.729)(1) = 0.729$

9. $P(0) = {}_6C_0 (0.5)^0 (0.5)^{6-0}$
$= \dfrac{6!}{0!6!}(1)(.0156252) = 0.015625$

10. $P(3) = {}_5C_3 (0.4)^3 (0.6)^{5-3} = \dfrac{5!}{3!2!}(.064)(.36) = 0.2304$

11. $p = 0.14,\ q = 1 - p = 1 - 0.14 = 0.86$
    a) $P(x) = {}_nC_x (0.15)^x (0.85)^{n-x}$
    b) $n = 12, x = 2, p = 0.14, q = 0.86$
    $P(2) = {}_{12}C_2 (0.14)^2 (0.86)^{12-2}$

12. a) $P(x) = {}_nC_x (0.0237)^x (0.9763)^{n-x}$
    b) $P(5) = {}_{20}C_5 (0.0237)^5 (0.9763)^{20-5}$

13. $P(1) = {}_6C_4 (0.3)^4 (0.7)^{6-4}$
$= \dfrac{6!}{4!2!}(.0081)(.49) = 0.05954$

14. $P(5) = {}_8C_5 (0.6)^5 (0.4)^{8-5}$
$= \dfrac{8!}{5!3!}(.07776)(.064) = 0.27869$

15. $P(2) = {}_3C_2 (0.96)^2 (0.04)^{3-2}$
$= \dfrac{3!}{2!1!}(.9216)(.04) = 0.1106$

16. $P(4) = {}_6C_4 (0.8)^4 (0.2)^{6-4}$
$= \dfrac{6!}{4!2!}(.4096)(.04) = 0.24576$

17. $P(4) = {}_6C_4 (0.92)^4 (0.08)^{6-4}$
$= \dfrac{6!}{4!2!}(.7164)(.0064) = 0.06877$

18. $P(2) = {}_4C_2 (0.01)^2 (0.99)^{4-2}$
$= \dfrac{4!}{2!2!}(.0001)(.9801) = 0.000588$

19.
$$P(4) = {_5}C_4 (.8)^4 (.2)^{5-4}$$
$$= \frac{5!}{1!4!}(.4096)(.2) = 0.4096$$

20. a)
$$P(0) = {_4}C_0 (.25)^0 (.75)^{4-0}$$
$$= \frac{4!}{4!}(1)(.3164) = 0.3164$$

b) P(at least 1) = 1 – P(0) = 1 – 0.3164 = 0.6836

21. a)
$$P(0) = {_5}C_0 (0.6)^0 (0.4)^{5-0}$$
$$= \frac{5!}{5!}(1)(.01024) = 0.01024$$

b) P(at least 1) = 1 – P(0) = 0.98976

22. a)
$$P(3) = {_5}C_3 \left(\frac{40}{80}\right)^3 \left(\frac{40}{80}\right)^2$$
$$= \frac{5!}{3!2!}(.125)(.25) = 0.3125$$

b)
$$P(3) = {_5}C_3 \left(\frac{20}{80}\right)^3 \left(\frac{60}{80}\right)^2$$
$$= \frac{5!}{3!2!}(.015625)(.5625) = 0.08789$$

23. a)
$$P(3) = {_6}C_3 \left(\frac{12}{52}\right)^3 \left(\frac{40}{52}\right)^3$$
$$\frac{6!}{3!3!}(.01229)(.45517) = 0.11188$$

b)
$$P(2) = {_6}C_2 \left(\frac{13}{52}\right)^2 \left(\frac{39}{52}\right)^4$$
$$= \frac{6!}{2!4!}(.0625)(.3164) = 0.29663$$

24. a)
$$P(3) = {_5}C_3 (0.7)^3 (0.3)^2$$
$$= \frac{5!}{3!2!}(.343)(.09) = 0.3087$$

b) P(at least 3) = P(3) + P(4) + P(5)
$$= 0.3087 + 0.3602 + 0.1681$$
$$= 0.8370$$

25. The probability that the sun would be shining would equal 0 because 72 hours later would occur at midnight.

Review Exercises

1. Relative frequency over the long run can accurately be predicted, not individual events or totals.
2. Roll the die many times then compute the relative frequency of each outcome and compare with the expected probability of 1/6.

3. P(mountain bike) = $\dfrac{8}{40} = \dfrac{1}{5}$

4. Answers will vary.

5. P(watches ABC) = $\dfrac{80}{200} = \dfrac{2}{5}$

6. P(even) = $\dfrac{5}{10} = \dfrac{1}{2}$

7. P(odd or > 5) = $\dfrac{5}{10} + \dfrac{4}{10} - \dfrac{2}{10} = \dfrac{7}{10}$

8. P(> 2 or < 5) = $\dfrac{7}{10} + \dfrac{5}{10} - \dfrac{2}{10} = \dfrac{10}{10} = 1$

9. P(even and > 4) = $\dfrac{2}{10} = \dfrac{1}{5}$

10. P(Grand Canyon) = $\dfrac{50}{240} = \dfrac{5}{24}$

11. P(Yosemite) = $\dfrac{40}{240} = \dfrac{1}{6}$

12. P(Rocky Mtn. or Smoky Mtn.) = $\dfrac{35}{240} + \dfrac{45}{240} = \dfrac{80}{240} = \dfrac{1}{3}$

13. P(not Grand Canyon) = $\dfrac{190}{240} = \dfrac{19}{24}$

16. P(wins Triple Crown) = $\dfrac{3}{85}$

14. a) 9:1   b) 1:9

15. 5:3

17. 7:3

18. a) E = P(win $200)•$198 + P(win $100)•$98
         + P(lose)•(–$2)
      = (.003)(198) + (.002)(98) – (.995)(2)
      = .594 + .196 – 1.990 = – 1.200 → -$1.20

   b) The expectation of a person who purchases three
      tickets would be 3(–1.20) = –$3.60.

   c) Expected value = Fair price – Cost
      –1.20 = Fair price – 2.00 $.80 = Fair price

19. a) $E_{Cameron}$ = P(pic. card)($9) + P(pic. card)(–$3)
       $= \left(\dfrac{12}{52}\right)(9) - \left(\dfrac{40}{52}\right)(3) = \approx -\$0.23$

   b) $E_{Lindsey}$ = P(pic. card)(–$9) + P(pic. card)($3)·
       $= \dfrac{-27}{13} + \dfrac{30}{13} = \dfrac{3}{13} \approx \$0.23$

   c) Cameron can expect to lose $(100)\left(\dfrac{3}{13}\right) \approx \$23.08$

20. E = P(sunny)(1000) + P(cloudy)(500) + P(rain)(100) = 0.4(1000) + 0.5(500) + 0.1(100) = 400 + 250+10
    = 660 people

21. a)
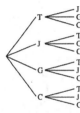

   b) Sample space:
      {TJ,TG,TC,JT,JG,JC,GT,GJ,GC,CT,CJ,CG}

   c) P(Gina is Pres. and Jake V.P.) = 1/12

22. a)

   b) Sample space:
      {H1,H2,H3,H4,T1,T2,T3,T4}

   c) P(heads and odd) = (1/2)(2/4) = 2/8 = ¼

   d) P(heads or odd) = (1/2)(2/4) + (1/2)(2/4)
                     = 4/8 + 2/8 = 6/8 = 3/4

23. P(even and even) = (4/8)(4/8) = 16/64 = 1/4

24. P(outer is greater than 5 and inner is greater than 5)
    = P(outer is > 5) · P(inner is > 5) = $\dfrac{3}{8} \cdot \dfrac{3}{8} = \dfrac{9}{64}$

25. P(outer odd and inner < 6)
    = P(outer odd) P(inner < 6) = $\dfrac{4}{8} \cdot \dfrac{5}{8} = \dfrac{1}{2} \cdot \dfrac{5}{8} = \dfrac{5}{16}$

26. P(outer is even or less than 6)
    = P(even) + P(< 6) – P(even and < 6)
    = $\dfrac{4}{8} + \dfrac{5}{8} - \dfrac{2}{8} = \dfrac{7}{8}$

27. P(inner even and not gold) = $\dfrac{1}{2} + \dfrac{6}{8} - \dfrac{2}{8} = \dfrac{1}{2} + \dfrac{4}{8} = 1$

28. P(outer gold and inner not gold)
    = $\left(\dfrac{2}{8}\right)\left(\dfrac{6}{8}\right) = \left(\dfrac{1}{4}\right)\left(\dfrac{3}{4}\right) = \dfrac{3}{16}$

29. P(all 3 are Hersheys) = $\dfrac{5}{12} \cdot \dfrac{4}{11} \cdot \dfrac{3}{10} = \dfrac{60}{1320} = \dfrac{1}{22}$

30. P(none are Nestle) = $\dfrac{8}{12} \cdot \dfrac{7}{11} \cdot \dfrac{6}{10} = \dfrac{336}{1320} = \dfrac{14}{55}$

31. P(at least one is Nestle) = 1 – P(none are Nestle)

$$= 1 - \frac{14}{55} = \frac{55}{55} - \frac{14}{55} = \frac{41}{55}$$

32. P(Hershey and Hershey and Reese)

$$= \frac{5}{12} \cdot \frac{4}{11} \cdot \frac{3}{10} = \frac{60}{1320} = \frac{1}{22}$$

33. P(yellow) = 1/4

34. Odds against yellow   3:1   Odds for yellow   1:3

35. $5 for red; $10 for yellow; $20 for green
    P(green) = ½; P(yellow) = ¼; P(red) = ¼
    EV = (1/4)(5) + (1/4)(10) + (1/2)(20) = $13.75

36. P(red, then green) = P(red)P(green)
    = (1/4)(1/2) = 1/8

37. P(not green) = 1/4 + 1/4 + 1/8 = 5/8

38. Odds in favor of green   3:5
    Odds against green   5:3

39. E = P(green)($10) + P(red)($5) + P(yellow)(–$20)
    = (3/8)(10) + (1/2)(5) – (1/8)(20)
    = (15/4) + (10/4) – (10/4) = 15/4 → $3.75

40. P(at least one red) = 1 – P(none are red)
    = 1 – (1/2)(1/2)(1/2) = 1-1/8 = 7/8

41. P(< 6 defects | American built) = 89/106 = 0.84

42. P(< 6 defects | foreign built) = 55/74 = 0.74

43. P(≥ 6 defects | foreign built) = 19/74 = 0.26

44. P(≥ 6 defects | American built) = 17/106 = 0.16

45. P(right handed) $= \frac{230}{400} = \frac{23}{40}$

46. P(left brained | left handed) $= \frac{30}{170} = \frac{3}{17}$

47. P(right handed | no predominance) $= \frac{60}{80} = \frac{3}{4}$

48. P(right brained | left handed) $= \frac{120}{170} = \frac{12}{17}$

49. a) 4! = (4)(3)(2)(1) = 24
    b) E = (1/4)(10K) + (1/4)(5K) + (1/4)(2K)
        + (1/4)(1K) = (1/4)(18K) = $4,500.00

50. # of possible arrangements = $(_5C_2)(_3C_2)(_1C_1)$

$$= \left(\frac{5!}{3!2!}\right)\left(\frac{3!}{1!2!}\right)\left(\frac{1!}{1!}\right) = \frac{(5)(4)(3)}{(2)(1)} = 30$$

51. $_{10}C_3 = \left(_{10}P_3\right)\left(\frac{10!}{7!3!}\right) = \frac{(10)(9)(8)(7!)}{(7!)(3)(2)(1)} = 120$

52. $_9P_3 = \frac{9!}{6!3!} = \frac{(9)(8)(7)(6)}{(3)(2)(1)} = (9)(8)(7) = 504$

53. $_6C_3 = \frac{6!}{3!3!} = \frac{(6)(5)(4)}{(3)(2)(1)} = 20$

54. a) $_{15}C_{10} = \frac{15!}{5!10!} = \frac{(15)(14)(13)(12)(11)}{(5)(4)(3)(2)(1)} = 3003$

    b) number of arrangements = 10! = 3,628,800

55. a) P(match 5 numbers) = $\dfrac{1}{_{52}C_5}$

$$= \dfrac{1}{\dfrac{50!}{45!5!}} = \dfrac{45!5!}{50!} = \dfrac{1}{2,118,760}$$

b) P(Big game win) = P(match 5 #s and Big #)
= P(match 5 #s)⊔P(match Big #)

$$= \left(\dfrac{1}{2118760}\right)\left(\dfrac{1}{36}\right) = \dfrac{1}{76,275,360}$$

58. P(two aces) = $\dfrac{_4C_2}{_{52}C_2} = \dfrac{\dfrac{4!}{2!2!}}{\dfrac{52!}{50!2!}}$

$$= \left(\dfrac{4!}{2!2!}\right)\left(\dfrac{50!2!}{52!}\right) = \dfrac{1}{221}$$

61. P(1st red, 2nd white, 3rd blue)

$$= \left(\dfrac{5}{10}\right)\left(\dfrac{3}{9}\right)\left(\dfrac{2}{8}\right) = \dfrac{1}{24}$$

63. P(3 N&WRs) =

$$\dfrac{_5C_3}{_{14}C_3} = \dfrac{\dfrac{5!}{3!2!}}{\dfrac{14!}{3!11!}} = \dfrac{5!3!11!}{3!2!14!} = \dfrac{(5)(4)(3)}{(14)(13)(12)} = \dfrac{5}{182}$$

65. $\dfrac{_8C_3}{_{14}C_3} = \dfrac{\dfrac{8!}{3!5!}}{\dfrac{14!}{3!11!}} = \dfrac{8!3!11!}{3!5!14!}$

$$= \dfrac{(8)(7)(6)}{(14)(13)(12)} = \dfrac{336}{2184} = \dfrac{2}{13}$$

68. n = 5, x = 3, p = 1/5, q = 4/5

$$P(3) = {_5}C_3\left(\dfrac{1}{5}\right)^3\left(\dfrac{4}{5}\right)^2 = 10\cdot\left(\dfrac{1}{5}\right)^3\left(\dfrac{4}{5}\right)^2 = 0.0512$$

56. $(_8C_2)(_{10}C_4) =$

$$\left(\dfrac{8!}{6!2!}\right)\left(\dfrac{10!}{6!4!}\right) = \dfrac{(8)(7)(10)(9)(8)(7)}{(2)(1)(4)(3)(2)(1)} = 5880 \text{ combos.}$$

57. $(_8C_3)(_5C_2)=$

$$\left(\dfrac{8!}{5!3!}\right)\left(\dfrac{5!}{2!3!}\right) = \dfrac{(8)(7)(6)(5)(4)}{(3)(2)(1)(2)(1)} = 560$$

59. P(all three are red) = $\left(\dfrac{5}{10}\right)\left(\dfrac{4}{9}\right)\left(\dfrac{3}{8}\right) = \dfrac{1}{12}$

60. P(1st 2 are red/3rd is blue) = $\left(\dfrac{5}{10}\right)\left(\dfrac{4}{9}\right)\left(\dfrac{2}{8}\right) = \dfrac{1}{18}$

62. P(at least one red) = 1 − P(none are red)

$$= 1 - 1 - \left(\dfrac{5}{10}\right)\left(\dfrac{4}{9}\right)\left(\dfrac{3}{8}\right) = 1 - \dfrac{1}{12} = \dfrac{11}{12}$$

64. P(2 NWs & 1 Time) =

$$\dfrac{(_6C_2)(_3C_1)}{_{14}C_3} = \dfrac{\left(\dfrac{6!}{2!4!}\right)\left(\dfrac{3!}{1!2!}\right)}{\dfrac{14!}{3!11!}}$$

$$= \dfrac{(6)(5)(3)(3)(2)(1)}{(2)(1)(14)(13)(12)} = \dfrac{45}{364}$$

66. $1 - \dfrac{2}{13} = \dfrac{11}{13}$

67. a) P(x) = $_nC_x (0.6)^x (0.4)^{n-x}$

b) P(75) = $_{100}C_{75}(0.6)^{75}(0.4)^{25}$

69. a) n = 4, p = 0.6, q = 0.4

$$P(0) = {_4}C_0(0.6)^0(0.4)^4$$

$$= (1)(1)(0.4)^4 = 0.0256$$

b) P(at least 1) = 1 − P(0) = 1 − 0.0256 = 0.9744

Chapter Test

1. $P(\text{fishing for bass}) = \dfrac{22}{30} = \dfrac{11}{15}$

2. $(>7) = \dfrac{2}{9} \approx 0.22$

3. $P(\text{odd}) = \dfrac{5}{9} \approx 0.55$

4. $P(\geq 4) = \dfrac{7}{9} \approx 0.78$

5. $P(\text{odd and} > 4) = \dfrac{3}{9} = \dfrac{1}{3} \approx 0.33$

6. $P(\text{both} > 5) = \dfrac{4}{9} \cdot \dfrac{3}{8} = \dfrac{12}{72} = \dfrac{1}{6}$

7. $P(\text{both even}) = \dfrac{4}{9} \cdot \dfrac{3}{8} = \dfrac{1 \cdot 1}{3 \cdot 2} = \dfrac{1}{6}$

8. $P(\text{1st odd, 2nd even}) = \dfrac{5}{9} \cdot \dfrac{4}{8} = \dfrac{5}{9} \cdot \dfrac{1}{2} = \dfrac{5}{18}$

9. $P(\text{neither} > 6) = \dfrac{6}{9} \cdot \dfrac{5}{8} = \dfrac{1 \cdot 5}{3 \cdot 4} = \dfrac{5}{12}$

10. $P(\text{red or picture})$
    $= P(\text{red}) + P(\text{picture}) - P(\text{red and picture})$
    $= \dfrac{26}{52} + \dfrac{12}{52} - \dfrac{6}{52} = \dfrac{32}{52} = \dfrac{8}{13}$

11. 1 die $(6)(3) = 18$

12.

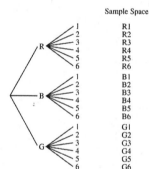

		Sample Space
R	1	R1
	2	R2
	3	R3
	4	R4
	5	R5
	6	R6
B	1	B1
	2	B2
	3	B3
	4	B4
	5	B5
	6	B6
G	1	G1
	2	G2
	3	G3
	4	G4
	5	G5
	6	G6

14. $P(\text{blue or 1}) = \dfrac{6}{18} + \dfrac{3}{18} - \dfrac{1}{18} = \dfrac{8}{18} = \dfrac{4}{9}$

15. $P(\text{not red or odd}) = \dfrac{12}{18} + \dfrac{9}{18} - \dfrac{6}{18} = \dfrac{15}{18} = \dfrac{5}{6}$

16. Number of codes $= (9)(26)(26)(10)(10) = 608{,}400$

17. a) $5:4$ b) $5:4$

18. odds against Aimee winning are $5:2$ or
    $$\dfrac{5}{2} = \dfrac{5/7}{2/7} = \dfrac{P(\text{not winning})}{P(\text{winning})}$$
    Therefore, $P(\text{Aimee wins}) = 2/7$

13. $P(\text{blue and 1}) = \dfrac{1}{18}$

19. $E = P(\text{club})(\$8) + P(\text{heart})(\$4)$
    $\quad + P(\text{spade or diamond})(-\$6)$
    $= \left(\dfrac{1}{4}\right)(8) + \left(\dfrac{1}{4}\right)(4) + \left(\dfrac{2}{4}\right)(-6)$
    $= \dfrac{8}{4} + \dfrac{4}{4} - \dfrac{12}{4} = \$0.00$

20. a) $P(\text{car}) = \dfrac{214}{456} = \dfrac{107}{228}$

    b) $P(\text{Golden Gate}) = \dfrac{230}{456} = \dfrac{115}{228}$

    c) $P(\text{SUV} \mid \text{Golden Gate}) = \dfrac{136}{230} = \dfrac{68}{115}$

20. d) $P(\text{GW Bridge} \mid \text{car}) = \dfrac{120}{214} = \dfrac{60}{107}$

21. $_6P_3 = \dfrac{6!}{(6-3)!} = \dfrac{6!}{3!} = 6 \cdot 5 \cdot 4 = 120$

22. P(neither is good) $= \dfrac{8}{20} \cdot \dfrac{7}{19} = \dfrac{2}{5} \cdot \dfrac{7}{19} = \dfrac{14}{95}$

23. P($\geq$ 1 good) $= 1 - $ P(neither -good) $= 1 - \dfrac{14}{95} = \dfrac{81}{95}$

24. $_7C_3 = \dfrac{7!}{4!\,3!} = \dfrac{(7)(6)(5)}{(3)(2)(1)} = 35$

$_5C_2 = \dfrac{5!}{3!\,2!} = \dfrac{(5)(4)}{(2)(1)} = 10$

$_{10}C_5 = \dfrac{12!}{7!\,5!} = \dfrac{(12)(11)(10)(9)(8)}{(5)(4)(3)(2)(1)} = 792$

P(3 red and 2 green) $= \dfrac{(35)(10)}{792} = \dfrac{350}{792} = \dfrac{175}{396}$

25. $(0.1)(0.1)(0.1) = 0.001$

$(0.1)(0.1)(0.1)(0.9)(0.9) = 0.00081$

$_5C_3 = \dfrac{5!}{3!\,2!} = \dfrac{(5)(4)}{(2)(1)} = 10$

$(10)(.00081) = 0.0081$

Group Projects

1. 0 because no measurement is exact.

2. a) 0.30199   b) 0.10737   c) 0.89263   d) 0.00000   e) 0.30199
    f) They should be the same.

3. a) $10^5 = 100,000$ b) $5^5 = 3125$ c) $\dfrac{1}{3125}$ d) 3125   e) 3125

    f) $\dfrac{1}{3125}$ g) same likelihood   h) More 5 digit codes are available.

CHAPTER THIRTEEN

STATISTICS

1. **Statistics** is the art and science of gathering, analyzing, and making inferences (predictions) from numerical information obtained in an experiment.
2. **Descriptive statistics** is concerned with the collection, organization, and analysis of data.
   Inferential statistics is concerned with making generalizations or predictions from the data collected.
3. Answers will vary.
4. Answers will vary.
5. Insurance companies, sports, airlines, stock market, medical profession
6. **Probability** is used to compute the chance of occurrence of a particular event when all possible outcomes are known.
   Statistics is used to draw conclusions about possible outcomes through observations of only a few particular events.
7. a) A **population** consists of all items or people of interest.
   b) A **sample** is a subset of the population.
8. a) A **systematic sample** is a sample obtained by selecting every n^{th} item on a list or production line.
   b) Use a random number table to select the first item, then select every n^{th} item after that.
9. a) A **random sample** is a sample drawn in such a way that each item in the population has an equal chance of being selected.
   b) Number each item in the population. Write each number on a piece of paper and put each numbered piece of paper in a hat. Select pieces of paper from the hat and use the numbered items selected as your sample.
10. a) A **cluster sample** is a random selection of groups of units.
    b) Divide a geographic area into sections. Randomly select sections or clusters. Either each member of the selected cluster is included in the sample or a random sample of the members of each selected cluster is used.
11. a) A **stratified sample** is one that includes items from each part (or strata) of the population.
    b) First identify the strata you are interested in. Then select a random sample from each strata.
12. a) A **convenience sample** uses data that is easily or readily obtained.
    b) For example, select the first 20 students entering a classroom.
13. a) An **unbiased sample** is one that is a small replica of the entire population with regard to income, education, gender, race, religion, political affiliation, age, etc.
14. a) No, the method used to obtain the sample is biased. In classes where students are seated alphabetically, brothers and sisters could be selected from different classes.
    b) The mean will be greater. Families with many children are more likely to be selected.

15. Stratified sample
16. Systematic sample
17. Cluster sample
18. Random sample
19. Systematic sample
20. Stratified sample
21. Convenience sample
22. Cluster sample
23. Random sample
24. Convenience sample

25. a) – c) Answers will vary.
26. Biased because the subscribers of *Consumer Reports* are not necessarily representative of the entire population.
27. President; four out of 42 U.S. presidents have been assassinated (Lincoln, Garfield, McKinley, Kennedy).
28. Answers will vary.

Exercise Set 13.2

1. Answers will vary.

2. Yes, the sum of its parts is 142%. The sum of the parts of a circle graph should be 100%. When the total percent of responses is more than 100%, a circle graph is not an appropriate graph to display the data. A bar graph is more appropriate in this situation.

3. There may have been more car thefts in Baltimore, Maryland than Reno, Nevada because many more people live in Baltimore than in Reno. But, Reno may have more car thefts per capita than Baltimore.

4. Mama Mia's may have more empty spaces and more cars in the parking lot than Shanghi's due to a larger parking lot or because more people may walk to Mama Mia's than to Shanghi's.

5. Although the cookies are fat free, they still contain calories. Eating many of them may still cause you to gain weight.

6. The fact that Morgan's is the largest department store does not imply it is inexpensive.

7. More people drive on Saturday evening. Thus, one might expect more accidents.

8. Most driving is done close to home. Thus, one might expect more accidents close to home.

9. People with asthma may move to Arizona because of its climate. Therefore, more people with asthma may live in Arizona.

10. We don't know how many of each professor's students were surveyed. Perhaps more of Professor Malone's students than Professor Wagner's students were surveyed. Also, because more students prefer a teacher does not mean that he or she is a better teacher. For example, a particular teacher may be an easier grader and that may be why that teacher is preferred.

11. Although milk is less expensive at Star Food Markets than at Price Chopper Food Markets, other items may be more expensive at Star Food Markets.

12. Just because they are the most expensive does not mean they will last the longest.

13. There may be deep sections in the pond, so it may not be safe to go wading.

14. Men may drive more miles than women and men may drive in worse driving conditions (like snow).

15. Half the students in a population are expected to be below average.

16. Not all students who apply to a college will attend that college.

17. a)

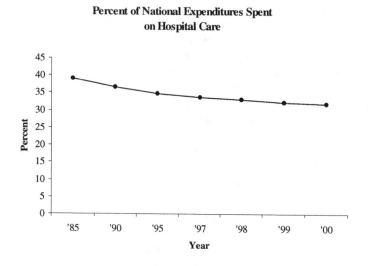

17.   b)

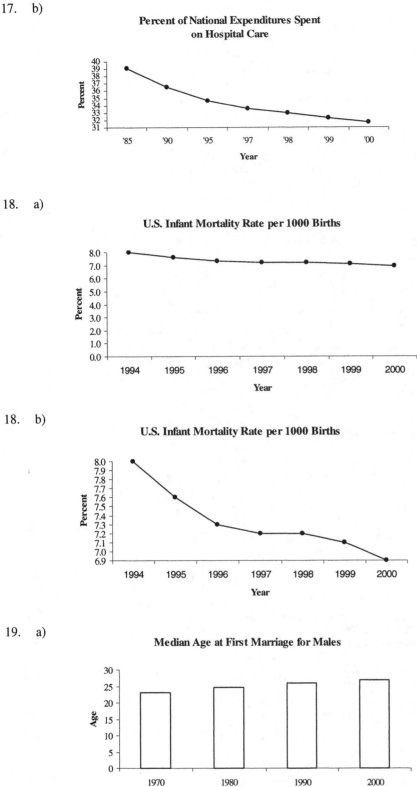

18.   a)

18.   b)

19.   a)

19.   b)

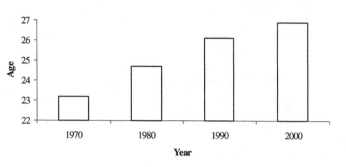

20.   a)

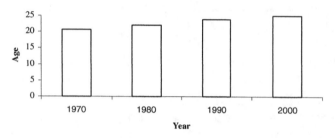

20.   b)

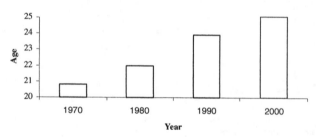

21.   a)

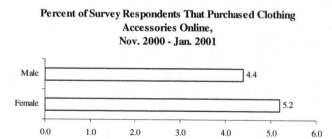

b) Yes. The new graph gives the impression that the percents are closer together.

22. a) $\dfrac{394,000,000-275,000,000}{275,000,000}=\dfrac{119,000,000}{275,000,000}$

     $=0.43\overline{27}\approx 43.3\%$ increase

b) Radius $=\dfrac{1}{4}$ in. $=0.25$ in.

     $A=\pi r^2=\pi(0.25)^2=0.0625\pi=0.196349541$

     ≈ 0.196 in.^{2}

c) Radius $=\dfrac{3}{8}$ in. $=0.375$ in.

     $A=\pi r^2=\pi(0.375)^2=0.140625\pi=0.441786467$

     ≈ 0.442 in.^{2}

d) $\dfrac{0.442-0.196}{0.196}=\dfrac{0.246}{0.196}=1.255102041$

     $\approx 125.5\%$ increase

e) Yes, the percent increase in the size of the area from the first circle to the second is greater than the percent increase in population.

23. A decimal point

Exercise Set 13.3

1. A **frequency distribution** is a listing of observed values and the corresponding frequency of occurrence of each value.

2. Subtract a lower class limit from the next lower class limit or subtract an upper class limit from the next upper class limit.

3. a) 7   b) 16-22   c) 16   d) 22

4. a) 9   b) 21-29   c) 21   d) 29

5. The **modal class** is the class with the greatest frequency.

6. The **class mark** is another name for the midpoint of a class. Add the lower and upper class limits and divide the sum by 2.

7. a) Number of observations = sum of frequencies = 18
   b) Width = $16-9=7$
   c) $\dfrac{16+22}{2}=\dfrac{38}{2}=19$
   d) The modal class is the class with the greatest frequency. Thus, the modal class is 16 - 22.
   e) Since the class widths are 7, the next class would be 51 - 57.

8. a) Number of observations = sum of frequencies = 25
   b) Width = $50-40=10$
   c) $\dfrac{50+59}{2}=\dfrac{109}{2}=54.5$
   d) 40 - 49 and 80 - 89 both contain 7 pieces of data. Thus, they are both modal classes.
   e) Since the class widths are 10, the next class would be $100-109$.

9.

Number Sold	Number of Days
0	3
1	8
2	3
3	5
4	2
5	7
6	2
7	3
8	4
9	1
10	2

10.

Number of Visits	Number of Families
20	3
21	2
22	0
23	3
24	4
25	2
26	6
27	2
28	2
29	1
30	1
31	2
32	2
33	1
34	1

11.

I.Q.	Number of Students
78 - 86	2
87 - 95	15
96 - 104	18
105 - 113	7
114 - 122	6
123 - 131	1
132 - 140	1

12.

I.Q.	Number of Students
80 - 88	4
89 - 97	17
98 - 106	15
107 - 115	8
116 - 124	4
125 - 133	1
134 - 142	1

13.

I.Q.	Number of Students
80 - 90	8
91 - 101	22
102 - 112	11
113 - 123	7
124 - 134	1
135 - 145	1

14.

I.Q.	Number of Students
80 - 92	11
93 - 105	24
106 - 118	9
119 - 131	5
132 - 144	1

15.

Placement test scores	Number of Students
472 - 492	9
493 - 513	9
514 - 534	5
535 - 555	2
556 - 576	3
577 - 597	2

16.

Placement test scores	Number of Students
470 - 486	4
487 - 503	9
504 - 520	8
521 - 537	2
538 - 554	2
555 - 571	2
572 - 588	2
589 - 605	1

17.

Placement test scores	Number of Students
472 - 487	4
488 - 503	9
504 - 519	7
520 - 535	3
536 - 551	2
552 - 567	2
568 - 583	2
584 - 599	1

18.

Placement test scores	Number of Students
472 - 496	9
497 - 521	12
522 - 546	4
547 - 571	2
572 - 596	3

19.

Circulation (thousands)	Number of Newspapers
209 - 458	36
459 - 708	8
709 - 958	3
959 - 1208	1
1209 - 1458	0
1459 - 1708	0
1709 - 1958	1
1959 - 2208	1

20.

Circulation (thousands)	Number of Newspapers
205 - 414	35
415 - 624	8
625 - 834	3
835 - 1044	1
1045 - 1254	1
1255 - 1464	0
1465 - 1674	0
1675 - 1884	1
1885 - 2094	0
2095 - 2304	1

21.

Circulation (thousands)	Number of Newspapers
209 - 408	34
409 - 608	9
609 - 808	3
809 - 1008	1
1009 - 1208	1
1209 - 1408	0
1409 - 1608	0
1609 - 1808	1
1809 - 2008	0
2009 - 2208	1

22.

Circulation (thousands)	Number of Newspapers
209 - 358	30
359 - 508	9
509 - 658	4
659 - 808	3
809 - 958	1
959 - 1108	0
1109 - 1258	1
1259 - 1408	0
1409 - 1558	0
1559 - 1708	0
1709 - 1858	1
1859 - 2008	0
2009 - 2158	1

23.

Population (millions)	Number of Counties
1.4 - 2.1	15
2.2 - 2.9	6
3.0 - 3.7	2
3.8 - 4.5	0
4.6 - 5.3	0
5.4 - 6.1	1
6.2 - 6.9	0
7.0 - 7.7	0
7.8 - 8.5	0
8.6 - 9.3	0
9.4 - 10.1	1

24.

Population (millions)	Number of Counties
1.0 - 2.7	19
2.8 - 4.5	4
4.6 - 6.3	1
6.4 - 8.1	0
8.2 - 9.9	1

25.

Population (millions)	Number of Counties
1.0 - 2.5	19
2.6 - 4.1	4
4.2 - 5.7	1
5.8 - 7.3	0
7.4 - 8.9	0
9.0 - 10.5	1

26.

Population (millions)	Number of Counties
1.4 - 2.9	21
3.0 - 4.5	2
4.6 - 6.1	1
6.2 - 7.7	0
7.8 - 9.3	0
9.4 - 10.9	1

27.

Price ($)	Number of States
0.35 - 0.44	6
0.45 - 0.54	10
0.55 - 0.64	11
0.65 - 0.74	3
0.75 - 0.84	2
0.85 - 0.94	4
0.95 - 1.04	1
1.05 - 1.14	2
1.15 - 1.24	2
1.25 - 1.34	1
1.35 - 1.44	0
1.45 - 1.54	1

28.

Price ($)	Number of States
0.35 - 0.45	7
0.46 - 0.56	13
0.57 - 0.67	8
0.68 - 0.78	2
0.79 - 0.89	5
0.90 - 1.00	2
1.01 - 1.11	2
1.12 - 1.22	1
1.23 - 1.33	2
1.34 - 1.44	0
1.45 - 1.55	1

29.

Price ($)	Number of States
0.35 - 0.54	16
0.55 - 0.74	14
0.75 - 0.94	6
0.95 - 1.14	3
1.15 - 1.34	3
1.35 - 1.54	1

30.

Price ($)	Number of States
0.35 - 0.48	12
0.49 - 0.62	12
0.63 - 0.76	6
0.77 - 0.90	6
0.91 - 1.04	1
1.05 - 1.18	3
1.19 - 1.32	2
1.33 - 1.46	1

31. February, since it has the fewest number of days

32. a) **Did You Know?, page 762:** There are 6 F's.
   b) Answers will vary.

Exercise Set 13.4

1. Answers will vary.

3. Answers will vary.

5. a) Answers will vary.
   b)

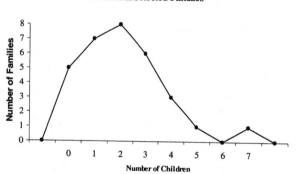

Children in Selected Families

2. a) Observed values
   b) Frequency

4. Answers will vary.

6. a) Answers will vary.
   b)

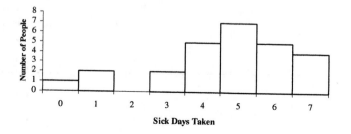

Number of Sick Days Taken Last Year

7. a) Answers will vary.

    b)
Observed Values	Frequency
45	3
46	0
47	1
48	0
49	1
50	1
51	2

8.
Observed Values	Frequency
16	1
17	2
18	1
19	1
20	0
21	1
22	2
23	1
24	1
25	2

9. Occasionally: $0.59(500) = 295$

    Most Times: $0.25(500) = 125$

    Every Time: $0.07(500) = 35$

    Never: $0.09(500) = 45$

10. Retail: $0.518(700) = 362.6 \approx 363$

    Services: $0.259(700) = 181.3 \approx 181$

    Other: $0.223(700) = 156.1 \approx 156$

11. Travelocity: $\dfrac{175}{500} = 0.35 = 35\%$ Priceline: $\dfrac{85}{500} = 0.17 = 17\%$

    Expedia: $\dfrac{125}{500} = 0.25 = 25\%$ Other: $\dfrac{115}{500} = 0.23 = 23\%$

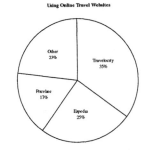

Using Online Travel Websites

12. 2 bedrooms: $\dfrac{182}{600} = 0.30\overline{3} \approx 30.3\%$ 4 bedrooms: $\dfrac{100}{600} = 0.1\overline{6} \approx 16.7\%$

    3 bedrooms: $\dfrac{230}{600} = 0.38\overline{3} \approx 38.3\%$ 5 or more bedrooms: $\dfrac{88}{600} = 0.14\overline{6} \approx 14.7\%$

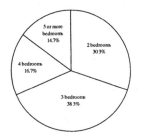

Permits for New Houses, Number of Bedrooms

13. a) and b)

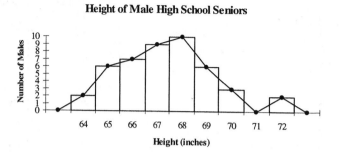

14. a) and b)

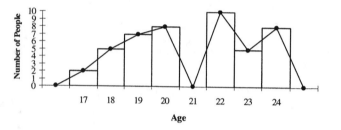

15. a) and b)

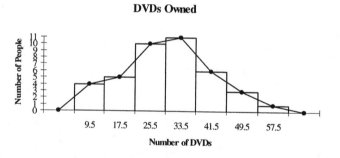

16. a) and b)

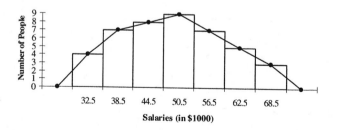

17. a) The total number of people surveyed:
2 + 7 + 8 + 5 + 4 + 3 + 1 = 30
   b) Four people purchased four soft drinks.
   c) The modal class is 2 because more people
   purchased 2 soft drinks than any other number of soft drinks.

e)

Number of Soft Drinks Purchased	Number of People
0	2
1	7
2	8
3	5
4	4
5	3
6	1

   d) Two people bought 0 soft drinks   0
   Seven people bought 1 soft drink   7
   Eight people bought 2 soft drinks   16
   Five people bought 3 soft drinks   15
   Four people bought 4 soft drinks   16
   Three people bought 5 soft drinks   15
   One person bought 6 soft drinks   6
   Total number of soft drinks purchased: 75

18) a) The total number of students surveyed: 2 + 4 + 6 + 8 + 7 + 3 + 1 = 31
   b) Since there are 51 units between class midpoints, each class width must also be 51 units.
   650 is the midpoint of the first class and there must be 25 units below it and 25 units above it.
   Therefore, the first class is 625 - 675. The second class will be 676 - 726.
   c) Six
   d) The class mark of the modal class is $803 because more students had an annual car insurance premium of
   $778 - $828 than any other annual car insurance premium.
   e)

Price	Number of Students
625 - 675	2
676 - 726	4
727 - 777	6
778 - 828	8
829 - 879	7
880 - 930	3
931 - 981	0
982 - 1032	1

19. a) 7 calls
   b) Adding the number of calls responded to in 6, 5, 4, or 3 minutes gives: 4 + 7 + 3 + 2 = 16 calls
   c) The total number of calls surveyed: 2 + 3 + 7 + 4 + 3 + 8 + 6 + 3 = 36

   d)

Response Time (min.)	Number of Calls
3	2
4	3
5	7
6	4
7	3
8	8
9	6
10	3

e)

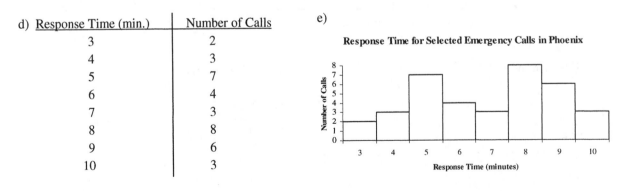

Response Time for Selected Emergency Calls in Phoenix

20. a) 8 families

b) At least six times means six or more times. Adding the families that went 6, 7, 8, 9, or 10 times gives $11 + 9 + 3 + 0 + 1 = 24$ families

c) Total number of families surveyed: $4 + 2 + 8 + 8 + 6 + 11 + 9 + 3 + 0 + 1 = 52$ families

d)

Number of Visits	Number of Families
1	4
2	2
3	8
4	8
5	6
6	11
7	9
8	3
9	0
10	1

e)

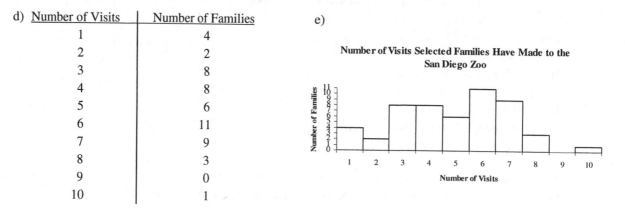

21.

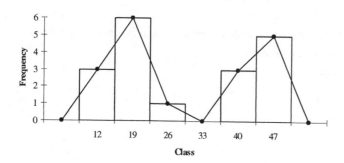

22.

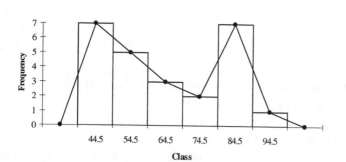

23. 1│5 represents 15

```
1│ 0   5   7
2│ 4   4
3│ 6   0   3
4│ 8   5   2   5   8
5│ 3   4
6│ 0   2   0
```

24. 1│2 represents 12

```
0│ 3   8   2   5
1│ 2   8   2   5   9   3   7   6
2│ 5   1   7   2   3
3│ 3   4
4│ 1
```

25. a)

Salaries (in $1000)	Number of Companies
27	1
28	7
29	4
30	3
31	2
32	3
33	3
34	2

b) and c)

Starting Salaries for 25 Different Social Workers

d) 2│3 represents 23

```
2│7  8  8  8  8  8  8  8  9  9  9  9
3│0  0  0  1  1  2  2  2  3  3  3  4  4
```

26. a)

Age	Number of People
20 - 24	9
25 - 29	6
30 - 34	10
35 - 39	6
40 - 44	5
45 - 49	4

b) and c)

Age of 40 People Attending a Broadway Show

d) 2│3 represents 23

```
2│0  1  1  2  3  3  3  4  4  5  5  6  7  8  8
3│0  0  0  1  1  2  3  4  4  4  5  5  5  7  8  9
4│0  0  0  2  4  5  5  6  7
```

27. a)

Advertising Spending (millions of dollars)	Number of Companies
597 - 905	19
906 - 1214	14
1215 - 1523	7
1524 - 1832	3
1833 - 2141	2
2142 - 2450	3
2451 - 2759	1
2760 - 3068	0
3069 - 3377	1

b) and c)

Advertising Spending

28. a)

Age	Number of Ambassadors
40 - 44	9
45 - 49	6
50 - 54	10
55 - 59	6
60 - 64	5
65 - 69	4

b) and c)

Ages of U.S. Ambassadors

29. a) - e) Answers will vary.
30. a) - e) Answers will vary.

Exercise Set 13. 5

1. **Ranked data** are data listed from the lowest value to the highest value or from the highest value to the lowest value.
2. The **mean** is the balancing point of a set of data. It is the sum of the data divided by the number of pieces of data.
3. The **median** is the value in the middle of a set of ranked data. To find the median, rank the data and select the value in the middle.
4. The **midrange** is the value half way between the lowest and highest values. To find the midrange, add the lowest and highest values and divide the sum by 2.
5. The **mode** is the most common piece of data. The piece of data that occurs most frequently is the mode.
6. The mode may be used when you are primarily interested in the most popular value, or the one that occurs most often, for example, when buying clothing for a store.
7. The median should be used when there are some values that differ greatly from the rest of the values in the set, for example, salaries.
8. The midrange should be used when the item being studied is constantly fluctuating, for example, daily temperature.
9. The mean is used when each piece of data is to be considered and "weighed" equally, for example, weights of adult males.
10. a) $\bar{x}$
    b) μ

	mean	median	mode	midrange
11.	$\dfrac{99}{9} = 11$	10	10	$\dfrac{5+23}{2} = 14$
12.	$\dfrac{550}{10} = 55$	$\dfrac{15+15}{2} = 15$	15	$\dfrac{9+370}{2} = 189.5$
13.	$\dfrac{485}{7} \approx 69.3$	72	none	$\dfrac{42+90}{2} = 66$
14.	$\dfrac{58}{7} \approx 8.3$	8	8	$\dfrac{5+12}{2} = 8.5$
15.	$\dfrac{64}{8} = 8$	$\dfrac{7+9}{2} = 8$	none	$\dfrac{1+15}{2} = 8$
16.	$\dfrac{510}{7} \approx 72.9$	60	none	$\dfrac{30+140}{2} = 85$

	mean	median	mode	midrange
17.	$\dfrac{118}{9} \approx 13.1$	11	1	$\dfrac{1+36}{2} = 18.5$
18.	$\dfrac{92}{14} \approx 6.6$	$\dfrac{4+4}{2} = 4$	1 and 4	$\dfrac{1+21}{2} = 11$
19.	$\dfrac{95}{8} \approx 11.9$	$\dfrac{12+13}{2} = 12.5$	13	$\dfrac{6+17}{2} = 11.5$
20.	$\dfrac{60}{6} = 10$	$\dfrac{5+15}{2} = 10$	5 and 15	$\dfrac{5+15}{2} = 10$
21.	$\dfrac{65}{10} = 6.5$	$\dfrac{5+5}{2} = 5$	3 and 5	$\dfrac{2+19}{2} = 10.5$
22.	$\dfrac{\$469}{7} = \$67$	$59	none	$\dfrac{\$25+\$140}{2} = \$82.50$
23. a)	$\dfrac{34}{7} \approx 4.9$	5	5	$\dfrac{1+11}{2} = 6$
b)	$\dfrac{37}{7} \approx 5.3$	5	5	$\dfrac{1+11}{2} = 6$
c)	Only the mean			
d)	$\dfrac{33}{7} \approx 4.7$	5	5	$\dfrac{1+10}{2} = 5.5$

The mean and the midrange

24. Answers will vary. The National Center for Health uses the median for averages in this exercise.

25. A 79 mean average on 10 quizzes gives a total of 790 points. An 80 mean average on 10 quizzes requires a total of 800 points. Thus, Jim missed a B by 10 points not 1 point.

26. a) Mean: $\dfrac{\$361,000}{10} = \$36,100$ b) Median: $\dfrac{\$27,000+\$28,000}{2} = \$27,500$

c) Mode: $26,000

d) Midrange: $\dfrac{\$24,000+\$81,000}{2} = \$52,500$

e) The median, since it is lower   f) The mean, since it is higher

27. a) Mean: $\dfrac{87.7}{10} \approx 8.8$ million   b) Median: $\dfrac{7.8+8.2}{2} = 8.0$ million

c) Mode: none

d) Midrange: $\dfrac{4.6+19.7}{2} \approx 12.2$ million

28. a) Mean: $\dfrac{\$14,810}{12} \approx \$1234.17$ b) Median: $\dfrac{\$1230+\$1250}{2} = \$1240$

c) Mode: $850

d) Midrange: $\dfrac{\$850+\$1900}{2} = \$1375$

29. a) Mean: $\dfrac{\$55.9}{11} \approx \$5.1$ billion

     b) Median: $2.3 billion

    c) Mode: $2.3 billion and $1.5 billion

     d) Midrange: $\dfrac{\$1.5 + \$26.5}{2} = \$14$ billion

    e) Answers will vary.

30. Let $x =$ the sum of his scores

    $\dfrac{x}{5} = 76$

    $x = 76(5) = 380$

31. Let $x =$ the sum of his scores

    $\dfrac{x}{6} = 85$

    $x = 85(6) = 510$

32. One example is 1, 1, 2, 5, 6. Mode = 1, Median = 2, Mean $= \dfrac{15}{5} = 3$

33. One example is 72, 73, 74, 76, 77, 78.

    Mean: $\dfrac{450}{6} = 75$, Median: $\dfrac{74 + 76}{2} = 75$, Midrange: $\dfrac{72 + 78}{2} = 75$

34. One example is 80, 82, 84, 88, 94, 100.

    Mean: $\dfrac{528}{6} = 88$

35. a) Yes

    b) Cannot be found since we do not know the middle two numbers in the ranked list

    c) Cannot be found without knowing all of the numbers

    d) Yes

    e) Mean: $\dfrac{24{,}000}{120} = 200$; Midrange: $\dfrac{50 + 500}{2} = 275$

36. A total of $80 \times 5 = 400$ points are needed for a grade of B. Jorge earned $73 + 69 + 85 + 80 = 307$ points on his first four exams. Thus, he needs 400 - 307 = 93 or higher to get a B.

37. a) For a mean average of 60 on 7 exams, she must have a total of $60 \times 7 = 420$ points. Sheryl presently has $49 + 72 + 80 + 60 + 57 + 69 = 387$ points. Thus, to pass the course, her last exam must be 420 - 387 = 33 or greater.

    b) A C average requires a total of $70 \times 7 = 490$ points. Sheryl has 387. Therefore, she would need 490 - 387 = 103 on her last exam. If the maximum score she can receive is 100, she cannot obtain a C.

    c) For a mean average of 60 on 6 exams, she must have a total of $60 \times 6 = 360$ points. If the lowest score on an exam she has already taken is dropped, she will have a total of $72 + 80 + 60 + 57 + 69 = 338$ points. Thus, to pass the course, her last exam must be 360 - 338 = 22 or greater.

    d) For a mean average of 70 on 6 exams, she must have a total of $70 \times 6 = 420$ points. If the lowest score on an exam she has already taken is dropped, she will have a total of 338 points. Thus, to obtain a C, her last exam must be 420 - 338 = 82 or greater.

38. The mode is the only measure which must be an actual piece of data since it is the most frequently occurring piece of data.

39. One example is 1, 2, 3, 3, 4, 5 changed to 1, 2, 3, 4, 4, 5.

    First set of data: Mean: $\dfrac{18}{6} = 3$, Median: $\dfrac{3+3}{2} = 3$, Mode: 3

    Second set of data: Mean: $\dfrac{19}{6} = 3.1\overline{6}$, Median: $\dfrac{3+4}{2} = 3.5$, Mode: 4

40. The mean changes from $\frac{9}{6}=1.5$ to $\frac{10}{6}=1.\overline{6}$. The mode changes from no mode to a mode of 1.

    The midrange changes from $\frac{3}{2}=1.5$ to $\frac{4}{2}=2$.

41. No, by changing only one piece of the six pieces of data you cannot alter both the median and the midrange.

42. Let x = sum of the values

    $\dfrac{x}{12}=85.20$

    $x = 85.20(12) = \$1022.40$

    $\$1022.40 - \$47 + \$74 = \$1049.40$

    $\dfrac{1049.40}{12} = \$87.45$ is the correct mean

43. The data must be arranged in either ascending or descending order.
44. She scored above approximately 73% of all the students who took the test.
45. He is taller than approximately 35% of all kindergarten children.
46. About 25% of the workers earn $20,750 or less.

47. a) Q_2 = Median = \$430

    b) $290, $300, $300, $330, $350, $350, $350, $350, $350, $400

    Q_1 = Median of the data listed below = $\dfrac{\$350+\$350}{2}=\$350$

    c) $450, $450, $500, $600, $650, $650, $700, $700, $750, $800

    Q_3 = Median of the data listed above = $\dfrac{\$650+\$650}{2}=\$650$

48. a) Q_2 = Median = $\dfrac{27+28}{2}=27.5$ ¢

    b) 17¢, 17¢, 20¢, 21¢, 24¢, 25¢, 27¢, 27¢, 27¢, 27¢

    Q_1 = Median of the data listed below = $\dfrac{24+25}{2}=24.5$ ¢

    c) 28¢, 28¢, 28¢, 28¢, 31¢, 33¢, 38¢, 74¢, 80¢, 81¢

    Q_3 = Median of the data listed above = $\dfrac{31+33}{2}=32$ ¢

49. Second quartile, median
50. a) No, the percentile only indicated relative position of the score and not the value of it.
    b) Yes, a higher percentile indicates a higher relative position in the respective population.
    Thus, Kendra was in a better relative position.

51. a) $490   b) $500   c) 25%   d) 25%   e) 17%   f) $100 \times \$510 = \$51,000$

52. a) $\dfrac{56}{7}=8$, $\dfrac{26}{4}=6.5$, $\dfrac{10}{5}=2$, $\dfrac{50}{5}=10$, $\dfrac{396}{6}=66$

    b) $\dfrac{92.5}{5}=18.5$ c) $\dfrac{538}{27}\approx 19.926$ d) No

53. a) Ruth: $\approx 0.290, 0.359, 0.301, 0.272, 0.315$

    Mantle: $\approx 0.300, 0.365, 0.304, 0.275, 0.321$

    b) Mantle's is greater in every case.

    c) Ruth: $\dfrac{593}{1878} \approx 0.316$; Mantle: $\dfrac{760}{2440} \approx 0.311$; Ruth's is greater.

    d) Answers will vary.

    e) Ruth: $\dfrac{1.537}{5} \approx 0.307$; Mantle: $\dfrac{1.565}{5} = 0.313$; Mantle's is greater.

    f) and g) Answers will vary.

54. a) $\dfrac{707,000}{25} = \$28,280$

    b) $\$21,000$

    c) $\$17,000$

    d) $\dfrac{17,000 + 100,000}{2} = \$58,500$

    e) The median because there are pieces of data that are much greater and much smaller than the rest of the data.

55. $\Sigma xw = 84(0.40) + 94(0.60) = 33.6 + 56.4 = 90$

    $\Sigma w = 0.40 + 0.60 = 1.00$

    weighted average $= \dfrac{\Sigma xw}{\Sigma w} = \dfrac{90}{1.00} = 90$

56. $\Sigma xw = 3.0(4) + 4.0(3) + 2.0(3) + 4.0(3) = 12 + 12 + 6 + 12 = 42$

    $\Sigma w = 4 + 3 + 3 + 3 = 13$

    weighted average $= \dfrac{\Sigma xw}{\Sigma w} = \dfrac{42}{13} = 3.230769231 \approx 3.23$

57. a) – c) Answers will vary.
58. a) Answers will vary. One example is 2, 3, 5, 7, 7.

    b) Answers will vary. The answers for the example given in part a) above are as follows:

    Mean: $\dfrac{24}{5} = 4.8$, Median = 5, Mode = 7

Exercise Set 13.6

1. To find the **range**, subtract the lowest value in the set of data from the highest value.

2. The **standard deviation** measures the spread of the data about the mean.

3. Answers will vary.

4. Zero since the mean is the same value as all of the data values. The spread about the mean is 0.

5. It may be important to determine the consistency of the data.

6. s

7. σ

8. Where one expects to find a large variability such as test scores

9. In manufacturing or anywhere else where a minimum variability is desired

10. The first set of data will have the greater standard deviation because the scores have a greater spread about the mean.

11. They would be the same since the spread of data about each mean is the same.

12. The sum of the values in the (Data − Mean)^2 column will always be greater than or equal to 0.

13. a) The grades will be centered about the same number since the mean, 75.2, is the same for both classes.

    b) The spread of the data about the mean is greater for the evening class since the standard deviation is greater for the evening class.

14. Answers will vary.

15. Range = 13 − 2 = 11

$$\overline{x} = \frac{35}{5} = 7$$

x	$x - \overline{x}$	$(x - \overline{x})^2$
7	0	0
5	-2	4
2	-5	25
8	1	1
13	6	36
	0	66

$$\frac{66}{4} = 16.5, s = \sqrt{16.5} \approx 4.06$$

16. Range = 16 − 8 = 8

$$\overline{x} = \frac{66}{6} = 11$$

x	$x - \overline{x}$	$(x - \overline{x})^2$
10	-1	1
10	-1	1
14	3	9
16	5	25
8	-3	9
8	-3	9
	0	54

$$\frac{54}{5} = 10.8, s = \sqrt{10.8} \approx 3.29$$

17. Range = 126 − 120 = 6

$$\overline{x} = \frac{861}{7} = 123$$

x	$x - \overline{x}$	$(x - \overline{x})^2$
120	-3	9
121	-2	4
122	-1	1
123	0	0
124	1	1
125	2	4
126	3	9
	0	28

$$\frac{28}{6} \approx 4.67, s = \sqrt{4.67} \approx 2.16$$

18. Range = 12 − 0 = 12

$$\overline{x} = \frac{70}{10} = 7$$

x	$x - \overline{x}$	$(x - \overline{x})^2$
3	-4	16
7	0	0
8	1	1
12	5	25
0	-7	49
9	2	4
11	4	16
12	5	25
6	-1	1
2	-5	25
	0	162

$$\frac{162}{9} = 18, s = \sqrt{18} \approx 4.24$$

19. Range = $15 - 4 = 11$

$\bar{x} = \dfrac{60}{6} = 10$

x	$x - \bar{x}$	$(x - \bar{x})^2$
4	-6	36
8	-2	4
9	-1	1
11	1	1
13	3	9
15	5	25
	0	76

$\dfrac{76}{5} = 15.2, \quad s = \sqrt{15.2} \approx 3.90$

20. Range = $9 - 9 = 0$

Since all pieces of data are identical,

the standard deviation is 0.

21. Range = $12 - 7 = 5$

$\bar{x} = \dfrac{63}{7} = 9$

x	$x - \bar{x}$	$(x - \bar{x})^2$
7	-2	4
9	0	0
7	-2	4
9	0	0
9	0	0
10	1	1
12	3	9
	0	18

$\dfrac{18}{6} = 3, s = \sqrt{3} \approx 1.73$

22. Range = $64 - 40 = 24$

$\bar{x} = \dfrac{424}{8} = 53$

x	$x - \bar{x}$	$(x - \bar{x})^2$
52	-1	1
50	-3	9
54	1	1
59	6	36
40	-13	169
43	-10	100
64	11	121
62	9	81
	0	518

$\dfrac{518}{7} = 74, s = \sqrt{74} \approx 8.60$

23. Range = $50 - 18 = \$32$

$\bar{x} = \dfrac{360}{10} = \$36$

x	$x - \bar{x}$	$(x - \bar{x})^2$
28	-8	64
28	-8	64
50	14	196
45	9	81
30	-6	36
45	9	81
48	12	144
18	-18	324
45	9	81
23	-13	169
	0	1240

$\dfrac{1240}{9} \approx 137.78, s = \sqrt{137.78} \approx \$11.74$

24. Range = $28 - 1 = 27$

$\bar{x} = \dfrac{84}{7} = 12$

x	$x - \bar{x}$	$(x - \bar{x})^2$
10	-2	4
23	11	121
28	16	256
4	-8	64
1	-11	121
6	-6	36
12	0	0
	0	602

$\dfrac{602}{6} \approx 100.33, \quad s = \sqrt{100.33} \approx 10.02$

25. Range = 200 − 50 = $150

$$\bar{x} = \frac{1100}{10} = \$110$$

x	$x - \bar{x}$	$(x - \bar{x})^2$
50	-60	3600
120	10	100
130	20	400
60	-50	2500
55	-55	3025
75	-35	1225
200	90	8100
110	0	0
125	15	225
175	65	4225
	0	23,400

$$\frac{23,400}{9} = 2600, \quad s = \sqrt{2600} \approx \$50.99$$

26. Range = 300 − 35 = $265

$$\bar{x} = \frac{980}{7} = \$140$$

x	$x - \bar{x}$	$(x - \bar{x})^2$
60	-80	6400
100	-40	1600
85	-55	3025
35	-105	11,025
250	110	12,100
150	10	100
300	160	25,600
	0	59,850

$$\frac{59,850}{6} = 9975, \quad s = \sqrt{9975} \approx \$99.87$$

27. a) Range = 68 - 5 = $63

$$\bar{x} = \frac{204}{6} = \$34$$

x	$x - \bar{x}$	$(x - \bar{x})^2$
32	-2	4
60	26	676
14	-20	400
25	-9	81
5	-29	841
68	34	1156
	0	3158

$$\frac{3158}{5} = 631.6, s = \sqrt{631.6} \approx \$25.13$$

b) New data: 42, 70, 24, 35, 15, 78

The range and standard deviation will be the same. If each piece of data is increased by the same number, the range and standard deviation will remain the same.

c) Range = 78 - 15 = $63

$$\bar{x} = \frac{264}{6} = \$44$$

x	$x - \bar{x}$	$(x - \bar{x})^2$
42	-2	4
70	26	676
24	-20	400
35	-9	81
15	-29	841
78	34	1156
	0	3158

$$\frac{3158}{5} = 631.6, s = \sqrt{631.6} \approx \$25.13$$

The answers remain the same.

28. a) - c) Answers will vary.

   d) If each piece of data is increased, or decreased, by n, the mean is increased, or decreased, by n. The standard deviation remains the same.

   e) The mean of the first set of numbers is $\dfrac{63}{7} = 9$. The mean of the second set is $\dfrac{4193}{7} = 599$.

Standard deviation of first set

x	$x - \bar{x}$	$(x - \bar{x})^2$
6	-3	9
7	-2	4
8	-1	1
9	0	0
10	1	1
11	2	4
12	3	9
	0	28

$\dfrac{28}{6} = 4.67, s = \sqrt{4.67} \approx 2.16$

Standard deviation of second set

x	$x - \bar{x}$	$(x - \bar{x})^2$
596	-3	9
597	-2	4
598	-1	1
599	0	0
600	1	1
601	2	4
602	3	9
	0	28

$\dfrac{28}{6} = 4.67, s = \sqrt{4.67} \approx 2.16$

29. a) - c) Answers will vary.

   d) If each number in a distribution is multiplied by n, both the mean and standard deviation of the new distribution will be n times that of the original distribution.

   e) The mean of the second set is $4 \times 5 = 20$, and the standard deviation of the second set is $2 \times 5 = 10$.

30. a) Same   b) More

31. a) The standard deviation increases. There is a greater spread from the mean as they get older.

   b) ≈ 133 lb

   c) $\dfrac{175 - 90}{4} = 21.25 \approx 21$ lb

   d) The mean weight is about 100 pounds and the normal range is about 60 to 140 pounds.

   e) The mean height is about 62 inches and the normal range is about 53 to 68 inches.

   f) $100\% - 95\% = 5\%$

32. a) and b) Answers will vary.

   c) Baseball: $\dfrac{172}{10} = \$17.20$ million

   NFL: $\dfrac{1216}{10} = \$12.16$ million

32.   d) Baseball Mean ≈ $17.2 million

NFL Mean ≈ $12.2 million

	Baseball				NFL	
x	$x-\bar{x}$	$(x-\bar{x})^2$		x	$x-\bar{x}$	$(x-\bar{x})^2$
22	4.8	23.04		15.4	3.2	10.24
20	2.8	7.84		13.3	1.1	1.21
18.7	1.5	2.25		13.0	0.8	0.64
17.2	0	0		12.0	-0.2	0.04
16	-1.2	1.44		11.7	-0.5	0.25
15.7	-1.5	2.25		11.7	-0.5	0.25
15.7	-1.5	2.25		11.4	-0.8	0.64
15.6	-1.6	2.56		11.3	-0.9	0.81
15.6	-1.6	2.56		11.3	-0.9	0.81
15.5	-1.7	2.89		10.5	-1.7	2.89
		47.08				17.78

$$\frac{47.08}{9} \approx 5.23, \quad s = \sqrt{5.23} \approx \$2.29 \text{ million}$$

$$\frac{17.78}{9} \approx 1.98, \quad s = \sqrt{1.98} \approx \$1.41 \text{ million}$$

33.   a)

East	
Number of oil changes made	Number of days
15-20	2
21-26	2
27-32	5
33-38	4
39-44	7
45-50	1
51-56	1
57-62	2
63-68	1

West	
Number of oil changes made	Number of days
15-20	0
21-26	0
27-32	6
33-38	9
39-44	4
45-50	6
51-56	0
57-62	0
63-68	0

b)

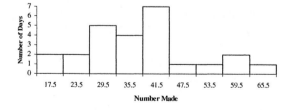

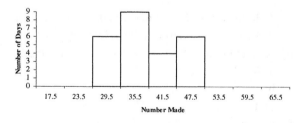

c) They appear to have about the same mean since they are both centered around 38.

d) The distribution for East is more spread out. Therefore, East has a greater standard deviation.

e) East: $\frac{950}{25} = 38$, West: $\frac{950}{25} = 38$

33. f)

	East			West	
x	$x-\bar{x}$	$(x-\bar{x})^2$	x	$x-\bar{x}$	$(x-\bar{x})^2$
33	-5	25	38	0	0
30	-8	64	38	0	0
25	-13	169	37	-1	1
27	-11	121	36	-2	4
40	2	4	30	-8	64
44	6	36	45	7	49
49	11	121	28	-10	100
52	14	196	47	9	81
42	4	16	30	-8	64
59	21	441	46	8	64
19	-19	361	38	0	0
22	-16	256	39	1	1
57	19	361	40	2	4
67	29	841	34	-4	16
15	-23	529	31	-7	49
41	3	9	45	7	49
43	5	25	29	-9	81
27	-11	121	38	0	0
42	4	16	38	0	0
43	5	25	39	1	1
37	-1	1	37	-1	1
38	0	0	42	4	16
31	-7	49	46	8	64
32	-6	36	31	-7	49
35	$\underline{-3}$	$\underline{9}$	48	$\underline{10}$	$\underline{100}$
	0	3832		0	858

$$\frac{3832}{24} \approx 159.67, \quad s = \sqrt{159.67} \approx 12.64 \qquad \frac{858}{24} = 35.75, \quad s = \sqrt{35.75} \approx 5.98$$

34. Answers will vary.

35. 6, 6, 6, 6, 6

Exercise Set 13.7

1. A **rectangular distribution** is one where all the values have the same frequency.
2. A **J-shaped distribution** is one where the frequency is either constantly increasing or constantly decreasing.
3. A **bimodal distribution** is one where two nonadjacent values occur more frequently than any other values in a set of data.
4. A **distribution skewed to the right** is one that has "a tail" on its right.
5. A **distribution skewed to the left** is one that has "a tail" on its left.
6. A **normal distribution** is a bell-shaped distribution.
7. a) *B*
   b) *C*
   c) *A*
8. a) Yes, 36
   b) *B*, since curve *B* is more spread out it has the higher standard deviation.
9. The distribution of outcomes from the roll of a die
10. Skewed left - a listing of test scores where most of the students did well and a few did poorly;
    Skewed right - number of cans of soda consumed in a day where most people consumed a few cans and a few people consumed many cans
11. J shaped right - consumer price index; J shaped left - value of the dollar
12. The distribution of heights of an equal number of males and females

13. Normal

14. Rectangular

15. Skewed right

16. Bimodal

17. The mode is the lowest value, the median is greater than the mode, and the mean is greater than the median. The greatest frequency appears on the left side of the curve. Since the mode is the value with the greatest frequency, the mode would appear on the left side of the curve (where the lowest values are). Every value in the set of data is considered in determining the mean. The values on the far right of the curve would increase the value of the mean. Thus, the value of the mean would be farther to the right than the mode. The median would be between the mode and the mean.

18. The mode is the highest value. The median is lower than the mode. The mean is the lowest value.

19. Answers will vary.

20. Answers will vary.

21. In a normal distribution the mean, median, and the mode all have the same value.

22. A **z-score** measures how far, in terms of standard deviation, a given score is from the mean.

23. A z-score will be negative when the piece of data is less than the mean.

24. Subtract the mean from the value of the piece of data and divide the difference by the standard deviation.

25. 0

26. a) $\approx 68\%$ b) $\approx 95\%$

27. 0.500

28. 0.500

29. $0.477 + 0.341 = 0.818$

30. $0.455 - 0.364 = 0.091$

31. $0.500 - 0.466 = 0.034$

32. $0.500 + 0.383 = 0.883$

33. $0.500 - 0.463 = 0.037$

34. $0.500 + 0.463 = 0.963$

35. $0.500 - 0.481 = 0.019$

36. $0.500 + 0.475 = 0.975$

37. $0.500 - 0.447 = 0.053$

38. $0.500 - 0.316 = 0.184$

39. $0.261 = 26.1\%$

40. $0.294 - 0.060 = 0.234 = 23.4\%$

41. $0.410 + 0.488 = 0.898 = 89.8\%$

42. $0.500 - 0.471 = 0.029 = 2.9\%$

43. $0.500 + 0.471 = 0.971 = 97.1\%$

44. $0.500 - 0.496 = 0.004 = 0.4\%$

45. $0.500 + 0.475 = 0.975 = 97.5\%$

46. $0.484 - 0.264 = 0.22 = 22.0\%$

47. $0.466 - 0.437 = 0.029 = 2.9\%$

48. $0.484 + 0.500 = 0.984 = 98.4\%$

49. a) Jake, Sarah, and Carol scored above the mean because their z-scores are positive.

   b) Marie and Kevin scored at the mean because their z-scores are zero.

   c) Omar, Justin, and Kim scored below the mean because their z-scores are negative.

50. a) Sarah had the highest score because she had the highest z-score.

   b) Omar had the lowest score because he had the lowest z-score.

51. $0.500 = 50\%$

52. $z_{14} = \dfrac{14 - 18}{4} = \dfrac{-4}{4} = -1.00$

   $z_{26} = \dfrac{26 - 18}{4} = \dfrac{8}{4} = 2.00$

   $0.341 + 0.477 = 0.818 = 81.8\%$

53. $z_{23} = \dfrac{23 - 18}{4} = \dfrac{5}{4} = 1.25$

   $0.500 - 0.394 = 0.106 = 10.6\%$

54. 10.6% of college students work at least 23 hours per week. (See Exercise 53.)

$0.106(500) = 53$ students

55. $z_{1650} = \dfrac{1650-1600}{100} = \dfrac{50}{100} = 0.50$

$0.500 + 0.192 = 0.692 = 69.2\%$

56. $z_{1750} = \dfrac{1750-1600}{100} = \dfrac{150}{100} = 1.50$

$0.500 - 0.433 = 0.067 = 6.7\%$

57. $z_{1650} = 0.50$ and $z_{1750} = 1.50$

(See Exercises 55 and 56.)

$0.433 - 0.192 = 0.241 = 24.1\%$

58. $z_{1400} = \dfrac{1400-1600}{100} = \dfrac{-200}{100} = -2.00$

$0.500 - 0.477 = 0.023 = 2.3\%$

59. $z_{1500} = \dfrac{1500-1600}{100} = \dfrac{-100}{100} = -1.00$

$z_{1625} = \dfrac{1625-1600}{100} = \dfrac{25}{100} = 0.25$

$0.341 = 0.099 = 0.44 = 44.0\%$

60. $z_{1480} = \dfrac{1480-1600}{100} = \dfrac{-120}{100} = -1.20$

$0.385 + 0.500 = 0.885 = 88.5\%$

61. $z_{7.4} = \dfrac{7.4-7.6}{0.4} = \dfrac{-0.2}{0.4} = -0.50$

$z_{7.7} = \dfrac{7.7-7.6}{0.4} = \dfrac{0.1}{0.4} = 0.25$

$0.192 + 0.099 = 0.291 = 29.1\%$

62. $z_{7.0} = \dfrac{7.0-7.6}{0.4} = \dfrac{-0.6}{0.4} = -1.50$

$0.500 - 0.433 = 0.067 = 6.7\%$

63. $z_{7.7} = 0.25$ (See Exercise 61.)

$0.500 + 0.099 = 0.599 = 59.9\%$

64. The 8-oz cup will overflow when the machine dispenses more than 8 oz of coffee.

$z_{8.0} = \dfrac{8.0-7.6}{0.4} = \dfrac{0.4}{0.4} = 1.00$

$0.500 - 0.341 = 0.159 = 15.9\%$

65. $0.500 = 50.0\%$

66. $z_{197} = \dfrac{197-206}{12} = \dfrac{-9}{12} = -0.75$

$z_{215} = \dfrac{215-206}{12} = \dfrac{9}{12} = 0.75$

$0.273 + 0.273 = 0.546 = 54.6\%$

67. $z_{191} = \dfrac{191-206}{12} = \dfrac{-15}{12} = -1.25$

$0.500 - 0.394 = 0.106 = 10.6\%$

68. $z_{224} = \dfrac{224-206}{12} = \dfrac{18}{12} = 1.50$

$0.500 - 0.433 = 0.067 = 6.7\%$

69. 10.6% of females have a cholesterol level less than 191. (See Exercise 67.)

$0.106(200) = 21.2 \approx 21$ women

70. 6.7% of females have a cholesterol level greater than 224. (See Exercise 68.)

$0.067(200) = 13.4 \approx 13$ women

71. $z_{30,750} = \dfrac{30,750-35,000}{2500} = \dfrac{-4250}{2500} = -1.70$

$z_{38,300} = \dfrac{38,300-35,000}{2500} = \dfrac{3300}{2500} = 1.32$

$0.455 + 0.407 = 0.862 = 86.2\%$

72. At least 39,000 miles means 39,000 miles or more.

$$z_{39,000} = \frac{39,000 - 35,000}{2500} = \frac{4000}{2500} = 1.60$$

$0.500 - 0.445 = 0.055 = 5.5\%$

73. The tires that last less than 30,750 miles will fail to live up to the guarantee.

$z_{30,750} = -1.70$ (See Exercise 71.)

$0.500 - 0.455 = 0.045 = 4.5\%$

74. 5.5% of tires will last at least 39,000 miles. (See Exercise 72.)

$0.055(200,000) = 11,000$ tires

75. $z_{3.1} = \frac{3.1 - 3.7}{1.2} = \frac{-0.6}{1.2} = -0.50$

$0.192 + 0.500 = 0.692 = 69.2\%$

76. $z_{2.5} = \frac{2.5 - 3.7}{1.2} = \frac{-1.2}{1.2} = -1.00$

$z_{4.3} = \frac{4.3 - 3.7}{1.2} = \frac{0.6}{1.2} = 0.50$

$0.341 + 0.192 = 0.533 = 53.3\%$

77. $z_{6.7} = \frac{6.7 - 3.7}{1.2} = \frac{3.0}{1.2} = 2.50$

$0.500 - 0.494 = 0.006 = 0.6\%$

78. $z_{6.7} = 2.50$ (See Exercise 77.)

$0.500 + 0.494 = 0.994 = 99.4\%$

79. 69.2% of the children are older than 3.1 years. (See Exercise 75.)

$0.692(120) = 83.04 \approx 83$ children

80. 53.3% of the children are between 2.5 and 4.3 years. (See Exercise 76.)

$0.533(120) = 63.96 \approx 64$ children

81. Customers will be able to claim a refund if they lose less than 5 lb.

$$z_5 = \frac{5 - 6.7}{0.81} = \frac{-1.7}{0.81} = -2.10$$

$0.500 - 0.482 = 0.018 = 1.8\%$

82. A motor will require repair or replacement if it breaks down in less than 8 years.

$$z_8 = \frac{8 - 10.2}{1.8} = \frac{-2.2}{1.8} \approx -1.22$$

$0.500 - 0.389 = 0.111 = 11.1\%$

83. The standard deviation is too large.
There is too much variation.

84. A z-score of 1.8 or higher is required for an A. The area from the mean to 1.8 is 0.464.
Thus, $0.500 - 0.464 = 0.036 = 3.6\%$ will receive an A.
A z-score between 1.8 and 1.1 is required for a B. The areas from the mean to these z-scores are
0.464 and 0.364, respectively. Thus, $0.464 - 0.364 = 0.100 = 10.0\%$ will receive a B.
A z-score between 1.1 and -1.2 is required for a C. The areas from the mean to these z-scores are
0.364 and 0.385, respectively. Thus, $0.364 + 0.385 = 0.749 = 74.9\%$ will receive a C.
A z-score between -1.2 and -1.9 is required for a D. The areas from the mean to these z-scores are
0.385 and 0.471, respectively. Thus, $0.471 - 0.385 = 0.086 = 8.6\%$ will receive a D.
A z-score of -1.9 or lower is required for an F. The area from the mean to -1.9 is 0.471.
Thus, $0.500 - 0.471 = 0.029 = 2.9\%$ will receive an F.

85. a) Katie: $z_{28,408} = \dfrac{28,408 - 23,200}{2170} = \dfrac{5208}{2170} = 2.4$

$\quad\quad$ Stella: $z_{29,510} = \dfrac{29,510 - 25,600}{2300} = \dfrac{3910}{2300} = 1.7$

$\quad$ b) Katie. Her z-score is higher than Stella's z-score. This means her sales are further above the mean than Stella's sales.

86. a) $\bar{x} = \dfrac{160}{30} = 5.\overline{3} \approx 5.33$

$\quad$ b)

x	$x - \bar{x}$	$(x - \bar{x})^2$	x	$x - \bar{x}$	$(x - \bar{x})^2$	x	$x - \bar{x}$	$(x - \bar{x})^2$
1	-4.33	18.75	4	-1.33	1.77	7	1.67	2.79
1	-4.33	18.75	4	-1.33	1.77	8	2.67	7.13
1	-4.33	18.75	4	-1.33	1.77	8	2.67	7.13
1	-4.33	18.75	5	-0.33	0.11	8	2.67	7.13
2	-3.33	11.09	6	0.67	0.45	8	2.67	7.13
2	-3.33	11.09	6	0.67	0.45	9	3.67	13.47
2	-3.33	11.09	6	0.67	0.45	9	3.67	13.47
2	-3.33	11.09	7	1.67	2.79	9	3.67	13.47
3	-2.33	5.43	7	1.67	2.79	10	4.67	21.81
3	-2.33	5.43	7	1.67	2.79	10	4.67	21.81
								260.70

$\quad\quad$ $260.70 \div 29 \approx 8.99$ $\quad\quad\quad\quad$ $s = \sqrt{8.99} \quad \approx 3.00$

$\quad$ c)
$\quad\quad$ $\bar{x} + 1.1s = 5.33 + 1.1(3) = 8.63$ $\quad\quad\quad$ $\bar{x} - 1.1s = 5.33 - 1.1(3) = 2.03$
$\quad\quad$ $\bar{x} + 1.5s = 5.33 + 1.5(3) = 9.83$ $\quad\quad\quad$ $\bar{x} - 1.15s = 5.33 - 1.5(3) = 0.83$
$\quad\quad$ $\bar{x} + 2.0s = 5.33 + 2.0(3) = 11.33$ $\quad\quad\quad$ $\bar{x} - 2.0s = 5.33 - 2.0(3) = -0.67$
$\quad\quad$ $\bar{x} + 2.5s = 5.33 + 2.5(3) = 12.83$ $\quad\quad\quad$ $\bar{x} - 2.5s = 5.33 - 2.5(3) = -2.17$

$\quad$ d) Between -1.1s and 1.1s or between scores of 2.03 and 8.63, there are 17 scores.

$\quad\quad$ $\dfrac{17}{30} = 0.5\overline{6} \approx 56.7\%$

$\quad\quad$ Between -1.5s and 1.5s, or between scores of 0.83 and 9.83, there are 28 scores.

$\quad\quad$ $\dfrac{28}{30} = 0.9\overline{3} \approx 93.3\%$

$\quad\quad$ Between -2.0s and 2.0s, or between scores of -0.67 and 11.33, there are 30 scores.

$\quad\quad$ $\dfrac{30}{30} = 1 = 100\%$

$\quad\quad$ Between -2.5s and 2.5s, or between scores of -2.17 and 12.83, there are 30 scores.

$\quad\quad$ $\dfrac{30}{30} = 1 = 100\%$

$\quad$ e)

Minimum %	K = 1.1	K = 1.5	K = 2.0	K = 2.5
(For any distribution)	17.4%	55.6%	75%	84%
Normal distribution	72.8%	86.6%	95.4%	99.8%
Given distribution	56.7%	93.3%	100%	100%

$\quad$ f) The percent between -1.1s and 1.1s is too low to be considered a normal distribution.

87. Answers will vary.
88. Using Table 13.7, the answer is 1.96.
89. Using Table 13.7, the answer is -1.18.
90. Answers will vary.

91. $\dfrac{0.77}{2} = 0.385$

Using the table in Section 13.7, an area of 0.385 has a z-score of 1.20.

$$z = \frac{x - \bar{x}}{s}$$

$$1.20 = \frac{14.4 - 12}{s}$$

$$1.20 = \frac{2.4}{s}$$

$$\frac{1.20s}{1.20} = \frac{2.4}{1.20}$$

$$s = 2$$

Exercise Set 13.8

1. The **correlation coefficient** measures the strength of the relationship between the quantities.

2. The purpose of **linear regression** is to determine the linear relationship between two variables.

3. 1   4. -1   5. 0

6. A negative correlation indicates that as one quantity increases, the other quantity decreases.

7. A positive correlation indicates that as one quantity increases, the other quantity increases.

8. The **line of best fit** represents the line such that the sum of the vertical distances between the points and the line is a minimum.

9. The **level of significance** is used to identify the cutoff between results attributed to chance and results attributed to an actual relationship between the two variables.

10. A **scatter diagram** is a plot of data points.

11. No correlation   12. Weak negative
13. Strong positive   14. Strong negative
15. Yes, $|\,0.76\,| > 0.684$ 16. No, $|\,0.43\,| < 0.537$
17. Yes, $|-0.73\,| > 0.707$ 18. No, $|-0.49| < 0.602$
19. No, $|-0.23| < 0.254$ 20. No, $|-0.49| < 0.590$
21. No, $|0.82| < 0.917$ 22. Yes, $|0.96| > 0.959$

Note: The answers in the remainder of this section may differ slightly from your answers, depending upon how your answers are rounded and which calculator you used.

23. a)

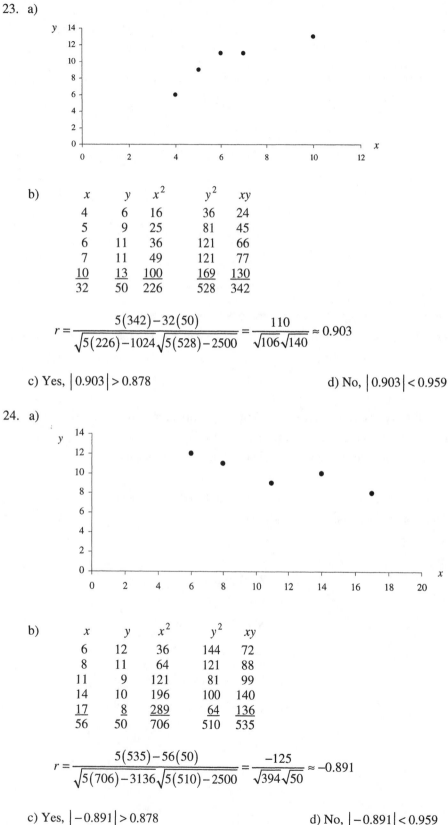

b)

x	y	x^2	y^2	xy
4	6	16	36	24
5	9	25	81	45
6	11	36	121	66
7	11	49	121	77
10	13	100	169	130
32	50	226	528	342

$$r = \frac{5(342) - 32(50)}{\sqrt{5(226) - 1024}\sqrt{5(528) - 2500}} = \frac{110}{\sqrt{106}\sqrt{140}} \approx 0.903$$

c) Yes, $|0.903| > 0.878$ d) No, $|0.903| < 0.959$

24. a)

b)

x	y	x^2	y^2	xy
6	12	36	144	72
8	11	64	121	88
11	9	121	81	99
14	10	196	100	140
17	8	289	64	136
56	50	706	510	535

$$r = \frac{5(535) - 56(50)}{\sqrt{5(706) - 3136}\sqrt{5(510) - 2500}} = \frac{-125}{\sqrt{394}\sqrt{50}} \approx -0.891$$

c) Yes, $|-0.891| > 0.878$ d) No, $|-0.891| < 0.959$

25. a)

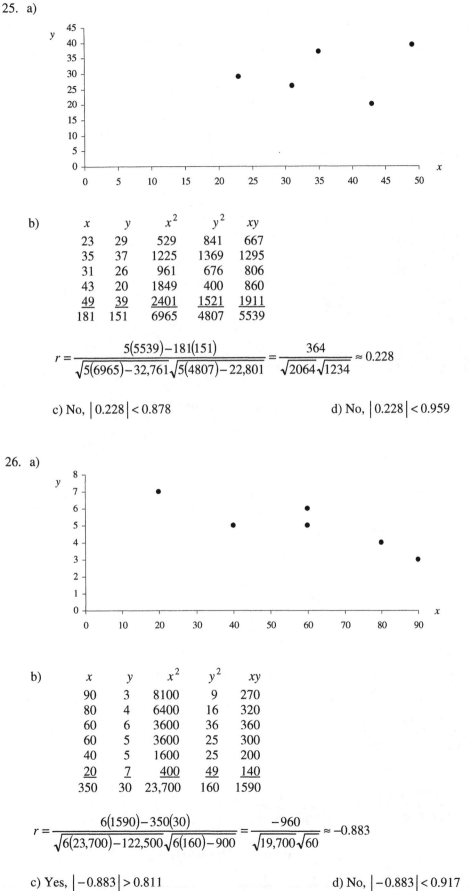

b)

x	y	x^2	y^2	xy
23	29	529	841	667
35	37	1225	1369	1295
31	26	961	676	806
43	20	1849	400	860
49	39	2401	1521	1911
181	151	6965	4807	5539

$$r = \frac{5(5539) - 181(151)}{\sqrt{5(6965) - 32{,}761}\sqrt{5(4807) - 22{,}801}} = \frac{364}{\sqrt{2064}\sqrt{1234}} \approx 0.228$$

c) No, $|0.228| < 0.878$

d) No, $|0.228| < 0.959$

26. a)

b)

x	y	x^2	y^2	xy
90	3	8100	9	270
80	4	6400	16	320
60	6	3600	36	360
60	5	3600	25	300
40	5	1600	25	200
20	7	400	49	140
350	30	23,700	160	1590

$$r = \frac{6(1590) - 350(30)}{\sqrt{6(23{,}700) - 122{,}500}\sqrt{6(160) - 900}} = \frac{-960}{\sqrt{19{,}700}\sqrt{60}} \approx -0.883$$

c) Yes, $|-0.883| > 0.811$

d) No, $|-0.883| < 0.917$

27. a)

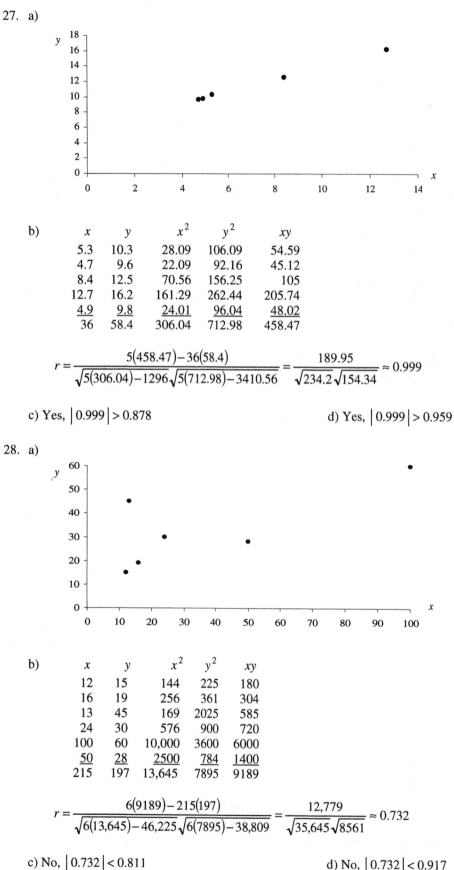

b)

x	y	x^2	y^2	xy
5.3	10.3	28.09	106.09	54.59
4.7	9.6	22.09	92.16	45.12
8.4	12.5	70.56	156.25	105
12.7	16.2	161.29	262.44	205.74
4.9	9.8	24.01	96.04	48.02
36	58.4	306.04	712.98	458.47

$$r = \frac{5(458.47)-36(58.4)}{\sqrt{5(306.04)-1296}\sqrt{5(712.98)-3410.56}} = \frac{189.95}{\sqrt{234.2}\sqrt{154.34}} \approx 0.999$$

c) Yes, $|0.999| > 0.878$ \hspace{2cm} d) Yes, $|0.999| > 0.959$

28. a)

b)

x	y	x^2	y^2	xy
12	15	144	225	180
16	19	256	361	304
13	45	169	2025	585
24	30	576	900	720
100	60	10,000	3600	6000
50	28	2500	784	1400
215	197	13,645	7895	9189

$$r = \frac{6(9189)-215(197)}{\sqrt{6(13,645)-46,225}\sqrt{6(7895)-38,809}} = \frac{12,779}{\sqrt{35,645}\sqrt{8561}} \approx 0.732$$

c) No, $|0.732| < 0.811$ \hspace{2cm} d) No, $|0.732| < 0.917$

29. a)

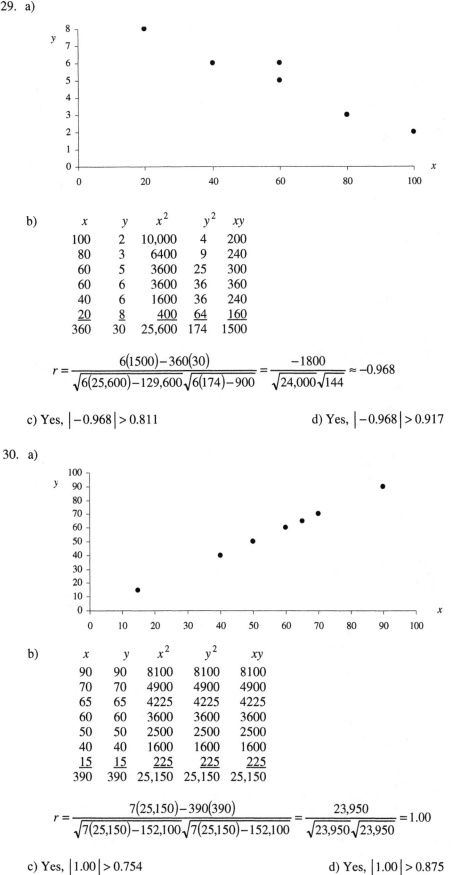

b)

x	y	x^2	y^2	xy
100	2	10,000	4	200
80	3	6400	9	240
60	5	3600	25	300
60	6	3600	36	360
40	6	1600	36	240
20	8	400	64	160
360	30	25,600	174	1500

$$r = \frac{6(1500)-360(30)}{\sqrt{6(25,600)-129,600}\sqrt{6(174)-900}} = \frac{-1800}{\sqrt{24,000}\sqrt{144}} \approx -0.968$$

c) Yes, $\left|-0.968\right| > 0.811$ 　　　　　　　　　d) Yes, $\left|-0.968\right| > 0.917$

30. a)

b)

x	y	x^2	y^2	xy
90	90	8100	8100	8100
70	70	4900	4900	4900
65	65	4225	4225	4225
60	60	3600	3600	3600
50	50	2500	2500	2500
40	40	1600	1600	1600
15	15	225	225	225
390	390	25,150	25,150	25,150

$$r = \frac{7(25,150)-390(390)}{\sqrt{7(25,150)-152,100}\sqrt{7(25,150)-152,100}} = \frac{23,950}{\sqrt{23,950}\sqrt{23,950}} = 1.00$$

c) Yes, $\left|1.00\right| > 0.754$ 　　　　　　　　　d) Yes, $\left|1.00\right| > 0.875$

31. From # 23: $m = \dfrac{5(342) - 32(50)}{5(226) - 1024} = \dfrac{110}{106} \approx 1.0$

$b = \dfrac{50 - \dfrac{110}{106}(32)}{5} \approx 3.4, \quad y = 1.0x + 3.4$

32. From # 24: $m = \dfrac{5(535) - 56(50)}{5(706) - 3136} = \dfrac{-125}{394} \approx -0.3$

$b = \dfrac{50 - \dfrac{-125}{394}(56)}{5} \approx 13.6, \quad y = -0.3x + 13.6$

33. From # 25: $m = \dfrac{5(5539) - 181(151)}{5(6965) - 32{,}761} = \dfrac{364}{2064} \approx 0.2$

$b = \dfrac{151 - \dfrac{364}{2064}(181)}{5} \approx 23.8, \quad y = 0.2x + 23.8$

34. From # 26: $m = \dfrac{6(1590) - 350(30)}{6(23{,}700) - 122{,}500} = \dfrac{-960}{19{,}700} \approx -0.05$

$b = \dfrac{30 - \dfrac{-960}{19{,}700}(350)}{6} \approx 7.8, \quad y = -0.05x + 7.8$

35. From # 27: $m = \dfrac{5(458.47) - 36(58.4)}{5(306.04) - 1296} = \dfrac{189.95}{234.2} \approx 0.8$

$b = \dfrac{58.4 - \dfrac{189.95}{234.2}(36)}{5} \approx 5.8, \quad y = 0.8x + 5.8$

36. From # 28: $m = \dfrac{6(9189) - 215(197)}{6(13{,}645) - 46{,}225} = \dfrac{12{,}779}{35{,}645} \approx 0.4$

$b = \dfrac{197 - \dfrac{12{,}779}{35{,}645}(215)}{6} \approx 20.0, \quad y = 0.4x + 20.0$

37. From # 29: $m = \dfrac{6(1500) - 360(30)}{6(25{,}600) - 129{,}600} = \dfrac{-1800}{24{,}000} \approx -0.1$

$b = \dfrac{30 - \dfrac{-1800}{24{,}000}(360)}{6} \approx 9.5, \quad y = -0.1x + 9.5$

38. From # 30: $m = \dfrac{7(25{,}150) - 390(390)}{7(25{,}150) - 152{,}100} = \dfrac{23{,}950}{23{,}950} = 1.0$

$b = \dfrac{390 - 1(390)}{7} = 0, \quad y = 1.0x$

39. a)

x	y	x^2	y^2	xy
8	15	64	225	120
20	28	400	784	560
9	20	81	400	180
15	25	225	625	375
16	28	256	784	448
2	5	4	25	10
70	121	1030	2843	1693

$$r = \frac{6(1693) - 70(121)}{\sqrt{6(1030) - 4900}\sqrt{6(2843) - 14,641}} = \frac{1688}{\sqrt{1280}\sqrt{2417}} \approx 0.960$$

b) Yes, $|\ 0.960\ | > 0.811$

c) $m = \dfrac{6(1693) - 70(121)}{6(1030) - 4900} = \dfrac{1688}{1280} \approx 1.3$, $b = \dfrac{121 - \dfrac{1688}{1280}(70)}{6} \approx 4.8$, $y = 1.3x + 4.8$

40. a)

x	y	x^2	y^2	xy
321	13	103,041	169	4173
380	23	144,400	529	8740
350	16	122,500	256	5600
358	14	128,164	196	5012
378	19	142,884	361	7182
391	19	152,881	361	7429
2178	104	793,870	1872	38,136

$$r = \frac{6(38,136) - 2178(104)}{\sqrt{6(793,870) - 4,743,684}\sqrt{6(1872) - 10,816}} = \frac{2304}{\sqrt{19,536}\sqrt{416}} \approx 0.808$$

b) No, $|\ 0.808\ | < 0.811$

c) $m = \dfrac{6(38,136) - 2178(104)}{6(793,870) - 4,743,684} = \dfrac{2304}{19,536} \approx 0.1$, $b = \dfrac{104 - \dfrac{2304}{19,536}(2178)}{6} \approx -25.5$, $y = 0.1x - 25.5$

41. a)

x	y	x^2	y^2	xy
20	40	400	1600	800
40	45	1600	2025	1800
50	70	2500	4900	3500
60	76	3600	5776	4560
80	92	6400	8464	7360
100	95	10,000	9025	9500
350	418	24,500	31,790	27,520

$$r = \frac{6(27,520)-350(418)}{\sqrt{6(24,500)-122,500}\sqrt{6(31,790)-174,724}} = \frac{18,820}{\sqrt{24,500}\sqrt{16,016}} \approx 0.950$$

b) Yes, $|0.950| > 0.917$

c) $m = \dfrac{6(27,520)-350(418)}{6(24,500)-122,500} = \dfrac{18,820}{24,500} \approx 0.8$, $\quad b = \dfrac{418-\dfrac{18,820}{24,500}(350)}{6} \approx 24.9$, $\quad y = 0.8x + 24.9$

42. a)

x	y	x^2	y^2	xy
765	119	585,225	14,161	91,035
926	127	857,476	16,129	117,602
1145	150	1,311,025	22,500	171,750
842	119	708,964	14,161	100,198
1485	153	2,205,225	23,409	227,205
1702	156	2,896,804	24,336	265,512
6865	824	8,564,719	114,696	973,302

$$r = \frac{6(973,302)-6865(824)}{\sqrt{6(8,564,719)-47,128,225}\sqrt{6(114,696)-678,976}} = \frac{183,052}{\sqrt{4,260,089}\sqrt{9200}} \approx 0.925$$

b) Yes, $|0.925| > 0.811$

c) $m = \dfrac{6(973,302)-6865(824)}{6(8,564,719)-47,128,225} = \dfrac{183,052}{4,260,089} \approx 0.04$, $\quad b = \dfrac{824-\dfrac{183,052}{4,260,089}(6865)}{6} \approx 88.2$, $\quad y = 0.04x + 88.2$

d) $y = 0.04(1500) + 88.2 = 148.2 \approx 148$ mountain lions

43. a)

x	y	x^2	y^2	xy
20	8	400	64	160
12	10	144	100	120
18	12	324	144	216
15	9	225	81	135
22	6	484	36	132
10	15	100	225	150
20	7	400	49	140
12	18	144	324	216
129	85	2221	1023	1269

$$r = \frac{8(1269)-129(85)}{\sqrt{8(2221)-16{,}641}\sqrt{8(1023)-7225}} = \frac{-813}{\sqrt{1127}\sqrt{959}} \approx -0.782$$

b) Yes, $|-0.782| > 0.707$

c) $m = \dfrac{8(1269)-129(85)}{8(2221)-16{,}641} = \dfrac{-813}{1127} \approx -0.7$, $\quad b = \dfrac{85-\dfrac{-813}{1127}(129)}{8} \approx 22.3$, $\quad y = -0.7x + 22.3$

d) $y = -0.7(14) + 22.3 = 12.5$ muggings

44. a)

x	y	x^2	y^2	xy
00	15.0	0	225	0
01	15.3	1	234.09	15.3
02	15.5	4	240.25	31
03	15.8	9	249.64	47.4
04	16.1	16	259.21	64.4
05	16.3	25	265.69	81.5
15	94	55	1473.88	239.6

$$r = \frac{6(239.6)-15(94)}{\sqrt{6(55)-225}\sqrt{6(1473.88)-8836}} = \frac{27.6}{\sqrt{105}\sqrt{7.28}} \approx 0.998$$

b) Yes, $|\,0.998\,| > 0.811$

c) $m = \dfrac{6(239.6)-15(94)}{6(55)-225} = \dfrac{27.6}{105} \approx 0.3$, $\quad b = \dfrac{94-\dfrac{27.6}{105}(15)}{6} \approx 15.0$, $\quad y = 0.3x + 15.0$

d) $y = 0.3(8) + 15.0 = 17.4$ million students

45. a)

x	y	x^2	y^2	xy
89	22	7921	484	1958
110	28	12,100	784	3080
125	30	15,625	900	3750
92	26	8464	676	2392
100	22	10,000	484	2200
95	21	9025	441	1995
108	28	11,664	784	3024
97	25	9409	625	2425
816	202	84,208	5178	20,824

$$r = \frac{8(20,824) - 816(202)}{\sqrt{8(84,208) - 665,856}\sqrt{8(5178) - 40,804}} = \frac{1760}{\sqrt{7808}\sqrt{620}} \approx 0.800$$

b) Yes, $|0.800| > 0.707$

c) $m = \dfrac{8(20,824) - 816(202)}{8(84,208) - 665,856} = \dfrac{1760}{7808} \approx 0.2$, $\quad b = \dfrac{202 - \dfrac{1760}{7808}(816)}{8} \approx 2.3$, $\quad y = 0.2x + 2.3$

d) $y = 0.2(115) + 2.3 = 25.3 \approx 25$ units

46. a)

x	y	x^2	y^2	xy
4	100	16	10,000	400
4	67	16	4489	268
3	80	9	6400	240
2	120	4	14,400	240
1	40	1	1600	40
3	90	9	8100	270
4	60	16	3600	240
2	60	4	3600	120
4	90	16	8100	360
1	100	1	10,000	100
28	807	92	70,289	2278

$$r = \frac{10(2278) - 28(807)}{\sqrt{10(92) - 784}\sqrt{10(70,289) - 651,249}} = \frac{184}{\sqrt{136}\sqrt{51,641}} \approx 0.069$$

b) No, $|0.069| < 0.632$

c) $m = \dfrac{10(2278) - 28(807)}{10(92) - 784} = \dfrac{184}{136} \approx 1.4$, $\quad b = \dfrac{807 - \dfrac{184}{136}(28)}{10} \approx 76.9$, $\quad y = 1.4x + 76.9$

47. a)

x	y	x^2	y^2	xy
1	80.0	1	6400.0	80.0
2	76.2	4	5806.4	152.4
3	68.7	9	4719.7	206.1
4	50.1	16	2510.0	200.4
5	30.2	25	912.0	151.0
6	20.8	36	432.6	124.8
21	326	91	20,780.7	914.7

$$r = \frac{6(914.7)-21(326)}{\sqrt{6(91)-441}\sqrt{6(20,780.7)-106,276}} = \frac{-1357.8}{\sqrt{105}\sqrt{18,408.2}} \approx -0.977$$

b) Yes, $\left|-0.977\right| > 0.917$

c) $m = \dfrac{6(914.7)-21(326)}{6(91)-441} = \dfrac{-1357.8}{105} \approx -12.9$, $b = \dfrac{326 - \dfrac{-1357.8}{105}(21)}{6} \approx 99.6$, $y = -12.9x + 99.6$

d) $y = -12.9(4.5) + 99.6 = 41.55 \approx 41.6\%$

48. Answers will vary.

49. a) and b) Answers will vary.

c)

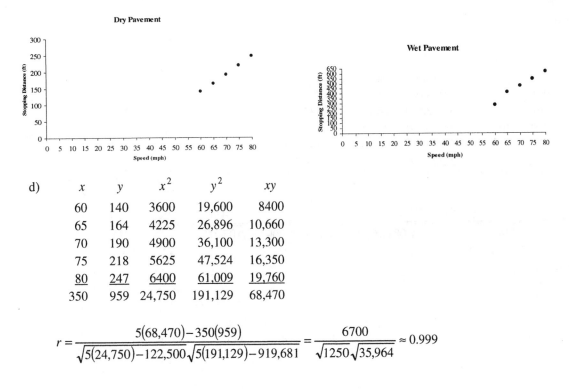

d)

x	y	x^2	y^2	xy
60	140	3600	19,600	8400
65	164	4225	26,896	10,660
70	190	4900	36,100	13,300
75	218	5625	47,524	16,350
80	247	6400	61,009	19,760
350	959	24,750	191,129	68,470

$$r = \frac{5(68,470)-350(959)}{\sqrt{5(24,750)-122,500}\sqrt{5(191,129)-919,681}} = \frac{6700}{\sqrt{1250}\sqrt{35,964}} \approx 0.999$$

49. e)

x	y	x^2	y^2	xy
60	280	3600	78,400	16,800
65	410	4225	168,100	26,650
70	475	4900	225,625	33,250
75	545	5625	297,025	40,875
80	618	6400	381,924	49,440
350	2328	24,750	1,151,074	167,015

$$r = \frac{5(167,015)-350(2328)}{\sqrt{5(24,750)-122,500}\sqrt{5(1,151,074)-5,419,584}} = \frac{20,275}{\sqrt{1250}\sqrt{335,786}} \approx 0.990$$

f) Answers will vary.

g) $m = \dfrac{5(68,470)-350(959)}{5(24,750)-122,500} = \dfrac{6700}{1250} \approx 5.4$, $b = \dfrac{959-\dfrac{6700}{1250}(350)}{5} = -183.4$, $y = 5.4x - 183.4$

h) $m = \dfrac{5(167,015)-350(2328)}{5(24,750)-122,500} = \dfrac{20,275}{1250} \approx 16.2$, $b = \dfrac{2328-\dfrac{20,275}{1250}(350)}{5} = -669.8$, $y = 16.2x - 669.8$

i) Dry: $y = 5.4(77)-183.4 = 232.4$ ft

  Wet: $y = 16.2(77)-669.8 = 577.6$ ft

50. a) The correlation coefficient will not change because $\sum xy = \sum yx$, $\left(\sum x\right)\left(\sum y\right)=\left(\sum y\right)\left(\sum x\right)$, and the square roots in the denominator will be the same.

  b) Answers will vary.

51. Answers will vary.

52. Answers will vary.

53. a)

x	y	x^2	y^2	xy
1996	157	3,984,016	24,649	313,372
1997	161	3,988,009	25,921	321,517
1998	163	3,992,004	26,569	325,674
1999	167	3,996,001	27,889	333,833
2000	172	4,000,000	29,584	344,000
2001	177	4,004,001	31,329	354,177
11,991	997	23,964,031	165,941	1,992,573

$$r = \frac{6(1,992,573)-11,991(997)}{\sqrt{6(23,964,031)-143,784,081}\sqrt{6(165,941)-994,009}} = \frac{411}{\sqrt{105}\sqrt{1637}} \approx 0.991$$

b) Should be the same.

53. c)

x	y	x^2	y^2	xy
0	157	0	24,649	0
1	161	1	25,921	161
2	163	4	26,569	326
3	167	9	27,889	501
4	172	16	29,584	688
5	177	25	31,329	885
15	997	55	165,941	2561

$$r = \frac{6(2561)-15(997)}{\sqrt{6(55)-225}\sqrt{6(165,941)-994,009}} = \frac{411}{\sqrt{105}\sqrt{1637}} \approx 0.991$$

The values are the same.

54. a) $SS(xy) = \sum xy - \frac{\left(\sum x\right)\left(\sum y\right)}{n} = 2335 - \frac{108(147)}{8} = 350.5$

$SS(x) = \sum x^2 - \frac{\left(\sum x\right)^2}{n} = 1866 - \frac{11,664}{8} = 408$

$SS(y) = \sum y^2 - \frac{\left(\sum y\right)^2}{n} = 3055 - \frac{21,609}{8} = 353.875$

$r = \frac{350.5}{\sqrt{408}\sqrt{353.875}} \approx 0.92$

b) Should be the same.

Review Exercises

1. a) A **population** consists of all items or people of interest.

   b) A **sample** is a subset of the population.

2. A **random sample** is one where every item in the population has the same chance of being selected.

3. The candy bars may have lots of calories, or fat, or sodium. Therefore, it may not be healthy to eat them.

4. Sales may not necessarily be a good indicator of profit. Expenses must also be considered.

5. a)   b)

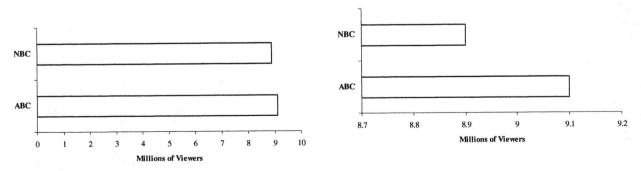

6. a)

Class	Frequency
35	1
36	3
37	6
38	2
39	3
40	0
41	4
42	1
43	3
44	1
45	1

b) and c)

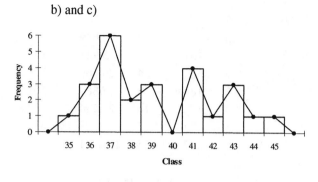

7. a)

High Temperature	Number of Cities
30 - 39	5
40 - 49	8
50 - 59	5
60 - 69	6
70 - 79	6
80 - 89	10

b) and c)

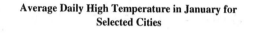

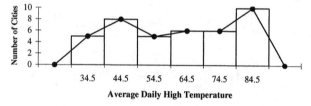

d)   3|6 represents 36

```
3|0  3  4  5  6
4|1  2  2  3  4  7  8  8
5|0  4  4  5  6
6|5  6  6  7  8  9
7|3  5  5  7  7  9
8|0  1  3  3  4  6  6  7  8  9
```

8. $\bar{x} = \dfrac{480}{6} = 80$

9. $\dfrac{79 + 83}{2} = 81$

10. None

11. $\dfrac{63 + 93}{2} = 78$

12. $93 - 63 = 30$

13.

x	$x - \bar{x}$	$(x - \bar{x})^2$
63	-17	289
76	-4	16
79	-1	1
83	3	9
86	6	36
93	13	169
	0	520

$$\dfrac{520}{5} = 104, \quad s = \sqrt{104} \approx 10.20$$

14. $\bar{x} = \dfrac{156}{12} = 13$

15. $\dfrac{12+14}{2} = 13$

16. 12 and 7

17. $\dfrac{4+23}{2} = 13.5$

18. $23 - 4 = 19$

19.

x	$x - \bar{x}$	$(x-\bar{x})^2$
4	-9	81
5	-8	64
7	-6	36
7	-6	36
12	-1	1
12	-1	1
14	1	1
15	2	4
17	4	16
19	6	36
21	8	64
23	10	100
	0	440

$\dfrac{440}{11} = 40, \; s = \sqrt{40} \approx 6.32$

20. $z_{37} = \dfrac{37-42}{5} = \dfrac{-5}{5} = -1.00$

$z_{47} = \dfrac{47-42}{5} = \dfrac{5}{5} = 1.00$

$0.341 + 0.341 = 0.682 = 68.2\%$

21. $z_{32} = \dfrac{32-42}{5} = \dfrac{-10}{5} = -2.00$

$z_{52} = \dfrac{52-42}{5} = \dfrac{10}{5} = 2.00$

$0.477 + 0.477 = 0.954 = 95.4\%$

22. $z_{50} = \dfrac{50-42}{5} = \dfrac{8}{5} = 1.60$

$0.500 + 0.445 = 0.945 = 94.5\%$

23. $z_{50} = \dfrac{50-42}{5} = \dfrac{8}{5} = 1.60$

$0.500 - 0.445 = 0.055 = 5.5\%$

24. $z_{39} = \dfrac{39-42}{5} = \dfrac{-3}{5} = -.60$

$0.500 + 0.226 = 0.726 = 72.6\%$

25. $z_{20} = \dfrac{20-20}{5} = \dfrac{0}{5} = 0$

$z_{25} = \dfrac{25-20}{5} = \dfrac{5}{5} = 1.00$

$0.341 = 34.1\%$

26. $z_{18} = \dfrac{18-20}{5} = \dfrac{-2}{5} = -0.40$

$0.500 - 0.155 = 0.345 = 34.5\%$

27. $z_{22} = \dfrac{22-20}{5} = \dfrac{2}{5} = 0.40$

$z_{28} = \dfrac{28-20}{5} = \dfrac{8}{5} = 1.60$

$0.445 - 0.155 = 0.29 = 29.0\%$

28. $z_{30} = \dfrac{30-20}{5} = \dfrac{10}{5} = 2.00$

$0.500 - 0.477 = 0.023 = 2.3\%$

29. a)

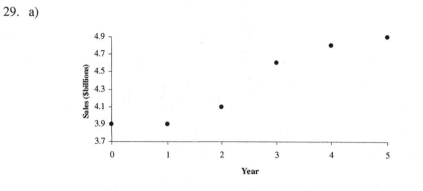

b) Yes; positive because generally as the year increases, the sales increase.

c)

x	y	x^2	y^2	xy
0	3.9	0	15.21	0
1	3.9	1	15.21	3.9
2	4.1	4	16.81	8.2
3	4.6	9	21.16	13.8
4	4.8	16	23.04	19.2
5	4.9	25	24.01	24.5
15	26.2	55	115.44	69.6

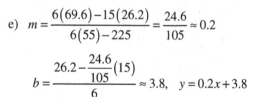

$$r = \frac{6(69.6)-15(26.2)}{\sqrt{6(55)-225}\sqrt{6(115.44)-686.44}} = \frac{24.6}{\sqrt{105}\sqrt{6.2}} \approx 0.964$$

d) Yes, $\mid 0.964 \mid > 0.811$

e) $m = \dfrac{6(69.6)-15(26.2)}{6(55)-225} = \dfrac{24.6}{105} \approx 0.2$

$b = \dfrac{26.2 - \dfrac{24.6}{105}(15)}{6} \approx 3.8, \quad y = 0.2x + 3.8$

30. a)

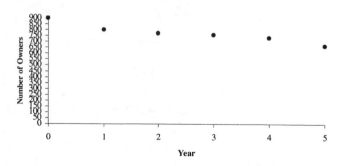

b) Yes; negative because generally as the year increases, the number of owners decreases.

30. c)

x	y	x^2	y^2	xy
0	897	0	804,609	0
1	800	1	640,000	800
2	770	4	592,900	1540
3	760	9	577,600	2280
4	735	16	540,225	2940
5	663	25	439,569	3315
15	4625	55	3,594,903	10,875

$$r = \frac{6(10,875)-15(4625)}{\sqrt{6(55)-225}\sqrt{6(3,594,903)-21,390,625}} = \frac{-4125}{\sqrt{105}\sqrt{178,793}} \approx -0.952$$

d) Yes, $|-0.952| > 0.811$

e) $m = \dfrac{6(10,875)-15(4625)}{6(55)-225} = \dfrac{-4125}{105} \approx -39.3$

$b = \dfrac{4625 - \dfrac{-4125}{105}(15)}{6} \approx 869.0, \quad y = -39.3x + 869.0$

31. a)

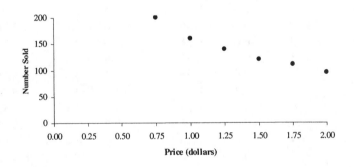

b) Yes; negative because generally as the price increases, the number sold decreases.

c)

x	y	x^2	y^2	xy
0.75	200	0.5625	40,000	150
1.00	160	1	25,600	160
1.25	140	1.5625	19,600	175
1.50	120	2.25	14,400	180
1.75	110	3.0625	12,100	192.5
2.00	95	4	9025	190
8.25	825	12.4375	120,725	1047.5

$$r = \frac{6(1047.5)-8.25(825)}{\sqrt{6(12.4375)-68.0625}\sqrt{6(120,725)-680,625}} = \frac{-521.25}{\sqrt{6.5625}\sqrt{43,725}} \approx -0.973$$

d) Yes, $|-0.973| > 0.811$

31. e) $m = \dfrac{6(1047.5) - 8.25(825)}{6(12.4375) - 68.0625} = \dfrac{-521.25}{6.5625} \approx -79.4$

$b = \dfrac{825 - \dfrac{-521.25}{6.5625}(8.25)}{6} \approx 246.7, \quad y = -79.4x + 246.7$

f) $y = -79.4(1.60) + 246.7 = 119.66 \approx 120$ sold

32. Mode = 175 lb
33. Median = 180 lb
34. 25%
35. 25%
36 100% - 86% = 14%
37. 100(187) = 18,700 lb
38. 187 + 2(23) = 233 lb
39. 187 - 1.8(23) = 145.6 lb
40. $\overline{x} = \dfrac{150}{42} \approx 3.57$
41. 2
42. $\dfrac{3+3}{2} = 3$
43. $\dfrac{0+14}{2} = 7$
44. 14 - 0 = 14

45.

x	$x - \overline{x}$	$(x - \overline{x})^2$	x	$x - \overline{x}$	$(x - \overline{x})^2$	x	$x - \overline{x}$	$(x - \overline{x})^2$
0	-3.6	12.96	2	-1.6	2.56	4	0.4	0.16
0	-3.6	12.96	2	-1.6	2.56	5	1.4	1.96
0	-3.6	12.96	3	-0.6	0.36	5	1.4	1.96
0	-3.6	12.96	3	-0.6	0.36	5	1.4	1.96
0	-3.6	12.96	3	-0.6	0.36	6	2.4	5.76
0	-3.6	12.96	3	-0.6	0.36	6	2.4	5.76
1	-2.6	6.76	3	-0.6	0.36	6	2.4	5.76
1	-2.6	6.76	3	-0.6	0.36	6	2.4	5.76
2	-1.6	2.56	4	0.4	0.16	6	2.4	5.76
2	-1.6	2.56	4	0.4	0.16	7	3.4	11.56
2	-1.6	2.56	4	0.4	0.16	8	4.4	19.36
2	-1.6	2.56	4	0.4	0.16	10	6.4	40.96
2	-1.6	2.56	4	0.4	0.16	14	10.4	108.16
2	-1.6	2.56	4	0.4	0.16			332.32
2	-1.6	2.56						

$\dfrac{332.32}{41} \approx 8.105, \quad s = \sqrt{8.105} \approx 2.85$

46.

# of Child.	# of Presidents
0 - 1	8
2 - 3	15
4 - 5	10
6 - 7	6
8 - 9	1
10 - 11	1
12 - 13	0
14 - 15	1

47. and 48.

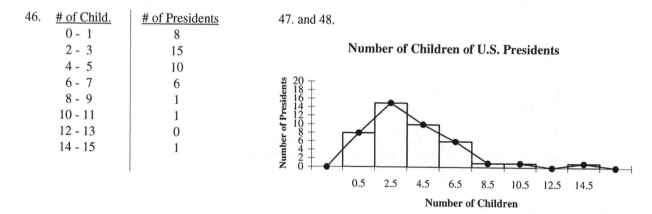

Number of Children of U.S. Presidents

49. No, it is skewed to the right.

50. No, some families have no children, more have one child, the greatest percent may have two children, fewer have three children, etc.

51. No, the number of children per family has decreased over the years.

Chapter Test

1. $\bar{x} = \dfrac{180}{5} = 36$

2. 37

3. 37

4. $\dfrac{21+46}{2} = 33.5$

5. $46 - 21 = 25$

6.

x	$x - \bar{x}$	$(x - \bar{x})^2$
21	-15	225
37	1	1
37	1	1
39	3	9
46	10	100
	0	336

$\dfrac{336}{4} = 84, \ s = \sqrt{84} \approx 9.17$

7.

Class	Frequency
25 - 30	7
31 - 36	5
37 - 42	1
43 - 48	7
49 - 54	5
55 - 60	3
61 - 66	2

8. and 9.

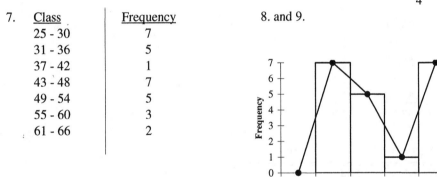

10. Mode = $695

11. Median = $670

12. 100% - 25% = 75%

13. 79%

14. 100(700) = $70,000

15. $700 + 1($40) = $740

16. $700 - 1.5($40) = $640

17. $z_{50,000} = \dfrac{50,000 - 75,000}{12,000} = \dfrac{-25,000}{12,000} \approx -2.08$

$z_{70,000} = \dfrac{70,000 - 75,000}{12,000} = \dfrac{-5000}{12,000} \approx -0.42$

$0.481 - 0.163 = 0.318 = 31.8\%$

18. $z_{60,000} = \dfrac{60,000 - 75,000}{12,000} = \dfrac{-15,000}{12,000} = -1.25$

$0.500 + 0.394 = 0.894 = 89.4\%$

19. $z_{90,000} = \dfrac{90,000 - 75,000}{12,000} = \dfrac{15,000}{12,000} = 1.25$

$0.500 - 0.394 = 0.106 = 10.6\%$

20. From #17 and #18,

$z_{60,000} = -1.25$ and $z_{70,000} \approx -0.42$

$0.394 - 0.163 = 0.231 = 23.1\%$

$0.231(300) = 69.3 \approx 69$ cars

21. a)

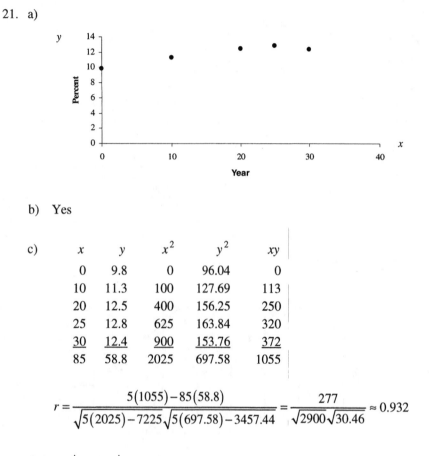

b) Yes

c)

x	y	x^2	y^2	xy
0	9.8	0	96.04	0
10	11.3	100	127.69	113
20	12.5	400	156.25	250
25	12.8	625	163.84	320
30	12.4	900	153.76	372
85	58.8	2025	697.58	1055

$$r = \frac{5(1055) - 85(58.8)}{\sqrt{5(2025) - 7225}\sqrt{5(697.58) - 3457.44}} = \frac{277}{\sqrt{2900}\sqrt{30.46}} \approx 0.932$$

d) Yes, $|\,0.932\,| > 0.878$

e) $m = \dfrac{5(1055) - 85(58.8)}{5(2025) - 7225} = \dfrac{277}{2900} \approx 0.1$

$$b = \frac{58.8 - \dfrac{277}{2900}(85)}{5} \approx 10.1, \quad y = 0.1x + 10.1$$

f) $y = 0.1(40) + 10.1 = 14.1\%$

Group Projects

1. a) – j) Answers will vary.
2. a) – g) Answers will vary.

CHAPTER FOURTEEN

GRAPH THEORY

1. A **graph** is a finite set of points, called **vertices**, that are connected with line segments, called **edges**.

2.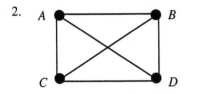

3.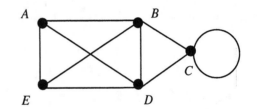

4. The **degree** of a vertex is the number of edges that connect to that vertex.

5. If the number of edges connected to the vertex is even, the vertex is **even**. If the number of edges connected to the vertex is odd, the vertex is **odd**.

6. Answers will vary. In the following graph, the edge *EF* is a bridge because if it were removed from the graph, the result would be a disconnected graph (i.e., there would be no path from vertices *A, B, E, H,* and *G* to vertices *C, D, J, I,* and *F*).

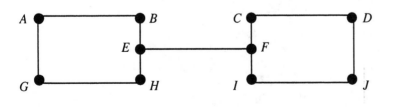

7. a) A **path** is a sequence of adjacent vertices and the edges connecting them.
   b) A **circuit** is a path that begins and ends at the same vertex.
   c)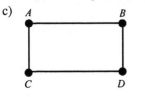

   The path *A, B, D, C* is a path that is not a circuit.
   The path *A, B, D, C, A* is a path that is also a circuit.

8. Answers will vary. In the graphs below, the graph on the right is disconnected since no path connects vertices
    A, *D*, and *E* to vertices *B* and *C*.

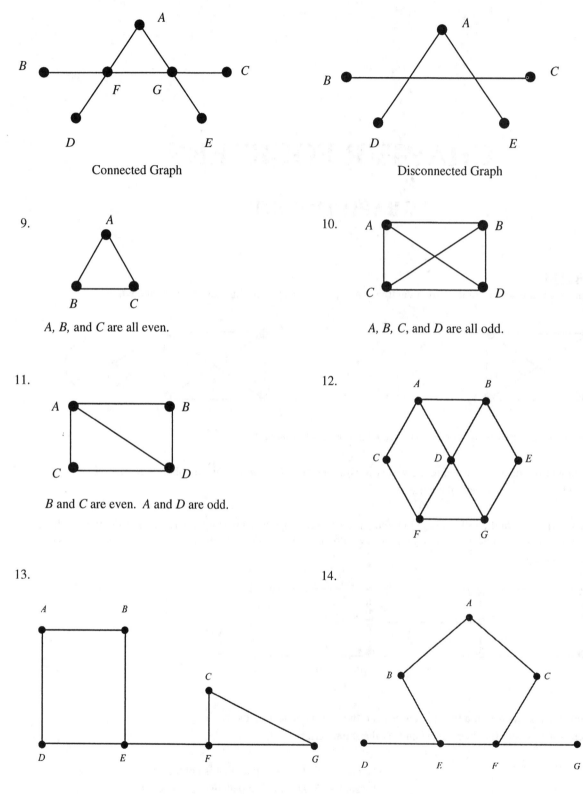

Connected Graph   Disconnected Graph

9.

A, *B*, and *C* are all even.

10.

A, *B*, *C*, and *D* are all odd.

11.

B and *C* are even. *A* and *D* are odd.

12.

13.

14.

15. No. There is no edge connecting vertices *B* and *C*. Therefore, *A, B, C, D, E* is not a path.

16. Edge *AC* (or *CA*) and edge *CD* (or *DC*)

17. Yes. One example is *A, C, E, D, B.*

18. Yes. One example is *C, D, E, C.*

19. Yes. One example is *C, A, B, D, E, C, D.*

20. Yes. One example is *A, B, D, E, C, A.*

21.
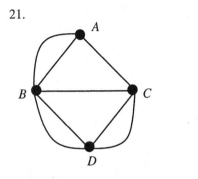

22.
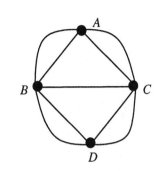

23.
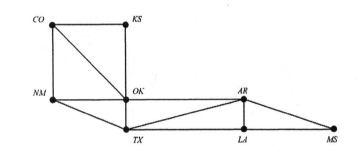

24.
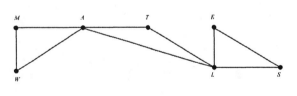

25.

26.
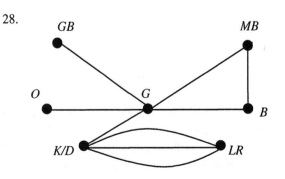

27.

28.
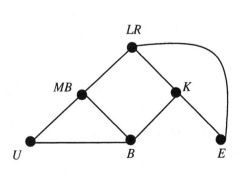

29.

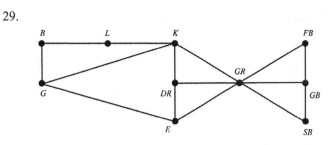

30.

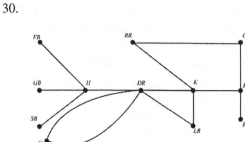

31.

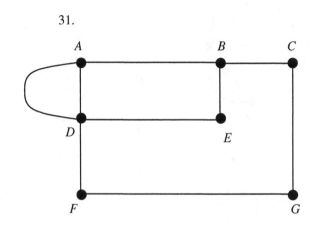

32.

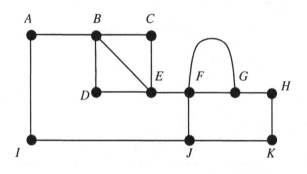

33. Disconnected. There is no path that connects
   A to *C*.

34. Connected

35. Connected

36. Disconnected. There is no path that
   connects *A* to *B*.

37. Edge *AB*

38. Edge *EF*

39. Edge *EF*

40. Edge *FK* and edge *HL*

41.

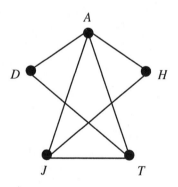

Other answers are possible.

42. Answers will vary.

43. It is impossible to have a graph with an odd number of odd vertices.

44. a) - c) Answers will vary.
   d) The sum of the degrees is equal to twice the number of edges. This is true since each edge
   must connect two vertices. Each edge then contributes two to the sum of the degrees.

45. a) and b) Answers will vary.

Exercise Set 14.2

1. a) An **Euler path** is a path that must include each edge of a graph exactly one time.
    b) and c)

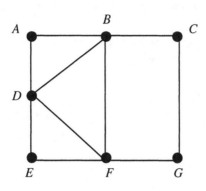

b) The path *A, B, E, D, C, A, D, B* is an Euler path.
c) The path *A, B, E, D, C* is a path that is not an Euler path.

2. a) An **Euler circuit** is a circuit that must include each edge of a graph exactly one time and return to the original vertex.
    b) and c)

b) The circuit *A, B, C, G, F, B, D, F, E, D, A* is an Euler circuit.
c) The path *A, B, C, G, F, E, D, A* is a circuit but not an Euler circuit.

3. a) Yes, according to Euler's Theorem.
    b) Yes, according to Euler's Theorem.
    c) No, according to Euler's Theorem.

4. a) Yes, according to Euler's Theorem.
    b) No, according to Euler's Theorem.
    c) No, according to Euler's Theorem.

5. If all of the vertices are even, the graph has an Euler circuit.

6. a) If all the vertices are even, then start with any vertex. If there are two odd vertices, then start with one of the odd vertices. Move from vertex to vertex without tracing any bridges until you have traced each edge of the graph exactly one time. You will finish at the other odd vertex.

    b) If there are any odd vertices, then there is no Euler circuit. If there are all even vertices, then start with any vertex. Move from vertex to vertex without tracing any bridges until you have traced each edge of the graph exactly one time. You will finish at the vertex you started from.

7. *A, B, C, D, E, B, E, D, A, C*; other answers are possible.

8. *C, A, B, E, D, C, B, E, D, A*; other answers are possible.

9. No. This graph has exactly two odd vertices. Each Euler path must begin with an odd vertex. *B* is an even vertex.

10. No. A graph with exactly two odd vertices has no Euler circuits.

11. *A, B, A, C, B, E, C, D, A, D, E*; other answers are possible.

12. *E, D, A, B, E, C, D, A, B, C, A*; other answers are possible.

13. No. A graph with exactly two odd vertices has no Euler circuits.

14. No. This graph has exactly two odd vertices. Each Euler path must begin with an odd vertex. *C* is an even vertex.

15. *A, B, C, E, F, D, E, B, D, A*; other answers are possible.

16. *B, D, F, E, B, C, E, D, A, B*; other answers are possible.

17. *C, B, A, D, F, E, D, B, E, C*; other answers are possible.

18. *D, A, B, C, E, B, D, E, F, D*; other answers are possible.

19. *E, F, D, E, B, D, A, B, C, E*; other answers are possible.

20. *F, D, E, C, B, A, D, B, E, F*; other answers are possible.

21. a) Yes. There are zero odd vertices.
    b) Yes. There are zero odd vertices.

22. a) Yes. There are two or fewer odd vertices.
    b) No. There are more than zero odd vertices.

23. a) No. There are more than two odd vertices.
    b) No. There are more than zero odd vertices.

24. a) No. There are more than two odd vertices.
    b) No. There are more than zero odd vertices.

25. a) Yes. Each island would correspond to an odd vertex. According to item 2 of Euler's Theorem, a graph with exactly two odd vertices has at least one Euler path, but no Euler circuit.
    b) They could start on either island and finish at the other.

26. a) Yes. The land at the top and the island on the left would each correspond to an odd vertex. According to item 2 of Euler's Theorem, a graph with exactly two odd vertices has at least one Euler path, but no Euler circuits.
    b) They could start either on the land at the top of the picture or on the island on the left. If they started on the island, then they would end on the land at the top, and vice versa.

In Exercises 27-32, one graph is shown. Other graphs are possible.

27. a)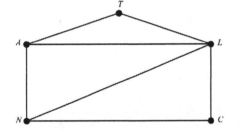

    b) Vertices *A* and *N* are both odd. According to item 2 of Euler's Theorem, since there are exactly two odd vertices, at least one Euler path, but no Euler circuits exist.
    Yes; *A, T, L, C, N, L, A, N*
    c) No. (See part b) above.)

28. a)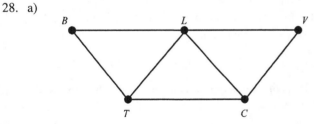

    b) Vertices *T* and *C* are both odd. According to item 2 of Euler's Theorem, since there are exactly two odd vertices, at least one Euler path, but no Euler circuits exist.
    Yes; *T, B, L, V, C, L, T, C*
    c) No. (See part b) above.)

29. a)

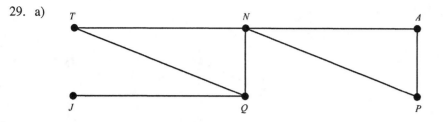

b) Vertices *J* and *Q* are both odd. According to item 2 of Euler's Theorem, since there are exactly two odd vertices, at least one Euler path, but no Euler circuits exist.

Yes; *J, Q, T, N, A, P, N, Q*

c) No. (See part b) above.)

30. a)

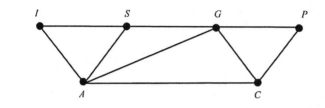

b) Vertices *S* and *C* are both odd. According to item 2 of Euler's Theorem, since there are exactly two odd vertices, at least one Euler path, but no Euler circuits exist.

Yes; *S, I, A, S, G, A, C, G, P, C*

c) No. (See part b) above.)

31. a)

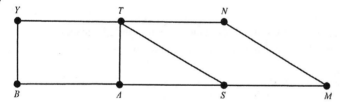

b) Vertices *A* and *S* are both odd. According to item 2 of Euler's Theorem, since there are exactly two odd vertices, at least one Euler path, but no Euler circuits exist.

Yes; *A, S, M, N, T, Y, B, A, T, S*

c) No. (See part b) above.)

32. a)

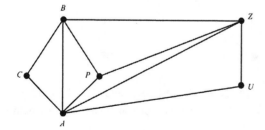

b) Vertices *A* and *P* are both odd. According to item 2 of Euler's Theorem, since there are exactly two odd vertices, at least one Euler path, but no Euler circuits exist.

Yes; *P, B, Z, P, A, B, C, A, U, Z, A*

c) No. (See part b) above.)

33. a) No. The graph representing the floor plan:

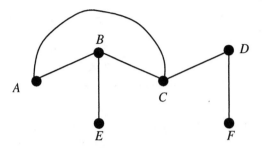

The wood carver is seeking an Euler path or an Euler circuit. Note that vertices *B, C, E,* and *F* are all odd. According to item 3 of Euler's Theorem, since there are more than two odd vertices, no Euler path or Euler circuit can exist.

b) No such path exists.

34. a) Yes. The graph representing the floor plan:

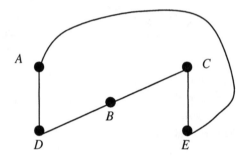

The wood carver is seeking an Euler path or an Euler circuit. Note that there are no odd vertices. According to item 1 of Euler's Theorem, since there are no odd vertices, at least one Euler path (which is also an Euler circuit) must exist.

b) One path (which is also a circuit) is *A, D, B, C, E, A.*

35. a) Yes. The graph representing the floor plan:

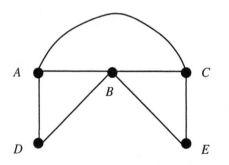

The wood carver is seeking an Euler path or an Euler circuit. Note that vertices *A* and *C* are both odd. According to item 2 of Euler's Theorem, since there are exactly two odd vertices, at least one Euler path, but no Euler circuits exist.

b) One path is *A, D, B, E, C, B, A, C.*

36. a) Yes. The graph representing the floor plan:

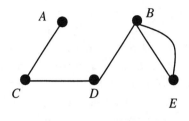

The wood carver is seeking an Euler path or an Euler circuit. Note that there are no odd vertices.
According to item 1 of Euler's Theorem, since there are no odd vertices, at least one Euler path (which is
also an Euler circuit) must exist.
b) One path is *A, C, D, B, E, B*.

37. a) Yes. The graph representing the map:

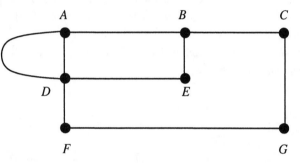

They are seeking an Euler path or an Euler circuit. Note that vertices *A* and *B* are both odd. According
to item 2 of Euler's Theorem, since there are exactly two odd vertices, at least one Euler path, but no
Euler circuits exist.
b) The residents would need to start at the intersection of Maple Cir., Walnut St., and Willow St. or at the
intersection of Walnut St. and Oak St.

38. a) Yes. The graph representing the map:

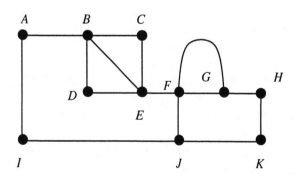

They are seeking an Euler path or an Euler circuit. Note that vertices *G* and *J* are both odd. According
to item 2 of Euler's Theorem, since there are exactly two odd vertices, at least one Euler path, but no
Euler circuits exist.
b) The residents would need to start at the intersection of Spring Blvd. and Lake St. or at the rightmost
intersection of Stream Cir. and Ocean Blvd.

39. *F, G, E, F, D, E, B, D, A, B, C, E*; other answers are possible.
40. *H, I, F, C, B, A, D, G, H, F, E, D, H, E, B*; other answers are possible.

41. *H, I, F, C, B, D, G, H, E, D, A, B, E, F*; other answers are possible.
42. *D, A, B, E, I, H, D, C, G, K, L, H, M, I, N, O, J, F, E*; other answers are possible.
43. *A, B, E, F, J, I, E, D, H, G, C, D, A*; other answers are possible.
44. *A, B, C, E, H, G, F, D, B, E, G, D, B, E, G, D, A*; other answers are possible.
45. *A, E, B, F, C, G, D, K, G, J, F, I, E, H, A*; other answers are possible.
46. *A, B, C, D, F, C, B, E, F, H, G, E, A*; other answers are possible.
47. *A, B, C, E, B, D, E, F, I, E, H, D, G, H, I, J, F, C, A*; other answers are possible.
48. *A, B, C, E, B, D, E, F, D, A, C, A*; other answers are possible.
49. *F, C, J, M, P, H, F, M, P*; other answers are possible.
50. *B, A, E, H, I, J, K, D, C, G, G, J, F, C, B, F, I, E, B*; other answers are possible.
51. *B, E, I, F, B, C, F, J, G, G, C, D, K, J, I, H, E, A, B*; other answers are possibe.
52. *J, G, G, C, F, J, K, D, C, B, F, I, E, B, A, E, H, I, J*; other answers are possible.
53. *J, F, C, B, F, I, E, B, A, E, H, I, J, G, G, C, D, K, J*; other answers are possible.
54. a) No.
    b) California, Nevada, and Louisiana (and others) have an odd number of states bordering them. Since a graph of the United States would have more than two odd vertices, no Euler path and no Euler circuit exist.
55. It is not possible to draw a graph with an Euler circuit that has a bridge. Therefore, a graph with an Euler circuit has no bridge.
56. a)   b)   c)

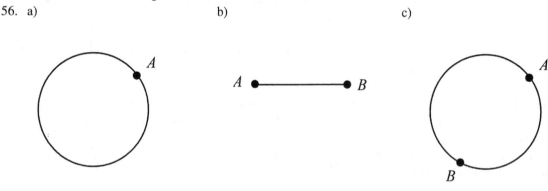

Exercise Set 14.3

1. a) A **Hamilton circuit** is a path that begins and ends with the same vertex and passes through all other vertices exactly one time.
   b) Both **Hamilton** and **Euler circuits** begin and end at the same vertex. A **Hamilton circuit** passes through all other *vertices* exactly once, while an **Euler circuit** passes through each *edge* exactly once.
2. a) A **Hamilton path** is a path that passes through each vertex exactly one time.
   b) A **Hamilton path** passes through each *vertex* exactly once; an **Euler path** passes through each *edge* exactly once.
3. a) A **weighted graph** is a graph with a number, or weight, assigned to each edge.
   b) A **complete graph** is a graph in which there is an edge between each pair of vertices.
   c) A **complete, weighted graph** is a graph in which there is an edge between each pair of vertices and each edge has a number, or weight, assigned to it.
4. a) The **factorial** of a number is computed by multiplying the given number by each natural number less than the given number.
   b) $7! = 7 \cdot 6 \cdot 5 \cdot 4 \cdot 3 \cdot 2 \cdot 1 = 5040$
   c) $8! = 8 \cdot 7 \cdot 6 \cdot 5 \cdot 4 \cdot 3 \cdot 2 \cdot 1 = 40,320$
   d) $10! = 10 \cdot 9 \cdot 8 \cdot 7 \cdot 6 \cdot 5 \cdot 4 \cdot 3 \cdot 2 \cdot 1 = 3,628,800$

5. a) The number of unique Hamilton circuits in a complete graph with n vertices is found by computing $(n-1)!$

b) $n = 4; (n-1)! = (4-1)! = 3! = 3 \cdot 2 \cdot 1 = 6$

c) $n = 9; (n-1)! = (9-1)! = 8! = 8 \cdot 7 \cdot 6 \cdot 5 \cdot 4 \cdot 3 \cdot 2 \cdot 1 = 40,320$

6. The **optimal solution** to a traveling salesman problem is the least expensive or shortest way to visit each location exactly one time and return to the starting location.

7. To find the optimal solution using the **Brute Force method**, write down all possible Hamilton circuits and then compute the cost or distance associated with each Hamilton circuit. The one with the lowest cost or shortest distance is the optimal solution to the traveling salesman problem.

8. Starting from your current position, choose the cheapest or shortest route to get to the next location. From there choose the cheapest or shortest route to a location you have not already visited. Continue this process until you have visited each location. The path found is the path found using the **Nearest Neighbor method** for approximating the optimal solution.

9. *A, B, C, G, F, E, D* and *E, D, A, B, F, G, C*; other answers are possible.

10. *F, B, C, A, D, E, G* and *E, G, D, A, C, F, B*; other answers are possible.

11. *A, B, C, D, G, F, E, H* and *E, H, F, G, D, C, A, B*; other answers are possible.

12. *A, B, C, D, H, G, F, E, I, J, K, L* and *A, E, I, J, F, B, C, G, K, L, H, D*; other answers are possible.

13. *A, B, C, E, D, F, G, H* and *F, G, H, E, D, A, B, C*; other answers are possible.

14. *A, D, F, G, H, E, B, C, I* and *I, C, B, A, D, E, F, H, G*; other answers are possible.

15. *A, B, D, E, G, F, C, A* and *A, C, F, G, E, D, B, A*; other answers are possible.

16. *A, B, C, D, H, L, K, G, F, J, I, E, A* and *A, E, I, J, K, L, H, D, C, G, F, B, A*; other answers are possible.

17. *A, B, C, F, I, E, H, G, D, A* and *A, E, B, C, F, I, H, G, D, A*; other answers are possible.

18. *A, B, F, G, H, I, E, D, C, A* and *A, C, D, E, I, H, G, F, B, A*; other answers are possible.

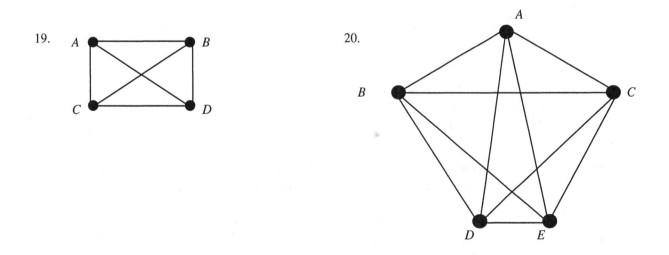

19.

20.

21. The number of unique Hamilton circuits within the complete graph with eight vertices representing this situation is $(8-1)! = 7! = 7 \cdot 6 \cdot 5 \cdot 4 \cdot 3 \cdot 2 \cdot 1 = 5040$ ways

22. The number of unique Hamilton circuits within the complete graph with thirteen vertices representing this situation is $(13-1)! = 12! = 12 \cdot 11 \cdot 10 \cdot 9 \cdot 8 \cdot 7 \cdot 6 \cdot 5 \cdot 4 \cdot 3 \cdot 2 \cdot 1 = 479,001,600$ ways

23. The number of unique Hamilton circuits within the complete graph with eleven vertices representing this situation is $(11-1)! = 10! = 10 \cdot 9 \cdot 8 \cdot 7 \cdot 6 \cdot 5 \cdot 4 \cdot 3 \cdot 2 \cdot 1 = 3,628,800$ ways

   (The vertices are the 10 different farms he has to visit and his starting point.)

24. The number of unique Hamilton circuits within the complete graph with twelve vertices representing this situation is $(12-1)! = 11! = 11 \cdot 10 \cdot 9 \cdot 8 \cdot 7 \cdot 6 \cdot 5 \cdot 4 \cdot 3 \cdot 2 \cdot 1 = 39,916,800$ ways

In Exercises 25-32, other graphs are possible.

25. a)

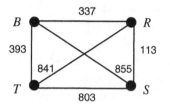

b)

Hamilton Circuit	First Leg/Cost	Second Leg/Cost	Third Leg/Cost	Fourth Leg/Cost	Total Cost
S, R, B, T, S	113	337	393	803	$1646
S, R, T, B, S	113	841	393	855	$2202
S, T, B, R, S	803	393	337	113	$1646
S, T, R, B, S	803	841	337	855	$2836
S, B, R, T, S	855	337	841	803	$2836
S, B, T, R, S	855	393	841	113	$2202

   The least expensive route is S, R, B, T, S or S, T, B, R, S

c) $1646

26. a)

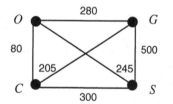

b)

Hamilton Circuit	First Leg/Distance	Second Leg/Distance	Third Leg/Distance	Fourth Leg/Distance	Total Distance
C, O, G, S, C	80	280	500	300	1160 miles
C, O, S, G, C	80	245	500	205	1030 miles
C, G, O, S, C	205	280	245	300	1030 miles
C, G, S, O, C	205	500	245	80	1030 miles
C, S, G, O, C	300	500	280	80	1160 miles
C, S, O, G, C	300	245	280	205	1030 miles

   The shortest route is C, O, S, G, C or C, G, O, S, C or C, G, S, O, C or C, S, O, G, C

c) 1030 miles

27. a)

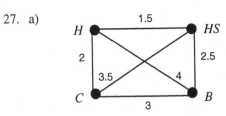

b)

Hamilton Circuit	First Leg/Distance	Second Leg/Distance	Third Leg/Distance	Fourth Leg/Distance	Total Distance
H, HS, B, C, H	1.5	2.5	3	2	9 miles
H, HS, C, B, H	1.5	3.5	3	4	12 miles
H, B, HS, C, H	4	2.5	3.5	2	12 miles
H, B, C, HS, H	4	3	3.5	1.5	12 miles
H, C, HS, B, H	2	3.5	2.5	4	12 miles
H, C, B, HS, H	2	3	2.5	1.5	9 miles

The shortest route *is H, HS, B, C, H* or *H, C, B, HS, H*

c) 9 miles

28. a)

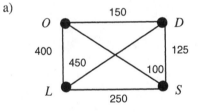

b)

Hamilton Circuit	First Leg/Distance	Second Leg/Distance	Third Leg/Distance	Fourth Leg/Distance	Total Distance
O, D, S, L, O	150	125	250	400	925 feet
O, D, L, S, O	150	450	250	100	950 feet
O, L, S, D, O	400	250	125	150	925 feet
O, L, D, S, O	400	450	125	100	1075 feet
O, S, D, L, O	100	125	450	400	1075 feet
O, S, L, D, O	100	250	450	150	950 feet

The shortest route is *O, D, S, L, O* or *O, L, S, D, O*

c) 925 feet

29. a)

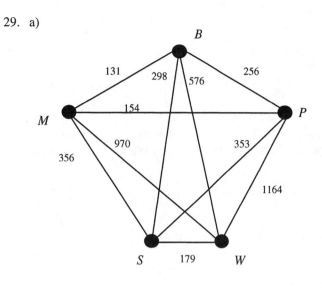

b) *B, M, P, S, W, B* for 131 + 154 + 353 + 179 + 576 = $1393
c) Answers will vary.

30. a)

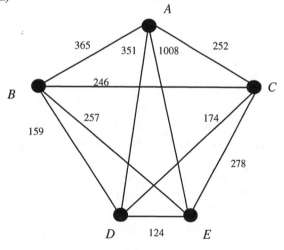

b) *A, C, D, E, B, A* for 252 + 174 + 124 + 257 + 365 = $1172
c) Answers will vary.

31. a)

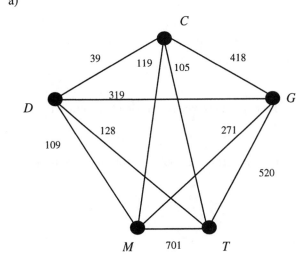

b) *C, D, M, G, T, C* for 39 + 109 + 271 + 520 + 105 = $1044
c) Answers will vary.

32. a)

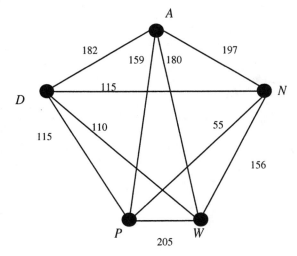

b) *N, P, D, W, A, N* for 55 + 115 + 110 + 180 + 197 = $657
c) Answers will vary.

33. a) – d) Answers will vary.

34. a)

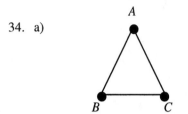

There are two choices for moving to the second vertex. There is one choice for moving to a third vertex.

$2(1) = 2$

$(3 - 1)! = 2! = 2(1) = 2$

The number obtained is the same as the number of Hamilton circuits in a complete graph with 3 vertices.

b)

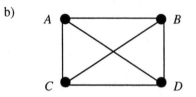

There are three choices for moving to the second vertex. There are two choices for moving to the third vertex. There is one choice for moving to the fourth vertex.

$3(2)(1) = 6$

$(4 - 1)! = 3! = 3(2)(1) = 6$

The number obtained is the same as the number of Hamilton circuits in a complete graph with 4 vertices.

c)

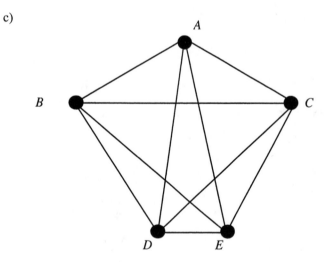

There are four choices for moving to the second vertex. There are three choices for moving to the third vertex. There are two choices for moving to the fourth vertex. There is one choice for moving to the fifth vertex.

$4(3)(2)(1) = 24$

$(5 - 1)! = 4! = 4(3)(2)(1) = 24$

The number obtained is the same as the number of Hamilton circuits in a complete graph with 5 vertices.

34. c)

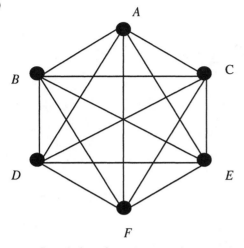

There are five choices for moving to the second vertex. There are four choices for moving to the third vertex. There are three choices for moving to the fourth vertex. There are two choices for moving to the fifth vertex. There is one choice for moving to the sixth vertex.

$5(4)(3)(2)(1) = 120$

$(6 - 1)! = 5! = 5(4)(3)(2)(1) = 120$

The number obtained is the same as the number of Hamilton circuits in a complete graph with 6 vertices.

d) When starting at a vertex in a complete graph with n vertices, you have $n - 1$ choices. At your second vertex, you have one less choice, or $n - 2$ choices. This process continues until you only have one vertex to choose from.

35. *A, E, D, N, O, F, G, Q, P, T, M, L, C, B, J, K, S, R, I, H, A*; other answers are possible.

Exercise Set 14.4

1. A **tree** is a connected graph in which each edge is a bridge.
2. In a tree, each edge is a bridge. In a graph that is not a tree, there is at least one edge that is not a bridge.
3. Yes, because removing the edge would create a disconnected graph.
4. A **spanning tree** is obtained by removing the edges of a graph one at a time, while maintaining a path to each vertex, until the graph is reduced to a tree.
5. A **minimum-cost spanning tree** is a spanning tree that has the lowest cost or shortest distance of all spanning trees for a given graph.
6. To find a minimum-cost spanning tree from a weighted graph, choose the edge with the smallest weight first. Continue to choose the edge with the smallest weight that does not lead to a circuit until a spanning tree is found.

7.

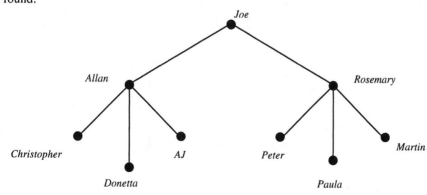

8.

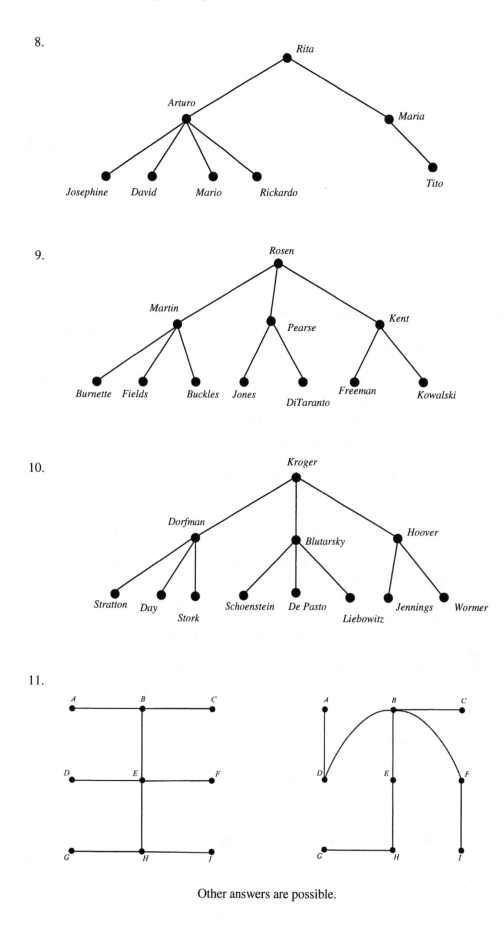

9.

10.

11.

Other answers are possible.

12.

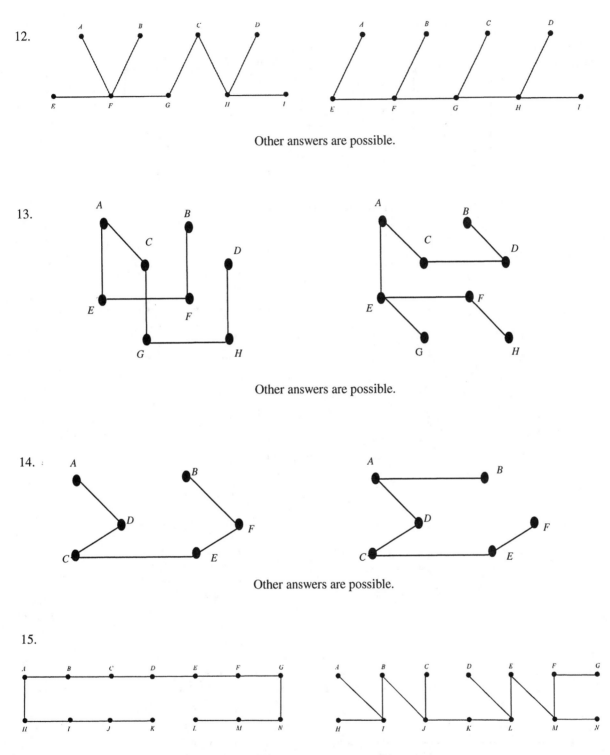

Other answers are possible.

13.

Other answers are possible.

14.

Other answers are possible.

15.

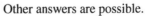

Other answers are possible.

16.

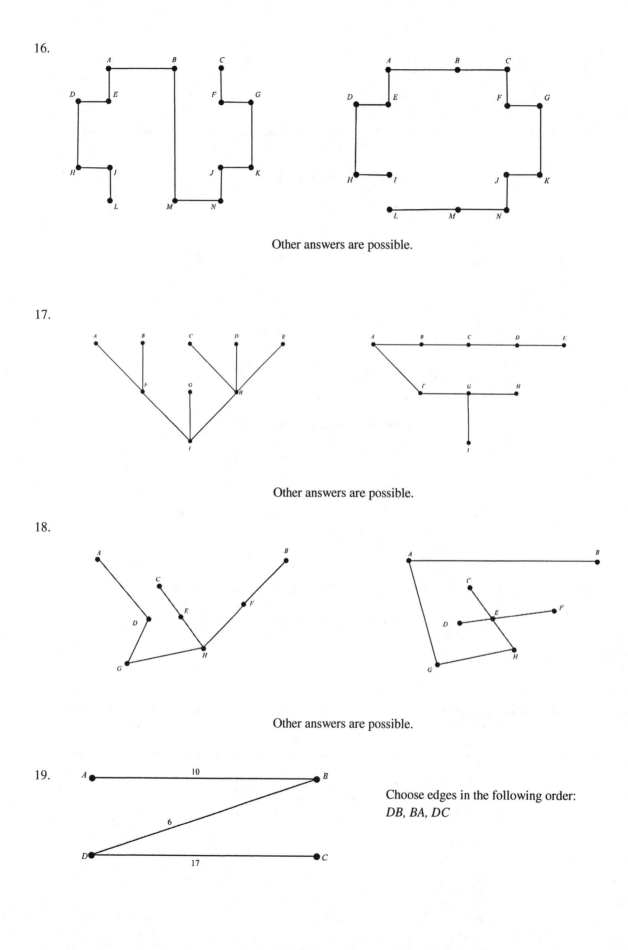

Other answers are possible.

17.

Other answers are possible.

18.

Other answers are possible.

19.

Choose edges in the following order:
DB, BA, DC

20.

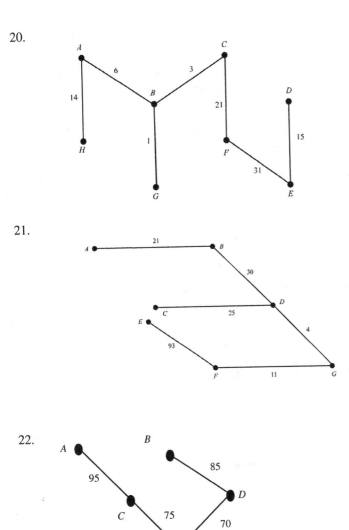

Choose edges in the following order:
GB, BC, BA, AH, DE, CF, FE

21.

Choose edges in the following order:
DG, GF, AB, CD, BD, EF

22.

Choose edges in the following order:
EF, FD, FC, BD, AC

23.

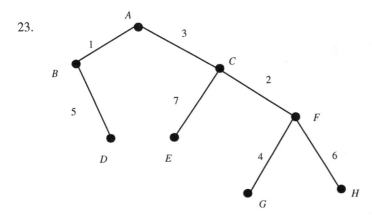

Choose edges in the following order:
AB, CF, AC, FG, BD, FH, EC

24.

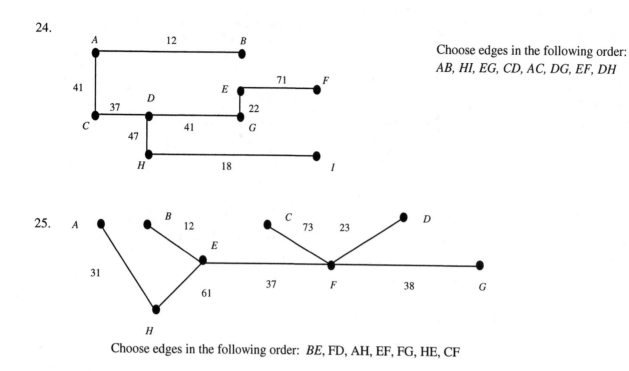

Choose edges in the following order:
AB, HI, EG, CD, AC, DG, EF, DH

25.

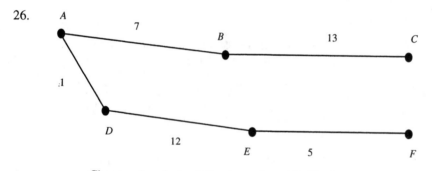

Choose edges in the following order: *BE*, FD, AH, EF, FG, HE, CF

26.

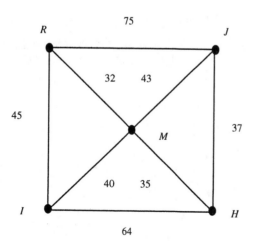

Choose edges in the following order: *AD, EF, AB, DE, BC*

27. a)

R   75   J
32   43
45
37
M
40   35
I   H
64

Other answers are possible.

27. b)

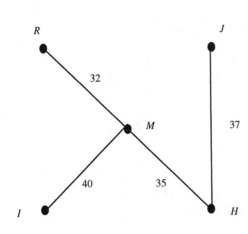

Choose edges in the following order:
RM, MH, JH, IM

c) $15(32 + 35 + 37 + 40) = 15(144) = \$2160$

28. a)

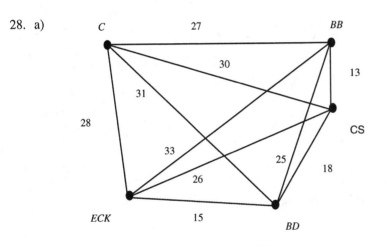

Other answers are possible.

b)

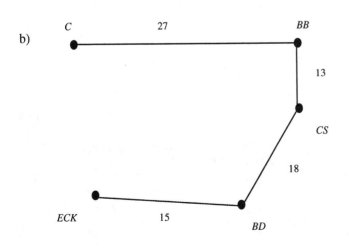

Choose edges in the following order:
BB CS, ECK BD, BD CS, C BB

c) $0.75(13 + 15 + 18 + 27) = 0.75(73) = \$54.75$

29. a)

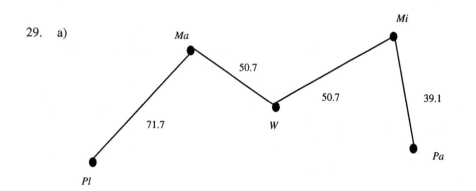

Choose edges in the following order: *Mi Pa, W Mi, Ma W, Ma Pl*

b) 895(39.1 + 50.7 + 50.7 + 71.7) = 895(212.2) = $189,919

30. a)

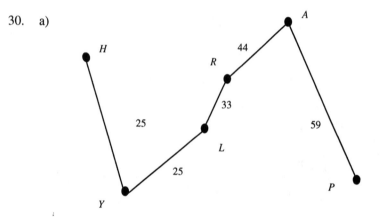

Choose edges in the following order: *HY, YL, LR, RA, AP*

b)   6800(25 + 25 + 33 + 44 + 59) = 6800(186) = $1,264,800

31. a)

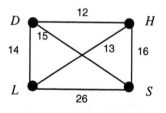

Other answers are possible.

b)

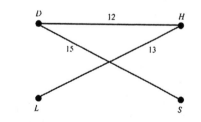

Choose edges in the following order:
DH, HL, DS

c) 3500 (12 + 13 + 15) = 3500 (40) = $140,000

32. a)

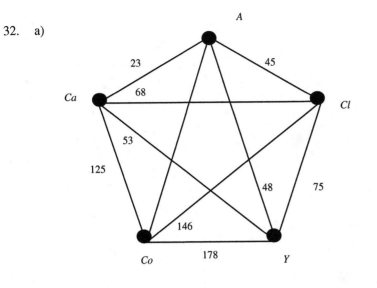

Other answers are possible.

b)

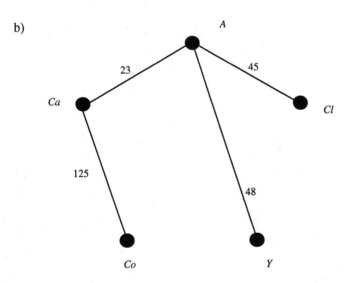

Choose edges in the following order:
ACa, AC, AY, CaCo

c) 2300(23 + 45 + 48 + 125) = 2300(241) = $554,300

33. a)

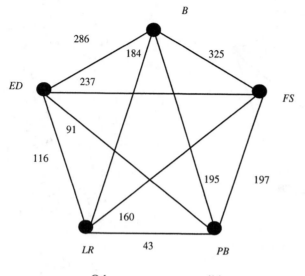

Other answers are possible.

b)

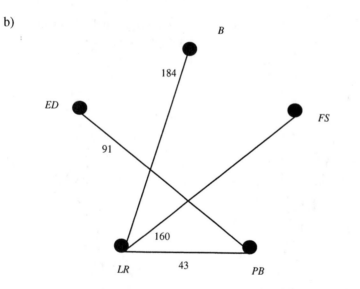

Choose edges in the following order: *LR PB, ED PB, LR FS, B LR*

c) 2500(43 + 91 + 160 + 184) = 2500(478) = $1,195,000

34. a)

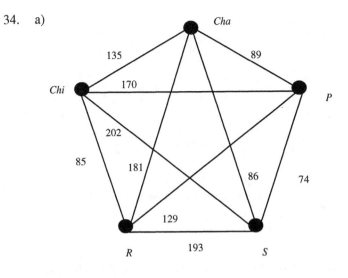

Other answers are possible.

b)

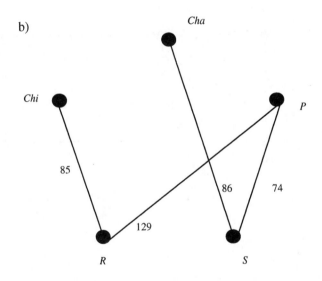

Choose edges in the following order: *PS, Chi R, Cha S, RP*

c) 74 + 85 + 86 + 129 = 374 miles

35. Answers will vary.
36. Answers will vary.
37. Answers will vary.
38. a) EULER
     b) FLEURY
     c) HAMILTON
     d) KRUSKAL

Review Exercises

1.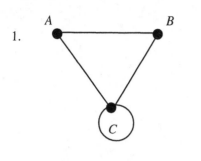

A, B, and C are all even. There is a
loop at vertex C.
Other answers are possible.

2.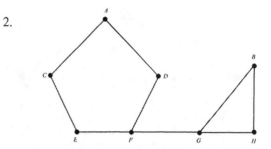

Edge *FG* is a bridge.
Other answers are possible.

3. *A, B, C, A, D, C, E, D*; other answers are possible.

4. No. To trace each edge in the graph with a path would require you to trace at least one edge twice (the graph has more than two odd vertices).

5.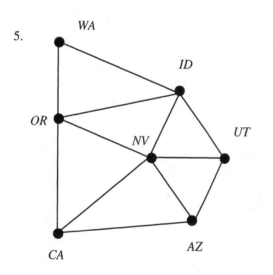

6.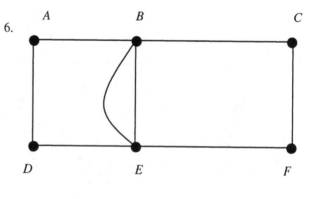

7. Connected

8. Disconnected. There is no path that connects *A* to *C*.

9. Edge *CD*

10. *C, B, A, F, E, D, C, G, B, A, G, E, D, G, F*; other answers are possible.

11. *F, E, G, F, A, G, D, E, D, C, B, A, B, G, C*; other answers are possible.

12. *B, C, A, D, F, E, C, D, E, B*; other answers are possible.

13. *E, F, D, E, C, D, A, C, B, E*; other answers are possible.

14. a) No. The graph representing the map:

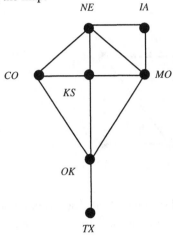

b) Vertices *CO* and *TX* are both odd. According to item 2 of Euler's Theorem, since there are exactly two odd vertices, at least one Euler path, but no Euler circuits exist.

Yes; *CO, NE, IA, MO, NE, KS, MO, OK, CO, KS, OK, TX*; other answers are possible.

c) No. (See part b) above.)

15. a) Yes. The graph representing the floor plan:

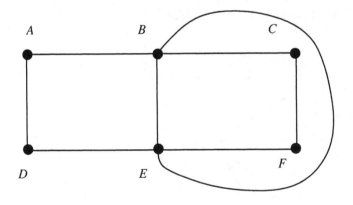

We are seeking an Euler path or an Euler circuit. Note that there are no odd vertices. According to item 1 of Euler's Theorem, since there are no odd vertices, at least one Euler path (which is also an Euler circuit) must exist.

b) The person may start in any room and will finish in the room where he or she started.

16. a) Yes. The graph representing the map:

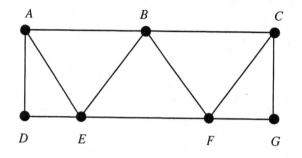

The officer is seeking an Euler path or an Euler circuit. Note that vertices *A* and *C* are both odd. According to item 2 of Euler's Theorem, since there are exactly two odd vertices, at least one Euler path but no Euler circuits exist.

b) The officer would have to start at either the upper left-hand corner or the upper right-hand corner. If the officer started in the upper left-hand corner, he or she would finish in the upper right-hand corner, and vice versa.

17. *A, B, F, A, E, F, G, C, D, G, H, D*; other answers are possible.
18. *A, B, C, D, H, G, C, F, G, B, F, E, A*; other answers are possible.
19. *A, C, B, F, E, D, G* and *A, C, D, G, F, B, E*; other answers are possible.
20. *A, B, C, D, F, E, A* and *A, E, F, B, C, D, A*; other answers are possible.

21.

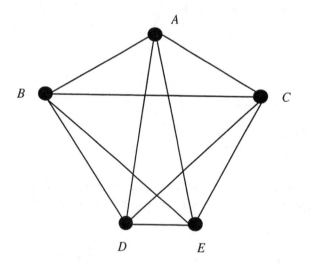

22. The number of unique Hamilton circuits within the complete graph with 5 vertices representing this situation is $(5-1)! = 4! = 4 \cdot 3 \cdot 2 \cdot 1 = 24$ ways

23. a)

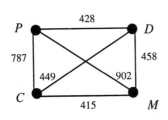

b)

Hamilton Circuit	First Leg/Cost	Second Leg/Cost	Third Leg/Cost	Fourth Leg/Cost	Total Cost
P, D, C, M, P	428	449	415	902	$2194
P, D, M, C, P	428	458	415	787	$2088
P, C, M, D, P	787	415	458	428	$2088
P, C, D, M, P	787	449	458	902	$2596
P, M, D, C, P	902	458	449	787	$2596
P, M, C, D, P	902	415	449	428	$2194

The least expensive route is *P, D, M, C, P* or *P, C, M, D, P*

c) $2088

24. a)

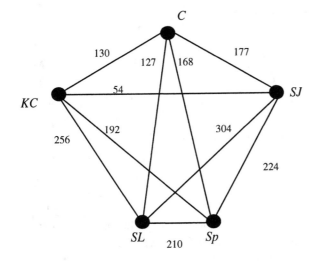

b) *SJ, KC, C, SL, Sp, SJ* traveling a total of 54 + 130 + 127 + 210 + 224 = 745 miles

c) *Sp, C, SL, KC, SJ, Sp* traveling a total of 168 + 127 + 256 + 54 + 224 = 829 miles

25.

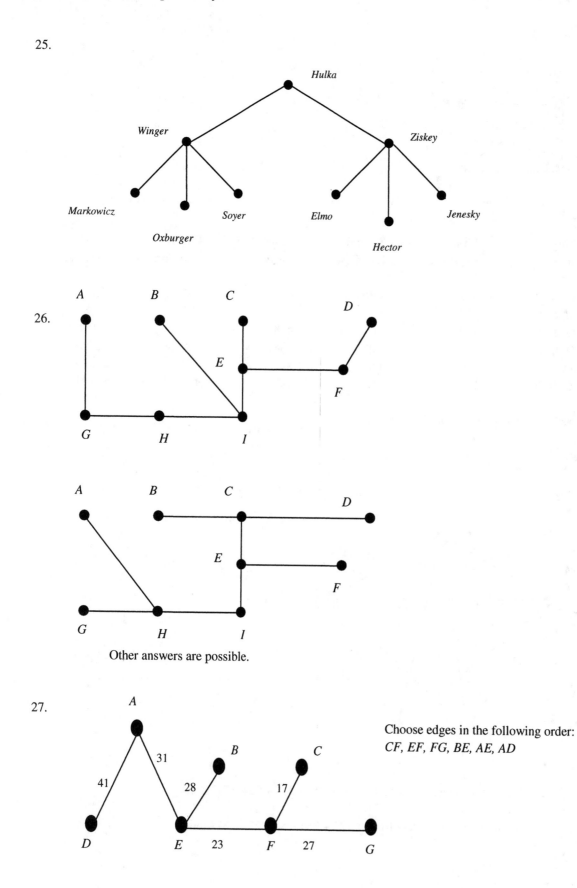

26.

Other answers are possible.

27.

Choose edges in the following order:
CF, EF, FG, BE, AE, AD

28. a)

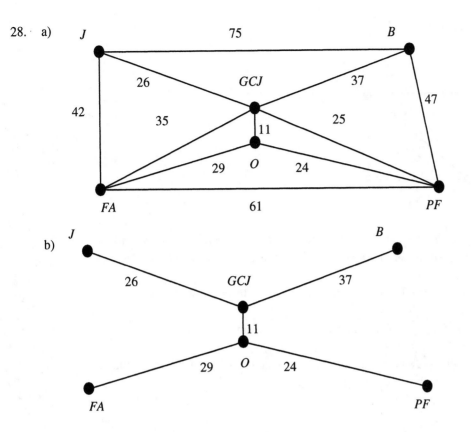

b)

Choose edges in the following order:
O GCJ, O PF, J GCJ, FA O, GCJ B

c) 2.50 (11 + 24 + 26 + 29 + 37) = 2.50 (127) = $317.50

Chapter Test

1.

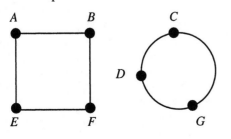

Edge *AB* is a bridge. There is a loop at vertex *G*.
Other answers are possible.

2.

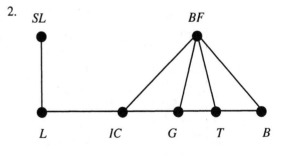

3. One example:

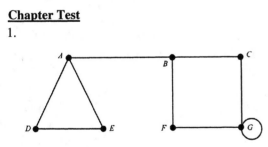

4. *D, A, B, C, E, B, D, E*;
other answers are possible.

5. Yes. The graph representing the floor plan:

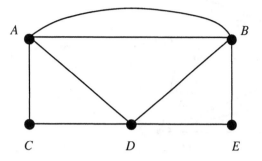

We are seeking an Euler path or an Euler circuit. Note that there are no odd vertices. According to item 1 of Euler's Theorem, since there are no odd vertices, at least one Euler path (which is also an Euler circuit) must exist.

The person may start in any room and will finish in the room where he or she started.

6. *A, D, E, A, F, E, H, F, I, G, F, B, G, C, B, A*; other answers are possible.

7. *A, B, C, D, H, I, L, K, J, G, F, E, A*; other answers are possible.

8. The number of unique Hamilton circuits within the complete graph with 8 vertices representing this situation is $(8-1)! = 7! = 7 \cdot 6 \cdot 5 \cdot 4 \cdot 3 \cdot 2 \cdot 1 = 5040$ ways

9. a)

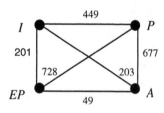

b)

Hamilton Circuit	First Leg/Cost	Second Leg/Cost	Third Leg/Cost	Fourth Leg/Cost	Total Cost
I, P, EP, A, I	449	728	49	203	$1429
I, P, A, EP, I	449	677	49	201	$1376
I, A, P, EP, I	203	677	728	201	$1809
I, A, EP, P, I	203	49	728	449	$1429
I, EP, A, P, I	201	49	677	449	$1376
I, EP, P, A, I	201	728	677	203	$1809

The least expensive route is *I, P, A, EP, I* or *I, EP, A, P, I* for $1376.

c) *I, EP, A, P, I* for $1376

10.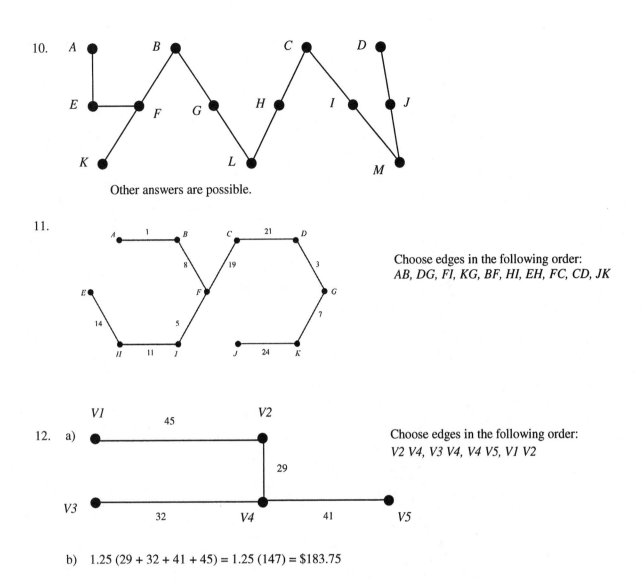

Other answers are possible.

11.

Choose edges in the following order:
AB, DG, FI, KG, BF, HI, EH, FC, CD, JK

12. a)

Choose edges in the following order:
V2 V4, V3 V4, V4 V5, V1 V2

b)   1.25 (29 + 32 + 41 + 45) = 1.25 (147) = $183.75

Group Projects
1. Answers will vary.
2. a) – d) Answers will vary.
3. a) – d) Answers will vary.
4. Answers will vary.

CHAPTER FIFTEEN

VOTING AND APPORTIONMENT

Exercise Set 15.1

1. When a candidate receives more than 50% of the votes.

2. Each voter votes for one candidate. The candidate receiving the most votes is declared the winner.

3. Voters rank candidates from most favorable to least favorable. Each last place vote is awarded one point, each next to last place vote is awarded two points, each third from last place vote is awarded three points, etc. The candidate receiving the most points is the winner.

4. Each voter votes for one candidate. If a candidate receives a majority of votes, that candidate is declared the winner. If no candidate receives a majority, eliminate the candidate with the fewest votes. (If there is a tie for the fewest votes, eliminate all tied candidates.) Repeat this process until a candidate receives a majority.

5. Voters rank the candidates. A series of comparisons in which each candidate is compared to each of the other candidates follows. If candidate A is preferred to candidate B, then A receives one point. If candidate B is preferred to candidate A, then B receives one point. If the candidates tie, each receives ½ point. The candidate receiving the most points is declared the winner.

6. Different systems can lead to a different winner.

7. A preference table summarizes the results of an election.

8. a) Pair-wise comparison method

$$1 \Leftarrow \begin{matrix} 2 \\ 3 \\ 4 \end{matrix} \qquad 2 < \begin{matrix} 3 \\ 4 \end{matrix} \qquad 3 \diagup 4 \qquad 3+2+1 = 6 \text{ groupings}$$

$$1 \Leftarrow \begin{matrix} 2 \\ 3 \\ 4 \\ 5 \\ 6 \end{matrix} \qquad 2 \Leftarrow \begin{matrix} 3 \\ 4 \\ 5 \\ 6 \end{matrix} \qquad 3 \Leftarrow \begin{matrix} 4 \\ 5 \\ 6 \end{matrix} \qquad 4 < \begin{matrix} 5 \\ 6 \end{matrix} \qquad 5 \diagup 6 \qquad 5+4+3+2+1 = 15 \text{ groupings}$$

9. a) Jeter is the winner; he received the most votes using the plurality method.

   b) No. $\dfrac{265128}{192827+210361+265128} = \dfrac{265128}{668316} \approx 0.40$ is not a majority. Majority is > 334,158 votes.

10. a) Felicia is the winner. Felicia received the most votes using the plurality method.

    b) No. $\dfrac{2863}{2192+2562+1671+2863+1959} = \dfrac{2863}{11247} \approx 0.25$ is not a majority. Majority is > 5,624 votes.

11.

Number of votes	3	1	2	2	1
First	B	A	C	C	A
Second	A	B	B	A	C
Third	C	C	A	B	B

12.

Number of votes	2	3	2	1
First	A	C	B	C
Second	B	A	A	B
Third	C	B	C	A

13. $9 + 5 + 3 + 2 = 19$ employees

14. No. Mop had the most with 9 votes, but $9/19 = 0.47$ which is not a majority. Majority is ≥ 10.

15. Votes – (M): 9, (V): $5+3 = 8$, (W): 2. Mop wins with the most votes.

16. M: 9 1st place votes $= (9)(3) = 27$
        5 2nd place votes $= (5)(2) = 10$
        5 3rd place votes $= (5)(1) = 5$
    V: 8 1st place votes $= (8)(3) = 24$
        11 2nd place votes $= (11)(2) = 22$
    B: 2 1st place votes $= (2)(3) = 6$
        3 2nd place votes $= (3)(2) = 6$
        14 3rd place votes $= (14)(1) = 14$
    M = 42 points; V = 46 points; B = 26 points
    Vacuum wins with 46 points.

17. A majority out of 19 votes is 10 or more votes.
    First choice votes: (M) 9, (V) 8, (B) 0
    None receives a majority, thus B with the least votes is eliminated.
    Second round: (M) 9, (V) $5+3+2 = 10$
    Vacuum wins with a majority of 10 votes.

18. M vs. V: M = 9 V = $5+3+2 = 10$ V gets 1 pt.
    M vs. B: M = $9+5 = 14$ B = $3+2 = 5$
              M gets 1 pt.
    V vs. B: V = $9+5+3 = 17$ B = 2 V gets 1 pt.
    Vacuum wins with 2 points.

19. B: 4 1st place votes $= (4)(3) = 12$
        2 2nd place votes $= (2)(2) = 4$
        3 3rd place votes $= (3)(1) = 3$
    G: 2 1st place votes $= (2)(3) = 6$
        4 2nd place votes $= (4)(2) = 8$
        3 3rd place votes $= (3)(1) = 3$
    M: 3 1st place votes $= (3)(3) = 9$
        3 2nd place votes $= (3)(2) = 6$
        3 3rd place votes $= (3)(1) = 3$
    B = 19 points; G = 17 points; M = 18 points
    Beach wins with 19 points.

20. Votes – (B): $3+1 = 4$, (G): 2, (M): $2+1 = 3$
    Beach wins with the most votes.

21. B vs. G: M = $3+2+1 = 6$ G = $2+1 = 3$
    B gets 1 pt.
    B vs. M: B = $3+1 = 4$ M = @+@+! = 5
    M gets 1 pt.
    G vs. M: G = $3+2 = 5$ M = $2+1+1 = 4$
    G gets 1 pt.
    All get 1 point, which indicates no winner.

22. A majority out of 9 votes is 5 or more votes.
First choice votes: (B) 4, (G) 2, (M) 3
None receives a majority, thus G with the least
votes is eliminated.
Second round: (B) 4, (M) 2+2+1 = 5
Mount Rushmore wins with a majority of 5 votes.

23. Votes: (S) 8+3+2 = 13   (L) 6+3 9
(H) 4+3+2 = 9   (T) 1
San Antonio wins with the most votes.

24. S: 9 1st place votes = (13)(4) = 52
5 2nd place votes = (5)(3) = 15
4 3rd place votes = (4)(2) = 8
10 4th place votes = (10)(1) = 10
L: 9 1st place votes = (9)(4) = 36
18 2nd place votes = (18)(3) = 54
4 3rd place votes = (4)(2) = 8
1 4th place vote = (1)(1) = 1
H: 9 1st place votes = (9)(4) = 36
9 2nd place votes = (9)(3) = 27
11 3rd place votes = (11)(2) = 22
3 4th place vote = (3)(1) = 3
T: 1 1st place votes = (1)(4) = 4
0 2nd place votes = 0
13 3rd place votes = (13)(2) = 26
18 4th place vote = (18)(1) = 18

S = 85 points; L = 99 points;
H = 88 points; T = 48 points
Los Angeles wins with 99 points.

25. A majority out of 32 votes is 16 or more votes.
First choice votes: (S) 13, (L) 9, (H) 9, (T) 1
None receives a majority, thus T with the least
votes is eliminated.
Second round: (S) 13, (L) 9, (H) 10
No majority, thus eliminate L.
Third round: (S) 16, (H) 16
Since S and H tied, there is no winner.

26. S vs. L: S = 8+3+2+1+2 = 16
L = 6+3+4+3 = 16   S / L get 0.5 pt.
S vs. H: S = 8+3+3+2 = 16
H = 6+4+3+1+2 = 16   S / H get 0.5 pt.
S and T: S = 8+3+2+1+2 = 16
T = 6+4+1 = 11   S gets 1 pt.
L and H: L = 8+6+3+3 = 20
H = 4+3+2+1+2 = 12   L gets 1 pt.
L and T: L = 8+6+3+4+3+3+2+2 = 31
T = 1   L gets 1 pt.
H and T: H = 8+6+3+4+3+2+2 = 28
T = 4   H gets 1 pt.
S = 2 H = 1.5 L = 2.5 T = 0 LA wins.

27. W: 5 1st place votes = (5)(3) = 15
4 2nd place votes = (4)(2) = 8
3 3rd place votes = (3)(1) = 3
D: 1 1st place votes = (1)(3) = 3
7 2nd place votes = (7)(2) = 14
4 3rd place votes = (4)(1) = 4
J: 6 1st place votes = (6)(3) = 18
1 2nd place votes = (1)(2) = 2
5 3rd place votes = (5)(1) = 5
W = 26 points; D = 21 points; J = 25 points
Williams wins with 26 points.

28. Votes: (W): 5, (D): 1, (J): 4 + 2 = 6
Johnson wins with the most votes.

29. W vs. D: W = 5+4 = 9 D = 1+2 = 3
W gets 1 pt.
W vs. J: W = 5 J = 1+4+2 = 7
J gets 1 pt.
D vs. J: D = 5+1 = 6 J = 4+2 = 6
D and J get 0.5 pt.
W = 1 pt. D = 1 pt. J = 1.5 pts.
Johnson wins with 1.5 points.

30. A majority out of 12 votes is 6 or more votes.
First choice votes: (W) 5, (D) 1, (J) 6
None receives a majority, thus D with the least votes is eliminated.
Second round: (W) 5, (J) 1+4+2 = 7
Johnson wins with a majority of 7 votes.

31. A majority out of 12 votes is 6 or more votes.
Most last place votes: (W) 3, (D) 4, (J) 5
Thus J with the most last place votes is eliminated.
Second round using the most last place votes:
(W) 1+2 = 3, (D) 5+4 = 9
Williams wins with the least last place votes.

32. Votes: (L): 5, (E): 2, (O): 4.   Lehigh Road wins with the most votes.

33. L: 5 1st place votes = (5)(3) = 15
6 3rd place votes = (6)(1) = 6
E: 2 1st place votes = (2)(3) = 6
9 2nd place votes = (9)(2) = 18
O: 4 1st place votes = (4)(3) = 12
2 2nd place votes = (2)(2) = 4
5 3rd place votes = (5)(1) = 5
L = 21 points; E = 24 points; O = 21 points
Erie Road wins with 24 points.

34. A majority out of 11 votes is 6 or more votes.
First choice votes: (L) 5, (E) 2, (O) 4
None receives a majority, thus E with the least votes is eliminated.
Second round: (L) 5, (O) 2+4 = 6
Ontario Road wins with a majority of 6 votes.

35. L vs. E: L = 5   E = 2+4 = 6   E gets 1 pt.
L vs. O: L = 5   O = 2+4 = 6   O gets 1 pt.
E vs. O: E = 5+2 = 6   O = 4   E gets 1 pt.
Erie Road wins with 2 points.

36. A majority out of 11 votes is 6 or more votes.
Most last place votes: (L) 2+4 = 6, (E) 0, (O) 5
Thus L with the most last place votes is eliminated.
Second round using the most last place votes:
(E) 0, (O) 4
Erie Road wins with the least last place votes.

37. a) Votes: (TI): 10, (C): 3, (HP): 2
Texas Instruments wins with the most votes.
b) TI: 10 1st place votes = (10)(4) = 40
5 2nd place votes = (5)(3) = 15
C: 3 1st place votes = (3)(4) = 12
6 2nd place votes = (6)(3) = 18
6 3rd place votes = (6)(2) = 12
S: 9 3rd place votes = (9)(2) = 18
6 4th place votes = (6)(1) = 6
9 4th place votes = (9)(1) = 9
HP: 2 1st place votes = (2)(4) = 8
4 2nd place votes = (4)(3) = 12
9 4th place votes = (9)(1) = 9
TI= 55 points; C = 42 points; S = 24 points,
HP = 29 points   TI wins with 55 points.

37. c) A majority out of 15 votes is 8 or more votes.
First choice votes: (TI) 10,   (C) 3
(S) 0,   (HP) = 2
Because TI already has a majority, TI wins.
d) TI vs. C: TI = 6+4+2 = 12   C = 3
TI gets 1 pt.
TI vs. S: TI = 6+4+3+2 = 15   TI gets 1 pt.
TI vs. HP: TI = 6+4+3 = 14   HP = 2
TI gets 1 pt.
C vs. S: C = 6+4+3+2 = 15   C gets 1 pt.
C vs. HP: C = 6+3 = 9   HP = 4+3 = 7
C = gets 1 pt.
S vs. HP: S = 6+3 = 9   HP = 4+2 = 6
S gets 1 pt.
TI wins with 3 points.

38. a) Votes: (L): 8, (M): 2, (S): 3, (H): 4
       I Love Lucy wins with the most votes.
    b) L: 8 1st place votes = (8)(4) = 32
          9 4th place votes = (9)(1) = 9
       M: 2 1st place votes = (2)(4) = 8
          15 2nd place votes = (15)(3) = 45
       S: 3 1st place votes = (3)(4) = 12
          2 2nd place votes = (2)(3) = 6
          12 3rd place votes = (12)(2) = 24
       H: 4 1st place votes = (4)(4) = 16
          5 3rd place votes = (5)(2) = 10
          8 4th place votes = (8)(1) = 8
       L= 41 points; M = 53 points; S = 42 points,
       H = 34 points   Mash wins with 53 points.

39. a) A: 6 1st place votes = (6)(4) = 24
          1 2nd place vote = (1)(3) = 3
          2 3rd place votes = (2)(2) = 4
          5 4th place votes = (5)(1) = 5
       B: 1 1st place vote = (1)(4) = 4
          4 2nd place vote = (4)(3) = 12
          9 3rd place votes = (9)(2) = 18
       C: 5 1st place votes = (5)(4) = 20
          6 2nd place vote = (6)(3) = 18
          1 3rd place vote = (1)(2) = 2
          2 4th place votes = (2)(1) = 2
       D: 2 1st place votes = (2)(4) = 8
          3 2nd place vote = (3)(3) = 9
          2 3rd place votes = (2)(2) = 4
          7 4th place votes = (7)(1) = 7
       A = 36 points; B = 34 points; C = 42 points;
       D = 28 points
       C wins with 42 points.

40. a) G vs. A: G = 69   A = 73   A gets 1 pt.
       G vs. C: G = 43   C = 99   C gets 1 pt.
       G vs. D: G = 43   D = 99   D gets 1 pt.
       A vs. C: A = 73   C = 69   A gets 1 pt.
       A vs. D: A = 73   D = 69   A gets 1 pt.
       C vs. D: C = 72   D = 70   C gets 1 pt.

    Apple wins with 3 points.

38. c) A majority out of 17 votes is 9 or more votes.
       First choice votes: (L) 8,   (M) 2
                           (S) 3,   (H) = 4
       None receives a majority, thus M with the least
        votes is eliminated.
       Second round: (L) 8, (S) 5, (H) 4
       No majority, thus eliminate H.
       Third round: (L) 8, (S) 9
       Seinfeld wins with 9 votes.
    d) L vs. M: L = 8   M = 9   M gets 1 pt.
       L vs. S: L = 8   S = 9   S gets 1 pt.
       L vs. H: L = 8   H = 9   H gets 1 pt.
       M vs. S: M = 14   S = 3   M gets 1 pt.
       M vs. H: M = 13   H = 4   M gets 1 pt.
       S vs. H: S = 13   H = 4   S gets 1 pt.
       Mash wins with 3 points.

39. b) Votes: (A): 6, (B): 1, (C): 5, (D): 2
       A wins with the most votes.
    c) A majority out of 14 votes is 7 or more votes.
       First choice votes: (A) 6,   (B) 1
                           (C) 5,   (D) = 2
       None receives a majority, thus B with the least
        votes is eliminated.
       Second round: (A) 7, (C) 5, (D) 2
       No majority, thus eliminate D.
       Third round: (A) 9, (C) 5
       A wins with 9 votes.
    d) A vs. B: A = 6   B = 8   B gets 1 pt.
       A vs. C: A = 9   C = 5   A gets 1 pt.
       A vs. D: A = 7   D = 7   A / D get 0.5 1 pt.
       B vs. C: B = 34   C = 11   C gets 1 pt.
       B vs. D: B = 9   D = 5   B gets 1 pt.
       C vs. D: C = 12   D = 2   C gets 1 pt.
       B and C tie with 2 points.

40. b) A majority out of 142 votes is 71 or more votes.
       First choice votes: G=43, A=30, C=29, D=40
       None receives a majority, thus C with the least
       votes is eliminated.
       Second round: (G) 43, (C) 30, (D) 69
       No majority, thus eliminate C.
       Third round: (G) 43, (C) 99
       Compaq wins with 99 votes.

40. c) G: 43 1st place votes = (43)(4) = 172
1 2nd place vote = (1)(3) = 3
26 3rd place votes = (26)(2) = 52
73 4th place votes = (73)(1) = 73
A: 30 1st place vote = (30)(4) = 120
43 2nd place vote = (43)(3) = 129
29 3rd place votes = (29)(2) = 58
26 4th place votes = (26)(1) = 26
C: 29 1st place votes = (29)(4) = 116
40 2nd place votes = (40)(3) = 120
73 3rd place vote = (73)(2) = 146
D: 40 1st place votes = (40)(4) = 160
59 2nd place vote = (59)(3) = 177
43 4th place votes = (43)(1) = 43
G = 300 points;   A = 333 points;
C = 382 points;   D = 380 points
Compaq wins with 380 points.

40. d) Votes: (G): 43, (A): 30, (C): 29, (D): 40
Gateway wins with the most votes.
e) You must choose the voting method prior to
the election.

41. a) If there were only two columns then only two of
the candidates were the first choice of the
voters. If each of the 15 voters cast a ballot,
then one of the voters must have received a
majority of votes because 15 cannot be split
evenly.
b) An odd number cannot be divided evenly so one
of the two first choice candidates must receive
more than half of the votes.

42. a) A: 1 + 2 + 3 + 1 = 8
B. 3 + 1 + 4 + 3 = 11
C: 4 + 4 + 1 + 1 = 10
D: 2 + 3 + 2 + 4 = 11
B and D tie with 11 points.
b) A: 0 + 1 + 3 + 1 = 5
B: 3 + 0 + 5 + 3 = 11
C: 5 + 5 + 0 + 0 = 10
D: 1 + 3 + 1 + 5 = 10
B wins with 11 points.
C wins with 42 points.

43. a) C: 4 + 1 +1 = 6   R: 4 + 4 + 3 = 11
W: 3 + 3 + 2 + 2 + 1 + 1 = 12
T: 4 + 3 + 2 + 2 = 11
The Warriors finished 1st, the Rams and the
Tigers tied for 2nd , and the Comets were 4th.
b) C: 5 + 0 = 5   R: 5 + 5 + 3 = 13
W: 3 + 3 + 1 + 1 + 0 + 0 = 8
T: 5 + 3 + 1 1 = 10
Rams - 1st, Tigers - 2nd, Warriors - 3rd, and
Comets - 4th.

44. a) Each voter casts 3+2+1 = 6 votes.
(20)(6) = 120 votes
b) 120 – (55+25) = 120 – 80 = 40 votes
c) No. Candidate B cannot win because the votes
for Candidate A > votes for Candidate B.

45. a) Each voter casts $+3+2+1 = 10 votes.
(15)(10) = 150 votes
b) 150 – (35+40+25) = 150 – 100 = 50 votes
c) Yes. Candidate D has more votes than each
of the other 3 candidates.

46. A = 10   B = 7   C = 5   D = 9
Candidates A and D will win.

Exercise Set 15.2
1. If a candidate receives a majority of first place votes, then that candidate should be declared the winner.
2. A candidate who wins a first election and then gains additional support without losing any of the original
support should also win a second election.
3. If a candidate is favored when compared individually with every other candidate, then that candidate should be
declared the winner.
4. If a candidate is declared the winner of an election, and in a second election, one or more of the other
candidates is removed, then the previous winner should still be declared the winner

5. A candidate that is preferred to all others will win each pairwise comparison and be selected with the pairwise comparison method.

6. A candidate that holds a majority of first place votes wins each pairwise comparison and is selected with the pairwise comparison method.

7. If a candidate receives a majority of first place votes, then that candidate should be declared the winner. Plurality counts only the 1st place votes.

8. If a majority is not reached on the 1st vote, then the candidate with the lowest vote total is eliminated and successive votes are taken until one of the candidates achieves a majority vote.

9. The plurality method yields Tacos are the winner with a majority of 8 1st place votes. However, if the Borda count method is used:
Tacos $(8)(3) + (3)(2) + (4)(1) = 24 + 6 + 4 = 34$
Pizza $(4+3)(4) + (8)(2) = 28 + 16 = 44$
Burgers $(4)(2) + (8+3)(1) = 8 + 11 = 19$
The winner is Pizza using the Borda count method, thus violating the majority criterion.

10. a) Total votes $= 2+4+2+3 = 11$
A vs. B: $A = 4+2 = 6$ $B = 2+3 = 5$ A gets 1 pt.
A vs. C: $A = 2+4 = 6$ $C = 2+3 = 5$ A gets 1 pt.
B vs. C: $B = 2+4 = 6$ $C = 2+3 = 5$ B gets 1 pt.
Plan A wins with 2 points.

b) C wins by a plurality of 5 votes. No, the head-to-head criterion is not satisfied.

11. Total votes $= 3+2+1+1 = 7$ Candidates A is the candidate of choice with a plurality of 4 votes.
A: 4 1st place votes $= (4)(4) = 16$
    3 4th place votes $= (3)(1) = 3$
B: 3 1st place vote $= (3)(4) = 12$
    4 2nd place vote $= (4)(3) = 12$
C: 2 2nd place vote $= (2)(3) = 6$
    4 3rd place vote $= (4)(2) = 8$
    1 4th place vote $= (1)(1) = 1$
D: 1 2nd place vote $= (1)(3) = 3$
    3 3rd place votes $= (3)(2) = 6$
    3 4th place votes $= (3)(1) = 3$
   G = 300 points;   A = 333 points;
A = 19 votes;   B = 24 votes;   C = 15 votes;
D = 12 votes
Candidate B is chosen with 24 votes, therefore the majority criterion is not satisfied.

12. a) Total votes $= 12+6+4+3 = 25$
B vs. W: $B = 12+6+4 = 22$ $W = 3$ B gets 1 pt.
B vs. S: $B= 12+3 = 15$ $S = 10$ B gets 1 pt.
B vs. R: $B = 12+6 = 18$ $R = 7$ B gets 1 pt.
W vs. S: $W = 12+3 = 15$ $S = 10$ W gets 1 pt.
W vs. R: $W = 12+6+3 = 21$ $R = 4$ W gets 1 pt.
S vs. R: $S = 12+6 = 18$ $S = 7$ S gets 1 pt.
Beach wins with 3 points.

b) B wins by a plurality of 12 votes. Yes, the head-to-head criterion is satisfied.

13. P: 4 1st place votes $= (4)(3) = 12$
    2 2nd place votes $= (2)(2) = 4$
    3 3rd place votes $= (3)(1) = 3$
L: 3 1st place vote $= (3)(3) = 9$
    5 2nd place vote $= (5)(2) = 10$
    1 3rd place vote $= (1)(1) = 1$
S: 2 1st place votes $= (2)(3) = 6$
    2 2nd place vote $= (2)(3) = 6$
    5 3rd place vote $= (5)(1) = 5$
P = 19 votes;   L = 20 votes;   S = 17 votes
P vs. L: $P = 4+1 = 5$ $L = 4$ P gets 1 pt.
P vs. S: $P = 4+1 = 5$ $S = 4$ P gets 1 pt.
L vs. S: $L = 4+1+2 = 7$ $S = 2$ L gets 1 pt.
Because Parking wins by head-to-head comparison and the Lounge Areas win by Borda count method, the head-to-head criterion is not satisfied.

14. A: 2 1st place votes = (2)(3) = 6
      7 2nd place votes = (7)(2) = 14
    B: 2 1st place vote = (2)(3) = 6
      2 2nd place vote = (2)(2) = 4
      5 3rd place vote = (5)(1) = 5
    C: 5 1st place votes = (5)(3) = 15
      4 3rd place vote = (4)(1) = 4
    A = 20 votes; B = 15 votes; C = 19 votes
    A vs. B: A = 2+2 = 4 B = 5 B gets 1 pt.
    A vs. C: A = 2+2 = 4 C = 5 C gets 1 p
    B vs. C: B = 2+2 = 4 C = 5 C gets 1 pt.
    Because C wins by head-to-head comparison
    and the A wins by the Borda count method,
    the head-to-head criterion is not satisfied.

15. A majority out of 25 votes is 13 or more votes.
    First choice votes: A=7, B=15, C=3
    Since B has > 13 votes, B wins by plurality with
    elimination.

    A vs. B: A = 7+3 = 10   B = 15   B gets 1 pt.
    A vs. C: A = 7   C = 25-7 = 18   C gets 1 pt.
    B vs. C: B = 15+7 = 22   C = 3   B gets 1 pt.

    Yes, because B wins by both methods, the
    head-to-head criterion is satisfied.

16. A majority out of 25 votes is 13 or more votes.
    First choice votes: (A) 10, (B) 2, (C) 8, (D) = 5
    None receives a majority, thus B with the least
    votes is eliminated.
    Second round: (A) 10, (C) 10, (D) 5
    Still no majority, thus eliminate D.
    Third round: (A) 10, (C) 15
    C wins with a majority of 15 votes.
    A vs. B: A = 10 B = 15 B gets 1 pt.
    A vs. C: A = 10 C = 15 C gets 1 pt.
    A vs. D: A = 12 D = 13 D gets 1 pt.
    B vs. C: B = 17 C = 8 B gets 1 pt.
    B vs. D: B = 20 D = 5 B gets 1 pt.
    C vs. D: C = 10 D = 15 D gets 1 pt.
    B wins with 3 points. Therefore, the head-to-head
    criterion is not satisfied.

17. Votes: A: 8, B: 4, C: 5; thus, A wins.
    If B drops out, we get the following:
    Votes: A: 8, C: 4 + 5 = 9, thus C would win.
    The irrelevant alternatives criterion is not satisfied.

18. Votes: A: 3, B: 4, C: 5; thus B wins
    If C drops out, we get the following:
    Votes: A: 3 + 5 = 8, B:6, thus A would win.
    The irrelevant alternatives criterion is not satisfied.

19. A receives 53 points, B receives 56 points, and
    C receives 53 points. Thus, B wins using the
    Borda count method. If A drops out, we get the
    following: B receives 37 points, and C receives
    44 points. Thus, C wins the second vote. The
    irrelevant alternatives criterion is not satisfied.

20. A receives 38 points, B receives 35 points, C
    receives 35 points. Thus, A wins using the Borda
    count method. If B drops out we get the following:
    A receives 25 points, and C receives 29 points.
    Thus, C wins the second vote.

    The irrelevant alternatives criterion is not satisfied.

21. A majority out of 32 voters is 16 or more votes.
    Votes: A: 8 + 3 = 11, B: 9, C: 12; none has a
    majority, thus eliminate B.
    Votes: A:8 + 3 = 11, C: 9 +12 = 21, thus C
    wins. If the three voters who voted for A,C,B
    change to C,A,B, the new set of votes becomes:
    Votes: A: 12, B: 9, C: 11; none has a
    majority, thus eliminate B.
    Votes: A: 9 + 12 = 21, C = 11, thus A wins.
    Thus, the monotonicity criterion is not satisfied.

22. A majority out of 29 voters is 15 or more votes.
Votes: A: 8,   B: 10,   C: 11; none has a majority,
 thus eliminate A.
Votes: B: 8 + 10 = 18,   C: 7 + 4 = 11, thus B
wins. After the four votes change their votes, the
the new set of votes is A: 8,   B: 14.,   C: 7;
none has a majority, thus eliminate C.
Votes: A: 7+ 8 = 15,   B:14; thus A wins.
Thus, the monotonicity criterion is not satisfied.

23. A majority out of 23 voters is 12 votes.
Votes: A: 10,   B: 8,   C: 5; none has a majority,
 thus eliminate C.
Votes: B: 10,   B: 8 + 5 = 13; thus B wins.
After A drops out, the new set of votes is
B: 8,   C: 10 + 5 = 15; thus C wins.

The irrelevant alternatives criterion is not satisfied.

24. A majority out of 13 voters is 7 votes.
Votes: A: 3,   B: 6,   C: 4; none has a majority, thus
eliminate A. Votes: B: 6,   C: 4 + 3 = 7; thus C
wins. After B drops out, the new set of votes is
Votes: A: 6 + 3 = 9,   C: 4; thus A wins.
The irrelevant alternatives criterion is not satisfied.

25. A receives 2 points, B receives 3 point, C receives
2 points, D receives 1 point, and E receives 2 pts.
B wins by pairwise comparison.
After A, C and E drop out, the new set of votes is
B: 2   D: 3, thus D wins. The irrelevant
alternatives criterion is not satisfied.

26. A receives 3 points, B receives 1 point, C receives
3 points, D receives 1 point, and E receives 2 points.
A and C tie, but when A vs. C, C wins and thus we
declare C the winner. After A, B and E drop out,
the new set of votes is table is C: 2 + 1 = 3,
D: 4, thus D wins.
The irrelevant alternatives criterion is not satisfied.

27. Total votes = 7   A wins with a majority of
4 votes.
A: 4 1st place votes = (4)(3) = 12
   3 3rd place votes = (3)(1) = 3
B: 2 1st place vote = (2)(3) = 6
   5 2nd place vote = (5)(2) = 10
C: 1 1st place votes = (1)(3) = 3
   2 2nd place votes = (2)(2) = 4
   4 3rd place vote = (4)(1) = 4
A = 15 points;   B = 16 points;   C = 11 points
B wins with 16 points. No. The majority
criterion is not satisfied.

28. Total votes = 11   B wins with a plurality of
5 votes.
A: 1 1st place votes = (1)(3) = 3
   5 2nd place votes = (5)(2) = 10
   5 3rd place votes = (5)(1) = 5
B: 6 1st place vote = (6)(3) = 18
   5 3rd place votes = (5)(1) = 5
C: 4 1st place votes = (4)(3) = 12
   6 2nd place votes = (6)(2) = 12
   1 3rd place vote = (1)(1) = 1
A = 21 points;   B = 23 points;   C = 25 points
C wins with 25 points. No. The majority
criterion is not satisfied.

29. Total votes = 31   Majority = 16 or more
a) Museum of Natural History
b) Museum of Natural History
c) Museum of Natural History
d) None of them

30. a) Total votes = 44   A majority is ≥ 22 votes.

    A: 8 1st place votes = (8)(5) = 40

      8 3rd place votes = (8)(3) = 24

      8 4th place votes = (8)(2) = 16

      20 5th place votes = (20)(1) = 20

    B: 20 1st place vote = (20)(5) = 100

      2 2nd place vote = (2)(4) = 8

      14 4th place votes = (14)(2) = 28

      8 5th place votes = (8)(1) = 8

    C: 4 1st place votes = (4)(5) = 20

      8 2nd place votes = (8)(4) = 32

      16 3rd place vote = (16)(3) = 48

      8 4th place votes = (8)(2) = 16

      8 5th place votes = (8)(1) = 8

    D: 4 1st place votes = (4)(5) = 20

      28 2nd place votes = (28)(4) = 112

      4 3rd place votes = (4)(3) = 12

      8 5th place votes = (8)(1) = 8

    E: 8 1st place votes = (8)(5) = 40

      2 2nd place votes = (2)(4) = 8

      16 3rd place votes = (16)(3) = 48

      14 4th place votes = (14)(2) = 28

    A = 100 pts.;   B = 136 pts.;   C = 124 pts.;

    D = 152 pts.;   E = 124 pts.

    Dow Chemical is chosen with 152 points.

    b) Burrows-Welcome will be chosen.

    c) Yes.

31. a) A majority out of 82 votes is 41 or more votes.

    First choice votes: (A) 28, (C) 30, (D) 24

    None receives a majority, thus D with the least

    votes is eliminated.

    Second round: (A) 52, (C) 30

    Thus, Jennifer Aniston is selected..

    b) No majority on the 1st vote; C is eliminated

    with the fewest votes.

    Second round: (A) 38, (D) 44

    Denzel Washington is chosen.

    c) Yes.

32. a) A receives 1 point, B receives 2½ points,

    C receives 1½ points, D receives 3 points,

    E receives 2 points. Thus, (D) wins.

    b) A receives 0 points, B receives 2½ points,

    D receives 2 points, E receives 1½ points.

    Thus, B wins.

    c) Yes.

33. A candidate who holds a plurality will only gain

    strength and hold and even larger lead if more

    favorable votes are added.

34. Answers will vary (AWV).

35. AWV   36. AWV   37. AWV

38. A majority out of 11 voters is 6 or more votes.

    a) Votes: A: 9,   B: 2; thus A wins.

    b) Votes: A: 4 + 2 = 6,   C: 5;   Yes, A wins.

    c) The five voters who favor C should vote C, B, A

      instead of C, A, B.

39. AWV

Exercise Set 15.3

1. If we divide the total population by the number of items to be apportioned we obtain a number called the standard divisor.

2. The standard quota is found by dividing each group's population by the standard divisor.

3. The standard quota rounded down to the nearest whole number.

4. The standard quota rounded up to the nearest whole number.

5. An apportionment should always be either the upper quota or the lower quota.

6. Hamilton's method

7. Jefferson's method, Webster's method, Adams's method

8. a) Jefferson's method   b) Adam's method   c) Webster's method

9. a) Webster's method   b) Adam's method   c) Jefferson's method

10. Jefferson's method, Webster's method, Adams's method

11. a) $\dfrac{7500000}{150} = 50,000$ = standard divisor

   b) and c)

State	A	B	C	D	Total
Population	1,222,000	2,730,000	857,000	2,693,000	7,500,000
Standard Quota	24.40	54.60	17.14	53.86	
Lower Quota	24	54	17	53	148
Hamilton's Apportionment	24	55	17	54	150

12. a) and b)

State	A	B	C	D	Total
Population	1,222,000	2,730,000	857,000	2,693,000	7,500,000
Modified Quota	24.65	55.15	17.31	54.40	
Jefferson's Apportionment (round down)	24	55	17	54	150

13. a) and b)

State	A	B	C	D	Total
Population	1,222,000	2,730,000	857,000	2,693,000	7,500,000
Modified Quota	24.70	55.26	17.35	54.51	
Jefferson's Apportionment (round down)	24	55	17	54	150

14. a) and b)

State	A	B	C	D	Total
Population	1,222,000	2,730,000	857,000	2,693,000	7,500,000
Modified Quota	24.11	53.95	16.94	53.22	
Adams' Apportionment (round up)	25	54	17	54	150

15. a) and b)

State	A	B	C	D	Total
Population	1,222,000	2,730,000	857,000	2,693,000	7,500,000
Modified Quota	24.06	53.85	16.90	53.12	
Adams' Apportionment (round up)	25	54	17	54	150

16. a) and b)

State	A	B	C	D	Total
Population	1,222,000	2,730,000	857,000	2,693,000	7,500,000
Standard Quota	24.40	54.60	17.14	53.86	
Webster's Apportionment (standard rounding)	24	55	17	54	150

17. a) and b)

State	A	B	C	D	Total
Population	1,222,000	2,730,000	857,000	2,693,000	7,500,000
Modified Quota	24.38	54.55	17.12	53.81	
Webster's Apportionment	24	55	17	54	150

18. a) Standard divisor $= \dfrac{\text{total}}{25} = \dfrac{675}{25} = 27$

b) and c)

Hotel	A	B	C	Total
Amount	306	214	155	675
Standard Quota	11.33	7.93	5.74	
Hamilton's Apportionment	11	8	6	25

19. a) and b)

Hotel	Al	Bob	Charlie	Total
Amount	350	530	470	1350
Modified Quota	8.05	12.18	10.84	
Jefferson's Apportionment (rounded down)	8	12	10	30

20. a) and b)

Hotel	Al	Bob	Charlie	Total
Amount	350	530	470	1350
Modified Quota	8.14	12.33	10.93	
Jefferson's Apportionment	8	12	10	30

21. a) and b)

Hotel	Al	Bob	Charlie	Total
Amount	350	530	470	1350
Modified Quota	7.45	11.28	10.00	
Adam's Apportionment (rounded up)	8	12	10	30

22. a) and b)

Hotel	Al	Bob	Charlie	Total
Amount	350	530	470	1350
Modified Quota	7.29	11.04	9.79	
Adam's Apportionment (rounded up)	8	12	10	30

23. a) and b)

Store	Al	Bob	Charlie	Total
Amount	350	530	470	1350
Standard Quota	7.78	11.78	10.44	
Webster's Apportionment (standard rounding)	8	12	10	30

24. a) and b)

Store	Al	Bob	Charlie	Total
Amount	350	530	470	1350
Modified Quota	7.61	11.52	10.22	
Webster's Apportionment	8	12	10	30

25. a) A standard divisor $= \dfrac{\text{total}}{30} = \dfrac{540}{30} = 18$

b)

Store	A	B	C	D	Total
Population	75	97	140	228	540
Standard Quota	4.177	5.39	7.78	12.67	30

26.

Store	A	B	C	D	Total
Population	123	484	382	271	1260
Standard Quota	5.86	23.05	18.19	12.90	
Lower Quota	5	23	18	12	58
Hamilton's Apportionment	6	23	18	13	60

27. A divisor of 20.5 was used.

Store	A	B	C	D	Total
Population	123	484	382	271	1260
Modified Quota	6.00	23.61	18.63	13.22	
Jefferson's Apportionment (round down)	6	23	18	13	60

28. A divisor of 21.5 was used.

Store	A	B	C	D	Total
Population	123	484	382	271	1260
Modified Quota	5.72	27.51	17.77	12.60	
Adams' Apportionment (round up)	6	23	18	13	60

29.

Store	A	B	C	D	Total
Population	123	484	382	271	1260
Standard Quota	5.86	23.05	18.19	12.90	
Webster's Apportionment	6	23	18	13	60

30. a) Standard divisor = $\dfrac{\text{total}}{250} = \dfrac{13000}{250} = 52$

b)

School	LA	Sci.	Eng.	Bus.	Hum	Total
Enrollment	1746	7095	2131	937	1091	13000
Standard Quota	33.58	136.44	40.98	18.02	20.98	

31.

School	LA	Sci.	Eng.	Bus.	Hum	Total
Enrollment	1746	7095	2131	937	1091	13000
Standard Quota	33.58	136.44	40.98	18.02	20.98	
Lower Quota	33	136	40	18	20	247
Hamilton's Apportionment	34	136	41	18	21	250

32. A divisor of 51.5 was used.

School	LA	Sci.	Eng.	Bus.	Hum	Total
Enrollment	1746	7095	2131	937	1091	13000
Modified Quota	33.90	137.77	41.38	18.19	21.18	
Jefferson's Apportionment (round down)	33	137	41	18	21	250

33. A divisor of 52.5 was used.

School	LA	Sci.	Eng.	Bus.	Hum	Total
Enrollment	1746	7095	2131	937	1091	13000
Modified Quota	33.26	135.14	40.59	17.85	20.78	
Adam's Apportionment (round up)	34	136	41	18	21	250

34.

School	LA	Sci.	Eng.	Bus.	Hum	Total
Enrollment	1746	7095	2131	937	1091	13000
Standard Quota	33.58	136.44	40.98	18.02	20.98	
Webster's Apportionment (standard rounding)	34	136	41	18	21	250

35. a) A standard divisor = $\dfrac{\text{total}}{150} = \dfrac{13500}{150} = 90$

Dealership	A	B	C	D	Total
Annual Sales	4800	3608	2990	2102	13500
Standard Quota	53.33	40.09	33.22	23.36	150.00

36.

Dealership	A	B	C	D	Total
Annual Sales	4800	3608	2990	2102	13500
Standard Quota	53.33	40.09	33.22	23.36	150.00
Hamilton's Apportionment	53	40	33	24	150

37.

Dealership	A	B	C	D	Total
Annual Sales	4800	3608	2990	2102	13500
Standard Quota	53.33	40.09	33.22	23.36	150.00
Jefferson's Apportionment	54	40	33	23	150

38.

Dealership	A	B	C	D	Total
Annual Sales	4800	3608	2990	2102	13500
Standard Quota	53.33	40.09	33.22	23.36	150.00
Adam's Apportionment	53	40	33	24	150

39.

Dealership	A	B	C	D	Total
Annual Sales	4800	3608	2990	2102	13500
Standard Quota	53.33	40.09	33.22	23.36	150.00
Webster's Apportionment	54	40	33	23	150

40. a) Standard divisor $= \dfrac{\text{total}}{210} = \dfrac{2940}{210} = 14$

b)

Precinct	A	B	C	D	E	F	Total
Crimes	743	367	432	491	519	388	2940
Standard Quota	53.07	26.21	30.86	35.07	37.07	27.71	

41.

Precinct	A	B	C	D	E	F	Total
Crimes	743	367	432	491	519	388	2940
Standard Quota	53.07	26.21	30.86	35.07	37.07	27.71	
Lower Quota	53	26	30	35	37	27	208
Hamilton's Apportionment	53	26	31	35	37	28	210

42. The divisor 3.8 as used.

Precinct	A	B	C	D	E	F	Total
Crimes	743	367	432	491	519	388	2940
Modified Quota	53.84	26.59	31.30	35.58	37.61	28.12	
Jefferson's Apportionment (round down)	53	26	31	35	37	28	210

43. The divisor 14.2 as used.

Precinct	A	B	C	D	E	F	Total
Crimes	743	367	432	491	519	388	2940
Modified Quota	52.32	22.85	30.42	34.58	36.55	27.32	
Adam's Apportionment (round up)	53	26	31	35	37	28	210

44.

Precinct	A	B	C	D	E	F	Total
Crimes	743	367	432	491	519	388	2940
Standard Quota	52.32	22.85	30.42	34.58	36.55	27.32	
Webster's Apportionment (standard rounding)	53	26	31	35	37	28	210

45. a) Standard divisor $= \dfrac{\text{total}}{200} = \dfrac{2400}{200} = 12$

b)

Shift	A	B	C	D	Total
Room calls	751	980	503	166	2400
Standard Quota	62.58	81.67	41.92	13.83	

46.

Shift	A	B	C	D	Total
Room calls	751	980	503	166	2400
Standard Quota	62.58	81.67	41.92	13.83	
Lower Quota	62	81	41	13	197
Hamilton's Apportionment	62	82	42	14	200

47. The divisor 11.9 was used.

Shift	A	B	C	D	Total
Room calls	751	980	503	166	2400
Modified Quota	63.11	82.35	42.27	13.95	
Jefferson's Apportionment (round down)	63	82	42	13	200

48. The divisor 12.1 was used.

Shift	A	B	C	D	Total
Room calls	751	980	503	166	2400
Modified Quota	62.07	80.99	41.57	13.72	
Adam's Apportionment (round up)	63	81	42	14	200

49. The divisor 12.02 was used.

Shift	A	B	C	D	Total
Room calls	751	980	503	166	2400
Modified Quota	62.48	81.53	41.85	13.81	
Webster's Apportionment (standard rounding)	62	82	42	14	200

50. Standard divisor $= \dfrac{3615920}{105} = 34437.33$

    a) Hamilton's Apportionment: 7, 2, 2, 2, 8, 14, 4, 5, 10, 10, 13, 2, 6, 2, 18

    b) Jefferson's Apportionment: 7, 1, 2, 2, 8, 14, 4, 5, 10, 10, 13, 2, 6, 2, 19

    c) States that Benefited: Virginia   States Disadvantaged: Delaware

Exercise set 15.4

1. The Alabama paradox occurs when an increase in the total # of items results in a loss of items for a group.

2. The new-states paradox occurs when the addition of a new group changes the apportionment of another group.

3. The population paradox occurs when group A loses items to group B, although group A's population grew at a higher rate than group B's.

4. Yes, it can produce the Alabama paradox, population paradox, and new-states paradox.

5. Hamilton's, Jefferson's

6. Adam's, Webster's

7. New divisor $= \dfrac{900}{51} = 17.65$

School	A	B	C	D	E	Total
Standard Quota	11.90	9.35	9.07	9.92	10.76	
Lower Quota	11	9	9	9	10	48
Hamilton's Apportionment	12	9	9	10	11	51

No. No school suffers a loss so the Alabama paradox does not occur.

8. a) Standard divisor $= \dfrac{2592}{144} = 18$

School	A	B	C	D	Total
Population	739	277	618	958	2592
Standard Quota	41.06	15.38	34.33	53.22	
Hamilton's Apportionment	41	16	34	53	144

    b) New divisor $= \dfrac{2592}{145} = 17.88$

School	A	B	C	D	Total
Population	739	277	618	958	2592
Standard Quota	41.33	15.49	34.56	53.57	
Hamilton's Apportionment	41	16	34	53	144

Yes. School B loses a monitor while schools C and D each gain a monitor.

9. a) Standard divisor $= \dfrac{900}{30} = 30$

State	A	B	C	Total
Population	161	250	489	900
Standard Quota	5.37	8.33	16.30	
Hamilton's Apportionment	6	8	16	30

9. b) New divisor = $\dfrac{900}{31} = 29.03$

State	A	B	C	Total
Population	161	250	489	900
Standard Quota	5.56	8.61	16.84	
Hamilton's Apportionment	5	9	17	31

Yes, state A loses 1 seat and states B and C each gain 1 seat.

10. a) Standard divisor = $\dfrac{1000000}{200} = 5000$

State	A	B	C	Total
Population	233,000	461,000	306,000	1,000,000
Standard Quota	46.60	92.20	61.20	
Lower Quota	46	92	61	199
Hamilton's Apportionment	47	92	61	200

10. b) New divisor = $\dfrac{1000000}{201} = 4975.12$

State	A	B	C	Total
Population	233,000	461,000	306,000	1,000,000
Standard Quota	46.83	92.66	61.51	
Lower Quota	46	92	61	199
Hamilton's Apportionment	47	93	61	201

No. None of the States lost a seat.

11. a) Standard divisor = $\dfrac{25000}{200} = 125$

City	A	B	C	Total
Population	8130	4030	12,840	25,000
Standard Quota	65.04	32.24	102.72	
Hamilton's Apportionment	65	32	103	200

b) New divisor = $\dfrac{25125}{200} = 125.625$

City	A	B	C	Total
New Population	8150	4030	12,945	25,125
Standard Quota	64.88	32.08	103.04	
Hamilton's Apportionment	65	32	103	200

No. None of the Cities loses a bonus.

12. a) Standard divisor = $\dfrac{900}{30} = 30$

College	A	B	C	Total
Faculty	162	249	489	900
Standard Quota	5.40	8.30	16.30	
Lower Quota	5	8	16	29
Hamilton's Apportionment	6	8	16	30

12. b) New divisor = $\dfrac{965}{30} = 32.167$

College	A	B	C	Total
Faculty	178	269	518	965
Standard Quota	5.53	8.36	16.10	
Lower Quota	5	8	16	29
Hamilton's Apportionment	6	8	16	30

No. The opportionment is the same.

13. a) Standard divisor = $\dfrac{5400}{54} = 100$

Division	A	B	C	D	E	Total
Population	733	1538	933	1133	1063	5400
Standard Quota	7.33	15.38	9.33	11.33	10.63	
Lower Quota	7	15	9	11	10	52
Hamilton's Apportionment	7	16	9	11	11	54

13. b) New divisor = $\dfrac{5454}{54} = 101$

Division	A	B	C	D	E	Total
Population	733	1539	933	1133	1116	
Standard Quota	7.26	15.238	9.238	11.22	11.05	
Lower Quota	7	15	9	11	11	53
Hamilton's Apportionment	8	15	9	11	11	54

Yes. Division B loses an internship Division A even though the population of division B grew faster than the population of division A.

14. a) Standard divisor = $\dfrac{30000}{250} = 120$

State	A	B	C	Total
Population	459	10551	18990	30000
Standard Quota	3.82	87.93	158.25	
Hamilton's Apportionment	4	88	158	250

b) Same divisor = $\dfrac{30000}{250} = 120$

State	A	B	C	Total
Population	464	10551	19100	30110
Standard Quota	3.87	87.93	159.17	
Hamilton's Apportionment	3	88	159	250

No. The opportionment is the same.

15. a) Standard divisor = $\dfrac{4800}{48} = 100$

Tech. Data	A	B	Total
Employees	844	3956	4800
Standard Quota	8.44	39.56	
Lower Quota	8	39	47
Hamilton's Apportionment	8	40	48

b) New divisor = $\dfrac{5524}{55} = 100.44$

Tech. Data	A	B	C	Total
Employees	844	3956	724	5524
Standard Quota	8.40	39.39	7.21	
Lower Quota	8	39	7	54
Hamilton's Apportionment	9	39	7	55

Yes. Group B loses a manager.

16. a) Standard divisor = $\dfrac{10000}{100} = 100$

State	A	B	Total
Population	1135	8865	10000
Standard Quota	11.35	88.65	
Hamilton's Apportionment	11	89	100

16. b) New divisor = $\dfrac{10625}{106} = 100.24$

State	A	B	C	Total
Population	1135	8865	625	10625
Standard Quota	11.32	88.44	6.24	
Hamilton's Apportionment	11	89	6	106

Yes. State C loses a seat to State B.

17. a) Standard divisor = $\dfrac{990000}{66} = 15,000$

State	A	B	C	Total
Population	68970	253770	667260	990000
Standard Quota	4.59	16.92	44.48	
Hamilton's Apportionment	5	17	44	66

b) New divisor = $\dfrac{1075800}{71} = 15,152.11$

State	A	B	C	D	Total
Population	68970	253770	667260	85800	1075800
Standard Quota	4.55	16.75	44.04	5.66	
Hamilton's Apportionment	4	17	44	6	71

Yes. State C loses a seat to State B.

18. a) Standard divisor $= \dfrac{3300}{33} = 100$

State	A	B	Total
Population	744	2556	3300
Standard Quota	7.44	25.56	
Lower Quota	7	25	32
Hamilton's Apportionment	7	26	33

b) New divisor $= \dfrac{4010}{40} = 100.25$

State	A	B	C	Total
Population	744	2556	710	4010
Standard Quota	7.42	25.50	7.08	
Lower Quota	7	25	7	39
Hamilton's Apportionment	7	26	7	40

No. The apportionment is the same.

Review Exercises

1. a) Robert Rivera wins with the most votes (12).

   b) A majority out of 24 voters is 13 or more votes. Robert Rivera does not have a majority.

2. a) Michelle MacDougal wins with the most votes (224).

   b) Yes. A majority out of 421 voters is 211 or more votes.

3.

# of votes	3	2	1	3	1
First	B	A	D	C	D
Second	A	C	C	B	A
Third	C	D	A	A	B
Fourth	D	B	B	D	C

4.

# of votes	2	2	2	1
First	C	A	B	C
Second	A	B	C	B
Third	B	C	A	A

5. Number of votes = 6 + 4 + 3 + 2 + 1 + 1 = 17

6. Park City wins with a plurality of 6 votes.

7. P: 50 points,   V: 47 points,   S: 35 points,
    A: 38 points. Park City wins with 50 points.

8. A majority out 17 voters is 9 or more votes.
    Votes: P: 6+1 = 7, V: 4, S: 3+2 = 5, A:1.
    None has a majority, thus eliminate A.
    Votes: P: 6+1 = 7, V: 4, S: 3+2+1 = 6
    None has a majority, thus eliminate V.
    Votes: P: 6+4+1 = 11, S: 3+2+1 = 6.
    Park City wins.

9. P: 3 pts.,   V: 2 pts.,   S: 0 pts.,   A: 1 pt.
    Park City wins with 3 points.

10. Votes: P: 7,   V: 4,   S: 5,   A: 1   None has a
    majority, thus eliminate S with most last place
    votes. Votes: P: 10, V: 4, A: 3; Park City wins.

11. 38+30+25+7+10 = 110 students voted

12. Volleyball wins with a plurality of 40 votes.

13. S: 223 pts.,   V: 215 pts.,   B: 222 pts.
    Soccer wins.

14. A majority out of 110 voters is 56 or more votes.
    Votes: S: 38, V: 40, B: 32; None has a majority,
    thus eliminate B. Votes: S: 45, V: 65
    Volleyball wins.

15. S: 1 pt.,   V: 1 pt.,   B: 1 pt.   A 3-way tie

16. Votes: S: 38,   V: 40,   B: 32   None has a
    majority, thus eliminate V with the most last place
    votes.   Votes: S: 68,   B: 42.   Soccer wins.

17. a) Votes: A: 161+134 = 295,   F: 45,   M: 12,
       P: 0   AARP wins.
    b) Yes. A majority out of 372 voters is 186 or more
       votes.   AARP receives a majority.
    c) A: 985 pts.,   F: 740 pts.,   M: 741 pts.,
       P: 852 pts.   AARP wins.
    d) 186 or more votes is needed for a majority.
       Votes: A: 295,   F: 45,   M: 12,   P: 0
       AARP wins.
    e) A: 3 pts.,   F: 1 pt.,   M: 1 pt.,   P: 1 pt.
       AARP wins.

18. Votes: (NO): 70,   (LV): 55,   (C): 30,   (SD): 45
    a) A majority out of 200 voters is 101 or more
       votes. None of the cities has a majority.
    b) New Orleans win a plurality of 70 votes.
    c) (NO):410 pts.,   (LV): 580 pts.,   (C): 505 pts.,
       (SD): 495 pts.   Las Vegas wins.
    d) Las Vegas wins with 130 pts. to 70 pts. for NO.
    e) NO: 0 pts.,   LV: 3 pts.,   C: 1 pt.,   SD: 1 pt.
       Las Vegas wins with points.

19. a) A majority out of 16 voters is 9 or more votes.
       Votes: (EB): 4+3+ = 7,   (FW): 1+1 = 2,
       (G): 0,   (WB): 6+1 = 7   None has a majority,
       thus eliminate G.   Votes: (EB): 4+3 = 7,
       (FW): 1+1 = 2,   (WB): 6 + 1 = 7   None has a
       majority, thus eliminate FW
       Votes: (EB): 4+3+1 = 8,   (WB): 6+1+1 = 8.
       Thus, EB and WB tie.
    b) Use the Borda count method to break the tie.
       (EB) = 46 points,   (WB) = 50 points;
       World Book wins.

19. c) (EB) vs. (WB): EB: 4+3+1 = 8 points,
       (WB): 6+1+1 = 8 points.
       EB and WB tie again.

20. A: 33 pts.,   B: 39 pts,   C: 28 pts.,   D: 20 pts.
    Using the Borda count, method B wins.
    However, B only has 3 first place votes, thus the
    majority criterion is not satisfied.

21. In a head-to-head comparison, B must win over all
    the others. For (B vs. A), A wins with 3 pts.
    The head-to-head criterion is not satisfied.

22. a) A majority out of 42 voters is 21 or more votes.
       Votes: A: 12,   B: 10+6 = 16,   C: 14
       None has the majority, thus eliminate A.
       Votes: B :10+6 = 16,   C: 14+12 = 26   C wins.

    b) The new preference table is

Number of votes	10	14	6	12
First	B	C	C	A
Second	A	B	B	C
Third	C	A	A	B

    Votes: A: 12,   B: 10,   C: 20;   None has a
    majority, thus eliminate B.
    Votes: A: 22,   C: 20   A wins.   When the
    order is changed A wins. Therefore, the
    monotonicity criterion is not satisfied.

22. c) If B drops out the new table is

Number of votes	10	14	6	12
First	A	C	C	A
Second	C	A	A	C

    Votes: A: 10+12 = 22,   C: 14+6 = 20   A wins.
    Since C won the first election and then after B
    dropped out A won, the irrelevant criterion is
    not satisfied.

23. a) M has 0 pts., S has 3 pts., F has 1 pt., and E has 1 pt. Thus, Starbucks wins.

b) Maxwell House wins w/a plurality of 33 votes.

c) M = 228 pts., S = 277 pts., F = 293 pts., and E = 292 pts. Thus, Folgers wins.

d) Eight O'clock wins over Maxwell House with 76 points.

e) Same results as in a), thus, Starbucks wins.

f) The plurality, plurality with elimination, and Borda count methods all violate the head-to-head criterion.

25. The Borda count method

26. Plurality and plurality w/elimination methods

27. Pairwise comparison and Borda count methods

24. a) Yes. Fleetwood Mac is favored when compared to each of the other bands.

b) Votes: A: 15, B: 34, C: 9+4 = 13, F: 25 Boston wins.

c) A: 217 points, B: 198 points, C: 206 points, F: 249 points Fleetwood Mac wins.

d) A majority out of 87 voters is 44 or more votes.
Votes: A: 15, B: 34, C: 13, F:25
None has a majority, thus eliminate C.
Votes: A: 15+9+4 = 28, B: 34, F: 25
None has a majority, thus eliminate F.
Votes: A: 28+25 = 53, B: 34 Abba wins.

e) A = 2 pts., B = 0 pts., C = 1 pt., F = 3 pts. Thus, Fleetwood Mac wins.

f) Plurality and plurality w/elimination methods

28. Standard divisor = $\dfrac{6000}{10} = 600$

Region	A	B	C	Total
Number of Houses	2592	1428	1980	6000
Standard Quota	4.32	2.38	3.30	
Lower Quota	4	2	3	9
Hamilton's Apportionment	4	3	3	10

29. Using the modified divisor 500.

Region	A	B	C	Total
Number of Houses	2592	1428	1980	6000
Modified Quota	5.18	2.86	3.96	
Jefferson's Apportionment (rounded down)	5	2	3	10

30. Using the modified divisor 700.

Region	A	B	C	Total
Number of Houses	2592	1428	1980	6000
Modified Quota	3.70	2.04	2.83	
Adam's Apportionment (rounded up)	4	3	3	10

31. Using the modified divisor 575.

Region	A	B	C	Total
Number of Houses	2592	1428	1980	6000
Modified Quota	4.51	2.48	3.4	
Webster's Apportionment (normal rounding)	5	2	3	10

32. Yes. Hamilton's Apportionment becomes 5, 2, 4.   Region B loses one truck.

33. Standard divisor $= \dfrac{690}{23} = 30$

Course	A	B	C	Total
Number of Students	311	219	160	690
Standard Quota	10.37	7.30	5.33	
Lower Quota	10	7	5	22
Hamilton's Apportionment	11	7	5	23

34. Use the modified divisor 28

Course	A	B	C	Total
Number of Students	311	219	160	690
Modified Quota	11.12	7.82	5.71	
Jefferson's Apportionment (round down)	11	7	5	23

35. Use the modified divisor 31.5

Course	A	B	C	Total
Number of Students	311	219	160	690
Modified Quota	9.87	6.95	5.08	
Adam's Apportionment (round up)	10	7	6	23

36. Use the modified divisor 29.5

Course	A	B	C	Total
Number of Students	311	219	160	690
Modified Quota	10.54	7.42	5.42	
Webster's Apportionment (standard rounding)	11	7	5	23

37. The new divisor is $\dfrac{698}{23} = 30.35$

Course	A	B	C	Total
Number of Students	317	219	162	698
Standard Quota	10.44	7.22	5.34	
Lower Quota	10	7	5	22
Hamilton's Apportionment	11	7	5	23

No. The apportionment remains the same.

38. The Standard divisor $= \dfrac{55000}{55} = 1000$

State	A	B	Total
Population	4862	50138	55,000
Standard Quota	4.86	50.14	
Hamilton's Apportionment	5	50	55

39. The apportionment is 4, 51.

41. The apportionment is 5, 50.

40. The apportionment is 5, 50.

42. The new divisor is $\dfrac{60940}{60} = 1015.67$

State	A	B	C	Total
Population	4862	50138	5940	60940
Standard Quota	4.79	49.36	5.85	
Hamilton's Apportionment	5	49	6	60

Yes. State A. gains a seat while State B loses a seat.

Chapter Test

1. $6+5+5+4 = 20$ members voted.

2. No candidate has a majority of ≥ 10 votes.

3. Chris wins with a plurality of 9 votes.

4. D = 41 pts., C = 44 pts., S = 35 pts. Chris wins.

5. Donyall wins with 11 pts.

6. D = 1.5 pts., C = 1 pt., S = 0.5 pt. Donyall wins.

7. a) Votes: H: $26+14 = 40$, I: 29, L: 30, S: 43
   Thus, the snail wins.

   b) (H) 1st $(40)(4) = 160$
   2nd $(59)(3) = 177$
   3rd $(0)(2) = 0$
   4th $(43)(1) = 43$ H receives 380 points.
   (I) 1st $(29)(4) = 116$
   2nd $(40)(3) = 120$
   3rd $(73)(2) = 146$
   4th $(0)(1) = 0$ I receives 382 points
   (L) 1st $(30)(4) = 120$
   2nd $(43)(3) = 129$
   3rd $(43)(2) = 86$
   4th $(26)(1) = 26$ L receives 361 points

7. b) (S) 1st $(43)(4) = 172$
   2nd $(0)(3) = 0$
   3rd $(26)(2) = 52$
   4th $(73)(1) = 73$ S receives 297 points.
   The iguana (I) wins with the most points.

   c) A majority out of 142 voters is 72 or more votes.
   Votes: H: 40, I: 29, L: 30, S: 43; None has a majority, thus eliminate I. Votes: H: 69, L: 30, S: 43 None has a majority, thus eliminate L. Votes: H: 99, S: 43
   The hamster wins.

   d) H vs. I: I gets 1 pt. H vs. L: L gets 1 pt.
   H vs. S: H gets 1 pt. I vs. L: L gets 1 pt.
   I vs. S: I gets 1 pt. L vs. S: L gets 1 pt.
   Ladybug wins with 3 points.

8. Plurality: Votes: W: 86, X: $52+28 = 80$, Y: 60, Z: 58 W wins.
   Borda count: W gets 594 points, X gets 760 points, Y gets 722 points, Z gets 764 points Z wins
   Plurality with elimination: A majority out of 284 voters is 143 or more votes.
   Votes: W: 86, X: 80, Y: 60, Z: 58
   None has a majority, thus eliminate Z.
   Votes: W: 86, X: $80+58 = 138$, Y: 60
   None has a majority, thus eliminate Y.
   Votes: W: 86, X: $138+60 = 198$ X wins.

8. Head-to-Head: When Y is compared to each of the others, Y is favored. Thus Y wins the head-to-head comparison.
   Plurality, Borda count and Plurality with elimination each violate the head-to-head criterion. The pairwise method never violates the head-to-head criterion.

9. A majority out of 35 voters is 18 or more votes. Louisiana (L) has a majority.
   However, Mississippi (M) wins using the Borda count method. Thus the majority criterion is violated.

10. a) The standard divisor $= \dfrac{33000}{30} = 1100$

State	A	B	C	Total
Population	6933	9533	16534	33,000
Standard Quota	6.30	8.67	15.03	
Hamilton's Apportionment	6	9	15	30

b)

State	A	B	C	Total
Population	6933	9533	16534	33,000
Modified Quota	6.30	8.67	15.03	
Jefferson's Apportionment (round down)	6	8	15	29

c) The new divisor 1064.52

State	A	B	C	Total
Population	6933	9533	16534	33,000
Standard Quota	6.51	8.96	15.53	
Hamilton's Apportionment	6	9	16	31

The Alabama paradox does not occur, sine none of the states loses a seat.

d) The divisor $= \dfrac{33826}{31} = 1091.16$

State	A	B	C	Total
Population	7072	9724	17030	33,826
Standard Quota	6.48	8.91	15.61	
Hamilton's Apportionment	6	9	16	31

The Alabama paradox does not occur, sine none of the states loses a seat.

10. e) The new divisor is $\dfrac{38100}{36} = 1058.33$

State	A	B	C	D	Total
Population	6933	9533	16534	5100	38100
Standard Quota	6.55	9.01	15.62	4.82	
Hamilton's Apportionment	6	9	16	5	36

The new states paradox does not occur, sine none of the existing states loses a seat.

APPENDIX

GRAPH THEORY

<u>**Exercise Set**</u>

1. A **vertex** is a designated point.
2. An **edge** (or an **arc**) is any line, either straight or curved, that begins and ends at a vertex.
3. To determine whether a vertex is odd or even, count the number of edges attached to the vertex. If the number of edges is odd, the vertex is **odd**. If the number of edges is even, the vertex is **even**.
4. Answers will vary.

5. 5 vertices, 7 edges
6. 6 vertices, 8 edges
7. 7 vertices, 11 edges
8. 5 vertices, 6 edges
9. Each graph has the same number of edges from the corresponding vertices.
10. Each graph has the same number of edges from the corresponding vertices.
11. Odd vertices: *C, D*
    Even vertices: *A, B*
12. Odd vertices: *A, C, E, F*
    Even vertices: *B, D*

13. Yes. The figure has exactly two odd vertices, namely *C* and *D*. Therefore, the figure is traversable. You may start at *C* and end at *D*, or start at *D* and end at *C*.
14. No. All four vertices are odd. There are more than two odd vertices. Therefore, the figure is not traversable.
15. Yes. The figure has no odd vertices. Therefore, the figure is traversable. You may start at any point and end where you started.
16. Yes. The figure has no odd vertices. Therefore, the figure is traversable. You may start at any point and end where you started.
17. No. The figure has four odd vertices, namely *A, B, E,* and *F*. There are more than two odd vertices. Therefore, the figure is not traversable.
18. Yes. The figure has exactly two odd vertices, namely *C* and *G*. Therefore, the figure is traversable. You may start at *C* and end at *G*, or start at *G* and end at *C*.
19. Yes. The figure has exactly two odd vertices, namely *A* and *C*. Therefore, the figure is traversable. You may start at *A* and end at *C*, or start at *C* and end at *A*.
20. Yes. The figure has no odd vertices. Therefore, the figure is traversable. You may start at any point and end where you started.
21. a) 0 rooms have an odd number of doors.
    5 rooms have an even number of doors.
    b) Yes because the figure would have no odd vertices.
    c) Start in any room and end where you began. For example: *A* to *D* to *B* to *C* to *E* to *A*.
22. a) 4 rooms have an odd number of doors.
    1 room has an even number of doors.
    b) No because the figure would have more than two odd vertices.

23. a) 2 rooms have an odd number of doors.

    4 rooms have an even number of doors.

    b) Yes because the figure would have exactly two odd vertices.

    c) Start at B and end at F, or start at F and end at B.

    For example: B to C to F to E to D to A to B to E to F

24. a) 2 rooms have an odd number of doors.

    4 rooms have an even number of doors.

    b) Yes because the figure would have exactly two odd vertices.

    c) Start at B and end at E, or start at E and end at B. For example: B to A to D to E to F to C to B to E

25. a) 4 rooms have an odd number of doors.

    1 room has an even number of doors.

    b) No because the figure would have more than two odd vertices.

26. a) 5 rooms have an odd number of doors.

    1 room has an even number of doors.

    b) No because the figure would have more than two odd vertices.

27. a) 3 rooms have an odd number of doors.

    2 rooms have an even number of doors.

    b) No because the figure would have more than two odd vertices.

28. a) 3 rooms have an odd number of doors.

    4 rooms have an even number of doors.

    b) No because the figure would have more than two odd vertices.

29. The door must be placed in room D. Adding a door to any other room would create two rooms with an odd number of vertices. You would then be unable to enter the building through the door marked "enter" and exit through the new door without going through a door at least twice.

30. The door must be placed in room D. Adding a door to any other room would create two rooms with an odd number of vertices. You would then be unable to enter the building through the door marked "enter" and exit through the new door without going through a door at least twice.

31. Yes because the figure would have exactly two odd vertices. Begin at either the island on the left or on the right and end at the other island.

32. Yes because the figure would have exactly two odd vertices. Begin at the island on the right and end on the land below the island, or vice versa.

33.

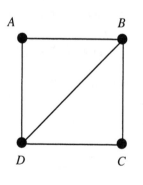

34.

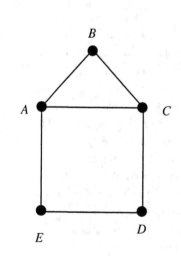

35. a) Kentucky, Virginia, North Carolina, Georgia, Alabama, Mississippi, Arkansas, Missouri
    b) Illinois, Arkansas, Tennessee

36. a) French Guiana, Surinam, Guyana, Venezuela, Columbia, Peru, Bolivia, Paraguay, Argentina, Uruguay
    b) Peru, Chile, Argentina, Paraguay, Brazil

37. a) 4
    b) 4
    c) 11

38.

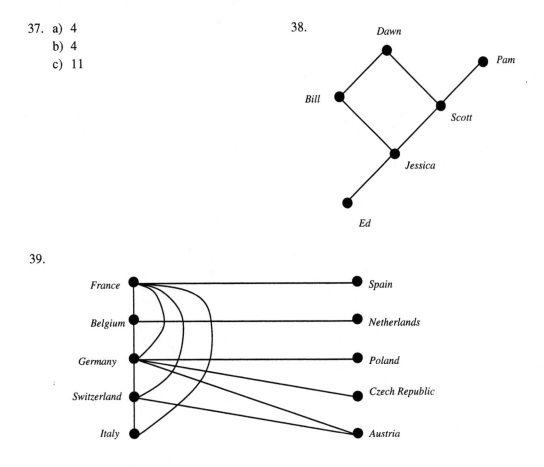

39.

40. No, it is not possible, assuming that your starting and ending points are considered vertices.

41. a) Yes, the graph has exactly two odd vertices, namely *C* and *G*.
    b) *C, A, B, E, F, D, G, C*

42. Number of Edges = Number of Vertices + Number of Regions - 2